AF823554

Studies in Systems, Decision and Control

Volume 252

Series Editor

Janusz Kacprzyk, Systems Research Institute, Polish Academy of Sciences, Warsaw, Poland

The series "Studies in Systems, Decision and Control" (SSDC) covers both new developments and advances, as well as the state of the art, in the various areas of broadly perceived systems, decision making and control–quickly, up to date and with a high quality. The intent is to cover the theory, applications, and perspectives on the state of the art and future developments relevant to systems, decision making, control, complex processes and related areas, as embedded in the fields of engineering, computer science, physics, economics, social and life sciences, as well as the paradigms and methodologies behind them. The series contains monographs, textbooks, lecture notes and edited volumes in systems, decision making and control spanning the areas of Cyber-Physical Systems, Autonomous Systems, Sensor Networks, Control Systems, Energy Systems, Automotive Systems, Biological Systems, Vehicular Networking and Connected Vehicles, Aerospace Systems, Automation, Manufacturing, Smart Grids, Nonlinear Systems, Power Systems, Robotics, Social Systems, Economic Systems and other. Of particular value to both the contributors and the readership are the short publication timeframe and the world-wide distribution and exposure which enable both a wide and rapid dissemination of research output.

** Indexing: The books of this series are submitted to ISI, SCOPUS, DBLP, Ulrichs, MathSciNet, Current Mathematical Publications, Mathematical Reviews, Zentralblatt Math: MetaPress and Springerlink.

More information about this series at http://www.springer.com/series/13304

Sergey Viktorovich Eliseev ·
Andrey Vladimirovich Eliseev

Theory of Oscillations

Structural Mathematical Modeling in Problems of Dynamics of Technical Objects

Springer

Sergey Viktorovich Eliseev
Irkutsk State Transport University
Irkutsk, Russia

Andrey Vladimirovich Eliseev
Irkutsk State Transport University
Irkutsk, Russia

ISSN 2198-4182 ISSN 2198-4190 (electronic)
Studies in Systems, Decision and Control
ISBN 978-3-030-31294-7 ISBN 978-3-030-31295-4 (eBook)
https://doi.org/10.1007/978-3-030-31295-4

© Springer Nature Switzerland AG 2020
This work is subject to copyright. All rights are reserved by the Publisher, whether the whole or part of the material is concerned, specifically the rights of translation, reprinting, reuse of illustrations, recitation, broadcasting, reproduction on microfilms or in any other physical way, and transmission or information storage and retrieval, electronic adaptation, computer software, or by similar or dissimilar methodology now known or hereafter developed.
The use of general descriptive names, registered names, trademarks, service marks, etc. in this publication does not imply, even in the absence of a specific statement, that such names are exempt from the relevant protective laws and regulations and therefore free for general use.
The publisher, the authors and the editors are safe to assume that the advice and information in this book are believed to be true and accurate at the date of publication. Neither the publisher nor the authors or the editors give a warranty, expressed or implied, with respect to the material contained herein or for any errors or omissions that may have been made. The publisher remains neutral with regard to jurisdictional claims in published maps and institutional affiliations.

This Springer imprint is published by the registered company Springer Nature Switzerland AG
The registered company address is: Gewerbestrasse 11, 6330 Cham, Switzerland

Preface

Ensuring safety of the operation of the modern machines and their reliable performance in the conditions of intensive dynamic loading is a topical area of focus in theoretical researches and constructional and technological developments in the sphere of the mechanical engineering, theory of machines and mechanisms, theory of vibrations, dynamics and strength, theory of control, mathematical modelling that are all parts of the interdisciplinary space combining the efforts of many specialists. Dynamic quality of different machines and the possibilities of their effective operation are studied and considered in detail and are in the zone of special attention at all the stages of making modern technical objects. In most cases, dynamic interactions of elements of different machines that are manifested through vibrations evoked by the internal and external elements become important trends in estimating the possibilities of machines while providing working efficiency and operation safety of technical systems. The basis of the exploratory researches and applied developments are most often the dynamic processes in mechanical vibration systems of different complexity that are considered as computational schemes, with the problems of dynamics of technical objects being converted into the corresponding research areas of the applied vibration theory.

In the presented monography, the generalized approaches in the researches of the properties of the mechanical vibratory systems based on the structural methods using the principles of dynamic analogies that characteristic for movements of elements of mechanical vibrating systems, including automatic control systems are developed. The commonality of approaches is especially well manifested through the problems of vibration protection of machines, equipment, instruments and equipment and implementations of vibration technological processes.

In the monography, the results of the researches that have been carried out during the recent years in the Irkutsk State Transport University are presented.

The area of the research is connected to the problems of machine dynamics and reflects the interests that have been formed during long-standing contacts with scientific schools in the sphere of theoretical and applied mechanics supported.

The generalized idea of the suggested research is the structural mathematical modelling in the dynamics of the mechanical vibratory systems in general and vibration protection systems in particular. The structural model (or scheme) within the developed methodology, which is regarded, is a graphical analogue of the source mathematical model in the form of a system of linear ordinary differential equations in operator form. In this regard, structural mathematical modelling based on the analogy with automated control systems, the principle of feedback and the equivalent transformations can be compared with the theory of circuits and theory of graphs.

The primary focus in the monography is the study of the peculiarities of the formation of extra connections in the mechanical vibratory systems. These extra connections are brought into the dynamic interaction by the elements of a different nature. In particular, the motion translation devices, lever mechanisms and lever linkages are closely regarded. The first three chapters convey general information and contain necessary survey material that specifies certain tendencies of formation of understanding on the expansion of the element base of the mechanical vibratory systems in general and vibration insulation systems in particular. There is wide diversity in constructive and technical forms of the elements' interconnection into different structures. That generally approves the necessity of separation of not typical elements only but more complex constructs defined as compacts, quasisprings etc., having the reduced mass-and-inertia and elastic dissipational properties.

The fourth chapter of the monography is dedicated to the specification of understanding ways and technologies of the transformation of structural diagrams that reflect the properties of the mechanical vibratory systems with one and more degrees of freedom.

The present work includes the definitions of automated control theory that are connected to the transfer functions of systems, frequency properties of systems and their particular aspects in the dynamic responses to different external perturbations. The specific feature is the capability of obtaining the constructs from a few interconnected elementary units that have the properties of interacting with each other using the same rules of transformations as elementary units.

The fifth chapter of the monography represents the further development of the methodological positions regarding the transformation of structural diagrams of mechanical vibratory systems. The specific aspect of the approaches is the particularization of ways of simplifying the systems by introducing concatenations. Some variations in changing the systems' properties by concatenating the elements are suggested, either based on the «zeroing» the relative coordinated of motion, or as well as by selecting the limiting values of mass and inertial or elastic characteristics in the certain parts of the system.

Chapter 6 contains the results of the researches dedicated to the studies of possible equivalent transformations of structural diagrams of mechanical vibratory systems. It describes the ways of defining the reduced stiffnesses and the reduced masses with regard to the computational schemes of technical objects that contain different mechanisms or motion translation devices.

In Chapter 7, the ideas of distinguishing the lever linkages and taking their special aspects into consideration have been further developed, being applied to mechanical vibratory systems with one or more degrees of freedom. Some dynamic effects that take place with the lever linkages are implemented in mechanical structures with partial systems that have different types of motions (translational, rotational and helicoidal).

The way different mechanisms influence the dynamic properties of the vibration protection systems, and the researches on the special aspects of this influence are represented in Chapter 8. Some original construction and technical solutions are considered. The approaches that are developed in the present monography make it possible to considerably expand the concepts of capabilities of systems while implementing the modes of dynamic absorbing of vibrations.

In the authors' opinion, the researches that have been done broaden the concepts about the capabilities of structural methods of mathematical modelling, due to their deep connection with control theory and system analysis, and also provide new opportunities in the solutions of problems of dynamic synthesis being applied to the dynamics of wide class of technical objects that are subject to intensive vibration loadings.

The authors would like to express their gratitude to N. K. Kuznetsov (D.Sc. in Engineering, Prof.) and P. A. Lontsikh (D.Sc. in Engineering, Prof.) from the Irkutsk National Research Technical University for their support, kind attention and help in finding solutions of scientific and technical questions.

Irkutsk, Russia
March, 2019

Sergey Viktorovich Eliseev
Andrey Vladimirovich Eliseev

Introduction

Ensuring reliability and safety of operation of machines is a modern problem solution of which is connected with interdisciplinary concepts of the dynamic interactions of the numerous elements of technical systems. Creating new machines and equipment is usually preceded by the stage of representational scientific researches and developments, in the course of which the capabilities of machines during the implementation of specified processes with the defined accuracy and the effectiveness, as well as their adaptive properties in case of any changes in conditions of their functioning.

Vibrations of machines and equipment are characteristic of the operation of the most technological and transport machines. That is specified by regulations and is formalized in the corresponding regulatory technical documents. The problems of vibration protection of machines, equipment and instruments are referred as widespread and sufficiently studied that have been reflected in the works of indigenous Russian scientists [1–3]. In the capacity of computational schemes, machines and mechanisms, in many cases mechanical vibratory systems with one, two and more degrees of freedom are considered. The greatest development in the evaluation of the dynamic properties of the mechanical vibratory systems is the analytical methods that are based on the mathematical models in the form of a system of a regular differential equations with constant coefficients. This approach is oriented to the models with linear properties and lumped parameters. Theoretical basics of the evaluation and studying of the dynamic properties of such systems found their implementation in the works dedicated to the theory of oscillations [4–8], which is acceptable enough to assume that the object of protection executes little vibrations with a relation to the static equilibrium position. These assumptions have been made during the preliminary and exploratory researches.

In more detail, nonlinear systems are considered, with special methods being applied to estimate their properties [9–11]. In theory and practice of vibration protection, there is an experience widely applied while solving the problems connected to the dynamics of transport systems [12, 13].

The development of technical means of restriction of vibration processes initiated the development of wide class of special devices in the form of suspension brackets, buffer springs, shock absorbers, dampers, vibration absorbers, etc. [14], which to the certain degree presupposed the interest towards the capabilities of expanding the set of typical elements of vibration protection systems and usage of different mechanisms to transform motion and implement the lever linkages ensuring the dynamic interactions in the certain sphere of interaction.

The ideas of active vibration control [15–19] had a considerable impact on the vibration protection theory. Within the framework of the active vibration control, the servo-actuators of different types, measuring devices and means of computer equipment became applicable within the structure of mechanical vibratory systems.

In their developed form, the vibration protection systems represent tailor-made automated control systems, which naturally presuppose the conditions of the generalized approach based on the concepts of the object state control and application of the feedback principles in the development of the corresponding mathematical models [1, 2, 20–22].

Regarding the problems of vibration protection in the linear setting satisfies only those research intentions that correspond to the preliminary evaluations, which is characteristic of the exploratory developments and approximate evaluations in case if there is not enough data on the structure and parameters of the technical object that is subject to the vibrations.

Structural mathematical modelling technology is based on the use of Laplace transformations in relation to the reference mathematical model in a form of a system of linear regular differential equations. Distinguishing the object of protection allows creating a structural diagram of a linear mechanical vibratory system which has the same form as an automated control system with an analytical mathematical model. Having functional analogues in relation to the elements of mechanical systems in the form of springs, dampers and mass-and-inertia units, it is easy to form a set of typical elements of the differential units of the first and second orders, and also integrating units of the first and second orders.

The obligatory condition of the inseparability of processes and continuity of interactions is the homogeneity of typical elements, when displacement is the input signal of each unit, and effort (or the force) is the output signal of a typical unit. The object of protection is a unit with a transfer function of the integration of the second order. The technology of these constructions and structural transformations has been reflected in [20, 23–24].

Transformation of structural diagrams is implemented on the basis of the rules of structural transformations of the automated control theory. Complication of computational schemes is related to the fact that different motion translation devices, as well as mechanisms expanding their functions of vibration protection, have been introduced into their structure.

Structural mathematical modelling has a certain convenience in distinguishing the constraints occurring during the introduction of elements, that is manifested while the structural transformations of the reference systems are being done, and also while taking into consideration the different factors arising during the additional constraints of a different nature being distinguished. In the physical signal, such additional constraints can be implemented in a form of special (lever, toothed, non-locking screw, etc.) mechanisms and devices. These kinds of approaches are characteristics of not only vibration protection systems but also of robotic science, suspension transport systems and vibrational technological machines [26–29].

References

1. Genkin, M.D., Elezov, V.G., Yablonsky, V.V.: Methods of active vibration damping mechanisms. In: Dynamics and Acoustics of Machines, pp. 70–87. Nauka Publ., Moscow (1971)
2. Kolovsky, M.Z.: Automatic Control of Vibration Protection Systems, 320 pp. Nauka Publ., Moscow (1976)
3. Frolov, K.V., Furman, F.A.: Applied Theory of Vibration Protection Systems, 286 pp. Mashinostroenie Publ., Moscow (1985)
4. Babakov, I.M.: Theory of Oscillations, p. 549. Nauka Publ., Moscow (1968)
5. Den-Hartog, J.P.: Mechanical Oscillations, p. 530. Publishing house of foreign literature, Moscow (1962)
6. Panovko, Ya.G.: Introduction to the Theory of Mechanical Vibrations, 255 p. Nauka Publication, Moscow (1991)
7. Timoshenko, S.P., Young, D.H., Weaver, U.: Theory of Oscillations in Engineering Business, 472 pp. Transl. from English by Korneychuk, L.G., edited by Grigolyuk, E.I. Mashinostroenie Publ., Moscow (1985)
8. Jie, F.S., Morse, I.E., Hinkle, R.T.: Mechanical Vibrations, p. 508. Mashinostroenie Publ., Moscow (1966)
9. Eliseev, S.V., Khomenko, A.P., Upyr, R.Yu.: Mechatronics of vibration protection systems with lever linkages. In: Modern Technologies. System Analysis. Modeling, vol. 3, no. (23), pp. 104–119. IrGUPS Publ., Irkutsk (2009)
10. Makhutov, N.A., Petrov V.P., Kuksova V.I., Moskvitin G.V.: Modern trends in the development of scientific research on problems of machine science and mechanical engineering. In: Problems of Mechanical Engineering and Automation, vol. 3, pp. 16–37. OJSC National Institute of Aviation Technologies, Moscow (2008)
11. Stoker, J.J.: Nonlinear Oscillations in Mechanical and Electrical Systems, p. 264. Publishing House of Foreign Literature, Moscow (1952)
12. Verigo, M.F., Kogan, A.Ya.: The Interaction of the Track and Rolling Stock, 559 pp. In: Verigo, M.F. (ed.). Transport Publ., Moscow (1986)
13. Kogan, A.Y.: Track Dynamics and its Interaction with Rolling Stock, 326 pp. Transport Publ., Moscow (1997)
14. Vibrations in technology: ref.: in 6 vols. In: Frolov, K.V. (ed.) Protection against Vibrations and Shocks, vol. 6, 586 pp. Mashinostroenie Publ., Moscow (1983)
15. Naucler, P.: Modelling and control of vibration in mechanical systems, 102 pp. Dissertation for the degree of Licentiate of Philosophy in Electrical Engineering, Uppsala University, Sweden (2005)
16. Nelson, P.A., Elliott, S.J.: Active Control of Sound, 562 pp. Academic Press Publ., Leningrad (1992)
17. Eliseev, S.V., Ol'kov, V.V., Zasyadko, A.A.: On the possibilities of active vibration protection using devices with motion conversion. In: Proceedings of Jubilee Science-Technical Conference of the Department of mechanics "Mechanics and Control Processes", pp. 89–97. Irkutsk (1971)
18. Kernop, D.: Principles of designing oscillation control systems using semi-active dampers. In: Dynamics of Systems, Mechanics and Control: Proceedings of ASME, vol. 112, 3, pp. 448–453 (1990)

19. Ruzhichka, D.J.: Active vibration protection systems. In: Testing Instruments and Stands, no. 10, pp. 14–25. Express-inform, VINITI, Moscow (1969)
20. Eliseev, S.V.: Structural Theory of Vibration Protection Systems, 224 pp. Nauka Publ., Novosibirsk (1978)
21. Eliseev, S.V.: The theory of active vibration protection systems and its applications: systems, 35 pp. Author's abstract of Dr. Sc. (Engineering) thesis, Tymoshenko Institute of Mechanics of AS Ukr. SSR, Kiev (1973)
22. Sinev, A.V., Safronov, Yu.G., Soloviev, V.S., et al.: Dynamic Properties of Linear Vibration Protection Systems. In Frolov K.V. (ed.), 208 p. Nauka Publ., Moscow (1982)
23. Eliseev, S.V., Reznik, Yu.N., Khomenko, A.P.: Mechatronic Approaches in the Dynamics of Mechanical Oscillatory Systems, 384 pp. Nauka Publ., Novosibirsk (2011)
24. Eliseev, S.V., Reznik, Yu.N., Khomenko, A.P., et al.: Dynamic Synthesis in Generalized Problems of Vibration Protection and Vibration Isolation of Technical Objects, 523 pp. ISU Publ., Irkutsk (2008)
25. Eliseev, S.V., Khomenko, A.P.: Dynamic Damping. Feedback Concept and Structural Methods of Mathematical Modeling, 357 pp. Nauka Publ., Novosibirsk (2014)
26. Goverdovsky, V.N.: Development of the theory and methods of designing machines with infra-frequency vibration protection systems, 42 pp. Author's abstract of Dr. Sc. (Engineering) thesis, SGUPS Publ., Novosibirsk (2006)
27. Gordeev, B.A., Erofeev, V.I., Sinev, A.V., Mugin, O.O.: Vibration protection systems using inertia and dissipation of rheological media, 176 p. Fizmatlit Publ., Moscow (2004)
28. Kiryukhin, A.V., Tikhonov, V.A., Chistyakov, A.G., Yablonsky, V.V.: Active vibration protection: purpose, principles, condition. 1. Purpose and principles of development. In: Problems of Mechanical Engineering and Automation, vol. 2, pp. 108–111. OJSC "National Institute of Aviation Technologies" Publication, Moscow (2011)
29. Kiryukhin, A.V., Tikhonov, V.A., Chistyakov, A.G., Yablonsky, V.V.: Active vibration protection: Purpose, principles, condition. 2. History of development and state. In: Problems of Mechanical Engineering and Automation, vol. 3, pp. 63–69. OJSC "National Institute of Aviation Technologies" Publ., Moscow (2011)

Contents

Chapter 1
Some Problems of Applied Theory of Vibrations

In some approaches of mathematical modeling, an computational scheme of machines, as well as transport ones, becomes a mechanical vibratory system with some degrees of freedom. As a part of such systems, elements and devices of different physical nature can be used. Computational schemes of objects of protection are defined by the aims of the research. In this regard, problems of vibration protection and vibration protection are typical ones, and the computational schemes are represented by mechanical vibratory systems [1].

1.1 Mechanical Vibratory Systems: Dynamics and Motion Control, Areas of Research

Control of dynamic state is based on the introduction of extra devices, elements, circuits, control loops, means of computer equipment into the mechanical system. Many technological machines are developed automated control systems which, it their turn, consist of autonomous subsystems [2–6]. As a result of the feedback theory development, structural mathematical models of the vibratory systems in a form of structural diagrams, dynamically equivalent to the automated control systems, have been introduced in the mechanics. A methodological basis of the transfer from mathematical models of mechanical systems in a form of a system of regular differential equations with constant coefficients, to mathematical models in a form of structural diagrams of automated control systems has been developed [7–10].

© Springer Nature Switzerland AG 2020

S. V. Eliseev and A. V. Eliseev, *Theory of Oscillations*, Studies in Systems, Decision and Control 252, https://doi.org/10.1007/978-3-030-31295-4_1

1.1.1 Structural Mathematical Modelling: Diagrams, Constraints, Transformations

Structural approaches in the theory of vibration protection systems are based on usage of the Laplace transformations in relation to the reference systems of linear differential equations with constant coefficients. In more detail, such approaches are regarded in [11–13]; the detailed concepts of the ways of transformation of structural diagrams and constructions of the corresponding transfer functions are represented in [14–16].

In this case, transfer function (TF) is perceived as carrier of information about the dynamic properties of the system and, in its physical sense, is a ratio of modules of a harmonic output signal to the input one and can be applied to construct different frequency characteristics, and also the system reactions to the typical impacts [17].

With all the diversity of construction and technical variants of construction of computational schemes in the problems of machine dynamics, most widespread are the mechanical vibratory systems with one, two and three degrees of freedom. Such computational schemes can be called basic. In Fig. 1.1, the basic model has mass and inertia element (m), and also constraints in a form of the elastic and dissipative elements (k_1, k_2—spring stiffnesses; resistance coefficients—b_1, b_2 with viscous friction).

The protection object can be associated with a fixed base (k_2, b_2p), and also rest against the base (k_1, b_1p), which has a known law of motion. The computational scheme can be represented with several graphical interpretations (Fig. 1.1a, b); their relationship is shown up in the structural diagram shown in Fig. 1.1c.

It follows from the structural diagram (Fig. 1.1c) that a mechanical system, if one has in mind its structural analog, consists of typical elementary units of three types:

- elastic elements or springs with stiffness coefficients k_1 and k_2 (input is a displacement, output is an elastic force), $W_i(p) = k_i$ ($i = 1, 2$);

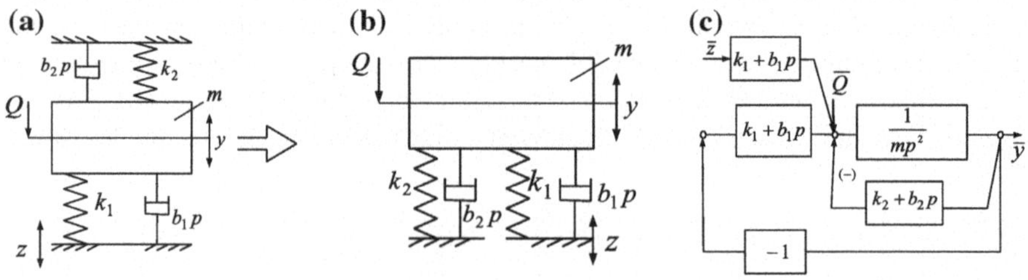

Fig. 1.1 The basic computational scheme in a form of mechanical system with one degree of freedom. **a** Model with a full set of elements; **b** a transformed model; **c** a structural model of the system ($p = j\omega$ is a complex variable; a sign «–» is a representation of the variable according to Laplace [11, 12])

- dissipative units (dampers) with transfer functions: $W_i(p) = b_i p$ (input is an displacement, output is a resistance force), that is, of a differentiating unit of the first order (b_i is a coefficient of resistance);
- a mass and inertial unit: transfer function $W_0(p) = \frac{1}{mp^2}$ is an integrating unit of the second order (input is the force, output is the displacement).

In typical elementary units of the structural theory of vibration protection systems, the input is the displacement, and the output is the force. As for the mass and inertial unit $W_0(p)$, it belongs to a different group, since the input in such a unit becomes the force, and the output is the displacement. Thus, the basic model (Fig. 1.1) consists of standard units that belong to two groups. The first group is characterized by a case where the input is the displacement, and the output is the force (such a group can be called generalized springs). The second group is the object of protection: an integrating units of the second order, but with the force as the input, and with the displacement as the output. It was shown in [11, 13, 16] that the first group of typical elementary units can be expanded by including a second-order differentiating unit, as well as integrating units of the first and second order (basically, and the unit of pure delay). The input signal of such units is the displacement, and the output is the force.

If we assume that in Fig. 1.1 $k_2 = 0$; $b_1 = 0$; $b_2 = 0$, then the basic computational scheme will take the simplest form. Then the introduction of any other elements (for example, $k_2 \neq 0$; $b_1 \neq 0$; $b_2 \neq 0$) can be considered as the introduction of additional negative feedback. Each additional constraint can be a more complex formation from the elementary units of the usual and extended set of typical elements [12]. Further changes in the structural diagram of the vibration protection system depend on the transformations in accordance with the rules of the circuits theory and the theory of automatic control.

Since the additional feedback is essentially the force that is generated when the input signal corresponds to the displacement, the structural transformations within the additional feedback branch (or simply an additional constraint) are made according to the rules of the theory of mechanical circuits (or the theory of electrical circuits, if one has in mind principles of electromechanical analogies). In its turn, the object of protection, within the framework of a structural diagram built on the notion that a mechanical oscillatory system, can be regarded as equivalent, in a dynamic sense, to an automatic control system, is an object of control. Therefore, in general, the construction of the transfer function of the vibration protection system, rather than its individual additional branches (or circuits), is performed according to the rules for transforming the structures of the theory of automatic control [16, 17].

Actually, the above-mentioned description of the methods of transformation can be considered in the same way as the idea of simplifying any computational schemes to the structure of the basic computational scheme. However, then typical elements form more complex structural formations of standard elements by a series of consecutive and parallel connections. In either case, the input signal to such an additional circuit or additional feedback is the displacement, and the output is the force. That is, the set of elementary units formally forms a generalized unit.

Materials on the methods of simplifying the computational schemes of vibration protection systems are given in [18, 19]. In our opinion, they could become the basis for a more general approach with the possibilities of considering the interaction of systems with elements of a different physical nature.

1.1.2 Changing Properties of Systems When Using Additional Constraints

The introduction of additional feedbacks (or expanding a set of typical elements of the first group) in a mechanical oscillatory system is shown in Fig. 1.2.

The basis of the structural diagram of the mechanical oscillatory system is the basic model of two elements: the mass and inertial $\frac{1}{mp^2}$ and the elastic k (these are the $W_0(p)$ and $W_i(p)$ mentioned above). As for the additional constraints $W_{\text{add1}}(p)$ and $W_{\text{add2}}(p)$, they can have the following structures:

$$-W_{add1}(p) = b'p + L'p^2 + \frac{A_1'}{p} + \frac{A_2'}{p^2}; \tag{1.1}$$

$$W_{\text{add2}}(p) = k'' + b''p + L''p^2 + \frac{A_1''}{p} + \frac{A_2''}{p^2}, \tag{1.2}$$

where $b', L', A_1', A_2', k'', b'', L'', A_1'', A_2''$ are the coefficients reflecting the properties of the corresponding typical elements of the first group, that is, an extended set of typical elements [12]. The expression for the transfer function of the system depends on the type of external influence. In the structural diagram under

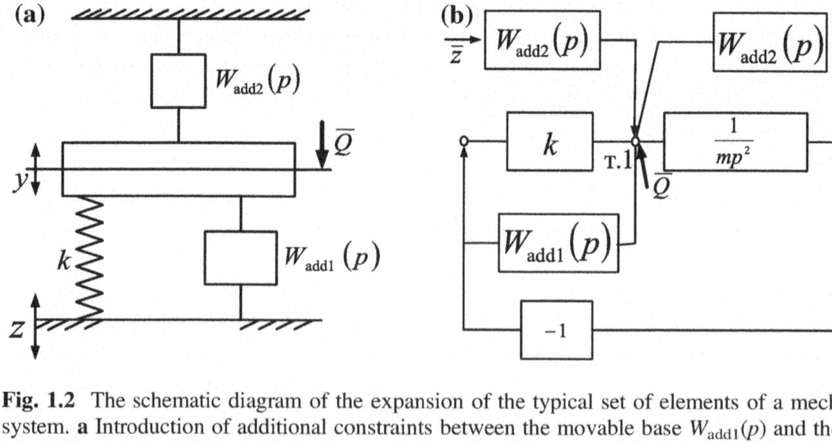

Fig. 1.2 The schematic diagram of the expansion of the typical set of elements of a mechanical system. **a** Introduction of additional constraints between the movable base $W_{\text{add1}}(p)$ and the fixed base $W_{\text{add2}}(p)$; **b** additional constraints on the structural diagram of the basic model

consideration (Fig. 1.2b), there are two of them ($\bar{z}$ and $\overline{Q}$) and they are applied at one point (point 1). Believing that $\bar{z} = 0$, and $\overline{Q} \neq 0$, we will find:

$$W_{\overline{Q}}(p) = \frac{\bar{y}}{\overline{Q}} = \frac{1}{mp^2 + k + W_{add1} + W_{add2}}; \tag{1.3}$$

with $\overline{Q} = 0$; $\bar{z} \neq 0$:

$$W_{\bar{z}}(p) = \frac{\bar{y}}{\bar{z}} = \frac{W_{add1}}{mp^2 + k + W_{add1} + W_{add2}}. \tag{1.4}$$

In cases where there are two perturbations, but they are related by some dependence, for example $\overline{Q} = a\bar{z}$ (a is some coupling coefficient), with $\overline{Q} \neq 0$; $\bar{z} \neq 0$, we get:

$$W_{z,Q} = \frac{\bar{y}}{\bar{z}} = \frac{W_{add1} + a}{mp^2 + k + W_{add1} + W_{add2}}. \tag{1.5}$$

Since the coupling coefficient a can be chosen quite arbitrarily (special systems are meant), the properties of the system may turn out to be unusual (up to obtaining invariant systems). Some results of investigations in this respect are given in [20, 21].

Changing the dynamic properties of a mechanical system due to the complication of its structure through the introduction of additional constraints makes it possible to obtain a number of interesting results, in particular, when introducing so-called active connections or constructing active vibration protection systems [22–24].

Complex computational schemes for systems with several degrees of freedom, provided that the protection object and equipment are represented by material points, are displayed by structural diagrams of dynamically equivalent automatic control systems. Such systems consist of several partial systems, each of which have one degree of freedom and can be considered as a basic one. If the protection object is associated with a certain material point, the remaining parts or branches of the mechanical system can be considered as additional feedbacks relative to the base model associated with the protection object. In this case, various simplifications in the representation of structural diagrams are possible, using the concepts of reduced mass of an object, reduced stiffness of the base elastic element, etc.

If the basic model and the protection object (i.e., the corresponding partial system) have an extended set of typical elements, then they can be arranged into a generalized spring: its reduced stiffness will depend on the frequency, which became the subject of research in [25–28].

When constructing structural diagrams of systems with several degrees of freedom, it is possible to construct a sufficient variety of mechanical circuits. The latter implies understanding circuits as a sequence of material points, connected by typical elements. Such structures can have many branches and closed contours.

The problems of estimating the dynamic properties of such circuits have been studied within a relatively narrow field: that is, taking into account the elastic elements of the traditional set of typical elements, rather than an extended one. Some concepts of changes in dynamic properties when introducing differentiators of the second order are given in [12]. In a number of cases, in circuit systems with several degrees of freedom, simultaneous dynamic absorbing modes can occur in several coordinates. The study of the properties of systems with several degrees of freedom showed that the selection of a system of generalized coordinates is of great importance. This allows us to introduce various effects of cooperative motions in several coordinates, as well as cases of equifrequent natural oscillations, coincidence of frequencies of resonance modes and natural oscillations. A peculiar feature of mechanical systems constructed from material points can be the fact that, most often, all motions in such systems occur along one straight line, on which the coordinates of points are laid off (one-dimensional construction).

It is important to note that in systems with several degrees of freedom, mass and inertial elements that are not the object of protection can, when it comes to building structures, transfer from the second group of integrating units of the second order, where the input is a force, and the output is a displacement, into the group of generalized springs via appropriate connections. In this case, in the structure of the system with respect to the protection object, other material points can form elementary generic units of a differentiating form of the second order. Some results of research in this direction are presented in [29, 30].

As a result of further comparative review of works and problems solved using mechanical systems with several material points, it is possible to propose a new vibration protection or vibration protection problem assignment when the protection object is two (or more) material points interconnected by elements of the expanded set of typical elements. We note that some concept of such an approach is inherent in problems of optimizing the parameters of dynamical absorbers [31, 32].

1.1.3 Special Aspects of the Motion of Solids in Vibrational Structures

The protection of a solid body that performs a planar motion (Fig. 1.3) is a more intricate problem. In the simplest variant, the basic model (Fig. 1.3a) is a beam with mass and inertia parameters M and J (M is the mass, J is the moment of inertia). The form of the structural diagram of a system with two degrees of freedom depends on the choice of generalized coordinates. In Fig. 1.3a, it is shown that in a system with two degrees of freedom, various additional feedbacks can also be introduced, as well as the elements of an extended set complementing the elements of the basic model.

So, the perturbation in the system can be of either kinematic (z_1 and z_2) nature or connected with force (Q_1 and Q_2), which is related to the consideration of the

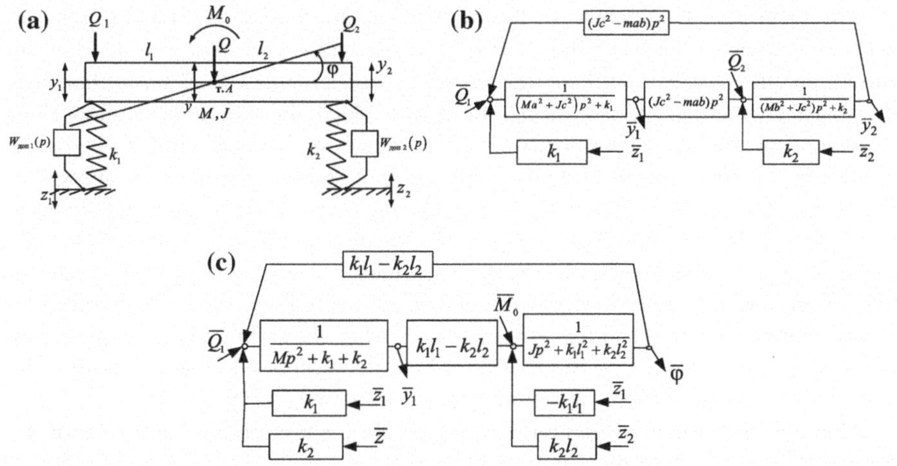

Fig. 1.3 The mechanical oscillatory system with additional feedbacks. **a** The computational scheme of a system with two degrees of freedom; **b** the structural diagram of a system in the coordinates y_1, y_2; **c** the structural diagram of systems in the coordinates y, φ

motion in the coordinates y_1, y_2 (relative to the conditionally fixed coordinate system). In the system there is a center of masses (barycentre) with the coordinate y, and the solid body can make angular vibrations of φ with respect to the barycentre A. Its position is determined by l_1 and l_2.

To assess the capabilities of this model, a number of ratios are used:

$$y = ay_1 + by_2,\ \varphi = c(\varphi_2 - \varphi_1),\ a = \frac{l_2}{l_1 + l_2},\ b = \frac{l_2}{l_1 + l_2},\ c = \frac{1}{l_1 + l_2},\qquad (1.6)$$
$$y_1 = y - y_1 l_1 \varphi,\ y_2 = y + l_2 \varphi.$$

From the computational scheme in Fig. 1.3a, it follows that the protection object in the form of a solid body (M, J) can be represented depending on the choice of the system of generalized coordinates (y_1, y_2 or y, φ) by two structural diagrams of dynamically equivalent automatic control systems (Fig. 1.3b, c). From these structural diagrams, it is possible to obtain the transfer functions necessary for estimating the dynamic properties, for example, $W'(p) = \frac{\bar{y}}{\bar{z}}$ (for $z = z_1 = z_2$, $Q_1 = 0$, $Q_2 = 0$) or $W''(p) = \frac{\bar{y}}{M_0}$ (for $z_1 = 0$, $z_2 = 0$, $Q = 0$) etc. An important circumstance is the need to relate the generalized forces to the coordinate system, which is done under conditions of equality of the work of generalized forces on virtual generalized coordinates [33]. The following properties are characteristic for systems in which the natural (but not artificial) connectivity of motions is implemented due to the fundamental difference between a material point and a solid body in a planar motion (Fig. 1.3b, c): in the basic model, only constraints with elastic elements (k_1 and k_2) exist between partial systems. They can be elastic or inertial. The selection of a system of generalized coordinates can change the nature of cross-coupling.

Introduction of the elements of the extended set to the system, in addition to the elastic elements of the basic model (Fig. 1.4a, k_1 and k_2), is quite simple [12]: by adding parallel elements of the extended set to the elements k_1 and k_2. In this case, the transfer functions of elementary units are added. Further transformations of structural diagrams are performed in accordance with known rules. Dynamic properties of mechanical systems with computational schemes, as shown in Fig. 1.3a, vary within a sufficiently wide spectrum, even when considering the basic model, if we introduce the coordinates of intermediate points into consideration, for example, when considering points that can be located between the gravity center of the system and the coordinates of the elastic elements [16]. When the number of elastic elements increases in the system, that is, when the solid body is supported by not two springs only, but by more of them with fixation at intermediate points, the concept of reduced stiffness is introduced, as discussed in [29, 34–36].

Although the structural diagrams in Fig. 1.3b, c show that the structure of the mechanical system with the computational scheme shown in Fig. 1.3a is a circuit structure, their direct representation as the interaction of two systems, each of which

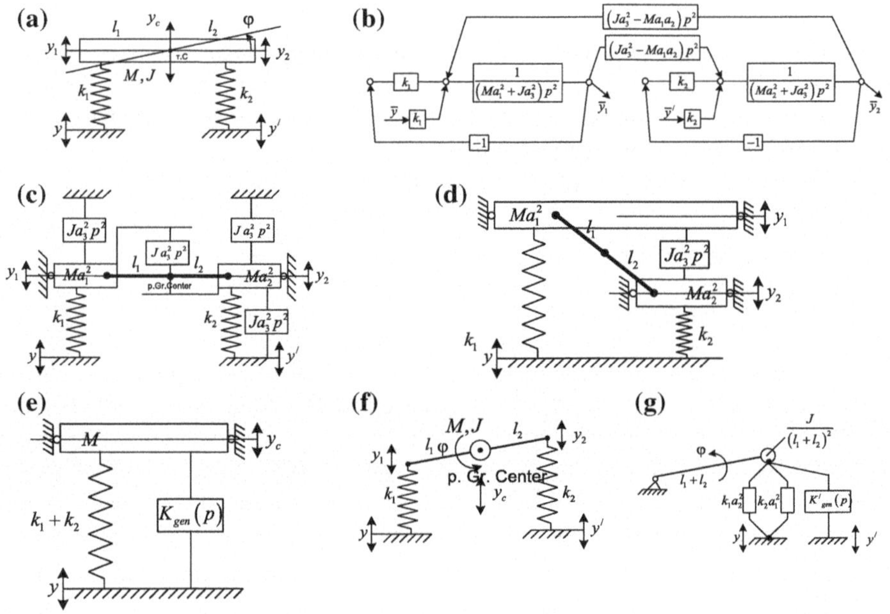

Fig. 1.4 Computational schemes of vibration protection systems (VPS) with two degrees of freedom. **a** The beam-type computational scheme; **b** the beam-type structural diagram with kinematic perturbation; **c** the beam-type computational scheme with the introduced standard units; **d** the computational scheme with the introduction of additional standard units and a lever linkage; **e** the computational scheme of the vibration protection system as a system with one degree of freedom with the introduced generalized spring; **f** the computational scheme of the VPS for deriving the differential equations of motion; **g** the computational scheme with the reduction of the partial rotational block to the translational one

has one degree of freedom, has its own essential aspects. They are in the fact that in the representation of a solid body having a mass M and a moment of inertia J in the coordinate system y_1 and y_2, the reduced masses are respectively of the form: $Ma^2 + Jc^2$ and $Mb^2 + Jc^2$; however, the motion of these mass and inertial elements is "constrained" by the presence of a weightless rod of the length $l_1 + l_2$, rotating around the barycentre. In the interpretations of the interaction of systems in the coordinates y and φ, there are similar features. This allows us to draw a conclusion, the essence of which is connected with the existence of the system of lever connections in systems with solid bodies as objects of protection or with the usual units. The lever linkage has various forms of physical representation, but it is characteristic that the lever transfer ratio is introduced into mathematical models. The lever linkage becomes the defining feature of mechanical oscillatory systems containing solid bodies, which in the end allows us to consider not one-dimensional dynamic interactions, but two-dimensional ones.

The results of studies in such representations were reflected in [37–41]. In particular, the relationships in the form of a ratio between the parameters of mechanical systems, for example, stiffnesses or masses of material points, can be found in [34], but these ratios do not reflect the geometric properties or the "metrics" of the system of forces arising in mechanical systems with solids.

The complication of these mechanical systems due to the introduction of additional constraints, including the units of the extended set, must be accompanied by consideration of such features of the connections as the coordinates of the points of attachment of the elementary units. Most often, the units have the form of dual elements, for which only two positions are controlled. In this respect, the lever is not considered an element of a typical set of elementary units of mechanical systems [42], since it is a unit with three characteristic points. This circumstance is very important, since it acts as a certain boundary, the transition of which should lead to the idea of expanding either the number of elements of a typical set, or a set of rules for connecting typical elements. Separate questions of such approaches of dynamic synthesis of mechanical circuits are considered in [43–46].

Figure 1.4 provides an overview of the transformation of the computational schemes of a system with two degrees of freedom in different coordinate systems and the special aspects of estimating dynamic properties.

In recent years, interest in lever mechanisms in the structures of oscillatory systems has increased noticeably, which was reflected in [17, 41, 44, 47, 48].

1.1.4 The Concept of a Basic Model

The inclusion of solids in the structure of mechanical oscillatory systems as protection objects with several controlled state parameters (at least two) is a research field in which joints play an important role [49, 50]. The joint of solid bodies can be regarded as a unique combination of two solid bodies (in the minimal form, two units), in which two elements having each their generalized coordinate can be described by a new coordinate of the relative motion, and that, in its turn, can tend to zero value. The concept of "joint" is more general in relation to the concept of a "kinematic pair." At the same time, the joint can be considered as one of the forms of constraints, in the understanding of the nature of a restraint in accordance with [1, 33]. The use of joints makes it possible to build mathematical models of a wide class of systems in which various modes of dynamic absorbing are implemented, created by the action of the inertial forces of moving space. Theoretical basics of the method of constructing mathematical models of mechanical systems with joints are given in [51].

The mechanical oscillatory system, the basic model of which includes a solid body, has three elastic supports ($k_1 \div k_3$), a solid body (M, J) in simplified form and executes a motion in three coordinates. Several systems of generalized coordinates are used to describe the motion. In particular, the coordinates of only the vertical motion of three points forming a triangle in the spatial system of Cartesian coordinates can be considered. Such a basic model is simplified with respect to the mathematical model of elastic oscillations of a solid body, which can be examined, for example, in the Eulerian coordinate system. There is sufficiently large number of mathematical models using different systems of generalized coordinates [52, 53]. In the application to the problems of vibration protection and vibration isolation of technical objects, spatial models with six generalized coordinates were developed by I. I. Blehaman, M. Z. Kolovsky; some questions of the simplification of models, taking into consideration the symmetry of forms, are given in [53–55]. The nature of mathematical models is determined not only by the large number of connections between partial subsystems, but also by the fact that, in the general case, the properties of the system are nonlinear. The problems of the dynamics of the rolling stock of railways and road transport are presented in [56–60].

The spatial dynamics of mechanical oscillatory systems is now more characteristic for machines of various purpose (technological robots, robotic complexes, walking machines, including anthropomorphic ones [61–63]), as well as for transport vehicles [64]. Mathematical models of complex mechanical systems related to space objects and rocket systems represent a special approach in the dynamics of machines, in which the main role is played by methods of numerical integration of differential equations.

Thus, mechanical oscillatory systems as the computational schemes of modern machines of various purpose are widely used in engineering practice to solve various problems. Mathematical models of systems reflect the specific character of the problems being solved and do not remain unchanged.

The development of mathematical models towards their complication is initiated by the processes of expanding the methods of mathematical modeling in electrical and electronic systems, which is typical for modern machines with automatic control systems. Integration of mathematical models, as the comparative survey shows, becomes possible, in connection with the sufficient proximity of the analytical tools of the theory of oscillations, the theory of mechanisms and machines, and the theory of automatic control.

1.2 Technical Objects and Special Aspects of Their Mathematical Modeling

Existing methods of modeling physical processes in technical systems of varying complexity make it possible, with a sufficiently clear formulation of the research problem, to offer a step-by-step description of the characteristics of its subsystems, including those that might contain elements of other physical nature. The complexity of the equations does not involve special difficulties for the numerical integration and formation of models of the development of processes under certain external influences. Such approaches have been developed in the dynamics of transport systems [56–60]. In such approaches, the question is how far it is possible to implement the opportunities that are able to reveal the regularities of the occurrence of the studied effects and their parameters that are accessible for measurement. Complex models make these approaches difficult to implement, which predetermines the relevance of research not only on simplifying models on the basis of the linearization methods, but also on the search for the distinguishing of the block pattern of the construction of mechanical systems.

1.2.1 Transport Vehicles and Technological Machines

The system of differential equations, for example, describing the spatial oscillations of the carriage, can be represented as a block of equations simulating the movement of wheel pairs in the track, and a block of equations describing the oscillations of the bolster structure. The selection of blocks facilitates the study and evaluation of the properties of the oscillatory mechanical system, offering the opportunities for specifying one's attention to the features of the mechanical systems themselves. As a result, the obtained effects and the opportunities of their implementation, the selection of rational parameters in the design concepts become the basis of technical developments [65, 66].

The dynamics of machines and transportation vehicles, in particular, is related to the study of the characteristics of vibration processes and their effect on reliability and safety of operation. At the same time, the safety of transport systems is of great importance, which is reflected in the development of methods and means of vibration protection, vibration isolation of suspension systems and cushioning.

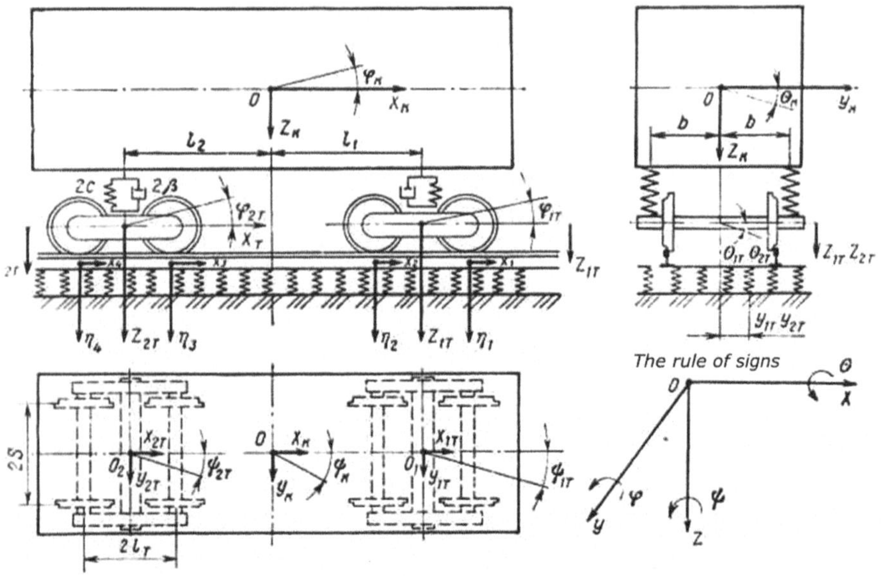

Fig. 1.5 The computational scheme of the four-axle freight car

As an example, Fig. 1.5 shows the computational scheme of a freight four-axle car, which reflects the main features of a mechanical system consisting of three solid bodies [67].

When the car moves along a straight, perfectly smooth path ($z_1 = 0$; $\varphi_1 = 0$; $\theta_1 = 0$; $z_2 = 0$; $\varphi_2 = 0$; $\theta_2 = 0$), eight holonomic constraints are superimposed on the system. Since a free solid body has six degrees of freedom, the system in question, as a whole, will have ten degrees of freedom.

Note that transport systems refer to complex technical objects, which, in turn, consist of equipment and assemblies. Locomotives are the same complex objects, therefore many problems related to the development of specific design and technical solutions and the evaluation of local properties are considered at an appropriate level using the computational schemes and mathematical models of individual locomotive and car blocks.

The computational schemes used to solve local problems of rolling stock dynamics are shown in Fig. 1.6. They are much simpler and are reduced to systems with several degrees of freedom, which can also be simplified.

In design solutions for locomotive suspension systems, it is characteristic to use rather complex connections implemented through various mechanisms. For example, Fig. 1.7a provides a structural diagram of the distribution of vibrations from the source to the operator of the road-building machine, as well as information on the solutions of measuring the level of the arising vibrations (Fig. 1.7b) [68].

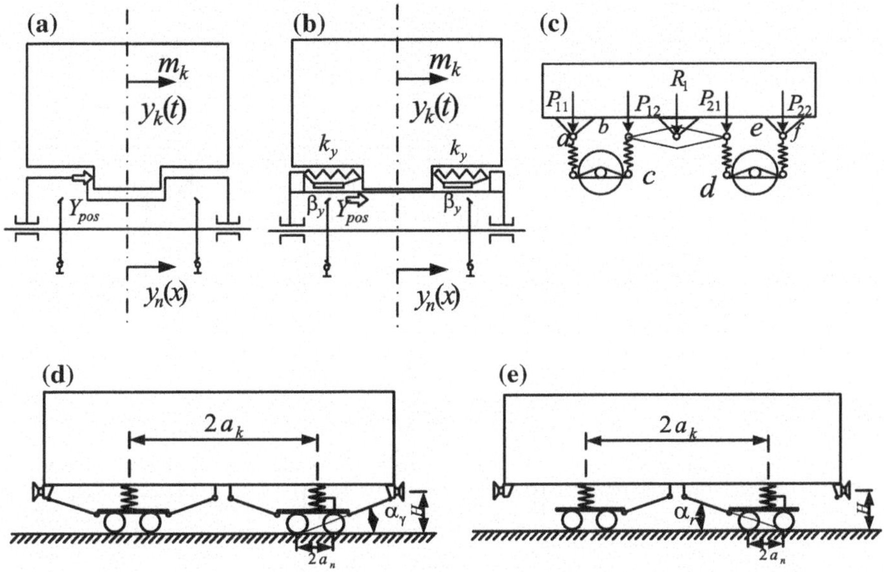

Fig. 1.6 The computational schemes for the study of oscillations of locomotives. **a–c** With lever-type joints; **d** considering lever linkages with traction; **e** considering traction and braking modes

1.2.2 Special Means of Protecting Objects from Vibration

One of the most important problems of ensuring safe working conditions for a human operator under vibration impact is the development and implementation of rational seat structures. Possible variants of constructive solutions are shown in Fig. 1.8, which shows the schematic diagrams reflecting the peculiar features of mechanical systems, manifested in various combinations of elastic, damping and lever elements [55, 69].

The variety of problems involves a variety of methods and means of protection that often take the form of specialized systems for automatically maintaining the parameters of the dynamic state of the protection object. In Fig. 1.9 schematic diagrams of structural implementations of active vibration protection systems [11, 69] are presented.

Such systems are multifunctional and contain, in addition to the mass-and-inertia, elastic and damping elements that are traditional for mechanical systems, devices of a different physical nature—elements of electro-, pneumo- and hydroautomatics, which predetermines interest in generalized methods of analysis and synthesis. The latter is connected with the development of the corresponding sections of the theory of circuits, in particular the theory of electrical circuits and impedance methods.

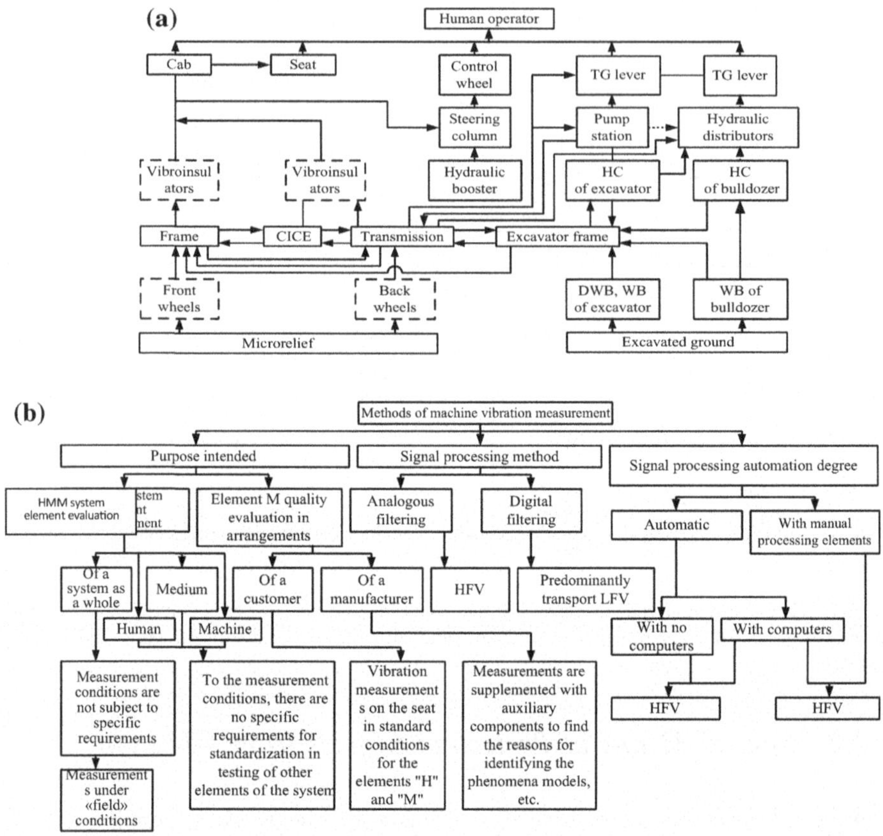

Fig. 1.7 **a** The structural diagram of the vibration propagation from its sources to the operator of a single-bucket excavator; **b** methods for assessing the vibrational background of machines (CICE—control of the internal combustion engine, TG—transmission gear, DWB—dynamic working body, WE—working equipment, HC—hydraulic cylinder, HFV—high-frequency vibration, LFV—low frequency vibration, H—human, M—machine, Med—medium)

The theory of automatic control, if one has in mind the development of multi-functional vibration protection systems (or active VPS), relies on structural methods of analysis and synthesis, and the computational schemes of systems in this case are represented in the form of the corresponding structural diagrams of automatic control systems [11–13, 16]. Elements of such systems are devices using external energy sources, which relate to a set of typical elements of automatic control systems. The capabilities of structural approaches are shown in the scheme shown in Fig. 1.10.

As the basic elements of systems of spring equalizing suspension of vehicles elastic elements are applied; as a rule, they are coil springs or leaf springs, as well as devices for dissipating energy of the vibrations in the form of hydraulic dampers of friction dampers and, in some cases, pneumatic shock absorbers. A protection

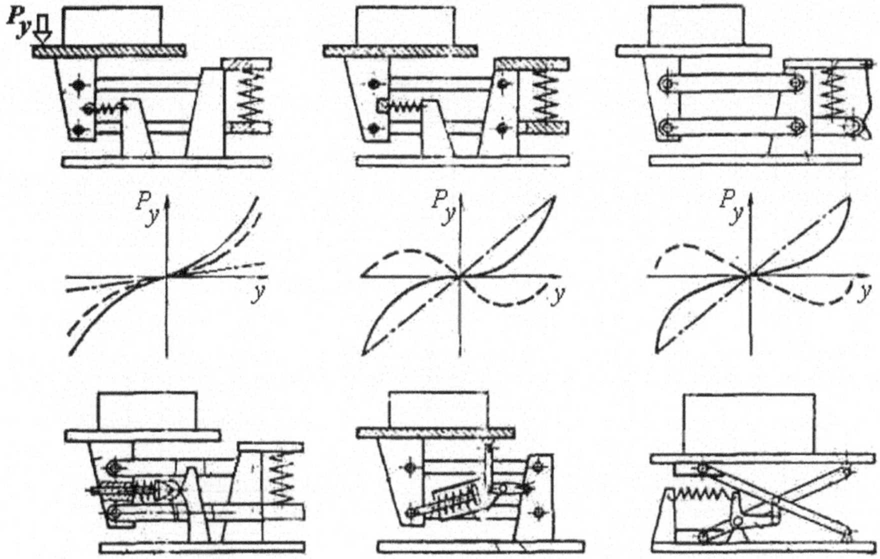

Fig. 1.8 Schematic diagrams of suspensions that allow providing stiffness of elastic elements with a progressive change in force

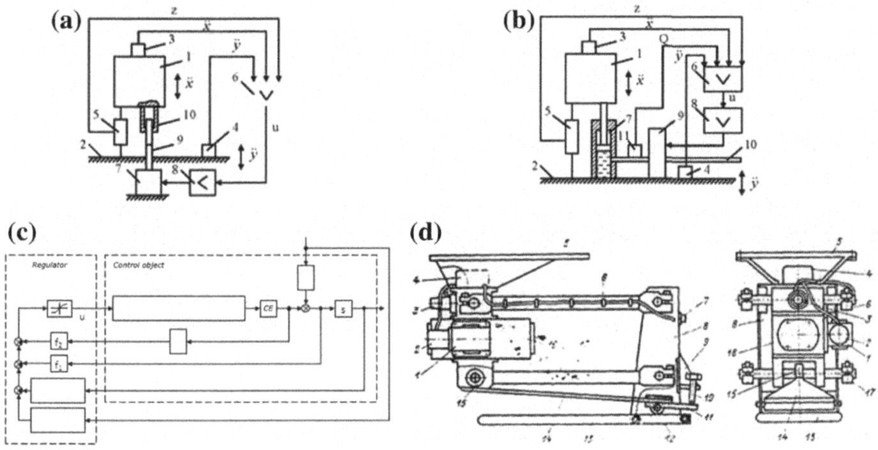

Fig. 1.9 Schematic diagrams of active vibration protection systems. **a** Diagram with electromechanical executive screw mechanism with a transmission “screw-nut” coupling; **b** system with electrohydraulic actuating mechanism; **c** structural diagram of vibration protection with electromechanical actuator and frequency regulator; **d** the scheme of passive suspension of a seat with the hydraulic shock absorber operated automatically on electric channels

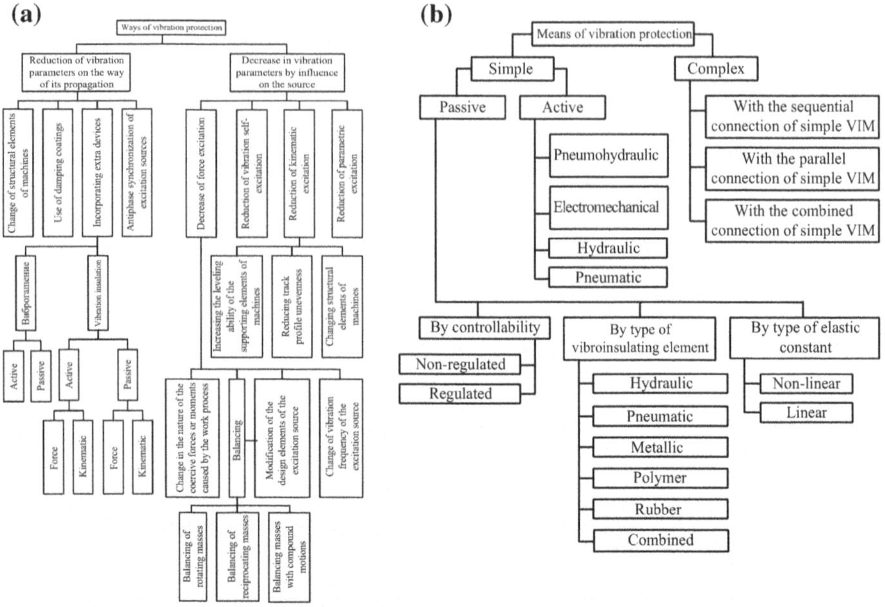

Fig. 1.10 General information on the structures of vibration protection systems. **a** Classification of vibration protection methods; **b** classification of means of vibration isolation taking into account the introduction of active elements

object occurs as backbone in the form of an oscillating loop; often the object has the form of a solid body of beam type or a solid body fixed at a point or having the ability to make more complex motions.

Thus, in most problems of transport dynamics, the mechanical oscillatory system as an computational scheme is a vibrational structure of various degrees of complexity, in which a set of traditional type elements consisting of elastic elements, dampers, mass-and-inertia units or jointed units and lever mechanisms (in a generalized sense), helping to form a spatial structure of interaction of system elements, is used.

The generalized notion of lever interactions is based on the fact that the location points for the attachment of springs, dampers, friction shock absorbers are spaced in a spatial scheme, which implies the possibility of the emergence of peculiar features in the dynamics of the system as a whole. In this direction in recent years, there have been some changes that are associated with the use of controlled forces to form the necessary dynamic state. This is due to the expansion of the set of traditional means of mechanical oscillating systems and is based on the introduction of servo or power actuators. This approach is reflected in the means of protection of rolling stock in the form of controlled pneumatic systems of spring suspension, mechatronic spring suspension systems and active vibration damping systems. Figure 1.11 presents the classification and generalized scheme of the oscillation control system in vibration protection systems.

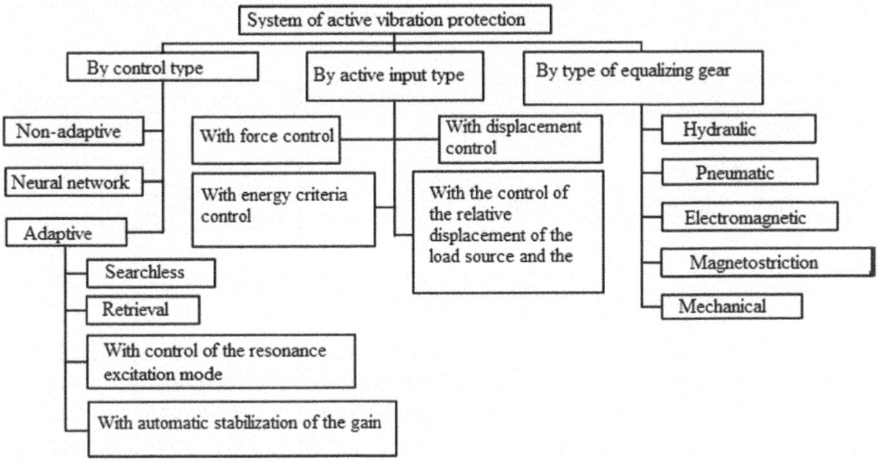

Fig. 1.11 The vibration control in vibration protection systems

Continuing the comparative analysis of assessments of methods, tools, approaches to solving problems of transport dynamics, we note that the complexity of the problem of estimating the dynamic state results in the need for dissection of the general problem to a system of local problems in which the features of dynamic interactions are detailed.

1.2.3 Lever Mechanisms and Devices in the Structures of Mechanical Systems

Lever devices are used in shaping the working space of the balanced spring equalizing suspension system, as well as in creating a system for transferring longitudinal forces from the trolley to the cargo body. The mechanical system through which the dynamic interaction of the trolley with the cargo body is implemented has a rather complex structure, in which the connection of elastic and dissipative elements using lever mechanisms can be noted. To ensure the stability of the mutual position of the body and the trolleys, diagrams are used to connect the mass-and-inertia elements through articulated leverages that determine the conditions for the connection between the cargo body and the trolleys. Suspension of the traction motor in the centerline support structural design ensures the relative movement of the traction motor, which implies the possibility of the existence of modes of dynamic absorbing of oscillations. It should be noted that the leverages providing the system of spatial arrangement of elastic and dissipative elements of spring equalizing suspension are also acting as mechanisms that implement the diagrams of force transmission and the connectivity of motions. A more detailed discussion on the possibilities of using lever linkages is presented in [48], which

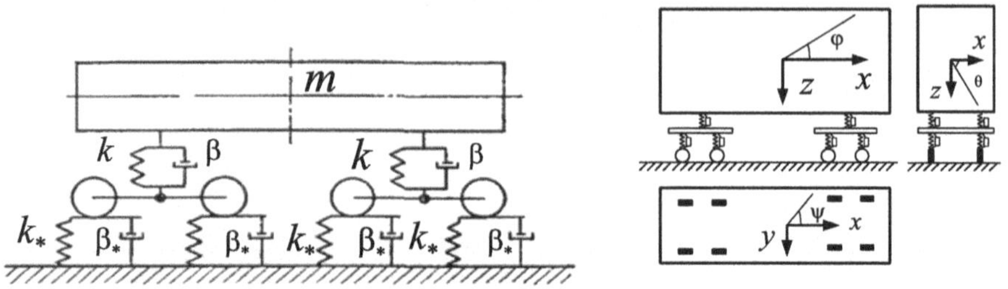

Fig. 1.12 Spatial oscillations of the four-axle freight car

served as a basis for transferring the lever linkages to the locomotive's computational schemes. More complex computational schemes in the dynamics of vehicles are associated with the consideration of objects of protection, which include the body of a mobile vehicle, supported by intermediate mass-and-inertia elements. The computational schemes for examining the spatial vibrations of a four-axle freight car are shown in Fig. 1.12.

Taking into account the features of the dynamic interaction, which involve a comprehensive examination of systems as a whole, has been reflected in [65]. The system of mathematical models for considering the spatial dynamics of cars is also correlated with the ideas of destructuring a complex model into a number of simplified or basic models. An interesting development of analytical approaches are proposals to use computational schemes in which the plane motion of the model is considered (with lateral rolling) taking into account elastic and dissipative elements; the actions of the mechanisms implemented by the system of absorbers and articulation constraints.

Taking the special aspects of the interaction of cars in the train into consideration seems to be promising for further studies of longitudinal dynamics problems. The methodological basis in this case is the investigation of a complex range of possible motions that are implemented in the development of a system of mathematical models, which eventually reveals the need to expand the set of basic models reflecting the most significant properties in the dynamics of vehicles. Basic models take the form of oscillatory systems with one or four (and more) degrees of freedom, in relation to which, with account of certain simplifications, it is possible to obtain analytical relations and to apply methods for the integral estimation of dynamic properties.

A system of models, which conveniently fixes and forms the algorithmic basis for an automated study of the generalized spectrum of the dynamic properties of vehicles, is presented in [9, 60].

Lever linkages, the use of which is quite common in the computational schemes of transport dynamics, are taken into account in the constructive and engineering study of spring equalizing suspension systems, stability of cargo body tilts and lateral oscillations, as well as in traction and braking modes. The areas of

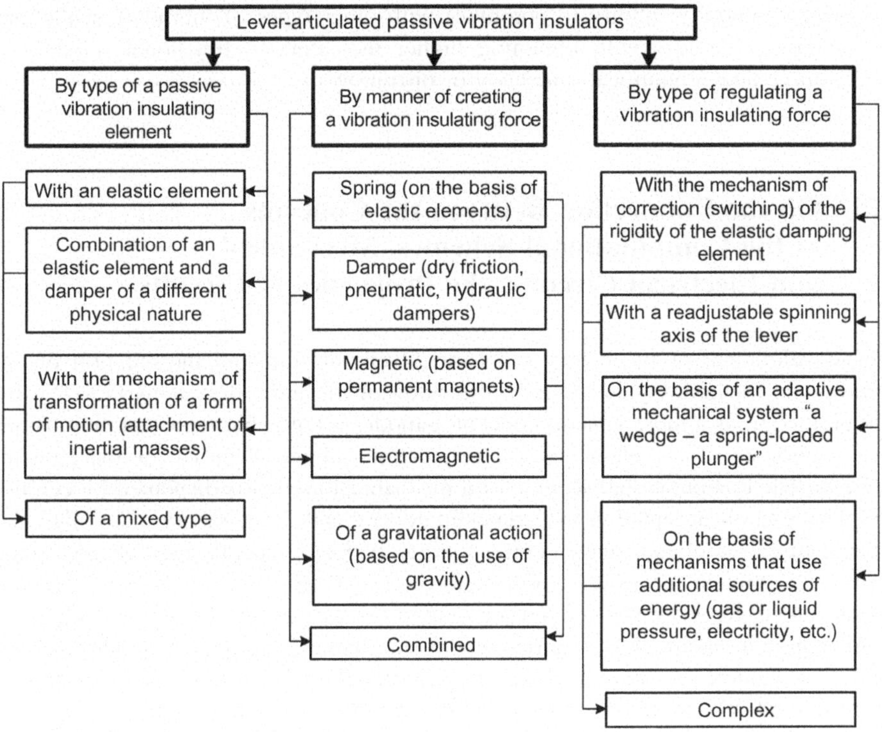

Fig. 1.13 Classification of passive articulation vibration insulators

application of articulation linkages in railway transport, presented in [40, 41], can be substantially expanded. Figure 1.13 presents a classification diagram of the articulation passive vibration insulators.

In recent years, separate development of passive lever vibration protection devices has been increasingly introduced into the practice of railway transport [70–72], which implement the useful aspects of lever mechanisms. In particular, it is believed that lever vibration protection devices have:

- a relatively low propensity to self-oscillations and the superimposition of resonant frequencies of oscillations of elastic links on the vibrations of the elastically suspended elements of the effector (which, apparently, is far from being the case);
- an ability to work in a wide range of external loads with small strokes and forces of elastic links with self-adjustment of these parameters during the reorganization of operating modes;
- an easy maintenance and correction of output parameters under a varied amplitude-frequency spectrum of external loads;

- good jointability (ability to combine) with other types of vibration insulating (damping) devices with obtaining higher indicators of functional reliability, resource and economical efficiency of operation.

1.3 On the Connection Between the Concepts of the Computational Schemes, Mechanical and Electrical Circuits, the Properties of Circuits

The development of the theory of electrical circuits initiated the creation of an appropriate methodological basis in the study of mechanical systems [73–75]. In circuit theory, as a more general concept, a model is a graphic representation using the symbolism of the elements and connections of the electrical or mechanical circuits. The computational scheme of a mechanical system is displayed by certain graphic symbols accepted in mechanics in general and in the theory of oscillations in particular. Graphical images correspond to typical elements of mechanical circuits and contain necessary information on the structure of constraints and parameters of the system. Such a representation of a mechanical system is called an computational scheme. It is assumed that the mechanical circuits are an integral part of the mechanical system as a whole. In microelectronics, the concepts of "electrical circuits" and "electronic circuits" are often identified with each other.

In the analysis of circuits in general and mechanical circuits in particular, matrix and topological methods are widely used. They are based on the representation of the circuit with the help of its graph. A graph of a circuit is the geometric system of lines (branches) connecting the set points (nodes). If the branches of the graph are oriented in a direction (for example, in the direction of the branch currents), then the graph is called oriented (directed). The graph contains information on the geometric structure of the computational scheme. In relation to mechanical systems, such approaches were developed in [9, 76]. Graph theory as a scientific discipline, offering analytical tools for evaluating the properties of systems, is now sufficiently developed that makes it possible to use formalized approaches for classifying structures and taking into account their principal features. However, the further development of the matrix and topological approach is associated with a detailed description of the typical elements of the chain, the physical nature of which is different. In particular, mechanical systems cannot always be represented by systems consisting only of material points; there are also solid bodies, which are characterized by interrelated motions. Interpretation of such motions with the help of a traditional set of elements is already becoming difficult to implement.

It can also be noted that the extension of the concepts of the typical elements (and in this case, the question is about their complication) also changes the idea of the nature of the links between the system elements. For the theory of circuits, transformations related to the simplification of circuits or the allocation of certain sections of circuits for detailed study are of great importance. In this respect, there are differences between the electrical and mechanical circuits. They are due to the different nature of physical processes. The analogy is based on the similarity of mathematical models, but not on their identity. The basis of various methods for transforming electrical circuits is the principle of equivalence, according to which the voltage and currents in the branches of the circuit not affected by the transformation remain unchanged [77, 78].

Structural approaches. In the structural theory of mechanical vibrational systems, in particular in the structural theory of vibration protection systems [11–13, 16], the basic laws of circuit theory have developed in the substantiation and detail of the notions of additional feedbacks. In the physical plane, the additional feedback is a mechanical circuit introduced into the structure of the oscillatory system. Note that the structural model of the vibration protection system (or mechanical oscillatory system) as a whole differs from the model in the form of a mechanical circuit. The main difference lies in the fact that in the structural model of a vibration protection system (VPS), an object is selected which dynamic state is provided by the effect of forces arising from the interaction of passive and active elements. The structural model is built in the conceptual space of the automatic control theory. The control object is a mass-and-inertia element in the form of a material point or solid body. The transfer function of the object taken separately corresponds to the transfer function of the second-order integral unit. The input of such a unit is the resultant force formed by the interacting elements. The structural model has direct and inverse relations (which can be positive and negative). The structural model of a vibration protection system with several degrees of freedom is built on the basis of the separation of partial systems with them to be subsequently linked into the overall structure. In this model, there are internal (with respect to the partial system) and interpartial constraints, which are called cross-couplings in the automatic control theory. The element base of the structural model of a mechanical system in a minimal set corresponds to a set of a mechanical circuit that is an analog of an electrical circuit.

The structural model of a mechanical system can be reduced to a mechanical circuit if there are no active elements in the system. In electrical circuits, these are dependent sources of current or voltage; in the structural models of active vibration protection systems, these are servo drives. Transitions from active vibration protection systems to passive ones are not sufficiently clear, therefore, a fairly large number of transitional forms are considered in the scientific literature.

One of the essential issues can be called the duality in the implementation of the function of a mass-and-inertia element in vibration protection systems with several degrees of freedom. If an element of mass m (in the system of material points) is acted upon as an object of control, then the remaining mass-and-inertia elements can be considered from two positions.

1. The first option assumes that other masses that are not objects of control, nevertheless form partial systems and a system of connections between themselves and the object of control (in the problems of vibration protection it is the object of protection). They can be called auxiliary control objects: they can be convolved with the help of equivalent transformations if necessary, and reduced to the main control object in the form of additional feedbacks.

A similar technique is used in the theory of electrical circuits, when a so-called active impedor is formed. As such an active impedor, the partial system of the protection object is "collapsed" to an active impedor.

This technique can be extended to systems in which the object of protection consists of two material points. Let's note that for cases with one and two degrees of freedom, this approach is correlated with the concepts of the choice of base models, that is, when the base model corresponding to the problem is selected, and the rest of the system is reduced to the basic way of transformations and simplifications. Such approaches are reflected in [12, 79].

2. The second option is that after selecting the control object (or protection object), the remaining masses are regarded as typical elements of the mechanical circuit at the level of electromechanical analogies in the theory of circuits. In this case, the mass-and-inertia element has a transfer function corresponding to the selected type of electromechanical analogies, and obeys the rules of equivalent transformations of the theory of circuits. Note that, being the active impedor, this element has its own peculiarities, since one of the ends of such an active impedor is connected to the base. This is an important circumstance that should be taken into account in the equivalent structural transformations of the structural models of mechanical oscillatory systems.

With the expansion of a typical set of elements of mechanical oscillatory systems, two types of inertia and mass units in the form of active impedors are distinguished in [11, 12]. The links of the first type have one end connected with the base, with corresponding features of the rules of equivalent transformations. The second type of elements corresponds to an active impedor with a similar transfer function, but the active impedor has two ends in this case, which have the same capabilities as the elastic element. If the output of an elastic normal element corresponds to a force proportional to the coordinate difference, then in the mentioned mass-and-inertia element the output corresponds to a force proportional to the difference of the accelerations of the input and output points. The physical form of the implementation of such an element can be represented by a non-locking screw pair without friction [30, 79].

Concluding the review of the theory of circuits in terms of a comparison of electrical and mechanical circuits, it is necessary to mention the many coincidences, which provide the basis for electromechanical analogies. At the same time, the differences in the physical nature of the chains predetermine many differences in the properties of the chains, especially in taking into account the features of the energy sources, the possibilities of dual matching, contour approaches and transformations.

One of the issues of the further development of the theory of mechanical circuits is the expansion of the element base of chains, in particular, due to the introduction of elements reflecting the properties of solids, etc.

1.4 Analytical Approaches to the Estimation of the Dynamic State

The variety of problems in the dynamics of machines and transportation vehicles has been reflected in a wide range of used computational schemes. The latter are based on mechanical oscillatory systems with one, two and several degrees of freedom, but their element base goes beyond the framework of the usual concepts typical of theoretical mechanics, the theory of oscillations and the theory of mechanisms and machines, which is associated with the use of technical systems, saturated with automation and complex mechanical devices, in the machines. To some extent, the theory of circuits [73–75] meets the emerging requirements of a unified methodological basis to be established for solving problems of analysis and synthesis. But between the electrical, mechanical and electronic circuits, there are great differences despite the obvious electromechanical analogies. This occurs due to many reasons, including the fact that the nature of the processes in the circuits themselves has not yet been sufficiently studied. The mechanical circuits and their theory, as expounded in [75, 77, 78], in many respects coincides with the theory of electrical circuits. At the same time, as the review shows, mechanical systems include solid bodies, lever mechanisms, motion translation devices (gear, screw mechanisms, etc.), electromagnetic, electrodynamic devices that make up an interconnected system. The further development of the theory of circuits depends on the expansion of concepts of the element base of systems, which leads to the consideration of complex elements in the form of four- and multiterminal networks, and to their usage in circuits of nonplanar connections. The interaction of the elements of the mechanical circuit with energy sources, which takes into account the limited power, is a topic that has been little studied so far. Compared with mechanical circuits, electrical circuits have a large set of typical elements and a more developed understanding of the interactions that are characteristic for the transformation and transmission of electrical energy (transformers, lines). Also, the analytical tools of the theory of electric circuits turned out to be more developed with respect to the possibilities of taking nonlinear factors into account.

Many devices of electrical nature have analogs in mechanical circuits, but nonetheless, the process of adapting new concepts, connections and properties lags behind the growth of the needs for solving the problems of the dynamics of complex technical systems. In this regard, the development of structural approaches based on the theory of circuits is quite relevant. Also worth noting that there are problems of agreement and compatibility of the structural methods of the theory of circuits and the automatic control theory. In both directions, there is the same theoretical basis in the form of the initial system of differential equations obtained on the basis of the corresponding laws. However, the structural interpretations of these two approaches are different due to the fact that the structural models of the automatic control theory are constructed using the principles of feedback. It should be noted that there is a control object in the structural diagrams of automatic control systems. There is no such element in the theory of circuits, which predetermines a number of differences in structural models, the rules for their construction and transformation. We believe that this area of research deserves special attention.

The existing analytical tools that used to solve problems of dynamics are characterized by a great diversity and is based on the methods of theoretical mechanics, the theory of oscillations, the theory of mechanisms and machines, and the control theory.

Traditional forms of mathematical models. Mathematical models in the form of systems of equations of the second and first order has become the most common in the assessment of the dynamic state of technical systems, including those of vibration protection. Systems with lumped parameters are considered; it is assumed that the system has linear properties. Problems of the nonlinearities taken into account, as a rule, require a separate approach and are connected with the study of local effects. The mathematical model can be developed in the form of a second-order matrix-vector equation

$$\theta\ddot{\varphi} + B\dot{\varphi} + G\varphi = F(t), \tag{1.7}$$

where $\bar{\varphi}$ is the n-dimensional vector of generalized coordinates; θ, B, G are symmetric matrices, respectively: inertial, dissipative and elastic; $F(t)$ is the n-dimensional vector of the external influence function. In many practical cases, the matrix θ is diagonal; in cases when it has a structure different from diagonal, it is always possible, by means of a nondegenerate modal transformation with respect to the initial inertial matrix of the transformation, to transform the system (1.7) to the form of a diagonal matrix.

If, for example, active units with non-linear properties are introduced in the system of vibration protection and vibration isolation, the system of equations is linearized. The mathematical model can be represented by a vector-matrix equation of the first order in the normal form

$$\dot{x} = Ax + F(t), \tag{1.8}$$

where x is the n-dimensional state vector (of the phase coordinates); A − $(n \times n)$ is the real or complex matrix; $F(t)$ is the n-dimensional vector function of external influences. In the computational terms, the fundamental difference between the models (1.7) and (1.8) lies in the difference of the parametric matrices: for the circuit models, the matrices Q, B, G are symmetric, and for models with directed couplings, the matrix A has an arbitrary structure (asymmetric in general or not reducible to the symmetric one). In those cases when the system has a more complex character, that is, the system is not simple circuit, the matrices B and G can become absolutely dense.

The class of mathematical models can be significantly expanded, taking into account the specifics of the problems of dynamics, and can include mathematical models in the form of systems of partial differential equations, integro-differential equations, etc.

Given the possibility of creating mathematical models that are convenient in terms of taking into account the properties of interaction of elements of systems of different physical nature, then, in this respect, structural models are of interest. Figure 1.14a–d shows the scheme of the relationship of the basic concepts of the structural method in the dynamics of mechanical oscillatory systems [12, 80]. In Fig. 1.14a, the computational scheme of a system consisting of traditional elements (such a model is called the base one) is shown. Its dynamic properties are described by a linear differential equation with constant coefficients (Fig. 1.14a, pos. I). The corresponding structural diagram of the dynamically equivalent automatic control system is shown in Fig. 1.14b. To describe the dynamic properties in this case, the transfer functions are used (Fig. 1.14b, pos. II). The structural diagram (Figs. 1.14b) reflects all the properties of the mathematical model (Fig. 1.14a, pos. I), and the transfer functions themselves (Fig. 1.14b, pos. II) can be obtained from the differential equation. However, structural diagrams carry information on many other properties of the system, which is of particular interest in problems of dynamics. The composition of the links reflected by the structural diagram (Fig. 1.14b) can be extended to implement different principles of dynamic state control in the system (Fig. 1.14c). In practical terms, this is related to the formation of additional links (or circuits taking the form of mechanical, electromechanical ones, etc.), specifically introduced into the system. However, the elements of the basic system alone can already serve as simplified forms of feedback by relative and absolute deviations. Figure 1.14d shows the structural diagram corresponding to the computational scheme in Fig. 1.14a. The transfer function of the system with the introduction of additional constraints (Fig. 1.14c) is represented by the expression $W(p)$, which is shown in Fig. 1.14f.

The introduction of additional constraints in practical terms is a rather complex issue, for the solution of which it is necessary to choose the places of application of the points of the active element, if the latter implements the controlling force. In this case, additional circuits should have devices for measuring, processing and amplifying signals (Fig. 1.14d).

Additional constraints. The introduction of additional constraints, as shown in Fig. 1.14, can be carried out by means of separate elements of mechanical circuits and mechanisms. Figure 1.15 shows that the introduction of an additional element in the circuit mechanical system with a second-order differentiation transfer function (a unit Lp^2) can change the values of the fundamental oscillation frequencies and the mutual position, as illustrated by the expressions (1)–(3) in Fig. 1.15. Figure 1.15b shows how the frequencies change with the variation of the parameter L. The properties of mechanical systems are regarded in [12, 30].

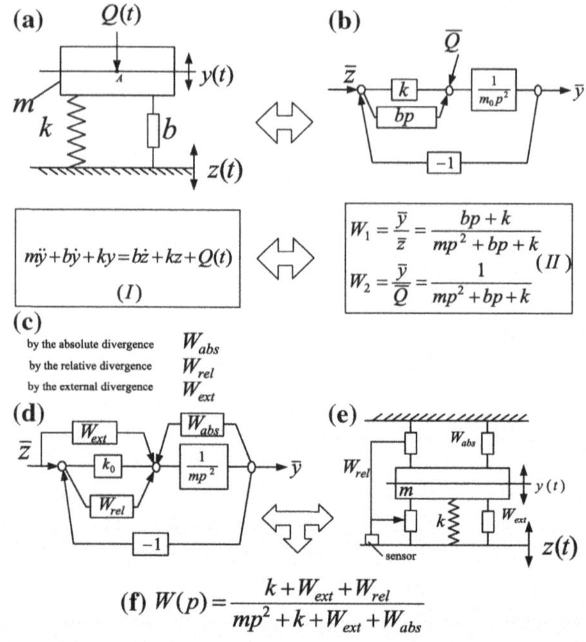

$$W_1 = \frac{\bar{y}}{\bar{z}} = \frac{bp + k}{mp^2 + bp + k} \qquad W_2 = \frac{\bar{y}}{\bar{Q}} = \frac{1}{mp^2 + bp + k} \quad (II)$$

$$\text{(f)}\ W(p) = \frac{k + W_{ext} + W_{rel}}{mp^2 + k + W_{ext} + W_{abs}}$$

Fig. 1.14 The diagram of the basic concepts of the structural method

Additional constraints can take the form of lever mechanisms, as shown in Fig. 1.16a. When constructing a mathematical model and during its linearization, the structural diagram of the system takes the form in accordance with Fig. 1.16b.

Variants of mechanical systems with geared lever linkages are shown in Figs. 1.17 and 1.18.

More complex constructive and engineering solutions in their linear interpretations are simply "integrated" into structural diagrams, creating conditions for solving problems of analytical and dynamic synthesis. The above examples confirm the need for further developments in the field of expanding the set of standard units, since formalized representations at the level of ordinary elastic and dissipative elements become difficult. The introduction of additional constraints in the form of electromechanical circuits in principle does not differ from the introduction of additional constraints the form of mechanisms. However, electrical circuits have a different element base, and structural models take the form of structures with dual elements [77, 78].

Real objects in the dynamics of machines are complex technical systems, consisting of blocks and elements of different physical nature, often electromechanical systems. As for the consideration of the design features of machines, in particular of vehicles, their computational schemes are mechanical oscillatory systems with many degrees of freedom. These systems consist, in their turn, of standard units, the joining of which determines the possibilities of dynamic interactions. At the same

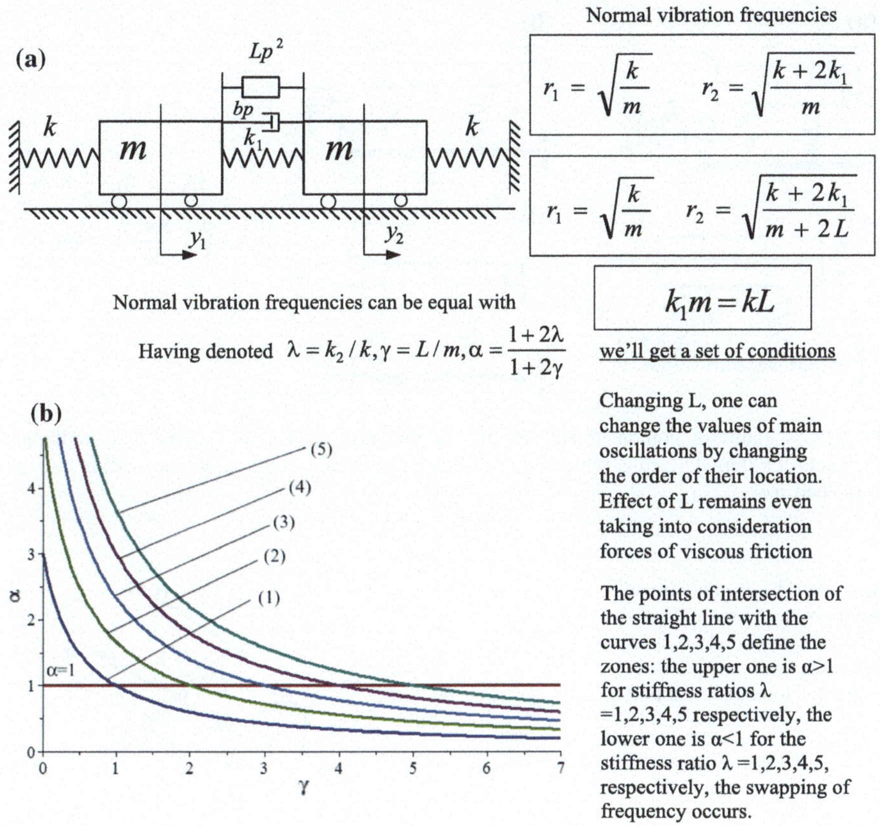

Fig. 1.15 The diagram and couplings between the system parameters when introducing additional element

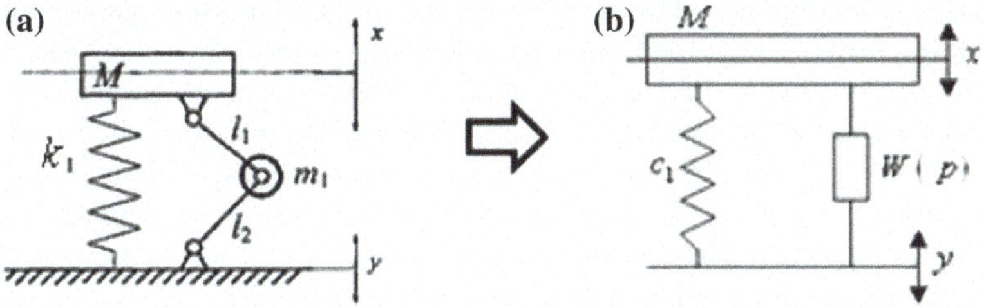

Fig. 1.16 The computational scheme of a vibration protection system with an extra constraint in a form of a lever mechanism

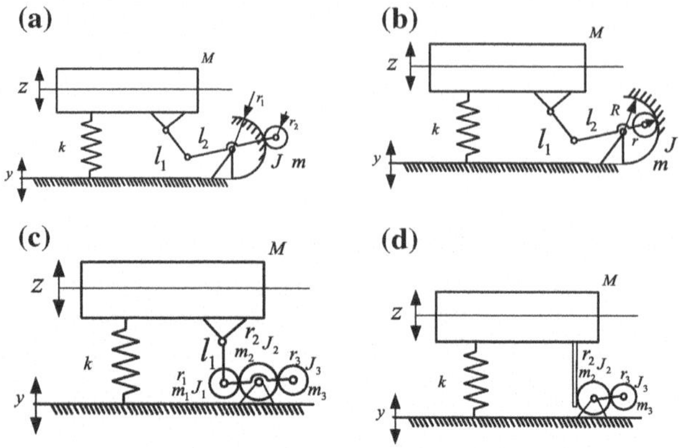

$$\omega_{dyn} = \sqrt{\frac{K + a^2 K_1 + \frac{1}{2} m\omega_0^2 a_2^2}{\frac{ma^2}{2}(1 - ctg\alpha_1)}}$$

Fig. 1.17 Variants of computational schemes of vibration protection systems with additional constraints. **a** External gearing; **b** internal gearing; **c** epicyclic transmission; **d** transmission through the toothed rack

Fig. 1.18 The computational scheme of a system with a two-drive group and an additional mass *m*, to which centrifugal forces are applied

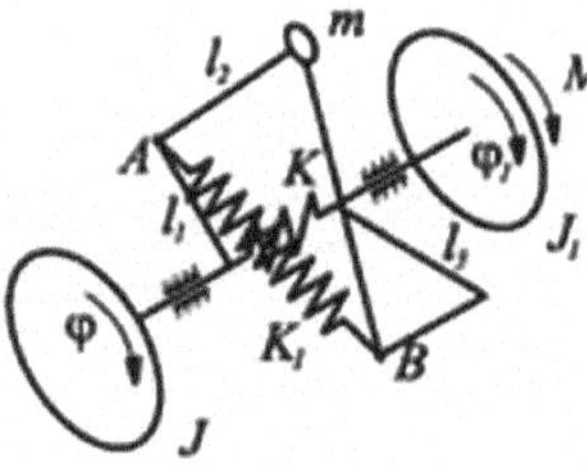

time, it can be noted that lever mechanisms play an important role in mechanical systems. At first glance, the lever mechanisms seem to fall out of the typical set of elements. However, their presence also predetermines the mechanical system of a new quality to be obtained. This is due to the formation of a certain geometric space in which the force vectors (and other parameters) are not in the same straight line. Similarly, in the theory of electrical circuits, transformers of various types are considered. In electrical circuits, transformers "break" the galvanic circuit, and their use often gives rise to nonplanarities in the structural models of circuits. The same can, apparently, be noted with respect to the elements of mechanical systems, including a solid body. Thus, in the dynamics of machines, mechanical oscillatory systems are the computational schemes for solving problems of vibration protection, control and management of the dynamic state of transport devices, robots, manipulators, etc.

Mathematical models in the form of circuits are widely used to evaluate the dynamic state, the selection of system parameters and specific modes of operation (dynamic absorbing, resonance, etc.). A detailed concept of the possibilities of changing the dynamic state by forming control forces makes it promising to use structural models in the form of dynamically equivalent structural diagrams of automatic control systems.

Peculiar features of mechanical systems. The element base of mechanical oscillatory systems is developed by adding links of double differentiation, single and double integration to the known elementary standard units in the form of elastic and dissipative beams typical elements with transfer functions. The list of typical elementary links can be extended. Dual elements are used as elementary model elements in mechanical circuits. The connection of the dual elements is carried out according to the rules of parallel and consecutive connection of the springs. The mechanical circuit can be transformed into a structural diagram of an equivalent automatic control system and vice versa.

Concepts of a typical set of elementary units of mechanical systems can be changed due to the fact that the development requires the interaction of not only material points, but also material points and solid bodies that perform both planar and spatial motions. Lever linkages, implemented by lever mechanisms, are typical of mechanical oscillatory systems that include solid bodies (plane motion). Depending on the forms of implementation, such mechanisms can take into account the stiffness of the levers and their mass-and-inertia parameters. Complex mechanical systems can be considered as integral forms formed by compounds of basic models, which presupposes the possibility of describing motion in various systems of generalized coordinates, as well as the estimation of the dynamic properties of systems through combinations of coordinates with obtaining the corresponding peculiar features of joint motions.

Thus, the solution of problems in the dynamics of machines is connected with the analysis of the dynamic interaction of elements and devices that are different in physical nature, which requires attention when constructing computational schemes and the selecting adequate mathematical models. With all the variety of mathematical models used, one of the most promising directions in the development of analytical forms for mapping the patterns of dynamic processes is structural methods. The latter is due to the fact that many machines and their equipment function as automatic control systems. When the control object coincides with specific problems of ensuring a dynamic state, especially such problems as vibration protection and vibration isolation, reliability and safety during periodic external impacts are effectively performed on the basis of operator methods of the automatic control theory and the theory of circuits. Transfer functions in mechanical oscillatory systems, which are the computational schemes of many machines and mechanisms, can be quite a universal and versatile carrier of information on the dynamic properties of systems. In this respect, the analogous properties are distinguished by the complex resistances of the electric circuits, which analytical tools for studying (the theory of circuits) have developed significantly in recent years.

Mechanical oscillatory systems can be regarded as mechanical circuits, and consequently can have the same properties as electric circuits, in accordance with the principle of electromechanical analogies. Therefore, they have analogous structural interpretations (or diagrams), formed by dual (bipolar) elements; the same rules for circuit transformation are used. At the same time, when considering the units in the mechanical circuit structures in the form of a solid body that performs a plane motion, the features conditioned by lever linkages occur during the interactions. This requires the development of specific approaches and techniques for structural transformation of circuits and the use of electromechanical analogies. The application of structural methods of the theory of automatic control makes it possible to compare mechanical systems with an automatic control system equivalent to them dynamically. In this case, a control object is allocated in the structural diagram of a mechanical system, and the mechanical system as a whole can have several degrees of freedom. In this case, the structural diagram of a mechanical system can contain in its composition both the control object and other types of links, including in the form of mechanical and electrical circuits. It can be assumed that such an approach can be extended to other types of circuits. Consideration of various circuits, taking into account the principle of electromechanical analogies, makes it possible to create a unified methodological basis for building a generalized technology for mathematical modeling of dynamic processes that arise in functional systems of machines, and to determine the parameters necessary for solving dynamic problems.

References

1. Theory of Mechanisms and Machines. Terminology, vol. 93, 34 pp. Nauka Publ., Moscow (1978)
2. Naucler, P.: Modelling and control of vibration in mechanical systems, 102 pp. Dissertation for the degree of Licentiate of Philosophy in Electrical Engineering, Uppsala University, Sweden (2005)
3. Nelson, P.A., Elliott, S.J.: Active Control of Sound, 562 pp. Academic Press Publ., Leningrad (1992)
4. Genkin, M.D., Eliseev, S.V., Migirenko, G.S., Frolov, K.V.: The principles of modern vibration protection. In: Collection of Scientific Papers Vibration Isolation Mechanisms and Machines, pp. 3–13. Novosibirsk (1984)
5. Kernop, D.: Principles of designing oscillation control systems using semi-active dampers. In: Dynamics of Systems, Mechanics and Control: Proceedings of ASME, vol. 112, 3, pp. 448–453 (1990)
6. Ruzhichka, D.J.: Active vibration protection systems. In: Testing Instruments and Stands, no. 10, pp. 14–25. Express-inform, VINITI, Moscow (1969)
7. Eliseev, S.V., Kuznetsov, N.K., Lukyanov, A.V.: Controlling Robots' Oscillations, 317 pp. Nauka Publ., Novosibirsk (1990)
8. Eliseev, S.V., Reznik, Yu.N., Upyr, R.Yu., et al.: Mechatronics Vibration Protection Systems. Elements of the Theory, 128 pp. IrGUPS Publ., Irkutsk (2009). Dep. in VINITI 27.11.09, No. 738-B 2009
9. Tarasik, V.P.: Mathematical Modeling of Technical Systems, 640 pp. Design-PRO Publ., Minsk (2004)
10. Frolov, K.V., Furman, F.A.: Applied Theory of Vibration Protection Systems, 286 pp. Mashinostroenie Publ., Moscow (1985)
11. Eliseev, S.V., Reznik, Yu.N., Khomenko, A.P.: Mechatronic Approaches in the Dynamics of Mechanical Oscillatory Systems, 384 pp. Nauka Publ., Novosibirsk (2011)

12. Eliseev, S.V., Reznik, Yu.N., Khomenko, A.P., et al.: Dynamic Synthesis in Generalized Problems of Vibration Protection and Vibration Isolation of Technical Objects, 523 pp. ISU Publ., Irkutsk (2008)
13. Khomenko, A.P., Eliseev, S.V., Ermoshenko, Yu.V.: System Analysis and Mathematical Modeling in Mechatronics of Vibration Protection Systems, 288 pp. IrGUPS Publ., Irkutsk (2012)
14. Eliseev, S.V., Artyunin, A.I., Ermoshenko, Yu.V., et al.: Methodological Approaches in Systems Analysis and Mathematical Modeling of Mechanical Oscillatory Systems, 96 pp. IrGUPS Publ., Irkutsk (2013). Dep. in VINITI 07.02.13, No. 37-B 2013
15. Eliseev, S.V., Artyunin, A.I., Ermoshenko, Yu.V., Moskovskikh, A.O., Kaimov, E.V.: The Method of Transformation of Mechanical Circuits Based on the Introduction into the Connections of Intermediate Devices, 68 pp. IrGUPS Publ., Irkutsk (2014). Dep. in VINITI 14.01.14, No. 19-B 2014
16. Eliseev, S.V., Khomenko, A.P.: Dynamic Damping. Feedback Concept and Structural Methods of Mathematical Modeling, 357 pp. Nauka Publ., Novosibirsk (2014)
17. Ivashchenko, N.N.: Automatic Adjustment. Theory and Elements of Systems, 4th edn., updated and revised, 736 pp. Mashinostroenie Publ., Moscow (1978)
18. Eliseev, S.V., Artyunin, A.I., Logunov, A.S., et al.: Dynamic Interactions of Machine Elements: Computational Schemes and Mathematical Models of Vibration States, 319 pp. IrGUPS Publ., Irkutsk (2013). Dep. in VINITI 08.11.13, No. 313-B 2013
19. Khomenko, A.P., Eliseev, S.V., Artyunin, A.I., et al.: Structural Mathematical Models in Problems of the Dynamics of Mechanical Oscillatory Systems, 231 pp. IrGUPS Publ., Irkutsk (2015). Dep. in VINITI 30.03.15, No. 62-B 2015
20. Eliseev, S.V.: Structural Theory of Vibration Protection Systems, 224 pp. Nauka Publ., Novosibirsk (1978)
21. Eliseev, S.V.: The theory of active vibration protection systems and its applications: systems, 35 pp. Author's abstract of Dr. Sc. (Engineering) thesis, Tymoshenko Institute of Mechanics of AS Ukr. SSR, Kiev (1973)
22. Eliseev, S.V., Moskovskikh, A.O., Bolshakov, R.S., Savchenko, A.A.: The Possibilities of Integrating the Methods of the Theory of Circuits and the Theory of Automatic Control in Problems of the Dynamics of Machines [Electronic Resource], 6th edn. technomag.edu.ru: Science and Education: Electronic Scientific and Technical Publication (2012). Access mode: URL: http://technomag.edu.ru/doc/378699.html. Access date: 10.06.2012
23. Logunov, A.S.: Dynamics of pneumatic elements and motion transformation devices in the systems of vibration protection of objects, 20 pp. Author's abstract of Ph.D. (Engineering) thesis, IrGUPS Publ., Irkutsk (2009)
24. Nasnikov, D.N.: Forms and features of dynamic interaction of units in vibration protection systems with an extended set of typical elements, 22 pp. Author's abstract of Ph.D. (Engineering) thesis, IrGUPS Publ., Irkutsk (2009)
25. Dimov, A.V.: Modeling and dynamic processes in generalized problems of vibration protection and vibration isolation of technical objects, 26 pp. Author's abstract of Ph.D. (Engineering) thesis, IrGUPS Publ., Irkutsk (2006)
26. Eliseev, S.V., Belokobylsky, S.V., Upyr, R.Yu.: The generalized spring in the tasks of machines and equipment. In: The Assistant of Scientific Works (Branch Mechanical Engineering, Construction), vol. 1, issue 3 (25), pp. 79–89. PoltSTU Publ., Poltava (2009)
27. Khomenko, A.P., Eliseev, S.V., Belokobylsky, S.V., Upyr, R.Yu., Trofimov, A.N., et al.: The method of regulating the rigidity of the vibration protection system and the device for its implementation. Pat. 2440523 Russian Federation, IPC F16F15/4 F16F 7/10. Applicant and patent holder: IrGUPS. Appl.: 01.02.2010; publ. 01/20/2012, Bull. No. 2, 5 pp
28. Khomenko, A.P., Eliseev, S.V., Artyunin, A.I., et al.: Mechanisms in Elastic Oscillatory Systems: Features of Taking into Account Dynamic Properties, Problems of Vibration Protection of Machines, Instruments and Equipment, 187 pp. IrGUPS Publ., Irkutsk (2013). Dep. in VINITI 08.15.13, No. 243-B 2013

29. Belokobylsky, S.V., Eliseev, S.V., Sitov, I.S.: Dynamics of Mechanical Systems. Lever Linkages and Inertial-Elastic Constraints, 319 pp. Polytekhnika Publ., St. Petersburg (2013)
30. Eliseev, S.V., Volkov, L.N., Kukharenko, V.P.: Dynamics of Mechanical Systems with Additional Constraints, 214 pp. Nauka Publ., Novosibirsk (1990)
31. Korenev, B.G., Reznikov, P.M.: Dynamic Vibration Dampers. Theory and Technical Applications, 535 pp. Nauka Publ., Moscow (1978)
32. Oleynik, A.I.: Development of the theory and constructive forms of multi-mass dynamic dampers and vibration protection devices for building structures and construction works, 341 pp. Dr. Sc. (Engineering) thesis. Moscow State University of Civil Engineering, Moscow (2002)
33. Lurie, A.I.: Analytical Mechanics, 516 pp. Nauka Publ., Moscow (1986)
34. Genkin, M.D., Ryaboy, V.M.: Elastic Inertia Vibration-Isolating Systems. Limit Possibilities, Optimal Structures, 191 pp. Nauka Publ., Moscow (1988)
35. Eliseev, S.V., Kaimov, E.V., Kinash, N.Zh.: Additional masses in the dynamics of vibrating systems. In: Bulletin of BelSUT: Science and Transport, vol. 2, pp. 70–76. BelSUT Publ., Gomel (2014)
36. Eliseev, S.V., Kashuba, V.B., Sitov, I.S.: The reduced stiffness of the feedback circuit. Determination of dynamic reactions in a mechanical oscillatory system. Syst. Methods Technol. BrSU Publ., Bratsk **2**(18), 15–22 (2013)
37. Eliseev, S.V., Belokobylsky, S.V., Kashuba, V.B., Sitov, I.S.: Lever linkages in mechanical circuits. Dynamic aspects. Syst. Methods Technol. BrSU Publ., Bratsk **3**(15), 7–16 (2012)
38. Eliseev, S.V., Belokobylsky, S.V., Upyr, R.Yu., Gozbenko, V.E.: Lever Linkages in the Problems of the Dynamics of Mechanical Oscillatory Systems. Theoretical Aspects, 159 pp. IrGUPS Publ., Irkutsk (2009). Dep. in VINITI 27.11.09, No. 737-B 2009
39. Eliseev, S.V., Kinash, N.Zh., Kaimov, E.V.: Lever linkages in mechanical oscillatory systems. Bull. All-Russ. Sci. Res. Des. Construct. Inst. VELNII Publ., Novocherkassk **1**(69), 112–126 (2015)
40. Ivanov, B.G.: Development of methods for calculating the dynamics and strength of units of transport equipment with articulation constraints, 48 pp. Dr. Sc. (Engineering) thesis, Korolev Samara Aerospace University Publ., Samara (2007)
41. Lavrus, V.V.: Improvement of pneumatic articulation railway transport systems, 20 pp. Author's abstract of Ph.D. (Engineering) thesis, Orlov State Technical University, Oryol (2006)
42. Vibrations in technology: ref.: in 6 vols. In: Genkin, M.D. (ed.) Measurements and Tests, vol. 5, 496 pp. Mashinostroenie Publ., Moscow (1981)
43. Belokobylsky, S.V., Eliseev, S.V., Kashuba, V.B.: Applied Problems of the Structural Theory of Vibration Protection Systems, 363 pp. Polytekhnika Publ., St. Petersburg (2013)
44. Eliseev, S.V., Artyunin, A.I., Kinash, N.Zh., Kaimov, E.V.: Lever linkages and mechanisms in vibration protection systems of technical objects. In: Theoretical and Applied Aspects of Modern Science: Collection of Scientific Papers, vol. 3, pp. 51–59. Belgorod (2014)
45. Eliseev, S.V., Ermoshenko, Yu.V.: Joints of Units in the Dynamics of Mechanical Oscillatory Systems, 156 pp. IrGUPS Publ., Irkutsk (2012)
46. Timoshenko, S.P., Young, D.H., Weaver, U.: Theory of Oscillations in Engineering Business, 472 pp. Transl. from English by Korneychuk, L.G. In: Grigolyuk, E.I. (ed.). Mashinostroenie Publ., Moscow (1985)
47. Belokobylsky, S.V., Eliseev, S.V., Kashuba, V.B., Nguen, D.H., Gypsy, V.V.: Relationships Between Motion Coordinates in Mechanical Oscillatory Systems with Lever Devices. Syst. Methods Technol. BrSU Publ., Bratsk **2**(26), 7–13 (2015)
48. Vargunin, V.N., Gusarov, V.N., Ivanov, B.G., et al.: Designing and Calculation of Articulation Devices and Assemblies, 86 pp. In: Mulyukin, O.P. (ed.). SamGAPS Publ., Samara (2006)
49. Fomina, I.V.: Development of methods for constructing mathematical models with joints of units, 18 pp. Author's abstract of Ph.D. (Engineering) thesis, IrGUPS Publ., Irkutsk (2011)

50. Khomenko, A.P., Eliseev, S.V.: Vibration protection systems with joints. Technology of construction of mathematical models. Mod. Technol. Syst. Anal. Model. IrGUPS Publ., Irkutsk **3**(27), 8–17 (2010)
51. Eliseev, S.V., Ermoshenko, Y.V., Fomina, I.V.: Opportunities for coupling of solids in chain mechanical systems. Mod. Technol. Syst. Anal. Model. IrGUPS Publ., Irkutsk **3**(27), 146–152 (2010)
52. Vibrations in technology: ref.: in 6 vols. In: Bolotin, V.V. (ed.) Fluctuations of Linear Systems, vol. 1, 352 pp. Mashinostroenie Publ., Moscow (1978)
53. Kruglov, Yu.A., Tumanov, Yu.A.: Shock Protection of Machines, Equipment and Equipment, 222 pp. Mashinostroenie Publ., Leningrad (1986)
54. Harris, C.M., Allan, G.: Shock and Vibration Handbook, 877 pp. McGraw-Hill, New York (2002)
55. Vibrations in technology: ref.: in 6 vols. In: Frolov, K.V. (ed.) Protection against Vibrations and Shocks, vol. 6, 586 pp. Mashinostroenie Publ., Moscow (1983)
56. Vershinsky, S.V., Danilov, V.N., Khusidov, V.D.: Dynamics of the Railway Car, 360 pp. Transport Publ., Moscow (1991)
57. Kogan, A.Y.: Track Dynamics and its Interaction with Rolling Stock, 326 pp. Transport Publ., Moscow (1997)
58. Kolchin, N.N.: Machine Mechanics, vol. 2, 486 pp. Mashinostroenie Publ., Leningrad (1965)
59. Khachaturov, A.A., Afanasyev, V.L., Vasilyev, V.S., et al.: Dynamics of the Road-Bus-Car-Driver System, 536 pp. In: Khachaturov, A.A. (ed.). Mashinostroenie Publ., Moscow (1976)
60. Khokhlov, A.A.: Dynamics of Complex Mechanical Systems, 172 pp. MIIT Publ., Moscow (2002)
61. Briskin, E.S., Zhoga, V.V., Chernyshev, V.V., et al.: Dynamics and Motion Control of Walking Machines with Cyclic Propellers, 191 pp. In: Briskin, E.S. (ed.). Mashinostroenie Publ., Moscow (2009)
62. Vukobratovich, M.: Walking Robots and Anthropomorphic Mechanisms, 544 pp. Mir Publ., Moscow (1976)
63. Lilov, L.K.: Modeling of Systems of Connected Bodies, 272 pp. Nauka Publ., Moscow (1993)
64. Silaev, A.A.: Spectral Theory of the Suspension of Transport Vehicles, 192 pp. Mashinostroenie Publ., Moscow (1972)
65. Kolesnikov, V.I., Vorobiev, V.B., Shapovalov, V.V., et al.: Improving the Interaction of the Track and Rolling Stock, 365 pp. Marshrut Publ., Moscow (2006)
66. Rothenberg, R.V.: Fundamentals of the "Driver-Car-Road-Environment" System Reliability, 216 pp. Mashinostroenie Publ., Moscow (1986)
67. Verigo, M.F., Kogan, A.Ya.: The Interaction of the Track and Rolling Stock, 559 pp. In: Verigo, M.F. (ed.). Transport Publ., Moscow (1986)
68. Kobzov, D.Yu., Zhmurov, V.V., Pleshivtseva, S.V.: Improving the Design of a Bucket Drive of Single-Bucket Hydroficated Road and Construction Machines. Mod. Technol. Syst. Anal. Model. IrGUPS Publ., Irkutsk **3**(15), 40–46 (2007)
69. Goverdovsky, V.N.: Development of the theory and methods of designing machines with infra-frequency vibration protection systems, 42 pp. Author's abstract of Dr. Sc. (Engineering) thesis, SGUPS Publ., Novosibirsk (2006)
70. Andreychikov, A.V.: Development of pneumatic systems for vibration isolation of a locomotive driver's seat using automated methods of search design, 295 pp. Ph.D. (Engineering) thesis, Bryansk (1984)
71. Eliseev, S.V., Khomenko, A.P., Logunov, A.S.: Dynamic synthesis in the tasks of building systems for protecting a human operator of vehicles from vibrations and impacts. Mod. Technol. Syst. Anal. Model. IrGUPS Publ., Irkutsk **4**(24), 64–74 (2009)
72. Eliseev, S.V., Khomenko, A.P., Logunov, A.S.: Use of pneumatic protection schemes for mechanisms or devices with movement conversion. Mod. Technol. Syst. Anal. Model. IrGUPS Publ., Irkutsk **1**(25), 8–13 (2010)

73. Atabekov, G.I.: Linear Electrical Circuits, 592 pp. Energia Publ., Moscow (1978)
74. Bakalov, V.P., Dmitrikov, V.F., Kruk, B.I.: Fundamentals of the Theory of Circuits. Textbook, 589 pp. Radio and Svyaz' Publ., Moscow (2000)
75. Druzhinsky, I.A.: Mechanical Chains, 240 pp. Mashinostroenie Publ., Moscow (1977)
76. Weits, V.L., Kochura, A.E., Martynenko, A.M.: Dynamic Calculations of Drive Machines, 352 pp. Mashinostroenie Publ., Leningrad (1971)
77. Gardner, M.F., Burns, J.L.: Transients in Linear Systems. Transl. from English, 3rd edn, 551 pp. Fizmatgiz Publ., Moscow (1961)
78. Lenk, A.: Electromechanical Systems: Systems with Distributed Parameters, 472 pp. Energoizdat Publ., Moscow (1982)
79. Eliseev, S.V., Lukyanov, A.V., Reznik, Yu.N., Khomenko, A.P.: Dynamics of Mechanical Systems with Additional Ties, 315 pp. Irkutsk State University, Irkutsk (2006)
80. Belokobylsky, S.V., Eliseev, S.V.: Mathematical Modeling of Dynamic Systems. Taking into Account the Features of the Physical Nature of Elements. In: Proceedings of XVIII All-Russian Conference "Information and Mathematical Technologies in Science and Management", vol. 1, pp. 150–160. Irkutsk (2013)

Chapter 2
Features of Representation of Mechanical Circuits Based on Equipment of the Theory of Circuits and Automatic Control Theory

The development of the theory of circuits, in particular, the theory of electric circuits and the principle of electromechanical analogies as a whole, has had a noticeable impact on the development of the analytical tools used to study the dynamic properties of mechanical oscillatory systems. The development of the theory of circuits gave rise to the formation of the theory of mechanical circuits and various applications of the feedback theory [1].

2.1 On the Question of the Development of the Equivalence Relations of Dynamical States in Mechanical Oscillatory Systems

In engineering calculations related to the problems of protecting equipment, instruments, equipment from the effects of vibrations and shocks, mechanical oscillating systems with lumped properties are considered as basic models. To describe the interaction between variables, the elements of the operational calculus are used, in which the time function t corresponds to the functions of the new variable p, the relationship between which is implemented by means of Laplace transforms [2]:

$$F(p) = \int_0^\infty e^{-pt} f(t)dt, \tag{2.1}$$

where $p = \delta + j\omega$, and the real part δ is positive and sufficiently large for the integral to be finite. Thereafter, the following relationships are used, which are applied to transform mathematical models in the form of systems of ordinary differential equations with constant coefficients:

© Springer Nature Switzerland AG 2020

S. V. Eliseev and A. V. Eliseev, *Theory of Oscillations*, Studies in Systems, Decision and Control 252, https://doi.org/10.1007/978-3-030-31295-4_2

(1) the transform of a constant quantity A:

$$F(A).\doteq\int\limits_0^\infty e^{-pt}Adt=\frac{Ae^{-pt}}{-p}\bigg|_0^\infty=-\frac{A}{p}(0-1)=\frac{A}{p};\tag{2.2}$$

(2) the transform of the first and second derivatives of the function f(t):

$$f'(t)\doteq pF(p);\ f''(t).\doteq p^2F(p);\tag{2.3}$$

(3) the transform of the integral $\int\limits_0^t f(t)dt$:

$$F(p).\doteq\frac{F(p)}{p}=\int\limits_0^t f(t)dt;\tag{2.4}$$

(4) the transform of the exponential function $f(t)=e^{-j\omega t}$:

$$F(p)\doteq\frac{1}{(p+j\omega)}\tag{2.5}$$

(5) the transform of the main trigonometric functions:

$$\sin\omega t\doteq\frac{\omega}{p^2+\omega^2};\ \cos\omega t\doteq\frac{p\omega}{p^2+\omega^2},\tag{2.6}$$

where $\doteq$ is the correspondence sign.

Note that the dynamic properties of simple systems are easily explained at the level of addition of vectors of different forces. In this case, force and kinematic perturbations are reduced under certain transformations to one another [3]. This fact reflects the reversibility of Newton's second law

$$Q=m\ddot{y},\ \dot{y}=\frac{1}{m\int Qdt}.\tag{2.7}$$

2.1.1 On the Relation of Ratios to Structural Diagrams

A system with one degree of freedom (Fig. 2.1) under the action of the force $Q(t)$, can be described by one of three equations depending on the functions y, v, a (displacement, velocity, acceleration):

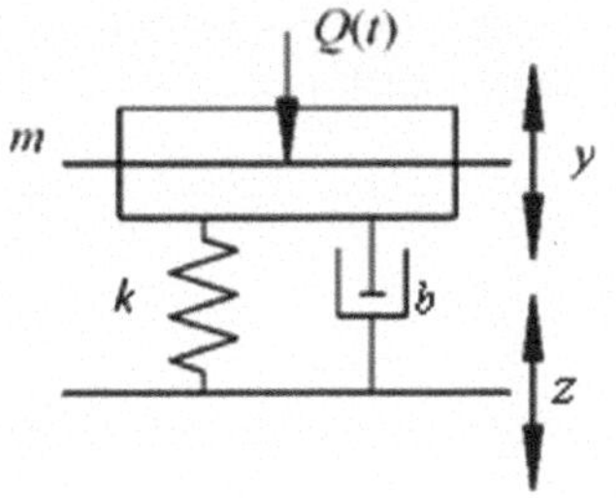

Fig. 2.1 The computational scheme of a system with one degree of freedom, taking into account elastic (k), dissipative (b) and mass-and-inertia (m) properties. y_1 is the kinematic perturbation; $Q(t)$ is the force action

$$\frac{md^2y}{dt^2}+b\frac{dy}{dt}+ky=Q(t); \tag{2.8}$$

$$\frac{mdv}{dt}+bv+k\int vdt=Q(t); \tag{2.9}$$

$$ma+b\int adt+k\iint adt^2=Q(t). \tag{2.10}$$

In the Laplace transforms, the expressions (2.8)–(2.10) take the following form:

$$mp^2\bar{y}+bp\bar{y}+k\bar{y}=\overline{Q}(p); \tag{2.11}$$

$$mp\bar{v}+b\bar{v}+\frac{k}{p}\bar{v}=\overline{Q}(p); \tag{2.12}$$

$$m\bar{a}+\frac{b}{p}\bar{a}+\frac{k}{p^2}\bar{a}=\overline{Q}(p). \tag{2.13}$$

Figure 2.2 represents structural diagrams corresponding to Eqs. (2.11)–(2.13), which are equivalent with respect to (2.8)–(2.10), but they differ because of the difference in the relationships between the elements.

Note that the points C and B are shown in the structural diagrams (see Fig. 2.2). At the point B, the input external impact (input signal) corresponds to the force perturbation, and at the point C, if necessary, a kinematic perturbation is applied.

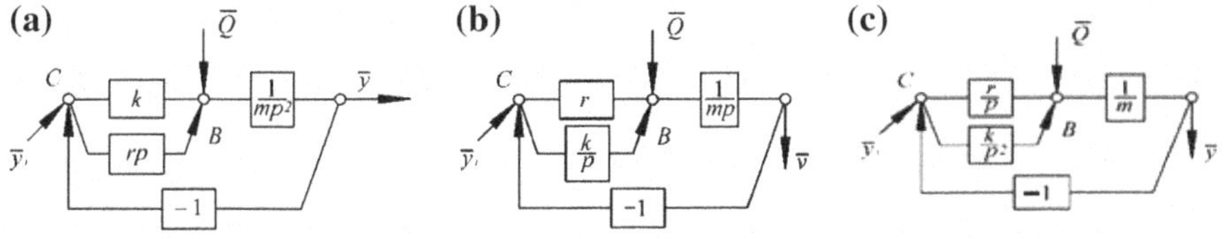

Fig. 2.2 The structural diagrams of the system shown in Fig. 2.1 that correspond to **a** equations of the form (2.8), (2.11); **b** equations of the form (2.9), (2.12), where $\bar{v}=\frac{d\bar{y}}{dt}=p\bar{y}$; **c** equations of the form (2.10), (2.13), where $\bar{a}=\frac{a\bar{v}}{dt}=p^2\bar{y}$

The developed approach can be used to write down Eqs. (2.8)–(2.10) in the integro-differential form:

$$mv + b\int vdt + k\iint vdt^2 = \int Q(t)dt; \tag{2.14}$$

$$m\bar{v} + \frac{b}{p}\bar{v} + \frac{k}{p^2}\bar{v} = \frac{\overline{Q}(p)}{p}. \tag{2.15}$$

Figure 2.3 shows the structural diagram of the system which corresponds to the equation of motion (2.15).

If we differentiate (2.8) with respect to time t, then we obtain an equation of the following form for the initial system (see Fig. 2.1)

$$m\frac{d^3y}{dt^3} + b\frac{d^2y}{dt^2} + k\frac{dy}{dt} = \frac{dQ(t)}{dt}; \tag{2.16}$$

$$mp^3\bar{y} + rp^2\bar{y} = p\overline{Q}(p). \tag{2.17}$$

and the corresponding structural diagram of the initial system (Fig. 2.4).

Tables 2.1 and 2.2 show the changes in displacement, speed and acceleration quantity for the main elements of the computational scheme (see Fig. 2.1). In these tables, inclined lines connect cells with identical values of parameters depending on the impact of forces and velocities.

The nomogram (Fig. 2.5) establishes a relationship between the parameters of possible transformations of the elements of the dynamic system.

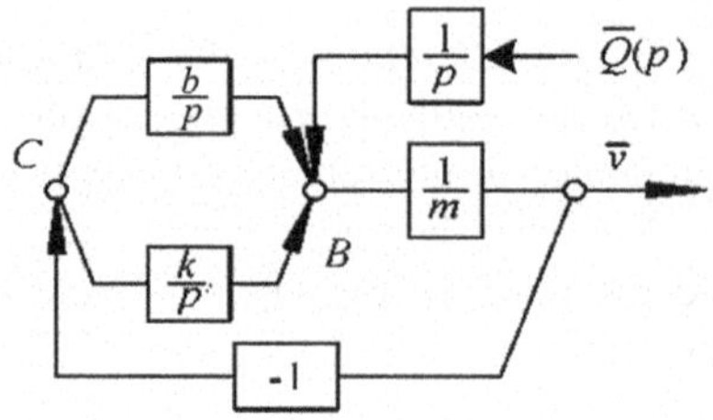

Fig. 2.3 The structural diagram of the system, which corresponds to Eq. (2.15)

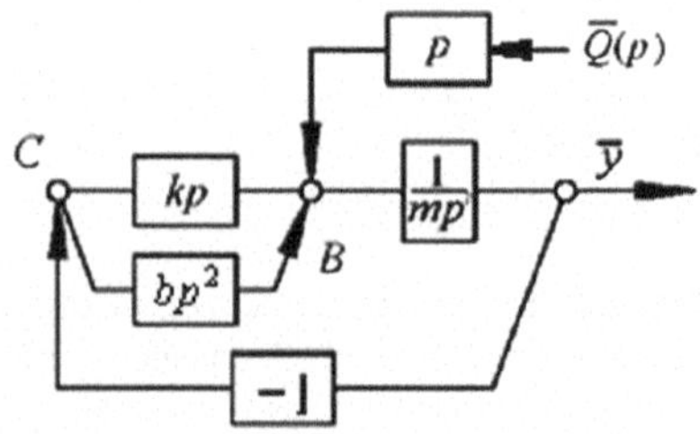

Fig. 2.4 The structural diagram of the system which corresponds to Eq. (2.17)

Table 2.1 Parameters of displacement, velocity and acceleration depending on the impact of the force or its derivatives

	$\frac{dQ(t)}{dt}$	$Q(t)$	$\int Q(t)dt$
m	$m\frac{da}{dt}$	ma	$m\int adt$
b	ba	$b\int adt$	$b\iint adt^2$
k	$k\int adt$	$k\iint adt^2$	$k\iiint adt^3$

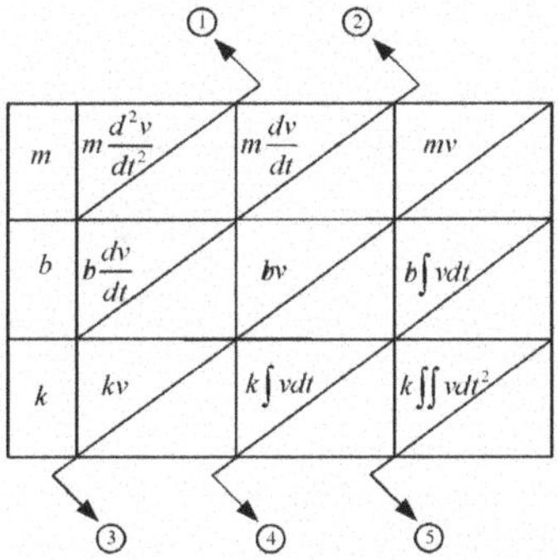

m	$m\frac{d^3y}{dt^3}$	$m\frac{d^2y}{dt^2}$	$m\frac{dy}{dt}$
b	$b\frac{d^2y}{dt^2}$	$b\frac{dy}{dt}$	by
k	$k\frac{dy}{dt}$	ky	$k\int ydt$

Note In Table 2.1 it is agreed that:

Point 1 corresponds to the differential of acceleration
Point 2 corresponds to acceleration
Point 3 corresponds to speed
Point 4 corresponds to displacement
Point 5 corresponds to the integral of the displacement

2.1.2 *The Relationship Between the Parameters of Electrical and Mechanical Circuits*

Thus, the initial system of differential equations (2.8)–(2.10), obtained by one of the known methods, can be transformed by applying differentiation or integration procedures followed by obtaining structural diagrams of systems analogous to automatic control systems. In prospect, using the analytical tools of the automatic control theory (ACT), it is possible to construct the transfer functions of the system and to evaluate the dynamic responses of the system to various types of perturbation. In this case, an important role is played by the choice of a system of electromechanical analogies.

Table 2.2 Parameters and its derivatives for passive elements depending on the effect of speed

	$a=\frac{dv}{dt}=\frac{d^2y}{dt^2}$	$\int adt=v=\frac{dy}{dt}$	$\iint adt^2=\int vdt=y$
m	$\frac{1}{m}Q$	$\frac{1}{m}\int Qdt$	$\frac{1}{m}\iint Qdt^2$
b	$\frac{1}{b}\frac{dQ}{dt}$	$\frac{1}{b}Q$	$\frac{1}{b}\int Qdt$
k	$\frac{1}{k}\frac{d^2Q}{dt^2}$	$\frac{1}{k}\frac{dQ}{dt}$	$\frac{1}{k}Q$

Note In Table 2.2 it is agreed that:

Point 1 corresponds to the force
Point 2 corresponds to the integral of the force
Point 3 corresponds to the double integral of the force
Point 4 corresponds to the differential of force
Point 5 corresponds to the double differential from the force

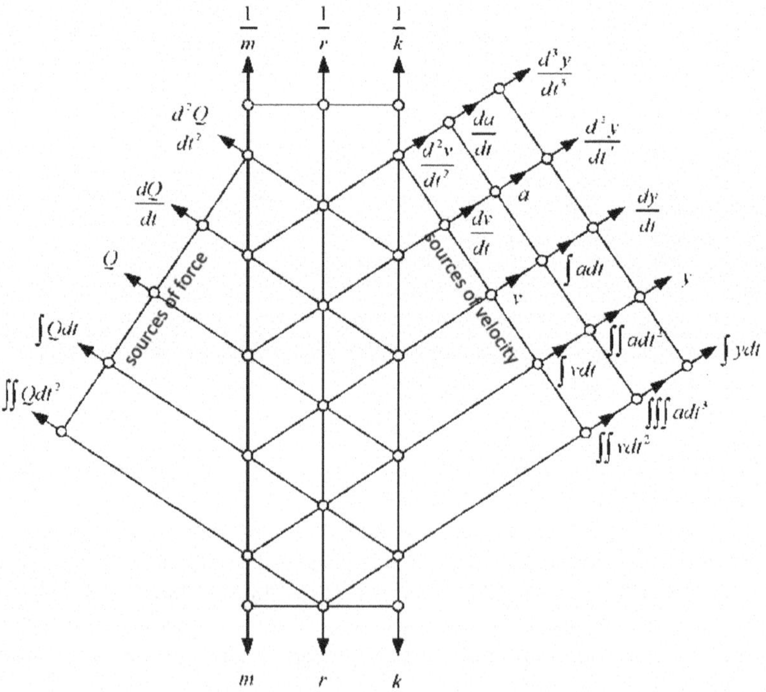

Fig. 2.5 The nomogram for selecting the parameters of the dynamic system

1. **The first system of analogy is the "tension-force" pair**. When the current i passes through the electric circuit consisting of the consecutively-connected passive elements—inductance L, resistance R and capacitance C (Fig. 2.6a), a voltage equal to (according to Kirchhoff's second law) the sum of the stress drops in individual elements

$$u_L + u_C + u_R = u(t) \tag{2.18}$$

or, replacing $u_L = L\frac{di}{dt}$, $u_R = Ri_R$, $i_C = C\frac{du_C}{dt}$, $u_C = \frac{1}{C}\int i_C dt$, we get

$$L\frac{di}{dt} + Ri + \frac{1}{C}\int idt = u(t). \tag{2.19}$$

For the steady-state mode, it can be assumed that the current changes according to the law, which, after transformations, leads to a relationship between the complex current and the voltage

$$I = U/\sqrt{R^2 + \left(\omega L - \frac{1}{\omega C}\right)^2} = U/Z, \tag{2.20}$$

where Z is the total complex resistance, or impedance, of the circuit [4].

It is known that the rate of change of the electric charge q in the capacitor C is equal to the current strength i: $i = \frac{dq}{dt}$; $\frac{di}{dt} = \frac{d^2q}{dt^2}$; $q = \int idt$. In this case, Eq. (2.19) takes the form

$$L\frac{d^2q}{dt^2} + R\frac{dq}{dt} + \frac{1}{C}q = u(t). \tag{2.21}$$

Comparing (2.21) and (2.9), one can find an analogy between the coefficients L and m, R and b, and k. However, for an electrical circuit with the serial connection of analogous elements in a mechanical system, their parallel connection is required.

2. **The second system of electromechanical analogies establishes a "current-to-force" relationship**. When passing through an electrical circuit consisting of passive parallel connected elements L, R, C, the current i is (according to the first Kirchhoff law) the sum of the currents in the individual elements

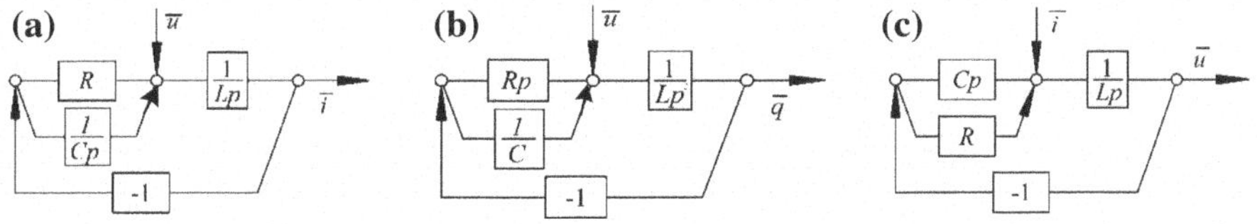

Fig. 2.6 Structural diagrams of electrical systems corresponding to: **a** Eq. (2.19): current–velocity; **b** Eq. (2.21): charge–displacement; **c** Eq. (2.23): stress–velocity

$$i_L + i_R + i_C = i(t), \tag{2.22}$$

which gives, after substitution, an equation of the form

$$\frac{1}{L}\int udt + \frac{1}{R}u + C\frac{du}{dt} = i(t). \tag{2.23}$$

In the steady-state mode, the $u = U_n \sin(\omega t + \xi)$ transformation leads to a relationship between the complex voltage and the current

$$U = F/\sqrt{\left(\frac{1}{R^2}\right) + \left(\frac{1}{\omega L} - \omega C\right)^2} = I/Y, \tag{2.24}$$

where Y is the complete complex compliance, or admittance, of the circuit.

If we compare (2.24) and (2.9), then we can find an analogy between the coefficients C and m, $\frac{1}{R}$ and b, $\frac{1}{L}$ and k. It should be noted that parallel connection of passive elements will be characteristic for both circuits (electrical and mechanical).

Let us construct, according to [5, 6], the structural diagrams of the electric systems corresponding to Eqs. (2.19), (2.21) and (2.23) (see Fig. 2.6).

2.1.3 Similarity Conditions

Thus, considering the mathematical models of oscillatory processes in systems whose circuit is shown in Fig. 2.1, we can take y, v, a as the "output signals"; at the same time, the external force Q, and also $\frac{dQ}{dt}$ and $\int Qdt$ can act as input signals. In addition, the consideration can be entered. If we take into account the possibility of a kinematic perturbation and its reducibility to a power perturbation, then it makes sense to treat $\int ydt$ as an "input signal". The relationship between the "signals" at the input and output can be presented in Table 2.3. In this case, y_1 is the kinematic action applied to the point C (see Fig. 2.2); $Q(t)$ is the force impact applied in the point B (see Fig. 2.2).

2.1.4 Circuit Transformation Rules

If we rely on structural representations, then the external effect is applied at the point D with the subsequent definition of the transfer function in its various forms (see Table 2.3, in which 14 items are marked, with some of them coinciding). In the

Table 2.3 Construction variants of the system transfer functions

Output signals		Input signals							
		Q	$\frac{dQ}{dt}$	$\int Qdt$	y_1	$\frac{dy_1}{dt}$	$\frac{d^2y_1}{dt^2}$	$\frac{d^3y_1}{dt^3}$	$\int y_1 dt$
	y	$\frac{\bar{y}}{\bar{Q}} = \frac{1}{[\,]}$	$\frac{\bar{y}}{\left(\frac{d\bar{Q}}{dt}\right)} = \frac{1}{p[\,]}$	$\frac{\bar{y}}{\left(\int \bar{Q}dt\right)} = \frac{p}{[\,]}$	$\frac{\bar{y}}{(\bar{y}_1)} = \frac{A}{[\,]}$	$\frac{\bar{y}}{\left(\frac{d\bar{y}_1}{dt}\right)} = \frac{A}{p[\,]}$	$\frac{\bar{y}}{\left(\frac{d^2\bar{y}_1}{dt^2}\right)} = \frac{A}{p^2[\,]}$	$\frac{\bar{y}}{\left(\frac{d^3\bar{y}_1}{dt^3}\right)} = \frac{A}{p^3[\,]}$	$\frac{\bar{y}}{\left(\int \bar{y}_1 dt\right)} = \frac{pA}{[\,]}$
	v	$\frac{\bar{v}}{\bar{Q}} = \frac{p}{[\,]}$	$\frac{\bar{v}}{\left(\frac{d\bar{Q}}{dt}\right)} = \frac{1}{[\,]}$	$\frac{\bar{v}}{\left(\int \bar{Q}dt\right)} = \frac{p^2}{[\,]}$	$\frac{\bar{v}}{(\bar{y}_1)} = \frac{pA}{[\,]}$	$\frac{\bar{v}}{\left(\frac{d\bar{y}_1}{dt}\right)} = \frac{A}{[\,]}$	$\frac{\bar{v}}{\left(\frac{d^2\bar{y}_1}{dt^2}\right)} = \frac{A}{p[\,]}$	$\frac{\bar{v}}{\left(\frac{d^3\bar{y}_1}{dt^3}\right)} = \frac{A}{p^2[\,]}$	$\frac{\bar{v}}{\left(\int \bar{y}_1 dt\right)} = \frac{Ap^2}{[\,]}$
	a	$\frac{\bar{a}}{\bar{Q}} = \frac{p^2}{[\,]}$	$\frac{\bar{a}}{\left(\frac{d\bar{Q}}{dt}\right)} = \frac{p}{[\,]}$	$\frac{\bar{a}}{\left(\int \bar{Q}dt\right)} = \frac{p^3}{[\,]}$	$\frac{\bar{a}}{(\bar{y}_1)} = \frac{p^2A}{[\,]}$	$\frac{\bar{a}}{\left(\frac{d\bar{y}_1}{dt}\right)} = \frac{pA}{[\,]}$	$\frac{\bar{a}}{\left(\frac{d^2\bar{y}_1}{dt^2}\right)} = \frac{A}{[\,]}$	$\frac{\bar{a}}{\left(\frac{d^3\bar{y}_1}{dt^3}\right)} = \frac{A}{p[\,]}$	$\frac{\bar{a}}{\left(\int \bar{y}_1 dt\right)} = \frac{Ap^3}{[\,]}$
	$\frac{da}{dt}$	$\frac{\left(\frac{d\bar{a}}{dt}\right)}{\bar{Q}} = \frac{p^3}{[\,]}$	$\frac{\left(\frac{d\bar{a}}{dt}\right)}{\left(\frac{d\bar{Q}}{dt}\right)} = \frac{p^2}{[\,]}$	$\frac{\left(\frac{d\bar{a}}{dt}\right)}{\left(\int \bar{Q}dt\right)} = \frac{p^4}{[\,]}$	$\frac{\left(\frac{d\bar{a}}{dt}\right)}{(\bar{y}_1)} = \frac{p^3A}{[\,]}$	$\frac{\left(\frac{d\bar{a}}{dt}\right)}{\left(\frac{d\bar{y}_1}{dt}\right)} = \frac{p^2A}{[\,]}$	$\frac{\left(\frac{d\bar{a}}{dt}\right)}{\left(\frac{d^2\bar{y}_1}{dt^2}\right)} = \frac{pA}{[\,]}$	$\frac{\left(\frac{d\bar{a}}{dt}\right)}{\left(\frac{d^3\bar{y}_1}{dt^3}\right)} = \frac{A}{[\,]}$	$\frac{\left(\frac{d\bar{a}}{dt}\right)}{\left(\int \bar{y}_1 dt\right)} = \frac{Ap^4}{[\,]}$
	$\int ydt$	$\frac{\left(\int \bar{y}dt\right)}{\bar{Q}} = \frac{1}{p[\,]}$	$\frac{\left(\int \bar{y}dt\right)}{\left(\frac{d\bar{Q}}{dt}\right)} = \frac{1}{p^2[\,]}$	$\frac{\left(\int \bar{y}dt\right)}{\left(\int \bar{Q}dt\right)} = \frac{1}{[\,]}$	$\frac{\left(\int \bar{y}dt\right)}{(\bar{y}_1)} = \frac{A}{p[\,]}$	$\frac{\left(\int \bar{y}dt\right)}{\left(\frac{d\bar{y}_1}{dt}\right)} = \frac{A}{p^2[\,]}$	$\frac{\left(\int \bar{y}dt\right)}{\left(\frac{d^2\bar{y}_1}{dt^2}\right)} = \frac{A}{p^3[\,]}$	$\frac{\left(\int \bar{y}dt\right)}{\left(\frac{d^3\bar{y}_1}{dt^3}\right)} = \frac{A}{p^4[\,]}$	$\frac{\left(\int \bar{y}dt\right)}{\left(\int \bar{y}_1 dt\right)} = \frac{A}{[\,]}$

Note $[\,] = mp^2 + bp + k$, $A = bp + k$

case of a kinematic perturbation (Fig. 2.8), the external action (the vibrations of the base y_1) will be applied to the point D or the point C.

If in the first case (see Fig. 2.7) the transfer function has the form

$$\overline{W} = \frac{\bar{y}}{\overline{Q}} = \frac{1}{mp^2 + bp + k}; \tag{2.25}$$

then in the second one (see Fig. 2.8):

$$\overline{W} = \frac{\bar{y}}{\bar{y}_1} = \frac{bp + k}{mp^2 + bp + k}. \tag{2.26}$$

The physical meaning of the transfer functions is also different (dynamic compliance and the transfer coefficient of the oscillation amplitude). From a comparison of the two approaches, one can see that the kinematic perturbation y_1 can be transferred from the point C to point D, but this is done in accordance with the transfer rules in structural transformations. At the point D, the transformed external impact will be defined as ky_1; the transfer function obtained by the force impact diagram will be the same as in the determination of the transfer function according to the kinematic perturbation diagram.

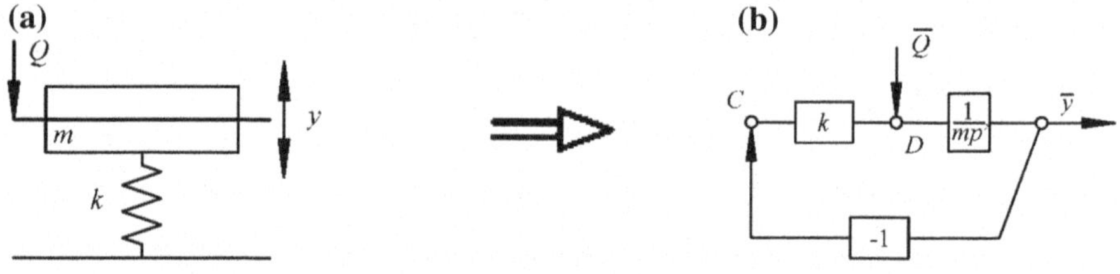

Fig. 2.7 The computational scheme (**a**) and structural diagram (**b**) of the system under the force perturbation (point D)

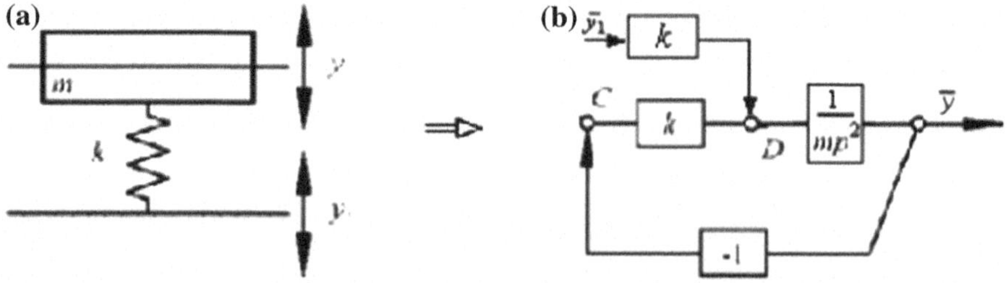

Fig. 2.8 The computational scheme (**a**) and structural diagram (**b**) of the system under kinematic impact (point C)

Let us consider the scheme of force loading through an elastic element (Fig. 2.9). However, the development of such a scheme will require a number of preliminary transformations of the computational scheme with two degrees of freedom, shown in Fig. 2.10.

We will formulate expressions for the kinetic and potential energy, use the Lagrange formalism of the second kind and obtain the following system of differential equations:

$$\left.\begin{array}{l} T = \frac{1}{2}m_1\dot{y}_1^2 + \frac{1}{2}m_2\dot{y}_2^2 \\ \Pi = \frac{1}{2}k_2y_1^2 + \frac{1}{2}k_1(y_2 - y_1)^2 \end{array}\right\}, \quad \left.\begin{array}{l} \frac{\partial \Pi}{\partial y_1} = k_2y_1 + k_1y_1 - k_1y_2 \\ \frac{\partial \Pi}{\partial y_2} = k_1y_2 - k_1y_1 \end{array}\right\}, \quad \left.\begin{array}{l} \frac{\partial T}{\partial \dot{y}_1} = m_1\dot{y}_1 \\ \frac{\partial T}{\partial \dot{y}_2} = m_2\dot{y}_2 \end{array}\right\},$$

$$\begin{cases} m_1\ddot{y}_1 + k_2y_1 + k_1y_1 - k_1y_2 = 0, \\ m_2\ddot{y}_2 + k_1y_2 - k_1y_1 = Q. \end{cases} \tag{2.27}$$

Figure 2.11 shows the successive positions of the transformation of the structural diagram of the automatic control system, which is equivalent in the dynamic relation to the computational scheme in Fig. 2.10. The structural diagram in

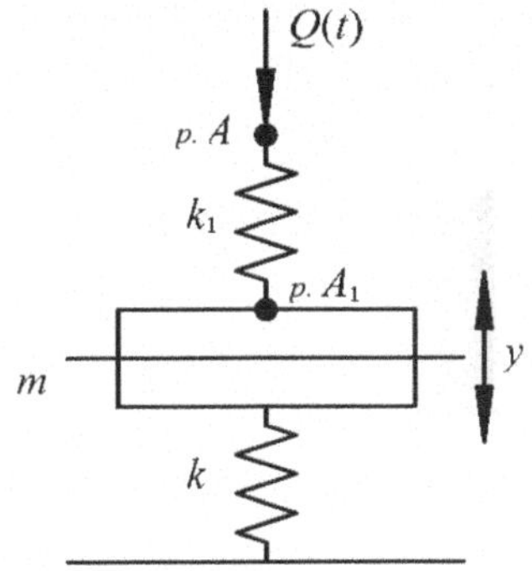

Fig. 2.9 Computational scheme of force perturbation through the elastic element k_1

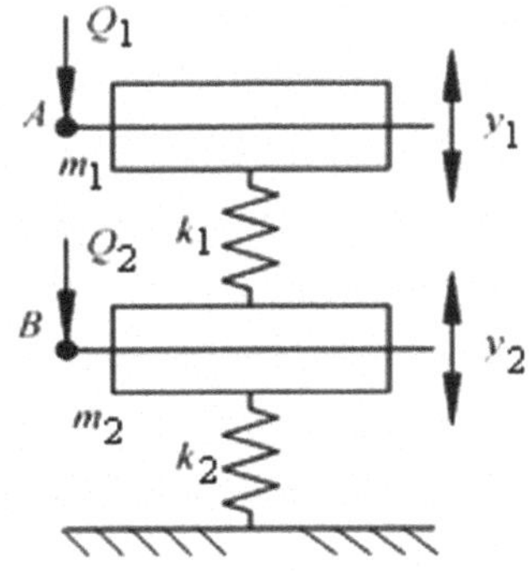

Fig. 2.10 The computational scheme with two degrees of freedom when force is applied to the mass m_1

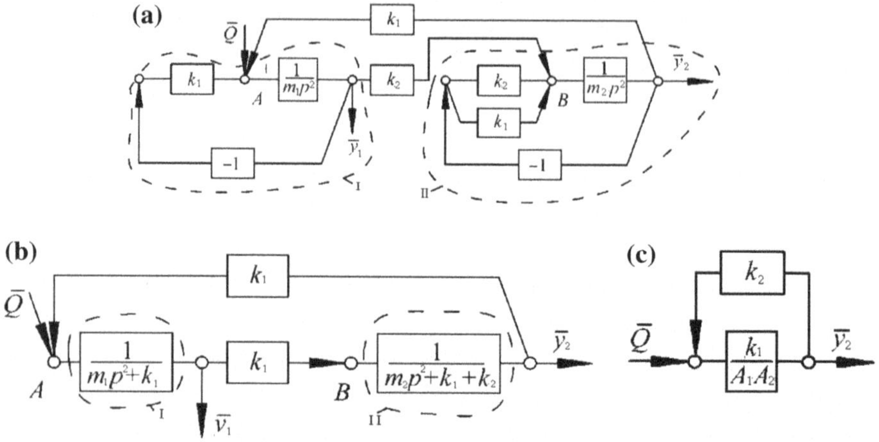

Fig. 2.11 Consecutive transformation of the structural scheme, corresponding to Fig. 2.10. **a** is detailed structural diagram, **b** is the structural diagram from blocks, corresponding to partial systems, **c** is the simplified system to define transfer function $\overline{Q}_1 = \bar{y}_2$ ($A_1 = m_1p^2 + k_1$; $A_2 = m_2p^2 + k_1 + k_2$)

Fig. 2.11a and can be called detailed, and the system itself consists of two blocks, which are partial systems interacting through elastic constraints (contours I, II).

If to convolve partial blocks I and II in Fig. 2.11a, the structural diagram will take the form as it is shown in Fig. 2.11b. Structural diagram of the system for determining the transfer function. The displacement y_2 is the force perturbation Q at the point A has the form shown in Fig. 2.11c, from which it follows that

$$\overline{W}_1 = \frac{\bar{y}_2}{\overline{Q}} = \frac{k_1}{p^4 m_1 m_2 + p^2(m_1k_2 + m_1k_1 + m_2k_1) + k_1k_2}, \tag{2.28}$$

and for $m_1 = 0$, respectively,

$$\overline{W}_2 = \frac{1}{m_2p^2 + k_2}. \tag{2.29}$$

Thus, if the force $\overline{Q}$ is applied at the point A, as it is shown in Fig. 2.9, then the force nature of the impact will not change when the force $\overline{Q}$ is transferred directly to the mass at the point A_1. If in the computational scheme in Fig. 2.10, *a* to assume that $m_2 = 0$, and the force Q is applied at the point A, then

$$\overline{W}_3 = \frac{\bar{y}_2}{\overline{Q}} = \frac{k_1}{m_1p^2(k_1 + k_2) + k_1k_2} = \frac{\frac{k_1}{k_1 + k_2}}{m_1p^2 + k_{sp}}, \tag{2.30}$$

where $k_{sp} = \frac{k_1k_2}{k_1 + k_2}$ is the stiffness of the consecutively connected springs.

A possible set of transfer functions for various combinations of parameters is given in Table 2.4.

From the comparative analysis it follows, in particular, that the structural diagrams of mechanical oscillatory systems are one of the forms of mathematical models that are used to evaluate the dynamic properties of oscillatory systems (both mechanical and electrical) based on the general principles of electromechanical analogies, and the algebra of structural diagram transformations makes it possible to obtain an additional number of characteristics that can be used in the analysis of dynamic properties (see Table 2.1).

It should also be noted that there are certain analogies in the construction of relations between the parameters of the state of systems, based on the possibilities of mutual reducibility of external kinematic and force impacts, which outlines generalized concepts about the results of dynamic transformations in the original systems.

Table 2.4 Expressions for intermediate transfer functions

Subitem No.	Place of force application	Values of parameters m_1, m_2	Transfer function	Notes
1	2	3	4	5
1	Q at the point A	$m_1 = 0$, $m_2 \neq 0$ $\frac{\bar{y}_2}{\bar{Q}}$	$\overline{W} = \frac{1}{m_2p^2 + k_2}$	Natural frequency does not depend on k_1
2	Q at the point A	$m_1 = 0$, $m_2 \neq 0$ $\frac{\bar{y}_1}{\bar{Q}}$	$\overline{W} = \frac{m_2p^2 + k_1 + k_2}{m_2p^2k_1 + k_1k_2}$	With this form of impact, a consecutive connection k_1 and k_2 is implemented
3	Q at point B	$m_1 = 0$, $m_2 \neq 0$ $\frac{\bar{y}_2}{\bar{Q}}$	$\overline{W} = \frac{1}{m_2p^2 + k_2}$	Impact at the point B is equivalent to the case of the force application at the point A (position 1)
4	Q at the point B	$m_1 = 0$, $m_2 \neq 0$ $\frac{\bar{y}_1}{\bar{Q}}$	$\overline{W} = \frac{1}{m_2p^2 + k_2}$	Element k_1 is not physically involved in the scheme
5	Q at the point A	$m_1 \neq 0$, $m_2 = 0$ $\frac{\bar{y}_2}{\bar{Q}}$	$\overline{W} = \dfrac{k_1}{m_1p^2(k_1 + k_2) + k_1k_2} = \dfrac{\frac{k_1}{k_1 + k_2}}{m_1p^2 + k_{\text{sp}}}, k_{sp} = \dfrac{k_1k_2}{k_1 + k_2}$	Case of parallel connection k_1 and k_2

(continued)

Table 2.4 (continued)

Subitem No.	Place of force application	Values of parameters m_1, m_2	Transfer function	Notes
6	Q at the point A	$m_1 \neq 0$, $m_2 = 0$ $\frac{y_1}{Q}$	$\overline{W} = \frac{k_1 + k_2}{m_1 p^2 (k_1 + k_2) + k_1 k_2} = \frac{1}{m_1 p^2 + k_{sp}}, k_{sp} = \frac{k_1 k_2}{k_1 + k_2}$	Case of parallel connection k_1 and k_2
7	Q at the point B	$m_1 \neq 0$, $m_2 = 0$ $\frac{y_2}{Q}$	$\overline{W} = \frac{m_1 p^2 + k_1}{m_1 p^2 (k_1 + k_2) + k_1 k_2} = \frac{\frac{m_1 p^2 + k_1}{k_1 + k_2}}{m_1 p^2 + k_{sp}}, k_{sp} = \frac{k_1 k_2}{k_1 + k_2}$	Case of parallel connection k_1 and k_2
8	Q at the point B	$m_1 \neq 0$, $m_2 = 0$ $\frac{y_1}{Q}$	$\overline{W} = \frac{k_1}{m_1 p^2 (k_1 + k_2) + k_1 k_2} = \frac{\frac{k_1}{k_1 + k_2}}{m_1 p^2 + k_{sp}}, k_{sp} = \frac{k_1 k_2}{k_1 + k_2}$	Case of parallel connection k_1 and k_2
9	Q at the point A, Q at the point B	$m_1 = 0$, $m_2 \neq 0$ $\frac{y_2}{Q}$	$\frac{y_2}{Q} = \overline{W}_{21} + \overline{W}_{22} = \frac{2}{m_2 p^2 + k_2}$	The case of simultaneous impact of two forces Q at the points A and B
10	Q at the point A, Q at the point B	$m_1 = 0$, $m_2 \neq 0$ $\frac{y_1}{Q}$	$\frac{y_1}{Q} = \overline{W}_{11} + \overline{W}_{12} = \frac{m_2 p^2 + 2k_1 + k_2}{m_2 p^2 k_1 + k_1 k_2}$	The case of simultaneous impact of two forces Q at the points A and B
11	Q at the point A	$m_1 = 0$, $m_2 = 0$ $\frac{y_2}{Q}$	$\overline{W} = \frac{1}{k_2}$	The loading scheme corresponds to the positions 1 and 5
12	Q at the point A	$m_1 = 0$, $m_2 = 0$ $\frac{y_1}{Q}$	$\overline{W} = \frac{k_1 + k_2}{k_1 k_2} = \frac{1}{k_{sp}}$	Case of parallel connection k_1 and k_2
13	Q at the point B	$m_1 = 0$, $m_2 = 0$ $\frac{y_2}{Q}$	$\overline{W} = \frac{1}{k_2}$	Spring k_1 does not work
14	Q at the point B	$m_1 = 0$, $m_2 = 0$ $\frac{y_1}{Q}$	$\overline{W} = \frac{1}{k_2}$	Spring k_1 does not work physically

2.2 On the Introduction of Additional Elements into Mechanical Systems Based on Structural Interpretations

In the simplest case, in problems of vibration protection and vibration isolation, as it was shown in Sect. 2.1, the object is an oscillatory system with one degree of freedom and it has two external perturbations (see Fig. 2.12a). The structural analogue (see Fig. 2.12b) and the equivalent mechanical circuit (Fig. 2.12c) form the basis for mathematical modeling of the processes occurring.

The transfer functions of the system can be determined from the structural diagram in Fig. 2.12b:

$$\overline{W}(p) = \frac{\bar{y}}{\overline{Q}} = \frac{1}{mp^2 + k}; \tag{2.31}$$

$$\overline{W}_1(p) = \frac{\bar{y}}{\bar{y}_1} = \frac{k}{mp^2 + k}. \tag{2.32}$$

2.2.1 *Introduction of Constraints*

Dynamic response (but not the complete one) in the absence of $\overline{Q}$ (only $\bar{y}_1$ is active) will be determined by the transfer function

$$\overline{W}_2(p) = \frac{\overline{Q}_{dyn}}{\bar{y}_1} = \frac{mp^2 \cdot k}{mp^2 + k}. \tag{2.33}$$

If we determine the value of the transfer function with $\bar{y}_1 = 0$ and $\overline{Q} \neq 0$, then the corresponding expression will have the form

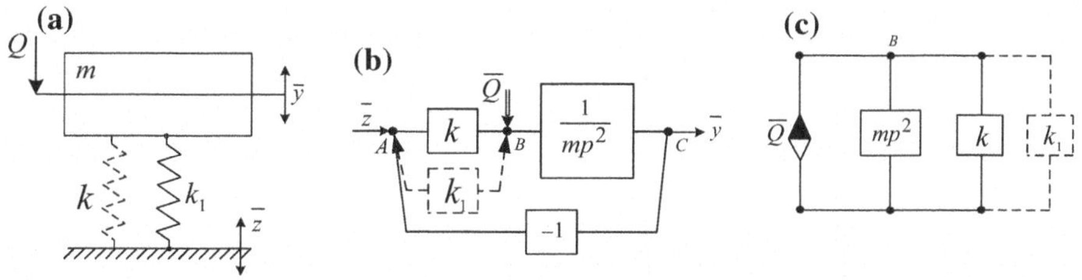

Fig. 2.12 The computational scheme (**a**), structural analog (**b**) and dual mechanical circuit (**c**) in the generalized problem of vibration protection and vibration isolation

$$\overline{W}_3(p) = \frac{\overline{Q}_{dyn}}{\overline{Q}} = \frac{mp^2}{mp^2 + k}. \quad (2.34)$$

The introduction of additional constraint by relative divergence [7], for example, of a spring with stiffness k_1, is accompanied by the following changes: in Fig. 2.12a, k_1 is introduced parallel to k; in Fig. 2.12b, $W_k'(p) = k + k_1$ is introduced instead of a unit with the transfer function $W_k(p)$; in Fig. 2.12c, k_1 is additionally introduced parallel to the branch k_1.

Correspondingly, the transfer functions (2.31)–(2.34) change:

$$\overline{W}(p) = \frac{\bar{y}}{\overline{Q}} = \frac{1}{mp^2 + k + k_1}; \quad (2.35)$$

$$\overline{W}_1(p) = \frac{\bar{y}}{y_1} = \frac{k + k_1}{mp^2 + k + k_1}; \quad (2.36)$$

$$\overline{W}_2(p) = \frac{\overline{Q}_{\text{dyn}}}{\overline{C}_1} = \frac{mp^2 \cdot (k + k_1)}{mp^2 + k + k_1}; \quad (2.37)$$

$$\overline{W}_3(p) = \frac{\overline{Q}_{\text{dyn}}}{\overline{Q}} = \frac{mp^2}{mp^2 + k + k_1}. \quad (2.38)$$

This same sort of situation will also be observed when introducing an additional constraint $\overline{W}_{\text{dyn}}(p)$ of a more complex type [7]. When introducing an additional constraint based on the principle of absolute deviation, the computational scheme (Fig. 2.12a) takes the form shown in Fig. 2.13a, and the structures in Fig. 2.12b and c, will take the form, respectively, as shown in Fig. 2.13b and c.

It should be noted that the introduction of an additional constraint in absolute deviation significantly changes both the computational scheme and the structural analogue (the structural diagram Fig. 2.13b) of the automatic control system of a dynamically equivalent mechanical oscillatory system (Fig. 2.13a).

But if we are dealing with a force perturbation $\overline{Q}$, then the mechanical circuit remains as if unchanged. (Note that this applies only to cases when the external impact is force.) With kinematic impact (or velocity perturbation), the mechanical circuit will look different. This is due to the possibility, in some cases, of an equivalent replacement of the kinematic perturbation by the force one. Thus, in Fig. 2.13b, the kinematic perturbation can be transferred from the point A to the point B, as shown in Fig. 2.14. In this case, the input kinematic perturbation is transformed separately, through the unit with the transfer function k. The reason for this is that an external force corresponding to an elastic force depending on the external displacement is applied to the point B.

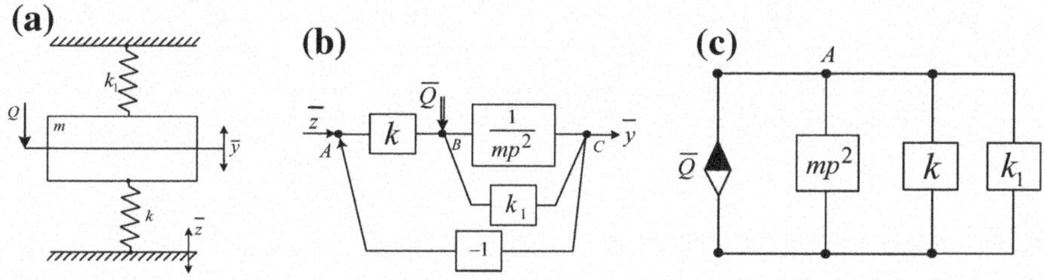

Fig. 2.13 The computational scheme in the problems of vibration protection when introducing an additional constraint k_1 according to the principle of absolute deviation control (**a**); **b** structural diagram—analog (automatic control system); **c** dual mechanical circuit—analog

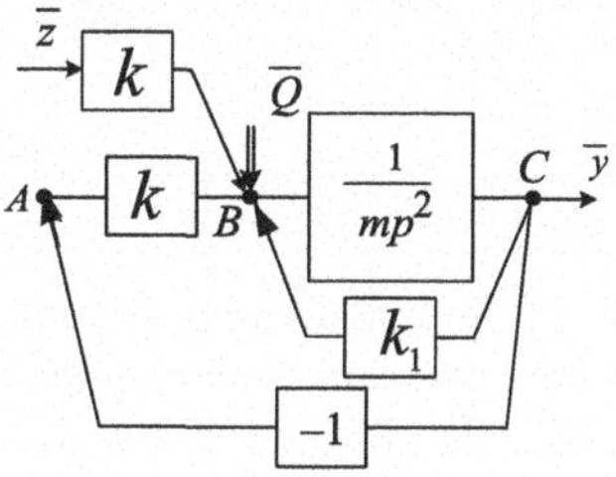

Fig. 2.14 The structural diagram of the system with the kinematic perturbation, reduced to the force one

2.2.2 *Equivalent Transformations*

In this connection, the question of the equivalence of certain interactions arises. Let us consider two diagrams of applying forces, as shown in Fig. 2.15a, b, which correspond to the structural diagrams of Fig. 2.15c, d.

Let us construct an equivalent structure of the automatic control system (ACS) (Fig. 2.16), taking into account that in Fig. 2.15a at the point C an element of mass m_1 will be involved, to which the force $\overline{Q}$ is applied.

From the diagram in Fig. 2.16b, it follows that the transfer function with respect to y from $\overline{Q}$ has the form

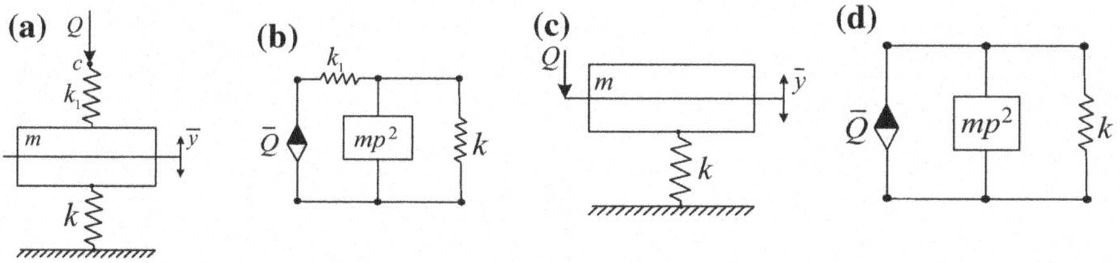

Fig. 2.15 Diagrams of equivalent relationships between structural circuits and dual mechanical circuits. The circuit in **a** corresponds to that in **c**; the circuit in **b** corresponds to that in **d**

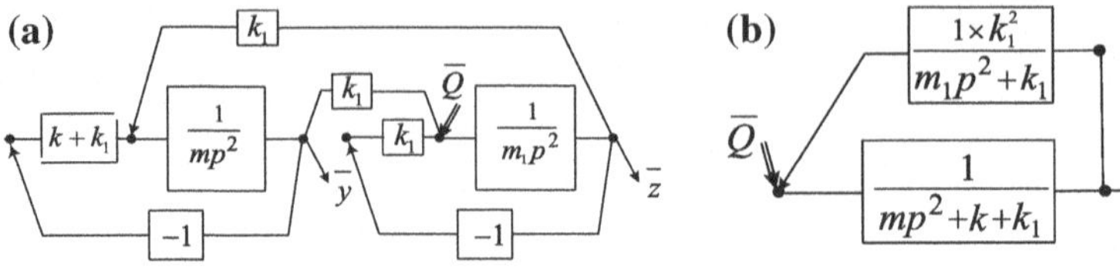

Fig. 2.16 Equivalent structural diagrams

$$W(p) = \frac{\bar{y}}{\overline{Q}} = \frac{mp^2 + k + k_1}{(mp^2 + k + k_1)(m_1p^2 + k_1) - k_1^2}, \tag{2.39}$$

which after transformations and for $m_1 = 0$ gives

$$\overline{W}(p) = \frac{\bar{y}}{\overline{Q}} = \frac{m_1p^2 + k + k_1}{mp^2 \cdot k_1 + kk_1}. \tag{2.40}$$

If the kinematic perturbation is replaced equivalently to the force one, according to the diagram shown in Fig. 2.14, then the mechanical circuit takes the form, as in Fig. 2.17, from which it follows that

$$\frac{\bar{y}}{\overline{Q}} = \frac{1}{mp^2 + k}; \qquad \frac{\bar{y}}{\bar{y}_1 k} = \frac{1}{mp^2 + k}; \qquad \frac{\bar{y}}{\bar{y}_1} = \frac{k}{mp^2 + k}. \tag{2.41}$$

Thus, the mechanical circuit used to determine the corresponding transfer functions (2.31) or (2.35) can be used to determine the transfer functions from the kinematic perturbation, but through the input force (transfer and equivalent) impact of a type $\overline{Q}_1 = \bar{y}_1 k$ (the transformation diagram results in the expression (2.41)). The external force $\overline{Q}$ and the displacement of the base exert, in the physical sense, the same impact—the force one. Therefore, the mechanical circuit (or the basis) is common, and the features of the transmission of impacts are determined through transfer functions that reflect compliance—the expression (2.31) or the transfer coefficient of the amplitude of the oscillations—expressions (2.32), (2.36), (2.41).

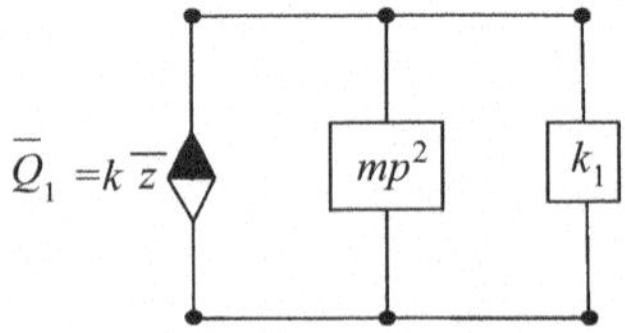

Fig. 2.17 The dual mechanical circuit for the problem of replacing the kinematic perturbation by an equivalent force

2.2.3 Introduction of a Typical Element

Let us note a characteristic feature of the construction of a mechanical circuit, taking into account in the computational scheme (according to Fig. 2.12a) the resistivity bp. In the mechanical circuit (Fig. 2.12c), the resistance (viscous friction) will be introduced as a branch bp parallel to the branch k (corresponds to the spring), then

$$\overline{W} = \frac{\bar{y}}{\overline{Q}} = \frac{1}{mp^2 + bp + k}. \tag{2.42}$$

If we consider the displacement of the base $z(t)$ instead of $\overline{Q}$, then expression (2.42) can be transformed according to the diagram:

$$\overline{Q} = \bar{z} \cdot [k + bp] = \bar{z} \cdot \overline{W}_0; \frac{\bar{y}}{\overline{Q}} = \frac{1}{mp^2 + bp + k}; \frac{\bar{y}}{\bar{z}} = \frac{bp + k}{mp^2 + bp + k}. \tag{2.43}$$

The use of the proposed approach has its special aspects in case if the additional constraint has the form of an oscillatory structure (with one or several degrees of freedom). This is due to the fact that the vibrational structure introduces additional force impacts generated by the intermediate masses (this is the inertia force of the intermediate mass, transmitted to the main mass m according to the diagram shown in Fig. 2.15 taking into account the expressions (2.33)). Let is consider an additional constraint in the form of a oscillating loop in the structural diagrams (Fig. 2.18) and the dual mechanical circuit (Fig. 2.19), which implies that

$$S(p) = \frac{\overline{Q}}{\bar{y}} = \frac{(m_1p^2 + k_3) \cdot k_2 + (mp^2 + k) \cdot (m_1p^2 + k_2 + k_3)}{m_1p^2 + k_2 + k_3}. \tag{2.44}$$

We obviously obtain this result from the structure of the automatic control system (ACS) (Fig. 2.18b, c)

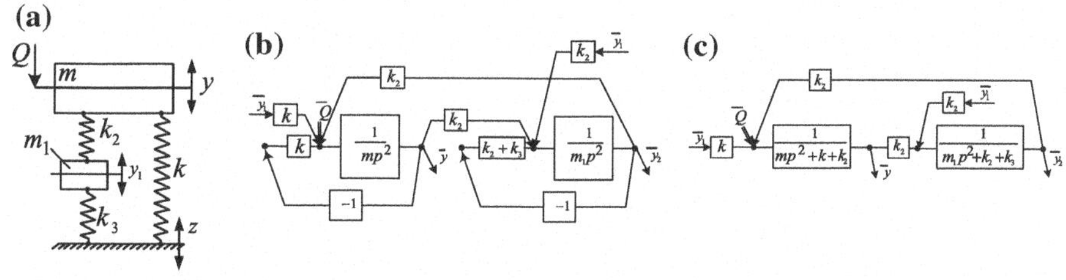

Fig. 2.18 The computational scheme in the problem of vibration protection (**a**) with additional constraint in the form of an oscillatory structure; **b** the structural diagram of a conventional system with two degrees of freedom; **c** the structural diagram using blocks (partial subsystems)

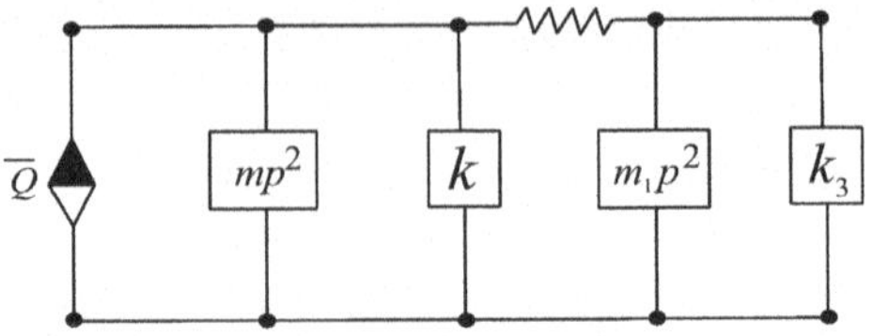

Fig. 2.19 The dual mechanical circuit, corresponding to the computational scheme in Fig. 2.18a

$$\overline{W}(p) = \frac{\bar{y}}{\overline{Q}} = \frac{m_1p^2 + k_2 + k_3}{(m_1p^2 + k + k_2)(mp^2 + k_2 + k_3) - k_2^2}, \quad (2.45)$$

Let us find the transfer function, taking into account the loading conditions and constraints of the computational scheme in Fig. 2.18a (we assume that $Q = 0$).

Since $\overline{W}^0 = \frac{\bar{y}}{\overline{Q}_1}$, and $\overline{Q} = \bar{y}_1\overline{W}_y$,

then

$$\overline{W}^0 = \frac{\bar{y}}{y_1} = \overline{W}_y \times \overline{W} = \overline{W}_y \times \frac{m_1p^2 + k_2 + k_3}{AB - k_2^2}, \quad (2.46)$$

where $A = m_1p^2 + k + k_2$; $B = m_1p^2 + k_2 + k_3$ ($\overline{W}_y = k$).

Let us find $\overline{W}_y$, taking into account that the displacement of the base forms a force impact on the element with mass m. It consists of $Q_{\text{cont}} = ky_1$, as well as the dynamic response from the action of the element with mass m_1, through springs with stiffness k_2 and k_3.

Let us use the dual mechanical circuit according to the computational scheme (Fig. 2.20), from which

$$S(p) = \frac{(m_1p^2 + k_3) \cdot k_2}{m_1p^2 + k_2 + k_3}. \quad (2.47)$$

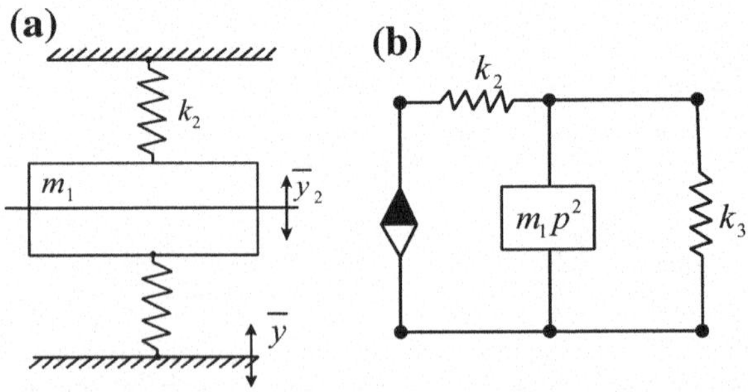

Fig. 2.20 The computational scheme (**a**) and the dual mechanical circuit (**b**) for taking into account the elastic properties of the base

Using (2.46) and (2.47), we find:

$$S(p) = \overline{W}_y = \frac{(m_1p^2 + k_3) \cdot k_2}{m_1p^2 + k_2 + k_3}. \quad (2.48)$$

2.2.4 *Specific Features of the Introduction of Constraints*

The circuit in Fig. 2.19 in the absence of branches k and mp^2 may be given as an explanation. The physical meaning is that this block generates a dynamic impact to the mass m on the side of the base.

Finally we get:

$$\overline{W^0} = \frac{\bar{y}}{\bar{y}_1} = \frac{(m_1p^2 + k_2 + k_3) \cdot \left[\frac{(m_1p^2 + k_3) \cdot k_2}{m_1p^2 + k_2 + k_3} + k\right]}{AB - k_2^2}. \quad (2.49)$$

After the transformations, the expression (2.49) takes the following form:

$$\overline{W^0} = \frac{\bar{y}}{\bar{y}_1} = \frac{k \cdot (m_1p^2 + k_2 + k_3) + k_2k_3 - k_2m_1p^2}{(m_1p^2 + k_2 + k_3) \cdot (AB - k_2^2)}. \quad (2.50)$$

If we assume that $A = m_1p^2 + k + k_2$; $B = m_1p^2 + k_2 + k_3$, and substitute into (2.50), we obtain

$$(m_1p^2 + k_3) \cdot k_2 + (mp^2 + k) \cdot (m_1p^2 + k_2 + k_3), \quad (2.51)$$

which coincides with the numerator of expression (2.45).

Thus, if an additional feedback is introduced into the system in the form of an oscillating unit, as it is shown in Fig. 2.10, then its transfer function has the form

$$\overline{W} = \frac{k \cdot (m_1p^2 + k_2 + k_3) + k_2k_3 - k_2m_1p^2}{(m_1p^2 + k_2 + k_3)(AB - k_2^2)}. \quad (2.52)$$

That is, the rule of introducing additional feedback in the form of an oscillating unit is confirmed by the diagram of relative control [7]. Therefore, the transfer function can be written in the following form

$$\overline{W}_{add} = \frac{m_1p^2 + k_3}{m_1p^2 + k_2 + k_3}. \quad (2.53)$$

In this case, k_3 is introduced parallel to the main spring. In terms of physical meaning, this should be dynamic stiffness, which is introduced into the main system, taking into account the spring stiffness, which binds both masses (in our case,

k_2). Then the transfer function (introduction by the principle of relative divergence) takes the form

$$\overline{W} = \frac{\bar{y}}{\bar{y}_1} = \frac{k + \left[\frac{m_1p^2 + k_3}{m_1p^2 + k_2 + k_3}\right]k_2}{mp^2 + k + \left[\frac{m_1p^2 + k_3}{m_1p^2 + k_2 + k_3}\right]}; \qquad (2.54)$$

$$\overline{W} = \frac{\bar{y}}{\bar{y}_1} = \frac{(m_1p^2 + k_2 + k_3)k + (m_1p^2 + k_3)k_2}{(m_1p^2 + k_2 + k_3)(mp^2 + k) + (m_1p^2 + k_3)k_2}. \qquad (2.55)$$

If the introduction of additional constraint is carried out by the principle of control on absolute deviation, then in contrast to (2.45) we obtain

$$\overline{W} = \frac{\bar{y}}{\bar{y}_1} = \frac{(m_1p^2 + k_2 + k_3) \cdot k}{(m_1p^2 + k_2 + k_3) \cdot (mp^2 + k) + (m_1p^2 + k_3) \cdot k_2}. \qquad (2.56)$$

Thus, if an oscillatory constraint is introduced into the computational schemes of a general form (Fig. 2.21) according to the principle of absolute deviation control, then the transfer takes the following form (which corresponds to the introduction of an additional constraint in absolute deviation):

$$\overline{W} = \frac{\bar{y}}{\overline{Q}} = \frac{1}{mp^2 + k + \overline{W}'_{add}}. \qquad (2.57)$$

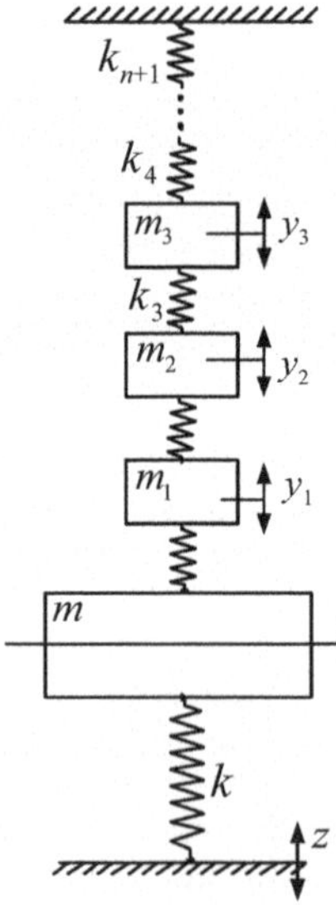

Fig. 2.21 The computational scheme in the problem of vibration protection, when an additional constraint in the form of an oscillatory structure of a general form is introduced according to the principle of absolute deviation control

The transfer function (kinematic perturbation) with an additional general constraint introduced by absolute deviation has the form similar to (2.57):

$$\overline{W} = \frac{\bar{y}}{\bar{y}_1} = \frac{k}{mp^2 + k + k_2 + \overline{W}'_{add}}. \tag{2.58}$$

The transfer function for an additional general constraint introduced by a relative deviation, in contrast to (2.58), is determined by the following expression:

$$\overline{W} = \frac{\bar{y}}{\bar{y}_1} = \frac{k + k_{n+1}\overline{W}'_{add}}{mp^2 + k + k_{n+1}\overline{W}'_{add}}, \tag{2.59}$$

where $\overline{W}'_{add} = \frac{k_1}{m_1p^2 + k_1 + k_2}$ is for the case with one intermediate mass m_1, and

$$\overline{W}'_{add} = \frac{k_1k_2}{(m_1p^2 + k_1 + k_2)\cdot(m_2p^2 + k_2 + k_3) - k_2^2} \tag{2.60}$$

for the case with two intermediate masses m_1 and m_2.

In the case of intermediate masses, when there are more than two of them, to calculate the transfer function of the additional constraint, it is best to use dual mechanical circuit, which for Fig. 2.21 has the form shown in Fig. 2.22.

2.2.5 Some Generalizations

Although the force and kinematic perturbations are quite different and have a different "structural" design, in theoretical studies the kinematic impact can be replaced by an equivalent force. In this substitution, it is interesting that in the transfer of the action from the bearing surface to the protection object, it is necessary to take into account how the object is influenced by the dynamic response which, in addition to the component from the elastic element, has additional parts.

This is, first of all, a component of the inertia forces, and then—of the viscous friction force. It was shown above how all factors are taken into account if the additional constraint acts as a general-type oscillating loop. The considered case

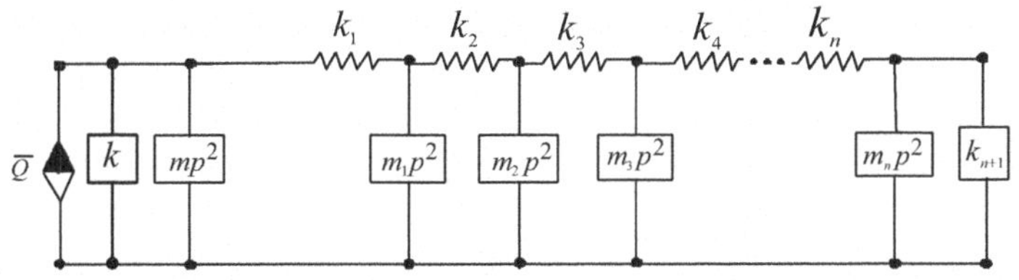

Fig. 2.22 The dual mechanical chain equivalent to the circuit in Fig. 2.21

with n-vibrational elements in the additional circuit is reduced to a closed mechanical circuit with n + 1-elements connected by elastic elements with stiffnesses k, k_1, k_2, …, k_{n+1}. By setting Q as the force applied to the element with the mass m, it is possible to determine the transfer function of the force from displacement, speed, etc. If we assume that the nth unit has a large mass, then y_n can be regarded as a parameter of the kinematic perturbation. It should also be noted that there may be other links between the elements m and m_n, considered as additional ones in this case. With an increase in the number of degrees of freedom, one can note an increase in the uncertainty in the combinations of the phase shift and the quantities of the divergence. Moreover, even at a fixed frequency of the external impact, due to the fact that as a matter of actual practice the number n (the number of intermediate masses when the computational scheme is chosen) is sufficiently indeterminate. In the future, it can be assumed, with respect to such elements, that the greater n is, the greater the degree of their approximation to the unit of the additional constraint as a stochastic unit (or to a unit with stochastic properties) will be. The smaller n is, the more determinism is. If n = 1, 2, 3, then we can obtain finite formulas for calculations.

If we select one mass m_1 (or m_1 and m_2) in a cluster and replace it with the whole block, that is, if we assume that the transfer function of the additional constraint has a structure

$$\overline{W}'_{add} = \frac{k'_1}{m'_1 p^2 + k'_1 + k'_2} + \text{remainder}, \tag{2.61}$$

then you can find the parameters k'_1, k'_2, m'_1, which would make the remainder "conditionally" small. The first step can be carried out using chain fractions, then find the main contour, and then evaluate the remainder. If there is a rapid decay in it, then it can be discarded. When estimating the remainder of a circuit by its influence upon the dynamical reaction or displacement, one must take into account that any resonance gives "infinity", so that the obtained conclusions are limited to the corresponding frequency range from 0 to some ω'.

If we introduce an additional constraint into the mechanical circuit, then, assuming that $\overline{W}'_{add}$ corresponds to the introduction rules described in [7], we will have a force factor. It should be noted that with the introduction of an additional constraint and the force nature of the perturbation, the dual mechanical circuit yields the same result as the approach based on the ACS. If we proceed to kinematic perturbation, then we need to introduce a new concept of some conditional force Q_1—this is the force that will influence the mass in the absence of $\overline{Q}$. That is, the impact on the system from the motion of the base (or kinematic perturbation) is replaced by an equivalent force perturbation Q_1, but it will already be applied to the object (mass m). Such an impact of Q_1 is formed in two directions: the elastic element κy and $k\overline{W}'_{add}$, and $\overline{W}'_{add}$ here acts as an element which generates a force when the input signal is a displacement. Thus, $\overline{W}'_{add}$ can be an additional spring, a damper, any set of these elements. But if there is an intermediate mass inside, then it makes the appropriate adjustments: it affects the input from the side of the base—on the object we should get a dynamic response

$$\frac{\bar{y}}{\overline{Q}_1} = S(p) = \frac{1}{\overline{W}(p)}; \tag{2.62}$$

$$\overline{W}(p) = \frac{\overline{Q}}{\bar{y}}; \tag{2.63}$$

$$\overline{Q}_1 = y_1(k + \overline{W}_{add}); \tag{2.64}$$

$$\frac{\bar{y}}{\bar{y}_1(k + \overline{W}_{add})} = \frac{k + \overline{W}_{add}}{mp^2 + k + \overline{W}_{add}}; \tag{2.65}$$

$$\frac{\bar{y}_1}{\bar{y}_1} = \frac{k + \overline{W}_{add}}{mp^2 + k + \overline{W}_{add}}. \tag{2.66}$$

All relations are defined similarly and correspondences are established between the structural theory of vibration protection systems and the theory of dual mechanical circuits. However, we note the differences in the understanding of the elements of the mechanical circuits: they connect the input signal—this is the displacement (coordinate) and the output signal in the form of a force factor. While in other works (for example [3, 8]), the input signal in the element is the velocity. Input signals in terms of physical meaning will be force factors, but their differences are that they will be determined by different transfer functions of the elements, namely:

- in the first case, the spring corresponds to the amplifying unit (k), the viscous friction resistance (bp) corresponds to the first-order differentiator, the mass corresponds to the second-order integrating unit $\left(\frac{1}{mp^2}\right)$;
- in the second case, the spring corresponds to the integrating unit of the first order $\left(\frac{k}{p}\right)$, the resistance of viscous friction corresponds to the amplifying link (b), and the mass corresponds to the differentiating link (mp).

2.3 Extension of a Typical Set of Elements

A comparative analysis of works devoted to electrical and mechanical circuits revealed sufficiently stable concepts about the elemental base of circuits, with their dual basis in mind. The theory of electrical circuits has been widely applied in electrical engineering in the study of transient processes, the evaluation of the dynamic properties of special operating modes. Numerous cases of application to specific electric circuits and electric machines are also known. A similar situation is observed with respect to mechanical systems, if they consist of linear passive elements. However, the technique for studying electric circuits seems to be more developed, since in structural interpretations it is possible to take into account internal limitations on the sources of electromotive forces and currents [6, 8].

2.3.1 Theory of Mechanical Circuits

In the theory of mechanical circuits [3, 9], the aforementioned questions are less developed. In the study of the above objects, the concepts of complex resistances and complex conductivities are used when using the same system of rules for coupling for the dual units; although in this case, in relation to circuits, as a rule, the problem of finding and calculating control forces is not posed; while the problems of dynamic synthesis are solved quite often. The development of methods for analyzing and synthesizing circuits occurs due to the complication of the original models, which is reflected in the development of the theory of four- and multiterminal networks in electrical circuits. One of the directions of the expansion of the element base was the use of the idea of lever linkages in electromechanical analogies. In electrical circuits, such elements are associated with transformers.

The element base of mechanical oscillatory systems, considered within the sphere of dynamic analogies with automatic control systems, naturally began to undergo changes, which was reflected in the structural theory of vibration protection systems. Structural models of mechanical systems differ from mechanical circuits and their analogues as they use the rule of transformation of structures with feedbacks. It should be noted that the process of expanding the element base is associated with the expansion of one group of elements that have a displacement at the input, and an effort at the output. In fact, these are varieties of a modified elastic element. Each of the elements has its own transfer function, which cannot be decomposed into simpler ones. The latter makes it possible to add three more to the known standard elements (an ordinary spring and a damper): a second-order differentiating unit and an integrating unit of the first and second order. In this regard, the elementary standard units of automatic control systems are complex and can be obtained from the elementary units of a mechanical system. The peculiarity of the approach developed for mechanical systems is that when selecting an object, which dynamic state is of interest, a basic model is formed, illustrated by the corresponding structural diagram equivalent in a dynamic relation to the automatic control system. In the basic model, the elastic element and the damper connecting the object to the support (or body) are used as the supporting or basic elastic elements. All other constraints introduced additionally to solve certain problems are elements of an extended type set, including additional elastic elements and dampers. To change the dynamic properties in a mechanical system, which is a computational scheme of a dynamic object, an additional constraint is introduced into its basic structural model, which in developed form can be represented by a mechanical circuit. Calculations of the parameters of such a circuit can be carried out to determine the total resistance (or reduced elasticity) on the basis of the rules of the theory of circuits. However, the input signal to such a circuit is the displacement, and the output is the force. As for additional constraints, they are introduced into the structural diagram in accordance with the principles known in the control theory. In the special aspects of the introduction of the additional constraint, the structural analogue of the mechanical system differs from the structural diagram of the dynamically equivalent automatic control system. The simplest form of a mechanical oscillatory system is an object of mass (m) and an elastic element (k). Such a system can be called a basic model. It corresponds to a dynamically equivalent automatic control system (Fig. 2.23).

The structural diagram gives an idea of the place and role of the elastic element, on the input of which the displacement (y) is supplied. The output of the link with the transfer function $k = W'$ is the force. This force is applied to the input of the second unit with the transfer function. Such a unit in the structural diagram is an "adder", in accordance with the d'Alembert principle at the input of the unit, the geometric sum of all forces is zero. This unit performs the basic function of the adder, it is not included in the number of elementary units and has a different transfer function in dimension, since the output signal is the displacement (y). In such a system of minimal complexity, the unit with the transfer function $k = W'$ creates the force of the elastic interaction. According to the automatic control theory (ACT), this unit is called amplifying. The second basic unit from the position of control theory is the second-order integrating unit, which "transforms" the elastic force into the displacement. Both diagrams in Fig. 2.23 correspond to the same differential equation

$$m\ddot{y} + ky = kz + P(t). \tag{2.67}$$

The structural diagram in Fig. 2.23b is the graphical form of the representation of the differential equation (2.67). Equation (2.66) can be obtained on the basis of the d'Alembert principle or by using the Lagrange equation. It becomes important to choose the "key" element in the basic model; this is an object of mass m. More

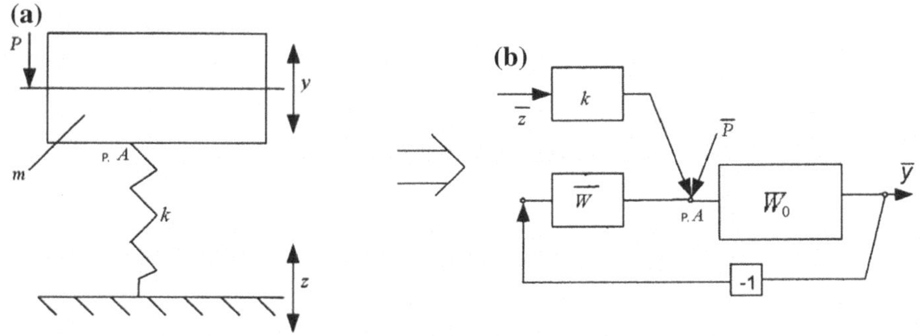

Fig. 2.23 The basic model of VPS (**a**) and its structural diagram (**b**)

precisely, the "key" element is a material point of mass m, against which the kinetostatics equation is generated. An elastic element with a stiffness coefficient k acts as an elementary unit, and two units with transfer functions W_1 and W_2 create a basic model of minimal complexity: here, $k = W'$ enters the extended set of elementary unit of the vibration systems. It should be noted that the point A (Fig. 2.23a) passes to the point of summation of forces (the point A in Fig. 2.23b), which makes it possible to find a dynamic response through the corresponding transfer function.

The VPS can be complicated via the introduction of an additional circuit parallel to the elastic element, the transfer function of which can be represented in the form of a fractional-rational expression

$$W_{add}(p) = \frac{a_0 + a_1 p + \cdots + a_n p^n}{b_0 + b_1 p + \cdots + b_m p^m}, \tag{2.68}$$

where m, n are integers ($n \leq m$); a_j, b_j are the coefficients determined by the constructional features of the VPS, $i = \overline{1, n}$, $j = \overline{1, m}$.

An additional circuit with the transfer function $W_{\text{add}}(p)$ is an additional feedback introduced into the structural diagram (Fig. 2.23b). In this case, it is possible to use principles that are well-known in ACT for introducing control by absolute, relative divergences, and also by external perturbation (Fig. 2.24).

If we assume that $W_1 = k_1$, and $W_2 = k_2$, then in Fig. 2.24b, you can see the form of physical implementation of such constraints. These are springs, which are introduced between the object of protection and the base: movable and non-movable. With regard to the introduction of constraint by the perturbation (W_3), its implementation requires the construction of a special circuit consisting of a series of units; this constraint can be shown in the structure of the transfer function of the VPS:

$$W = \frac{\bar{y}}{\bar{z}} = \frac{k + W_1 + W_3}{mp^2 + k + W_1 + W_2}. \tag{2.69}$$

If the system has several degrees of freedom, then the structural diagram of the VPS will consist of partial systems, each of them having one degree of freedom, but

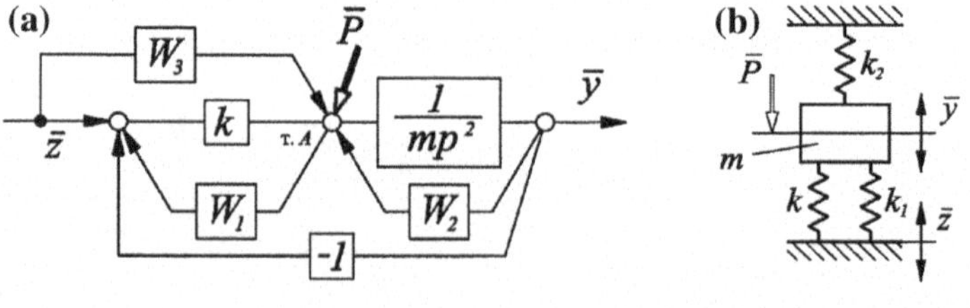

Fig. 2.24 Types of additional constraints in the structural diagram (**a**) and computational scheme (**b**)

these systems will be connected via some units (elastic ones, in this case). The latter demonstrates the "every action has a reaction" principle in the system. The development of structural diagrams can be carried out directly by the computational scheme. However, the preliminary generation of differential equations of motion can be useful. In particular, the matrices of the coefficients of the system of equations possess certain symmetry properties, which makes it possible to control the correctness of the generation of the equations of the VPS.

The various additional constraints discussed above are essentially different types of generalized springs, since the input for them is the coordinate of the object (or its displacement), the output is the force. The point *A* (see Fig. 2.24a) is an "adder" in which the sum of all applied forces is zero (in accordance with the d'Alembert principle). The number of elementary units can be expanded, and it is not limited to springs only, as shown in [7]. If by an elementary unit we mean a unit which transfer function cannot be obtained by means of the simplest transformations, then the number of elementary units can include: units of single and double differentiation, of single and double integration. Let us suppose that out of these elementary units that make up the extended elementary base of the VPS, one can get other standard units, but they will already turn out to be units of the second level, since they will consist of the mentioned elementary units. In this regard, the standard units in ACT, with a few exceptions, are the units of the second level, from the standpoint of the structural theory of the VPS.

2.3.2 Special Aspects of the Structures of the Automatic Control Theory

In the automatic control theory, the standard units are also used. However, their selection is based on other approaches. In particular, the transfer function of the vibration protection system is considered as a result of some transformations that allow the transfer function of the system to be perceived as a result of a consecutive and parallel connection of standard units. By taking into account this circumstance, differences in the methods of dynamic synthesis of the VPS and the ACS are determined, although they have a common basis. It should be noted that the additional constraint introduced (the auxiliary circuit) is essentially a generalized spring and it consists of elementary units, each of which also acts as a kind of spring: at the input there is a displacement, at the output there is a force. In all other cases, the ACS and the VPS follow the general rules of structural transformation.

Standard units in automation systems. In the theory of automatic control, typical elementary units are also used. The transfer function of the automatic control system in the form of an arbitrary polynomial can be decomposed into prime factors, that is, the transfer function of the system (of the unit) represented by the expression (2.68) can always be written as a product of prime factors and fractions of the form k, p, $\frac{1}{p}$, $Tp \pm 1$, $\frac{1}{Tp \pm 1}$, $T^2p^2 \pm 2\xi Tp + 1$, $\frac{1}{T^2p^2 \pm 2\xi Tp + 1}$. It is to be recalled that in ACT, k is called the transfer coefficient, T is the time constant and ξ ($0 < \xi < 1$) is the damping coefficient. In the context of the above positions in understanding the elementary units in (2.32), it can be seen that k, p, $\frac{1}{p}$ are elementary, and the rest can be obtained by using the rules of consecutive and parallel connections, as in the automatic control theory. The diagram of the correlation of the transformation rules is shown in Fig. 2.25.

A significant feature of the mechatronics of the VPS should be also mentioned. In this case, the additional constraint or additional circuit has the dimension of the unit, at the input of which there is an offset, and the output is force. That is, basically, we consider a mechanical circuit consisting of the same type of elements. There are five such elements. But a unit of pure delay can be added to them on the same conditions and in certain circumstances. Such elements act as dual elements

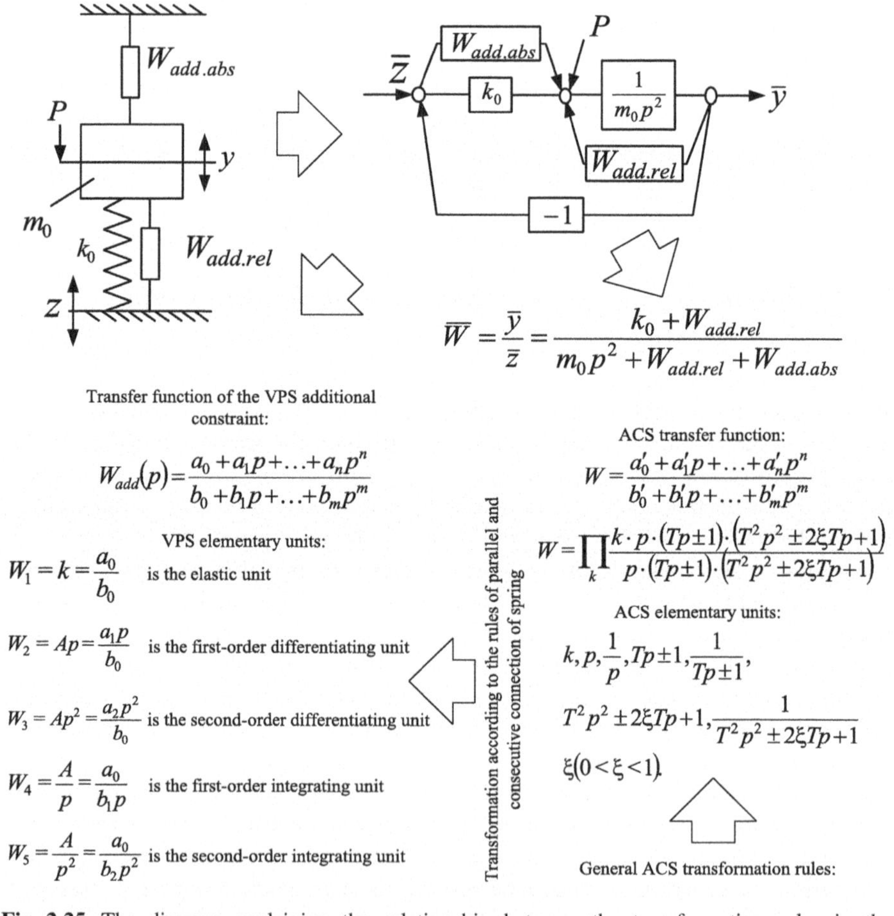

Fig. 2.25 The diagram explaining the relationship between the transformation rules in the supplementary feedback circuit and in the ACS structural diagram

and are interconnected to form more complex structures according to the rules of consecutive and parallel addition of springs [7, 10–12].

The transfer function (2.67), in our case, refers to an additional mechanical circuit introduced in a certain way into the structural diagram of a dynamically equivalent automatic control system. The complexity of this circuit predetermines the structure, the number of elementary units and the rules for their commutation.

2.4 Special Aspects of Circuit Structures

The study of mechanical oscillatory systems is an indispensable stage in the solution of various problems in the dynamics of machines. The formation of computational schemes, in which the concepts of the most essential properties of research objects are reflected, is predetermined by the features of the problems of dynamics. In this respect, vibration protection and vibration isolation of equipment and apparatus can be attributed to the problems of evaluation, control and management of the dynamic state of mechanical oscillating systems, which involves an interest in simplifying the initial systems and developing methods for simplifying mechanical oscillatory systems to basic structures with several degrees of freedom.

2.4.1 Opportunities for Simplification

Let us consider some general provisions that can be taken into account at the stage of justification and selection of computational schemes for vibration protection and vibration protection of machines and equipment, taking into account that the selected object of protection is only a fragment of more complex systems. The assumptions about the presence of a fixed base, racks, base models adopted in research technologies are conditional, which suggests the search for opportunities for developing new approaches. In this respect, studies of the simultaneous influences of several perturbations and the generalization of the concept of elastic elements are of particular interest [4, 13–15].

Let us consider a mechanical system with two degrees of freedom (Fig. 2.26a) and its structural diagram (Fig. 2.26b), based on the approaches outlined in [7]. In Fig. 2.26a, b, the following designations are accepted: y_1, y_2 are the coordinates in the stationary reference system, y_1, y_2 are the kinematic perturbations from the side of the base (or body); m_1, m_2, $k_1 \div k_3$ are the parameters of mass-and-inertia and elastic elements.

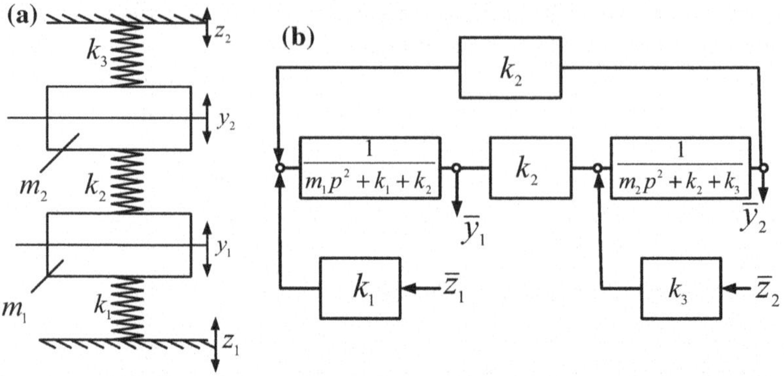

Fig. 2.26 The computational scheme (**a**) and structural diagram (**b**) of a mechanical oscillatory system with two degrees of freedom

Transfer functions of the system can be obtained on the basis of known techniques [3] and have the following form:

$$\overline{W}_1(p) = \frac{\bar{y}_1}{\bar{z}_{1(\bar{z}_3=0)}} = \frac{k_1(m_2p^2 + k_2 + k_3)}{A}; \tag{2.70}$$

$$W_2(p) = \frac{\bar{y}_2}{\bar{z}_{1(\bar{z}_3=0)}} = \frac{k_1k_2}{A}; \tag{2.71}$$

$$\overline{W}_3(p) = \frac{\bar{y}_1}{\bar{z}_{2\,(\bar{z}_1=0)}} = \frac{k_2k_3}{A}; \tag{2.72}$$

$$W_4(p) = \frac{\bar{y}_2}{\bar{z}_{2\,(\bar{z}_1=0)}} = \frac{k_2(m_1p^2 + k_1 + k_2)}{A}, \tag{2.73}$$

where $p = j\omega$;

$$A = (m_1p^2 + k_1 + k_2) \cdot (m_2p^2 + k_2 + k_3) - k_2^2. \tag{2.74}$$

If two kinematic perturbations act simultaneously in the system (see Fig. 2.26) and the condition $z = z_1 = z_2$ is fulfilled, then the transfer functions of the system will take the form

$$W_1'(p) = \frac{\bar{y}_1}{\bar{z}} = \frac{k_1(m_2p^2 + k_2 + k_3) + k_2k_3}{A}; \tag{2.75}$$

$$W_2'(p) = \frac{\bar{y}_2}{\bar{z}} = \frac{k_3(m_1p^2 + k_1 + k_2) + k_1k_2}{A}. \tag{2.76}$$

The joint impact of external perturbing factors leads to a change in the transfer functions, which can be interpreted physically as the occurrence of additional constraints in the system, which can lead to a change in the parameters of the dynamic absorbing of oscillations, change the structure of the dynamic interactions, but the denominator of the transfer functions remains unchanged. The latter can also be considered in such a way that, in the presence of a second perturbation channel, the processes in the system are formed on the basis of the principles of control by force perturbation [3]. At the computational schemes and structural diagrams, the kinematic perturbations $k_1\bar{z}_1$ and $k_2\bar{z}_2$ can be replaced by force perturbations $\overline{Q}_1 = k_1z_1$ and $\overline{Q}_2 = k_2z_2$, but at the same time it is necessary to take into account the corresponding changes in the numerators of the transfer functions [16].

If to consider the computational scheme in Fig. 2.26a as a particular case of a "triangle" connection type, the model in Fig. 2.27 can be used.

It can also be assumed that the circuit in Fig. 2.27 can be represented in the general form of a "triangle" type connection. If the support point O belongs to an element with infinitely large mass ($m_0 \to \infty$, where m_0 is the mass of the base or the body), then the circuit in Fig. 2.27 is converted to the form such as that in Fig. 2.28.

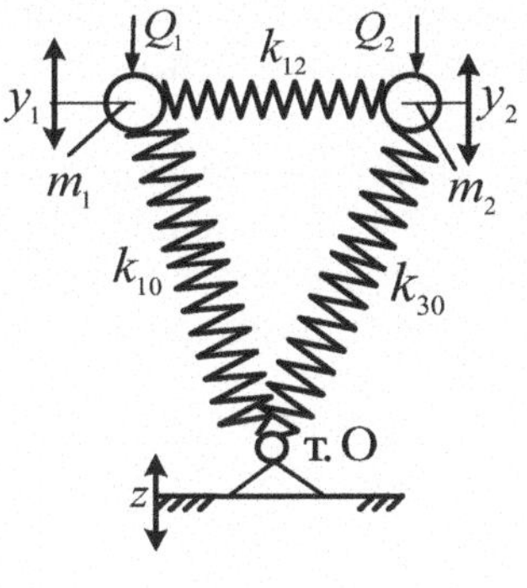

Fig. 2.27 The schematic diagram of the reference system in the form of a "triangle"

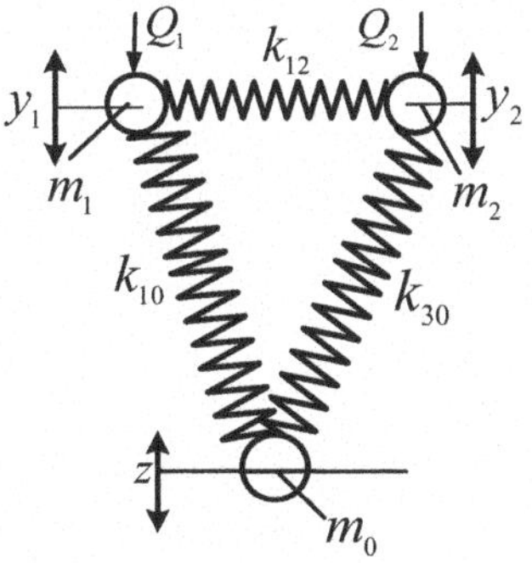

Fig. 2.28 The schematic diagram of the connection of the elements of the system taking into account the mass of the base or body

We write the expressions for the kinetic and potential energy of the system

$$\mathrm{T} = \frac{1}{2}m_0\dot{z}^2 + \frac{1}{2}m_1\dot{y}_1^2 + \frac{1}{2}m_2\dot{y}_2^2; \tag{2.77}$$

$$\Pi = \frac{1}{2}k_{10}(y_1 - z)^2 + \frac{1}{2}k_{12}(y_2 - y_1)^2 + \frac{1}{2}k_{30}(y_2 - z)^2 \tag{2.78}$$

and we obtain a system of equations of motion:

$$m_1\ddot{y}_1 + y_1(k_{10} + k_{12}) - k_{10}z - k_{12}y_2 = 0; \tag{2.79}$$

$$m_2\ddot{y}_2 + y_2(k_{12} + k_{30}) - k_2y_1 - k_{30}z - k_{12}y_1 = 0; \tag{2.80}$$

$$m_0\ddot{z} + z(k_{10} + k_{30}) - k_{10}y_1 - k_{30}y_2 = 0. \tag{2.81}$$

The structural diagram for the system shown in Fig. 2.28, has the form shown in Fig. 2.29.

As a result of the transformations, the initial system acquires one more degree of freedom in the motions, but its structure becomes closed. Let us suppose that $m_0 \to \infty$, then the output signal on the partial system (m_0p^2 + k_{10} + k_{30}) can be considered equal to 0. In this case, the structural diagram in Fig. 2.29 should be simplified (Fig. 2.30). The z coordinate will be the external impact at the points 1 and 2.

That is, there is a link between the points 1 and 6, but the link between the points 2 and 5 is not essential. The link between the points 4 and 5 (see Fig. 2.29) is also present; between the points 6 and 3 it is implemented as the force $k_{30}z$ (see Fig. 2.29). With a large difference between m_0 and m_2, the interaction in the

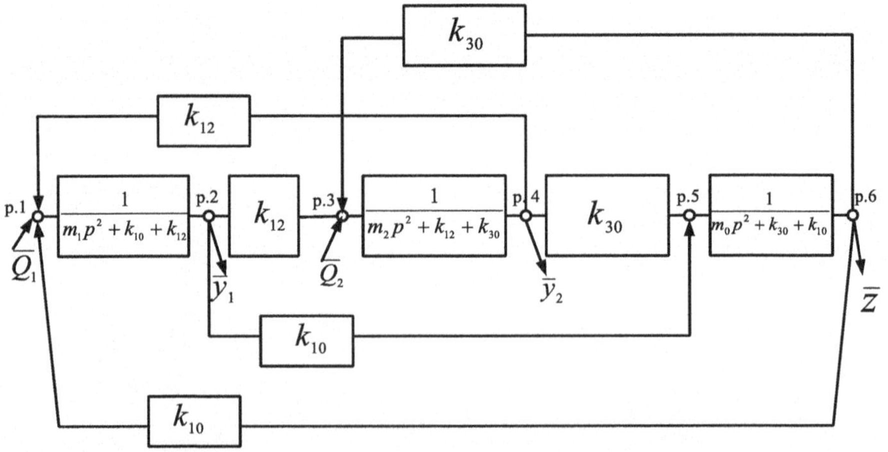

Fig. 2.29 The structural diagram of the system taking into account the mass of the base or body

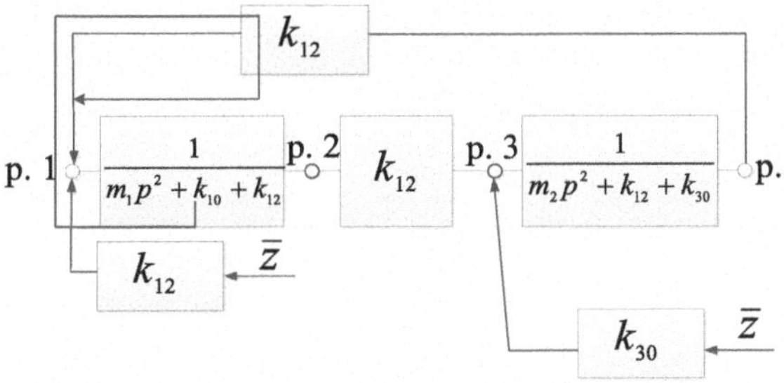

Fig. 2.30 The structural diagram of the system for $m_0 \to \infty$

physical sense can be considered on a case-by-case basis: the mass m_1 acts on the body in a "one-way" fashion, and then the constraints (2) ÷ (5) can be removed, and (6) ÷ (1) turn into $k_{10}z$. In turn, (4), (5) also "disappear", and (6), (3) are transformed into $k_{30}z$.

With simplifications, it is necessary to note the existence of asymmetric effects. Thus, after simplification, the structural diagram in Fig. 2.29 is transformed into the structural diagram shown in Fig. 2.30.

The base (fixed reference frame) in the computational schemes can be represented as some element with infinitely large mass ($m_0 \to \infty$). In the structural diagram (see Fig. 2.30), the influence of the partial systems ($m_1p^2 + k_{10} + k_{12}$) and ($m_2p^2 + k_{30} + k_{12}$) on the element m_0 can be considered inessential, so direct constraints k_{10} and k_{30} in the structural diagram (see Fig. 2.30) can be considered "insignificant". Physically, this means that the dynamic forces on the part of the elements m_1 and m_2 are very small to affect the change in motion along the z coordinate. At the same time, the motion of the element with mass m_0 is given and is defined as z, so the feedbacks m_0 by k_{10} and k_{30} remain and are transformed into kinematic perturbations $k_{10}z$ and $k_{30}z$, as can be seen for $z = z_1 = z_2$ in Fig. 2.30.

In this interpretation, the computational scheme in Fig. 2.26a can be represented as a joint of three masses of the "star" type. In the absence of other additional elastic elements, such a system can be considered under the assumption of free motion under the impact of known external forces.

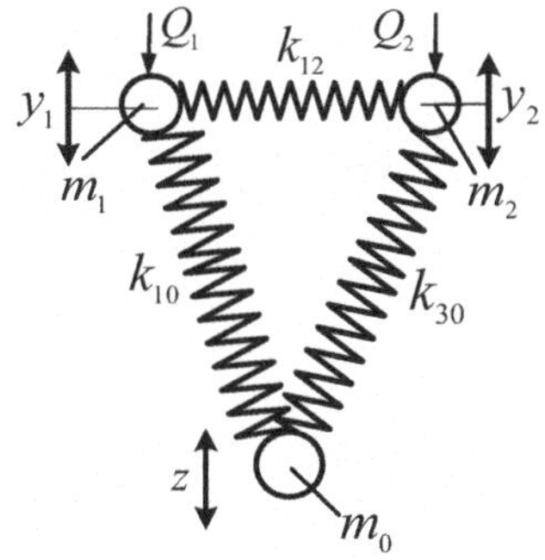

Fig. 2.31 The computational scheme in the form of a structure of a particular kind

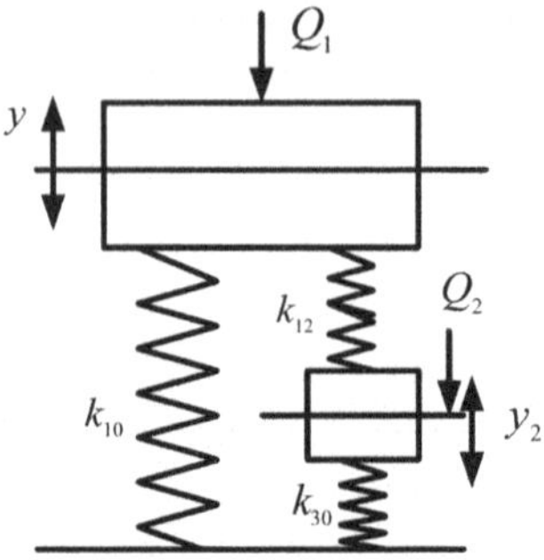

Fig. 2.32 The computational scheme of the transformed system

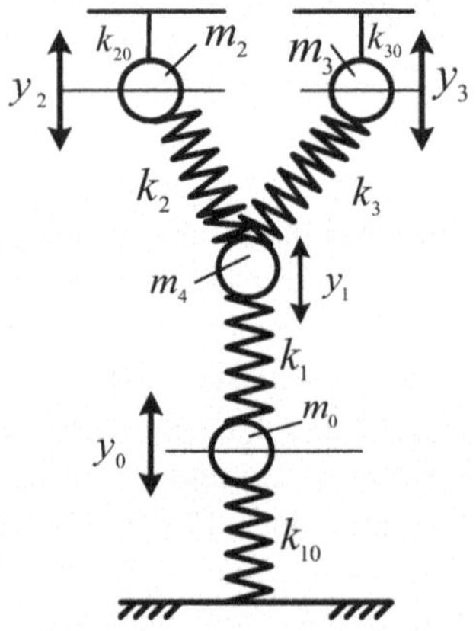

Fig. 2.33 The mechanical system of the "star" type

1. Of particular interest is the case when $z_1 = 0$, $z_2 \neq 0$ or $z_1 \neq 0$, $z_2 = 0$ is implemented. Let $z_2 = 0$, then the computational scheme is a structure of a particular kind (Fig. 2.31) or a circuit system at the input of which the external perturbation is applied (to the mass m_1), and the second element of the circuit is connected by the elastic element with the fixed base.
   By analogy with the foregoing, one can also assume that the elastic element k_3 (see Fig. 2.26a) will be related to an element of mass m_0 ($m_0 \to \infty$). In this case it will be a non-closed mechanical circuit.
2. If $z_1 = z_2 = 0$, then the mechanical circuit in Fig. 2.26a can be considered closed (see Fig. 2.28).
3. If $m_0 \to \infty$, then the "triangle" is transformed and takes the form, as shown in Fig. 2.32. Further calculations are carried out on the basis of known methods.

If we assume that a system with three degrees of freedom is, in a certain sense, basic and has the simplest configuration, then it can be obtained from a connection of the "star" type (Fig. 2.33).

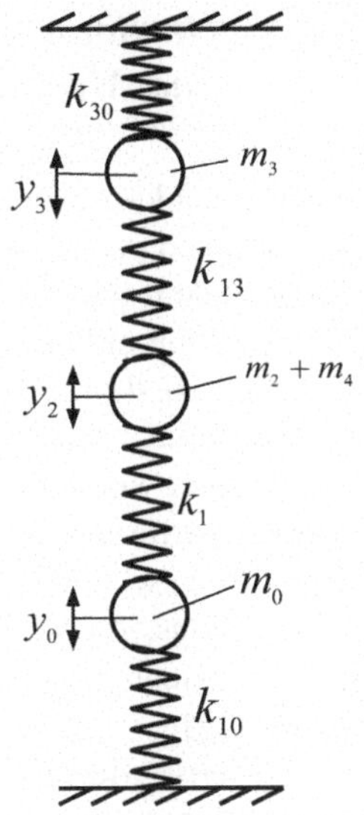

Fig. 2.34 The three-mass mechanical chain system

We will assume that the three-beam star has four masses. If we join m_2 and m_4, we obtain a chain system of three masses (Fig. 2.34), which is one of the known structures of mechanical systems. The same structure can be obtained if there is a fusion of $m_2 + m_3$, then $m_2 + m_3 \to m_4 \to m_1$.

If we connect the elements $m_1 + m_4$, then we can also obtain a chain system of the same type as in Fig. 2.34:

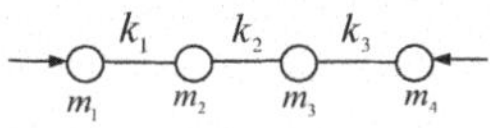

in which the necessary parameters are simply defined. Variants of possible simplifications in mechanical systems, which are reduced to basic models with two and three degrees of freedom, are considered in more detail in [16–18].

Thus, the notions that basic models with one to three degrees of freedom have a fixed basis, which allocates them to some particular class of systems, are not absolute. In fact, it can always be assumed that such representations are associated with the fulfillment of certain conditions. The practice of measuring the oscillations of mechanical systems shows that the basic models are correct in relation to a certain particular range. At the same time, basic models act as some simplifications of more complex structures. It is important that in the process of simplification, the asymmetry of the connections changes, or more precisely, is assumed, which in the physical sense corresponds to the difference in the parameters of the dynamic interaction between the elements of the system.

2.4.2 On Some Special Aspects of Electrical and Mechanical Circuits

When considering electrical circuits, two types of external influences are given. The first is the source of voltage; the second is the source of current. Sources of current can be ideal if their power is large enough. However, the limited sources of external energy in the circuit are taken into account. Real energy sources have an internal resistance. It should be noted that current and voltage sources are interchangeable, i.e., one source can be equivalently transformed into another [5, 6].

The same situation is typical for mechanical circuits. There are two sources of external perturbations: the source of force and the source of velocity [3]. In the physical sense, these external influences are considered as power and kinematic perturbations. As for kinematic perturbations, in the theory of mechanical circuits speed is usually used, although displacements can also be applied with the same reason [17].

The most widespread is the electromechanical analogy "current–speed". If the velocity V is the first derivative of the displacement S, then $V = dS/dt$. At the same time, the current i and the electric charge q are related by a ratio $dq/dt = i$. It follows that there is also an electromechanical analogy "charge–displacement" (q–S). In the theory of electromechanical circuits that uses the electromechanical analogy as a certain methodological basis, it is known that the number of variants of electromechanical analogies can be extended (for example, this is given in [19, 20]).

The selection of analog pairs with respect to electrical and mechanical circuits can be based on the idea of the invariance of certain estimates of the general properties of systems. If the power of the electrical circuits is defined by $W = UI$ as the product of the current by the voltage, then the following statement is quite logical: the electrical circuit may undergo certain transformations, but they will lead to an equivalent circuit only if the power equality is satisfied. Similar representations that include mechanical circuits yield $N = PV$, since power, in the mechanical sense, is the product of the force by speed.

If we take such considerations as a basis, then the pairs of analogues $U \sim P$ (voltage–force) and $i \sim V$ (current–velocity) can be considered quite justified. At the same time, taking into account that i and V are the first derivatives of the corresponding quantities characterizing the state, we can assume that the analogy is preserved even in the situation when the primitive functions differ by a constant quantity (we assume that the displacement, as $S(t) = S_0(t) + \text{const}$, can be considered as an primitive function with respect to $V(t)$, etc.). In other words, the choice of analog pairs is not absolute, but relative, and in practical problems requires the establishment of similarity relations, based on the specific conditions of the problem being solved. In the choice of analog pairs, the phase relations and the initial conditions in the occurrence of simultaneous processes can play a certain role.

Special aspects of mechanical circuits. In the theory of mechanical circuits, in this way, not only the force–velocity pair, but also others, can be considered as external perturbations. Traditionally, in mechanics, the external impact is associated with the external force applied to the mass (this approach is based on Newton's laws), this approach is common in the dynamics of machines. As for speed as a form of external impact, this kind of perturbation is used along with the displacement of the base or the preset acceleration of the base or support of the technical object. Such an impact is called kinematic, and is implemented, as a rule, via the elements by which the object is supported or is in contact with the base, the body, the fixed stand. Traditional ideas about external influences and their types come from the theory of machines and mechanisms and the corresponding problems of machine dynamics (in particular, vibration protection and vibration isolation of machines and equipment). If the law of motion is given, then the problems of determining the influencing forces are most often solved. Basically, such problems are called inverse problems of dynamics. If external forces are given or known, then the motion of the system is determined, that is, the direct problem of dynamics is solved.

Considering the possible replacement of energy sources in electric circuits, it can be assumed that equivalent substitutions of force perturbations for kinematic and vice versa are also feasible. In particular, such possibilities in mechanical oscillatory systems are considered in [21, 22]. The question of the joint impact of several perturbing factors in systems with several degrees of freedom is more complex. Linear theories of circuits of different nature, as well as the theory of linear oscillations, are based on the superposition principle, which is universally recognized. However, it is worth pointing out that the idea of an equivalent replacement of sources of perturbations in the circuits predetermines the possibility of implementing the process of obtaining the total perturbation, applied at the point chosen for this.

The proposed approach allow us to formulate problems that are oriented to the search for conditions for obtaining a cumulative effect of the desired type, for example, equalities of the sum to zero, which in a certain situation would provide the effect of dynamic absorbing of the oscillations. Naturally, in order to obtain such conditions, it is necessary to have quite definite links between external influences. These can be the coincidence of effects in frequency, certain phase shifts, the presence of certain common integral properties, etc.

Thus, the method of electromechanical analogies makes it possible to extend the system of equations relating to mechanical circuits in which other variables, such as displacement, acceleration, including derivatives and integral relations, can be used as state variables in the form of velocity points and forces. In this case, the conditions for preserving power, energy, and momentum are fulfilled. The introduction of additional constraints in the structural diagrams of equivalent automatic control systems used in the structural theory of vibration protection systems can be transferred to structural interpretations or diagrams of mechanical and electrical circuits that will allow establishing certain rules for the transformation of structural diagrams of one kind into structural diagrams of another one. The difference between the structural diagrams of mechanical systems (that are represented in the form of circuits) and the dynamically equivalent structural diagrams of automatic control systems lies in the fact that, within the apparatus of the theory of automatic control, the object of vibration protection is allocated as an object of control. The remaining parts of the system can be considered as mechanical circuits, interpreted as additional feedbacks.

The element base of the circuits can be expanded by incorporating new elements into it, for example, motion transformation devices in the form of various mechanisms (in particular, levers). Similar processes are also characteristic of electric circuits when introducing transformers of various types. It can be assumed that the introduction of transformers into an electrical circuit is equivalent to introducing a solid body in the mechanical circuit that performs a plane motion, which presupposes a significant change in the concepts of the rules for transforming circuits. It should be noted that the new elements of circuits, which are subject to the rules for transforming structures, can be constructed as some blocks possessing the properties of generalized springs. In special cases (statics), such elements are reduced to ordinary elements. For more complex structures, special transformation rules can be developed, for example, "star–triangle".

References

1. Eliseev, S.V., Khomenko, A.P.: Dynamic Damping. Feedback Concept and Structural Methods of Mathematical Modeling, 357 pp. Nauka Publ., Novosibirsk (2014)
2. Lurie, A.I.: Operational Calculus and Application in Engineering, 368 pp. Nauka Publ., Moscow (1959)
3. Druzhinsky, I.A.: Mechanical Chains, 240 pp. Mashinostroenie Publ., Moscow (1977)
4. Eliseev, S.V.: Impedance Methods in the Study of Mechanical Systems. Fundamentals of the Theory, 86 pp. IPI Publ., Irkutsk (1979)
5. Atabekov, G.I.: Linear Electrical Circuits, 592 pp. Energia Publ., Moscow (1978)
6. Bakalov, V.P., Dmitrikov, V.F., Kruk, B.I.: Fundamentals of the Theory of Circuits. Textbook, 589 pp. Radio and Svyaz' Publ., Moscow (2000)
7. Eliseev, S.V., Reznik, Yu.N., Khomenko, A.P., et al.: Dynamic Synthesis in Generalized Problems of Vibration Protection and Vibration Isolation of Technical Objects, 523 pp. ISU Publ., Irkutsk (2008)
8. Gardner, M.F., Burns, J.L.: Transients in Linear Systems. Transl. from English, 3rd edn., 551 pp. Fizmatgiz Publ., Moscow (1961)
9. Lenk, A.: Electromechanical Systems: Systems with Distributed Parameters, 472 pp. Energoizdat Publ., Moscow (1982)
10. Eliseev, S.V.: Structural Theory of Vibration Protection Systems, 224 pp. Nauka Publ., Novosibirsk (1978)
11. Eliseev, S.V., Reznik, Yu.N., Khomenko, A.P.: Mechatronic Approaches in the Dynamics of Mechanical Oscillatory Systems, 384 pp. Nauka Publ., Novosibirsk (2011)
12. Rothenberg, R.V.: Fundamentals of the "Driver-Car-Road-Environment" System Reliability, 216 pp. Mashinostroenie Publ., Moscow (1986)
13. Eliseev, S.V., Kashuba, V.B., Bolshakov, R.S.: The Possibility of the Influence of External Influences on the Reduced Rigidity of the System. Mashinostroenie and Life Safety, vol. 3

(13), pp. 46–52. Murom Institute (Branch) of the Alexander and Nikolai Stoletovs Vladimirsky State University, Murom (2012)
14. Eliseev, S.V., Kovyrshin, S.V., Bolshakov, R.S.: Design features of compacts of elastic elements in mechanical oscillatory systems. Interactions with the elements of the system and forms of connections. Mod. Technol. Syst. Anal. Model. IrGUPS Publ., Irkutsk **4**(36), 61–70 (2012)
15. Eliseev, S.V., Parshuta, E.A., Bolshakov, R.V.: Transfer functions of mechanical oscillatory systems. Possibilities for estimating reduced stiffness. Int. J. Appl. Basic Res. RAE Publ., Penza **1**, 11–18 (2013)
16. Belokobylsky, S.V., Eliseev, S.V., Kashuba, V.B.: Opportunities for dynamic interactions in mechanical oscillatory systems with coupled external forces. Syst. Methods Technol. BrSU Publ., Bratsk **4**(16), 7–13 (2012)
17. Eliseev, S.V., Lontsikh, P.A.: Evaluation of the forms of interaction between partial systems in mechanical circuits. Possible simplifications. Bull. Irkutsk State Tech. Univ. NI ISTU Publ., Irkutsk **6**(65), 17–21 (2012)
18. Khomenko, A.P., Eliseev, S.V.: Protection against vibration for an object with two degrees of freedom. Dynamic reactions. News Transsib. OmGUPS Publ., Omsk **3**(15), 74–85 (2013)
19. Moscovskikh, A.O.: Possibilities of integration of the tools of the theory of chains and the theory of automatic control in the problems of the dynamics of machines, 20 pp. Ph.D. (Engineering) thesis. IrGUPS Publ., Irkutsk (2011)
20. Tarasik, V.P.: Mathematical Modeling of Technical Systems, 640 pp. Design-PRO Publ., Minsk (2004)
21. Kolovsky, M.Z.: Automatic Control of Vibration Protection Systems, 320 pp. Nauka Publ., Moscow (1976)
22. Khomenko, A.P., Eliseev, S.V., Artyunin, A.I., Eliseev, A.V., Bol'shakov, R.S., Kaimov, E. V.: Elastic Elements in Mechanical Systems. Structural Interpretations, 460 pp. IrGUPS Publ., Irkutsk (2013). Dep. in VINITI 08.15.13, No. 230-B 2013

Chapter 3
Some Issues of the Methodology of the Approaches of the Structural Modeling in the Dynamics of Mechanical Oscillatory Systems

Evaluation of the dynamic properties of a quite complex system, in particular a vibration protection system, is ultimately reduced to establishing and investigating the relationship between the signal fed to its input and the signal received at its output. The physical nature of the quantities characterizing the signals can be very different and do not always coincide for the input and output signals.

When a signal is fed to the input of the system, there are processes occurring in it, the development of which in time can be denoted by the general concept—the dynamic state (motion) of the system, although often these processes are not connected with spatial displacements at all. However, in any case, there is a change in the system—its transition from one state to another.

3.1 Dynamic Systems, Signals, External Influences

During the system motion, the quantities that determine its state and are called *the generalized coordinates of the system* are interrelated and change with time. Relations that remain at each instant of time between the generalized coordinates of the system and its input signal form, in the general case, the system of equations of motion of the system under consideration. As the output signal of the system, one of its generalized coordinates or some function of these coordinates is usually used. Therefore, the notion of output signal is introduced as one of the unknown functions in the system of equations of motion.

The system of equations of motion must obviously be complete, and then, by eliminating the corresponding generalized coordinates, one can obtain the equation of motion that links the output signal to the input one. The number of generalized coordinates, which must be equal to the number of independent equations of motion, determines the number of degrees of freedom of the system.

© Springer Nature Switzerland AG 2020

S. V. Eliseev and A. V. Eliseev, *Theory of Oscillations*, Studies in Systems, Decision and Control 252, https://doi.org/10.1007/978-3-030-31295-4_3

If we denote the input signal by $y_0(t)$; and the output by $y(t)$, then from the mathematical point of view, the system's action is reduced to the fact that it causes some dependence between the quantities of $y(t)$ and $y_0(t)$, which can be written in a general way:

$$D[y(t)] = D_0[y_0(t)], \tag{3.1}$$

where D and D_0 are some operators applied to the functions $y(t)$ and $y_0(t)$. The certain form of these operators is determined from the equation of motion of the system under consideration.

The operator D is said to be linear and has the following basic properties:

$$\left.\begin{aligned} D\left[\sum_k y_k(t)\right] &= \sum_k D[y_k(t)], \\ D[cy(t)] &= cD[y(t)], c = \text{const.} \end{aligned}\right\} \tag{3.2}$$

The first of the equalities (3.2) reflects *the principle of superposition* (imposition). In this case, the general solution of the linear equation can be represented as the sum of linearly independent particular solutions. It is known that a derivative of any order and an integral of any multiplicity applied to a function are linear operators with respect to this function; linear combinations of such linear operators are also linear operators, from which linear differential equations are formed [1].

In the theory of vibration protection systems, linear processes are usually referred to only those processes that are described by linear differential equations with constant coefficients. Therefore, linear systems should satisfy certain conditions:

- the system parameters on which the coefficients in the equations of motion depend, should not depend on the generalized coordinates of the system, which are the sought-for functions of time, and also on their derivatives;
- the system parameters should not depend on time.

Satisfying the first condition is mandatory, since otherwise the equation becomes nonlinear and will describe the motion of the nonlinear system. Satisfying the second condition determines the class of linear systems, since if the condition is violated, the equation remains linear, but with variable coefficients. It is, however, expedient to select systems, the motion of which is described by linear differential equations with variable coefficients, into a separate class of so-called parametric systems. This is due to the fact that linear parametric systems have some specific properties that are of great practical importance. Thus, in a linear system described by an equation with constant coefficients, the signal spectrum is not transformed, i.e. in the spectrum of the output signal, the emergence of components of new frequencies that were absent in the spectrum of the input signal is impossible. When passing through such a system, only the amplitudes and phases of the individual frequency components can change. Meanwhile, parametric systems convert the spectrum of the input signal; at their output, new frequency components may occur.

In this respect, they are similar to nonlinear systems. In addition, the mathematical apparatus of the parametric systems investigating has its own characteristics and is much more complicated than the apparatus for studying linear systems with constant coefficients [2].

It should be noted that, along with very diverse systems that can be regarded as linear with a sufficient degree of accuracy for practice, cases are possible where a deliberately nonlinear system allows "linearization". In these cases, in the course of motion the generalized coordinates are characterized by only small deviations from their equilibrium or steady-state value; such equations for small deviations can be approximately considered linear.

3.1.1 Equations of Motion

Taking into account the equation of motion of a linear system, described above, is a linear differential equation with constant coefficients. In the general case, this equation can be either ordinary or in partial derivatives.

With respect to real systems, the ordinary differential equation corresponds to the description of processes, whose propagation time through the system (i.e., the time of signal transmission from the input to the output of the system) can be neglected. Moreover, the spatial extent of the system does not matter, it is regarded as a system with lumped parameters. Partial differential equations physically correspond to the accounting of wave processes occurring in special devices (for example, such as a drill string, a long shaft, etc. Such devices are called distributed parameter systems [3].) Let us consider an example of a linear system whose motion is described by the differential equation of the second order of the form (3.3)

$$a_0 \frac{d^2y}{dt^2} + a_1 \frac{dy}{dt} + a_2 y = f, \tag{3.3}$$

where y is the generalized coordinate; a_0, a_v, a_2 are constant coefficients, depending on the parameters of the system; f is a known function of time t. A system whose motion is described by an equation of the form (3.3) can be a physical pendulum making small oscillations in one plane. With reference to models of mechanical oscillating systems, the given time function f is an input signal. As an output signal, the general coordinate of the position of the object y (or a quantity proportional to it) is taken. As an example, the system of vibration protection of the object M under the force perturbation Q, a simplest diagram of which is shown in Fig. 3.1.

The differential equation of motion has the following form:

$$M \frac{d^2y}{dt^2} + B \frac{dy}{dt} + K = Q(t), \tag{3.4}$$

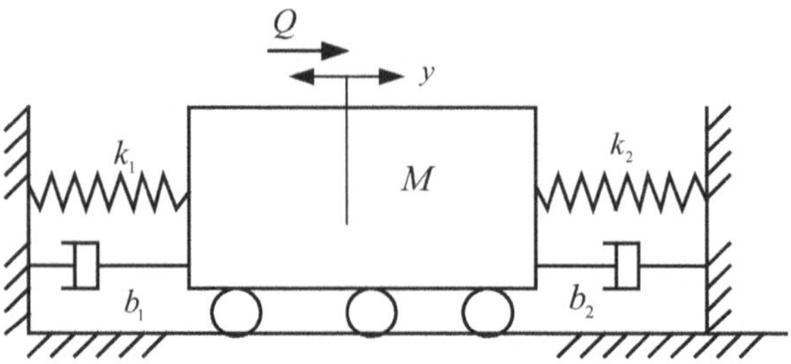

Fig. 3.1 The computational scheme of the oscillatory system with elastic (k_1, k_2) and dissipative (b_1, b_2) constraints

where $B = b_1 + b_2$; $K = k_1 + k_2$ are the dissipative and elastic parameters of the system, respectively. In the kinematic form of the external action, as well as in considering chain systems with several degrees of freedom, it is advantageous to have an equation of motion of the system that directly connects a certain output signal to the input signal. To obtain the equation, we use the properties of linear differential operators, which make it possible to write

$$D = \alpha_n \frac{d^n}{dt^n} + \alpha_{n-1} \frac{d^{n-1}}{dt^{n-1}} + \ldots \alpha_1 \frac{d}{dt} + \alpha_0 = \sum_{k=0}^{n} \alpha_k \frac{d^k}{dt^k}, a_n, a_{n-1},$$

where a_0, a_1, …, a_n are constant coefficients, and the order of the higher derivative determines the order of the operator. The operator D acquires a certain meaning when applied to some function of the variable t. At the same time, operators D have independent properties, due to which the use of operators can reduce the intermediate transformations:

- the product of linear operators is also a linear operator, i.e. $D_1 D_2 \ldots D_n = D$;
- the result of multiplication of operators does not change when the order of the sequence of factors is changed; $D_1 D_2 = D_2 D_1$;

$$D_2[D_1(y)] = D_2 D_1(y) = D(y),$$

  where $D = D_2 D_1 = D_1 D_2$;
- in the process of intermediate transformations, the operator expression $D(y)$ can be considered as the product of the operator D itself by the function y.

From mechanical systems of the simplest kind (see Fig. 3.1), more complex computational schemes can be developed. Note that the order of the highest derivative in the operator D_0 cannot be higher than the order of the highest derivative of the operator D, if y is the output of a physically implementable system. The operator of the left side of the equation of motion $D(y)$ remains unchanged with respect to any output quantity y that is a linear operator of the generalized coordinate of the system.

The state of a complex system at any time is determined by several generalized coordinates, the number of which is equal to the number of degrees of freedom of the system. At the same time, determining the number of degrees of freedom of the system and the corresponding convenient generalized coordinates is not always simple. However, a complex system can always be represented in the form of units, the simplest interconnected systems. Therefore, in the most general case, the motion of a complex linear system with lumped parameters is described by a system of differential equations of the form

$$\begin{aligned} D_{11}(y_1)+D_{12}(y_2)+\cdots+D_{1s}(y_s)&=y_{01};\\ D_{21}(y_1)+D_{22}(y_2)+\cdots+D_{2s}(y_s)&=y_{02};\\ &\vdots\\ D_{s1}(y_1)+D_{s2}(y_2)+\cdots+D_{ss}(y_s)&=y_{0s}. \end{aligned} \tag{3.5}$$

In these equations, y_1, y_2, …, y_s are the generalized coordinates of the system; $y_{01}, y_{02}, \ldots, y_{0s}$ are input signals (external forces acting on the system). In the most general case of links of the second order, the operators D_{ik} have the form

$$D_{ik}=\alpha_{ik}\frac{d^2}{dt^2}+\beta_{ik}\frac{d}{dt}+\gamma_{ik}, \tag{3.6}$$

where α_{ik}, β_{ik} and γ_{ik} are constant coefficients depending on the parameters of the system, some of which may have zero values. Members with indices $i\neq k$ correspond to existing internal links in the system.

Without loss of generality, we can take, for example, $y_{01}=y_0$; $y_{02}=y_{03}==\cdots=y_{0s}$, since, due to the linearity of Eq. (3.5), the impact of all external forces y_{0i} can be considered as the result of superposition of each of these forces separately. Taking into account the above properties of the operators D, we can rewrite Eq. (3.5) in the following form:

$$\begin{aligned} D_{11}y_1+D_{12}y_2+\cdots+D_{1s}y_s&=y_0;\\ D_{21}y_1+D_{22}y_2+\cdots+D_{2s}y_s&=0;\\ &\vdots\\ D_{s1}y_1+D_{s2}y_2+\cdots+D_{ss}y_s&=0. \end{aligned} \tag{3.7}$$

From the system of Eq. (3.7) it is easy to find an equation for any coordinate y_k by eliminating all the others. The elimination procedure is simplified if, in the course of intermediate calculations, the operators D_{ik} are regarded as some coefficients for the unknowns y_k. If we use the theory of determinants, we obtain

$$y_k=\frac{\Delta_k}{\Delta}=\frac{\Delta_{1k}y_0}{\Delta}, \tag{3.8}$$

where

$$\Delta = \begin{vmatrix} D_{11} & D_{12} & \ldots & D_{1s} \\ D_{21} & D_{22} & \ldots & D_{2s} \\ \vdots & \vdots & \vdots & \vdots \\ D_{s1} & D_{s2} & \ldots & D_{ss} \end{vmatrix}; \quad (3.9)$$

$$\Delta_k = \begin{vmatrix} D_{11} & D_{12} & \ldots & D_{1k-1} & x_0 & D_{1k+1} & \ldots & D_{1s} \\ D_{21} & D_{22} & \ldots & D_{2k-1} & 0 & D_{2k+1} & \ldots & D_{2s} \\ \vdots & \vdots & \vdots & \vdots & \vdots & \vdots & \vdots & \vdots \\ D_{s1} & D_{s2} & \ldots & D_{sk-1} & 0 & D_{sk+1} & \ldots & D_{ss} \end{vmatrix} = \Delta_{1k} x_0. \quad (3.10)$$

Obviously, the determinants Δ and Δ_{1k} are linear operators $\Delta = D$ and $\Delta_{ik} = D_{0k}$, so that, as a result, for the coordinate y_k we get a linear differential equation of the form

$$D(y_k) = D_{0k}(y_0). \quad (3.11)$$

This result shows that the left-hand member of the equation for any coordinate y_k is the same, and the order of the operator D is $n \leq 2s$. The operator D_{0k} on the right-hand member of the equation for each coordinate has its own meaning, but the order of the operator D_{0k} is always less than the order of the operator D [4].

It can be shown that with respect to the quantities $y'_k = D_k(y_k)$, where D_k is a linear operator of arbitrary order, Eq. (3.11) takes the form (3.11′)

$$D(y'_k) = D'_{0k}(y_0); D'_{0k} = D_k D_{0k}. \quad (3.11')$$

Thus, the left-hand member of the equation does not change, whereas the operator D'_{0k} on the right-hand member of the equation can acquire any order depending on the order of the operator D_k. However, for the quantity y'_k that is an output signal of a physically implementable system, the operator D_k cannot be of order higher than the second. Then the order of the operator D'_{0k} is not higher than the order of the operator D. Such representations are characteristic for signal interactions formed by measuring systems and elements of these systems. However, in systems of a particular kind, and mechanical systems with mass-and-inertia and elastic dissipative elements are related to them, other variants of constructing the model of interaction and links are possible [5]. In the general case, the equation of motion of a system of any complexity has the form

$$D(y) = D_0(y_0), \quad (3.12)$$

where

$$D = \alpha_0 \frac{d^n}{dt^n} + \alpha_1 \frac{d^{n-1}}{dt^{n-1}} + \cdots + \alpha_n, a_0, a_1 \ldots; \qquad (3.13)$$
$$D_0 = b_0 \frac{d^m}{dt^m} + b_1 \frac{d^{m-1}}{dt^{m-1}} + \cdots + b_m.$$

In this case, $m \leq n$ if y is the output signal of a physically implementable system and y_0 is the input signal. All the coefficients ai and bi are real, since they are functions of the parameters of the system. The form of the operator D does not depend on which of the generalized coordinates of the system or which linear operator of the system is used as an output signal; only the operator D_0 of the right-hand member of the equation changes. Moreover, the operator D remains the same when the input (the point of application of the external force) to the system changes; this follows from the invariance of the principal determinant Δ of the system of Eq. (3.5) when any of the functions x_{0i} is selected as the input signal [4].

These conclusions are consistent with the fact that the formal mathematical solution of the linear equation (3.12) is known to be determined by the sum of two solutions: the general solution of Eq. (3.12) without the right-hand member; of the general solution of the homogeneous equation $D(y) = 0$, and the particular solution of the inhomogeneous equation (3.12). With respect to a particular system, the input signal y_0 and output signal y have corresponding dimensions, which are not always the same. Therefore, the coefficients a_i and b_i also have the appropriate dimensions. This, however, does not prevent us from considering Eq. (3.12) as common to any linear system, independently of its physical nature and purpose. By introducing relative quantities of $y' = \frac{y}{y^0}$ and $y_0' = \frac{y_0}{x_0^0}$, where y^0 and y_0^0 are certain basic values of the quantities y and y_0, Eq. (3.12) can be reduced to a universal form with coefficients $a_i' = a_i y^0$ and $b_i' = b_i y_0^0$, which have the dimension of time in the $(n - i)$th power.

Such a generality of mathematical research became the basis for the field of applied mathematics that is currently widely developed—mathematical modeling. Indeed, the analytical solution of the linear equation can be replaced by an experimental study by modeling a preset equation using an appropriately constructed real linear system whose equation of motion coincides with the preset equation [6].

3.1.2 Unforced, Steady-State and Transient Motions of the System: Initial Conditions, Transition Function

In the case of a regular input signal, the quantity $y_0(t)$ is a preset function of time and, consequently, the expression $D_0(y_0)$ on the right-hand member of Eq. (3.12) is a known function of time. Thus, the investigation of a linear dynamical system reduces to solving the inhomogeneous differential equation (3.12) for the preset initial conditions

$$y(0), \left(\frac{dy}{dt}\right)_{t=0} = y^{(1)}(0), \left(\frac{d^2y}{dt^2}\right)_{t=0} = y^{(2)}(0), \ldots, \left(\frac{d^{n-1}y}{dt^{n-1}}\right)_{t=0} = y^{(n-1)}(0). \quad (3.14)$$

The general solution of Eq. (3.12) can be represented in the form

$$y(t) = y_{unf}(t) + y_{ss}(t). \quad (3.15)$$

In this expression, $y_{unf}(t)$ is the general solution of the homogeneous equation obtained from (3.12) by equating its right-hand member (that is, for $y_0 \equiv 0$) to zero, and $y_{ss}(t)$ is a particular solution of the inhomogeneous Eq. (3.12).

Thus, the solution $y_{unf}(t)$ corresponds to the motion of the system in the absence of an external signal, i.e. own free motion of the system. The general form of the function $y_{unf}(t)$ does not depend on the form of the function $y_0(t)$, but is determined by the properties of the system itself, which are manifested in the properties of the roots of the characteristic equation of the system. The characteristic equation is obtained from the operator D of the system (the left-hand member of Eq. (3.12)) by replacing $\frac{d^k}{dt^k}$ by p^k in it and has the form:

$$a_0p^n + a_1p^{n-1} + \cdots + a_n = 0. \quad (3.16)$$

If, for example, all the roots of the characteristic equation (3.16) are different, then

$$y_{unf}(t) = \sum_{k=1}^{n} C_k e^{p_k t}, \quad (3.17)$$

where p_k is the roots of Eq. (3.16) and C_k is arbitrary constants.

In the presence of multiple roots, the corresponding general expression for $y_{unf}(t)$ is also known and also contains n arbitrary constants C_k. The particular solution $y_{ss}(t)$ depends on the form of the function $y_0(t)$, which determines the external impact on the system, and corresponds to the forced or established motion (regime, state) of the system [7].

The solution (3.15) of Eq. (3.12) determines the dynamic process in the system, which takes place from the moment the external signal is applied to the input of the system. Usually, at the time of the input signal, the origin of the time count is taken, hence, all the variable quantities included into Eqs. (3.5), (3.12), and the solutions obtained from them are considered as a function of time t only for $t \geq 0$. For time $t < 0$ all of them can be taken identically equal to zero. It should be noted that, although the general solution of $y_{unf.}(t)$ has standard forms, as, for example, (3.17), completely independent of the $y_0(t)$ external signals. The particular form of the solution depends on the properties of the roots of the characteristic equation of system (3.16), as well as on the initial conditions (3.14) and the initial value $y_0(0)$, since the values of arbitrary constants C_k depend on them.

The solution (3.15), which is the result of the superposition of two solutions, one of which corresponds to the regime of free oscillations of the system $y_{unf}(t)$, and the other—$y_{ss}(t)$—to the steady-state regime, is called the transient process. The type of the transient process also depends on the form of the function $y_0(t)$. According to (3.17), the free oscillations of the system will become dull if the real parts of the roots p_k of the characteristic equation are negative. In this case, theoretically, the system with $t \to \infty$ goes into a steady-state regime, completely independent of the initial conditions.

It should be noted that in the automatic control theory, only the function $y_{unf}(t)$ can be sometimes called a transient process. The output signal $y(t)$, obtained during the course of the transient process, is the most complete characteristic of the dynamic properties of a system that are of great importance. From the previous presentation it is clear that the character of the function $y(t)$ depends on the series of conditions. However, an estimation of the dynamic properties of a system with sufficient completeness can be made as a result of its investigation under certain standard conditions [8, 9].

The first condition is that the system has zero initial conditions. In addition, the system is at rest, if all of its units are at rest. It should be noted that for the second-order unit, the initial values of its generalized coordinate and its first derivative are zero; for the first-order unit, the initial value of its generalized coordinate is zero.

This corresponds to the requirement of continuity of the indicated quantities. The second condition defines the standard form of the input signal, which is accepted as a unit function of time $y_0(t) = 1(t)$ [1].

The output signal of the system, obtained under observance of these standard conditions (the initial state of the system is the state of rest, the input signal is a unit function) is called the transition function of the system $h(t)$.

With reference to the analysis of diverse and complex systems, the classical method of solving equations is not unique. At present, methods of mathematical investigation have found wide application: the operational method (operational calculus), the convolution integral method (time method), and the frequency method. All these methods, like the classical ones, are based on the principle of superposition, which plays an important role in the theory of linear systems. All of them are related to each other, but each of them has its own characteristics, which are more adapted to solving various specific issues encountered in specific applications.

Mechanical oscillatory systems, consisting of mass-and-inertia, elastic (springs) and dissipative units (dampers), in many problems of dynamics are considered as computational schemes. With the preliminary stages of the investigation in mind mathematical models in the approaches are usually reduced to systems of ordinary linear differential equations with constant coefficients. A sufficiently detailed methodological base has been developed for such systems, which is reflected, for example, in [10–12]. Some sections of the theory of oscillations are used in the theory of circuits, in particular when using electromechanical analogies for studying the general properties of mechanical and electrical systems [13–15]. The theory of mechanical circuits develops in concert with the theory of electrical circuits, which is predetermined by the concept of elementary standard units [16]. The development of concepts on the properties of elements and devices of various mechanisms and machines in those aspects that are related to the analysis of the possibilities of integration of elastic dissipative mass-and-inertia properties in real constructive forms were considered in the first section of the monograph. Note that here, as in electrical circuits, mechanical elements are only mathematical abstractions. Real physical elements usually simultaneously possess several properties attributed to these idealized elements. Bodies that have a mass usually have elasticity, and springs usually have mass. Both in these and other bodies, internal energy losses occur, characteristic for elements with viscous friction and causing attenuation of oscillations. Finally, the damping elements are also not devoid of mass. The question is only the relative significance of these secondary properties of the physical elements. The question of the appropriateness of certain secondary properties of elements to be included in the study must be solved independently in each specific case.

The natural development of such representations has become the possibilities of constructing certain technologies of mathematical models, which, as mentioned, lead to systems of linear ordinary differential equations with constant coefficients. The solution of such equations in applications to the theory of mechanical circuits often relies on the use of an operational calculus giving results in a convenient form for both the theory of mechanical circuits and the automatic control theory, where operator methods are widely used [16–18]. Fundamentals of operational calculus, in the application to chain structures, presented in [8, 15, 19].

3.1.3 Elements of Operational Calculus. Laplace Pre-formation

At the heart of the operational calculus lies the linear Laplace transform, which is written in the following form:

$$L\{f(t)\} = \int_0^\infty e^{-pt} f(t)\,dt. \tag{3.18}$$

In this expression, $f(t)$ is some function of the real variable t, which is identically equal to zero for $t < 0$ and $p = \sigma + j\omega$ is a complex number [19]. The integral (3.18) is called the Laplace integral of the function $f(t)$. For a wide class of functions $f(t)$, the Laplace integral is a regular function of p in some range of values of p (that is, it defines a function having derivatives of all orders). Then outside this region there are singular points of this function, and the integral itself may not make sense. Let, for example, $f(t) = e^{\alpha t}$ для $t \geq 0$ and $\alpha > 0$. Then

$$\int_0^{\infty} e^{-pt} e^{\alpha t} dt = \int_0^{\infty} e^{-(p-\alpha)t} dt = \frac{1}{p-\alpha},$$

if Re $p = \sigma \geq \sigma_0 > \alpha$. Obviously, for Re $p < \alpha$ the Laplace integral becomes meaningless. Thus, in this case the Laplace integral transforms the function $f(t) = e^{\alpha t}$ of the independent variable t into a new function of the independent variable p; and this function is regular at the points of the complex plane $p = + j\omega$ lying to the right of the line Re $p = \alpha$. The pole of the function $\frac{1}{p-\alpha}$ is located to the left of this line at the point $p = \alpha$ on the real axis (Fig. 3.2).

The function $L\{f(t)\}$ defined this way is called the Laplace transform of the function $f(t)$ and is considered for all values of p, although the integral itself makes sense only in a certain range of values of p. The function $f(t)$, for which there exists a Laplace transform, is said to be initial. We note that not every function can correspond to the Laplace transform. For example, discontinuous functions $\frac{1}{t}, e^{t^2}$ cannot be Laplace transformed, since the corresponding Laplace integrals do not make sense for any values of p. Similarly, not every function of p corresponds to the initial function. For example, the function tg p does not have an initial function, since the tangent poles are located on the entire real axis, and not to the left of a straight line parallel to the imaginary axis of the complex plane $p = \sigma + j\omega$. For functions $f(t)$ encountered in technology, there always exists a Laplace transform, since for large t they grow more slowly than an exponential function $e^{\sigma_0 t}$, where σ_0 is some finite positive number [7, 8].

If $L\{f(t)\}$ is a Laplace transform, then the corresponding initial function $f(t)$ is unique in some sense: two piecewise continuous initial functions $f_1(t)$ and $f_2(t)$,

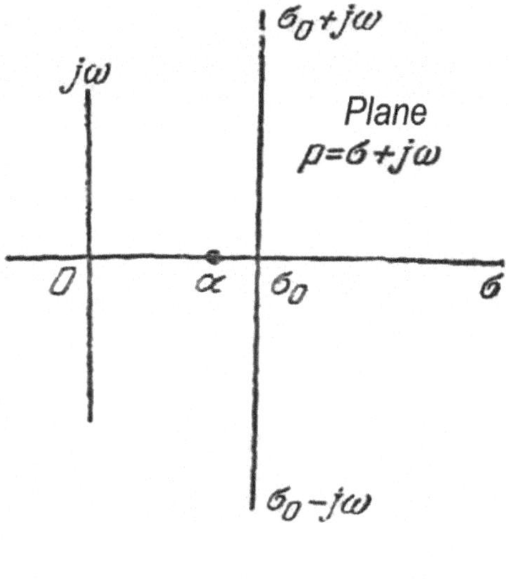

Fig. 3.2 The arrangement of the roots in the complex plane

corresponding to the same Laplace transformation, can differ only at the points of discontinuity; consequently, if $L\{f(t)\}$ corresponds to a continuous initial function $f(t)$, then the latter is uniquely determined.

Original and transform. The initial function $f(t)$, satisfying the condition $f(t) = 0$ for $t < 0$, is called the original. When the operational calculus is applied to the solution of technical problems [20], the Laplace transform is used in two variants. In one of them, the original $f(t)$ is associated with the function $L\{f(t)\}$, the Laplace transform, which is called the Laplace transform or the L-transform and is denoted. Another variant of the original is used, in which $f(t)$ is associated with the $pL\{f(t)\}$, the Laplace-Carson transformation, which is called the Laplace-Carson transform $\bar{f}_{LK}(p)$ [2]. According to (3.18), the following relation holds:

$$\bar{f}(p) = \frac{\bar{f}_{LK}(p)}{p} = \int_0^{\infty} e^{-pt} f(t)\,dt. \tag{3.19}$$

Thus, two transforms can be associated with this original: an L-transform and an LK-transform. Accordingly, in the process of transformations, the transform of the original can be obtained as the L- or LK-transform. Note that when using transformations, the type of the transform taken in the calculations is usually indicated, and the relation (3.19) between the L-, LK-transform and the corresponding original $f(t)$ is taken into account. Each of them has its own advantages. For example, in studying the dynamic properties of systems, as a standard signal, a unit function $1(t) = 1$ is widely used with $t \geq 0$, which is zero with $t > 0$. Substituting $f(t) = 1$ into the integral (3.19), we can obtain (under the condition $\mathrm{Re}\, p > 0$)

$$\bar{f}(p) = \frac{\bar{f}_{LK}(p)}{p} = \frac{1}{p}. \tag{3.20}$$

Thus, LK is the transform of the unit function equal to unity. In this case, as will be shown below, the fundamental characteristic of the system—the transfer function—obtains a simple interpretation, proving to be an LK-transform of the transition function.

However, when considering impulse systems, when the main role is played by a unit impulse function $\delta(t)$, it becomes more convenient to use the L-transform. In addition, the L-transform has a certain advantage also in establishing a relationship between the operational and frequency methods. In the further presentation of the operational calculus, the Laplace transform is used. Correspondence of L-transforms to the original (and vice versa) can be written in a symbolic form

$$\bar{f}(p) \div f(t) \text{ or } f(t) \div \bar{f}(p). \tag{3.21}$$

Thus, the correspondence (3.21) indicates that the relation (3.19) exists between $\bar{f}(p)$ and $f(t)$, that the Laplace transform (3.18) is linear, that is,

$$L\{cf(t)\} = cL\{f(t)\}, c = \text{const}, L\left\{\sum_{k=1}^{n} f_k(t)\right\} = \sum_{k=1}^{n} L\{f_k(t)\}.$$

This implies two basic properties of the original and its transform (both Laplace and Laplace-Carson).

3.1.4 Algebraization of Differential Equations

The method of operational calculus consists in the fact that, in the general case, the solution of the initial system of differential equations, to which the desired originals match, is replaced by the solution of the corresponding system of equations for the transforms. When the basic equations for the originals are ordinary linear differential equations with constant coefficients, the corresponding equations for the transforms turn out to be algebraic, so that the procedure for moving to equations for transforms is called the algebraization of differential equations for originals.

It should be noted that the operational calculus is applicable to solving more complicated linear equations, for example, differential equations with variable coefficients and equations in partial derivatives. However, in the first case, the equation for transforms is also a linear differential equation with variable coefficients, and whether it is simpler than the original one depends on the form of the variable coefficients of the original equation. Undoubtedly, the application of operational calculus to solving equations in partial derivatives with constant coefficients, especially in the one-dimensional case, simplifies the problem [2, 8].

From algebraic equations, the transform of the required original is elementary. In this case, the cumbersome calculations arising in the classical method are eliminated in connection with the determination of arbitrary integration constants in accordance with the preset initial conditions, since in the operational method the initial conditions are automatically introduced into the solution transform. The original solution with its known transform can usually be easily found by using extensive correspondence tables and numerous properties of transforms in special manuals and reference books; they also give elementary and general formulas that allow us to calculate the original from a well-known transform. On the other hand, the study of the properties of the transform of the desired solution of the differential equations of motion of the system already makes it possible to establish very general and practical properties of the system itself.

To pass from the differential equation for the originals to the equation for transforms, it is necessary to find transforms of the derivatives of the original. It is possible to derive a general formula for the L-transform of the derivative of the nth order of the original. According to (3.19),

$$\bar{f}(p) = \int\limits_0^\infty e^{-pt} f(t)\,dt. \tag{3.22}$$

It is required to find the L-transform of the derivative

$$\frac{d^n f}{dt^n} \doteq f^{(n)}(t) \doteq \bar{f}^{(n)}(p).$$

Applying (3.22) to the first-order derivative, we obtain, by integrating by parts

$$\bar{f}^{(1)}(p) = \int\limits_0^\infty e^{-pt} \frac{df}{dt} dt = e^{-pt} f(t) \Big|_0^\infty + p \int\limits_0^\infty e^{-pt} f(t) dt = -f(0) + p\bar{f}(p). \tag{3.23}$$

Here it is assumed that Re $p > \sigma_0$, where σ_0 is chosen in such a way that $\lim\limits_{t\to\infty} e^{-\sigma_0 t} f(t) = 0$. Then

$$f^{(1)}(t) \doteq \bar{f}^{(1)}(p) = p\bar{f}(p) - f(0), \tag{3.24}$$

where $f(0) = [f(t)]_{t=0}$.

To calculate the L-transform of the second derivative, we can write

$$\bar{f}^{(2)}(p) = \int\limits_0^\infty e^{-pt} \frac{d^2 f}{dt^2} dt = e^{-pt} \frac{df}{dt} \Big|_0^\infty + p \int\limits_0^\infty e^{-pt} \frac{df}{dt} dt = -f^{(1)}(0) + p\bar{f}^{(1)}(p), \tag{3.25}$$

where $f^{(1)}(0) = \left(\frac{df}{dt}\right)_{t=0}$ and it is assumed that $\lim\limits_{t\to\infty} e^{-\sigma_0 t} \frac{df}{dt} = 0$, due to the proper choice of the value of Re $p > \sigma_0$. Therefore,

$$f^{(2)}(t) \doteq \bar{f}^{(2)}(p) = p\bar{f}^{(1)}(p) - f^{(1)}(0) = p^2 \bar{f}(p) - pf(0) - f^{(1)}(0). \tag{3.26}$$

From the obtained results it follows that for the L-transform of the derivative of the nth order of the original $f(t)\dot{}\bar{f}(p)$ there is a recurrence formula

$$\frac{d^n f}{dt^n} \doteq \bar{f}^{(n)}(p) = p\bar{f}^{(n-1)}(p) - f^{(n-1)}(0), \quad n = 1, 2, \tag{3.27}$$

where $f^{(n-1)}(0) = \left(\frac{d^{n-1}f}{dt^{n-1}}\right)_{t=0}$; $\bar{f}^{(0)}(p) = \bar{f}(p), f^{(0)}(0) = f(0)$. Since $\bar{f}_{LK}(p) = p\bar{f}(p)$ and, consequently, instead of (3.27) for the *LK*-transform we obtain

$$\frac{d^n f}{dt^n} \doteq \bar{f}_{LK}^{(n)}(p) = p\bar{f}_{LK}^{(n-1)}(p) - pf^{(n-1)}(0), \quad n = 1, 2. \tag{3.28}$$

In the case when the initial conditions are zero, i.e.

$$f(0) = f^{(1)}(0) = f^{(2)}(0) = \cdots = f^{(n-1)}(0) = 0,$$

for both types of transforms, the same type of correspondence is satisfied

$$\frac{d^n f}{dt^n} \doteq p^n \bar{f}(p). \tag{3.29}$$

Let us find the transform of the integral of the initial function $\int_0^t f(t)\,dt$ via the transform $\bar{f}(p)$ of the very initial function. Let

$$F(t) = \int_0^t f(t)dt, \tag{3.30}$$

Besides,

$$\bar{f}(p) = \int_0^\infty e^{-pt} f(t)\,dt.$$

By the definition of transforms, we have

$$\begin{aligned}\overline{F}(p) &= \int_0^\infty e^{-pt}\left[\int_0^t f(t)dt\right]dt = -\frac{1}{p}e^{-pt}\left[\int_0^t f(\tau)d\tau\right]\Bigg|_0^\infty + \frac{1}{p}\int_0^\infty e^{-pt}f(t)(dt)\\ &= \frac{1}{p}\bar{f}(p),\end{aligned} \tag{3.31}$$

if Re $p > \sigma_0$, where σ_0 is chosen so that $\lim\limits_{t\to\infty} e^{-\sigma_0 t}\int_0^t f(t)\,dt = 0$. The result obtained shows that, regardless of the type of transform, there is a correspondence

$$\int_0^t f(t)dt \doteq \frac{1}{p}\bar{f}(p). \tag{3.32}$$

Similarly, it can be shown that

$$\underbrace{\int_0^t \cdots \int_0^t}_{} f(t)_n (dt)^n \doteq \frac{1}{p^n} \bar{f}(p). \tag{3.33}$$

Suppose given a differential equation of the form

$$\sum_{k=0}^{n} a_{n-k} \frac{d^k y}{dt^k} = \sum_{k=0}^{m} b_{m-k} \frac{d^k y_0}{dt^k}, \tag{3.34}$$

the solution of which must be found under preset initial conditions

$$y(0), y^{(1)}(0), \ldots, y^{(n-1)}(0). \tag{3.35}$$

According to Formula (3.27), Eq. (3.33) in L-transforms will have the form

$$D(p)\bar{y}(p) - N(p) = Q(p)\bar{y}_0(p) - M(p), \tag{3.36}$$

where

$$D(p) = \sum_{k=0}^{n} a_{n-k} p^k, \quad N(p) = \sum_{k=1}^{n} a_{n-k} \sum_{s=1}^{k} p^{s-1} y^{(k-s)}(0),$$
$$Q(p) = \sum_{k=0}^{m} b_{m-k} p^k, \quad M(p) = \sum_{k=1}^{m} b_{m-k} \sum_{s=1}^{k} p^{s-1} y_0^{(k-s)}(0).$$

The expressions obtained show that $N(p)$ is a polynomial in p of order at most $n - 1$, and its coefficients are proportional to the values of the initial conditions (3.35); $M(p)$ is a polynomial in p of order at most $m - 1$, and its coefficients are proportional to the values at $t = +0$ of the function x_0 and its derivatives up to $(m - 1)$th inclusive.

From (3.36) we can find

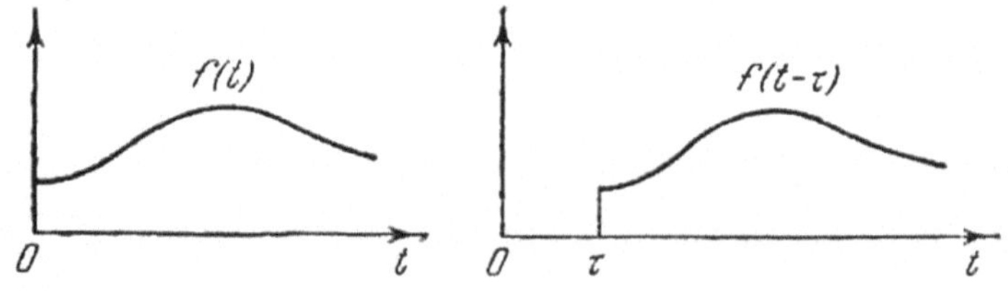

Fig. 3.3 The diagram for determining the delay

$$\bar{x}(p) = \frac{Q(p)}{D(p)}\bar{y}_0(p) + \frac{N(p) - M(p)}{D(p)}. \tag{3.37}$$

Thus, if at the initial moment the system is at rest, then the algebraization of the equation of motion of the system is reduced to a simple replacement of the original $y(t)$ by its transforms, and the symbols of the derivatives of $\frac{d^k}{dt^k}$ – by the complex number p^k. In this case, the initial conditions (3.36) can be found from the expression obtained for the transform $\bar{y}(p)$.

3.1.5 Some Correspondences and Equalities of the Operational Calculus

Let us list some of the basic correspondences and equalities of the operational calculus, used at a later stage.

The delay theorem. If $\bar{f}(p) \doteq f(t)$, then

$$\begin{aligned} e^{-p\tau}\bar{f}(p) &\doteq 0 && \text{for } t < \tau, \\ e^{-p\tau}\bar{f}(p) &\doteq f(t-\tau) && \text{for } t < \tau. \end{aligned} \tag{3.38}$$

The original $f_\tau(t) = f(t - \tau)$ repeats the values of the original $f(t)$ with a delay for a time interval τ, as explained in Fig. 3.3.

Since $f(t) = 0$ at $t < 0$, the function $f(t - \tau)$ for $t < \tau$. Assuming, according to (3.19), we obtain

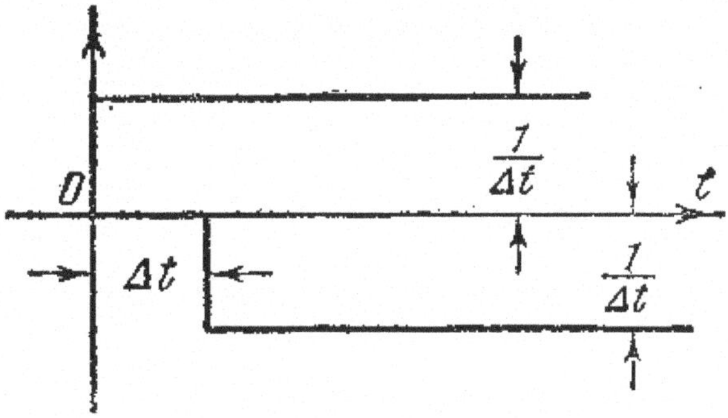

Fig. 3.4 The graph of the shift of a single impulse function

$$\bar{f}_\tau(p) = \int_0^\infty e^{-pt} f_\tau(t)dt = \int_\tau^\infty e^{-pt} f(t-\tau)dt$$

$$= \int_0^\infty e^{-p(t'+\tau)} f(t')dt' = e^{-pt} \int_0^\infty e^{-pt} f(t)dt = e^{-p\tau} f(p),$$

which proves the theorem.

L is the transform of the unit function 1(t). According to (3.19), we have

$$\bar{1}(p) = \int_0^\infty e^{-pt} 1(t)dt = \int_0^\infty e^{-pt} dt = \frac{1}{p}, \quad \mathrm{Re}\, p > 0$$

Consequently,

$$\bar{1}(p) = \frac{1}{p} \doteqdot 1(t). \tag{3.39}$$

L is the transform of the unit impulse function δ(t). To determine the transform of a single impulse function, we represent it as a difference of two step functions shifted relative to each other by an interval Δt and having a step height equal to $\frac{1}{\Delta t}$ (Fig. 3.4) at $\Delta t \to 0$, i.e.

$$\delta(t) = \lim_{\Delta t \to 0} \frac{1}{\Delta t} [1(t) - 1(t - \Delta t)]. \tag{3.40}$$

On the basis of (3.39) and the delay theorem (3.38), we obtain

$$\lim_{\Delta t \to 0} \frac{1}{\Delta t} [1(t) - 1(t-\tau)] \doteqdot \lim_{\Delta t \to 0} \frac{1}{\Delta t} \left[\frac{1}{p} - e^{-p\Delta t} \frac{1}{p}\right] = 1.$$

Consequently,

$$\bar{\delta}(p) = 1 \doteqdot \delta(t). \tag{3.41}$$

From the relation (3.40) formally follows

$$\delta(t) = \frac{d}{dt} 1(t). \tag{3.42}$$

The link between the L-transform for p = ∞ and p = 0 and the original for t = 0 and t = ∞. If

$\bar{f}(p) \doteq f(t)$, then

$$\lim_{p\to\infty} p\bar{f}(p) = f(0); \tag{3.43}$$

$$\lim_{p\to 0} p\bar{f}(p) = f(\infty). \tag{3.44}$$

Equality (3.44) is valid only if $f(t)$ tends to a certain limit as t tends to infinity. For example, $\frac{1}{p-j\omega} \doteq e^{j\omega t}$ but equality (3.44) does not make sense for this correspondence, since $e^{j\omega t}$ does not tend to a certain limit at $t \to \infty$.

The convolution theorem (for L-transforms). If

$$\bar{f}_1(p) \doteq f(t) \text{ and } \bar{f}_2(p) \doteq f_2(t),$$

then

$$\bar{f}_1(p)\bar{f}_2(p) \doteq \int_0^t f_1(t-\tau)f_2(\tau)d\tau = \int_0^t f_1(\tau)f_2(t-\tau)\,d\tau. \tag{3.45}$$

Since the original, in accordance with (3.45), is zero for $t = 0$, then

$$p\bar{f}_1(p)\bar{f}_2(p) \doteq \frac{d}{dt}\int_0^t f_1(t-\tau)f_2(\tau)d\tau = \frac{d}{dt}\int_0^t f_1(\tau)f_2(t-\tau)\,d\tau. \tag{3.46}$$

3.1.6 The Definition of the Original from the Well-Known L-Transforms

In most cases, the presence of correspondence tables in numerous manuals and reference books on operational calculus makes it possible to find the finished form of the desired original corresponding to the *L*-transform obtained from the operational equation of the system [19]. In some manuals and tables, the original $f(t)$ is put in correspondence to an *LK*-transform. When using them, one just needs to remember that.

The transform obtained from the operating equation does not always completely coincide with one of the table transforms. Then the corresponding transform of the result can bring it to a form that makes it possible to use more complex correspondences between transforms and originals. Particularly useful is the correspondence (3.45) proposed by the convolution theorem. The rule for calculating the original yields the Heaviside decomposition formulas, which exist, for example, in [19] and can be used in cases usually encountered in practice. We turn our attention to the general method for determining the original from a preset transform, since it is of fundamental importance and, in particular, allows us to establish a connection between the operational and frequency methods of investigating communication and control systems. The Laplace transform formula

$$\bar{f}(p) = \int_0^\infty e^{-pt} f(t)\, dt \tag{3.47}$$

for its preset left-hand member as a function of the independent variable $p = \sigma + j\omega$ is an integral equation with respect to the unknown function $f(t)$. Note that the originals are functions $f(t)$ of the real variable t that satisfy certain conditions. These are functions that vanish for $t < 0$, piecewise-continuous for $t > 0$, and their modulus satisfies condition

$$|f(t)| < Me^{\sigma_0 t},$$

where M and σ_0 are finite numbers. Under these conditions, the Laplace integral (3.47) converges absolutely and represents a function of the complex variable p regular in the half-plane Re $p > \sigma_0$. Besides,

$$\lim_{|p|\to\infty} \bar{f}(p) = 0. \tag{3.48}$$

The solution of the integral equation (3.47) is given by the so-called Riemann-Mellin inversion formula

$$f(t) = \frac{1}{2\pi j} \int_{c-j\infty}^{c+j\infty} e^{pt} \bar{f}(p)\, dp, \tag{3.49}$$

where the number c can be arbitrary, but greater than σ_0 [19]. Thus, the integration takes place in the plane of the complex variable $p = \sigma + j\omega$ along any line parallel to the imaginary axis and located in the semi-plane Re $p \geq c > \sigma_0$, i.e. to the right of the straight line σ_0 shown in Fig. 3.4.

The derivation of Formula (3.49) is based on the use of the Fourier integral formula (see above). If the function $f(t)$ satisfies the conditions specified above, which determines the original, and, in addition, is differentiable at points where it is continuous, then the function

$$\psi(t) = e^{-\sigma t} f(t), \ \sigma > \sigma_0, \tag{3.50}$$

can be represented in the form of a double Fourier integral:

$$\psi(t) = \frac{1}{2\pi} \int_{-\infty}^{\infty} d\omega \int_{0}^{\infty} \psi(\tau)\, e^{-j\omega(\tau - t)}\, d\tau. \tag{3.51}$$

Substituting the expression for $\psi(t)$ from (3.50) in (3.51) and multiplying both sides of the equality by $e^{\sigma t}$, we obtain

$$\begin{aligned} f(t) &= \frac{1}{2\pi} \int_{-\infty}^{\infty} d\omega \int_{0}^{\infty} e^{-\sigma\tau} f(\tau) e^{\sigma t} e^{-j\omega(\tau - t)} d\tau \\ &= \frac{1}{2\pi} \int_{-\infty}^{\infty} e^{(\sigma + j\omega)t} d\omega \int_{0}^{\infty} e^{-(\sigma + j\omega)\tau} f(\tau) d\tau. \end{aligned} \tag{3.52}$$

Suppose that p = σ + jω in (2.52), and replace the internal integral according to (2.46). For fixed $\sigma = c$, the path of integration represents a line parallel to the imaginary axis, where $dp = jd\omega$. Formula (3.52) then becomes (3.49). Thus, from the representation of $\bar{f}(p)$ in the form of the Laplace integral (3.46), the inversion formula (3.49) follows. It can be shown that, conversely, from (3.49), under the conditions indicated above (which the function $\bar{f}(p)$ must satisfy), supplemented by the requirement of absolute convergence of the integral

$$\int_{c - j\infty}^{c + j\infty} \bar{f}(p)\, dp, \tag{3.53}$$

For $c > \sigma_0$, it follows that $\bar{f}(p)$ can be represented in the form of the Laplace integral (3.46). It should be noted that the value of the integral (3.49) does not depend on the choice of the constant c in the semi-plane Re $p > \sigma_0$, and that for $t < 0$ this integral becomes zero [19].

So, if the function $f(t)$ has properties inherent in the original, then its transform is determined by Formula (3.46), conversely, if the function is an L-transform, then the corresponding original is determined by Formula (3.49). To calculate the integral (3.49), we usually use the method known from the theory of functions of a complex variable method, which consists in the fact that the given path of integration is replaced by other paths and contours that allow the use of Cauchy's residue theorem. On the basis of the Jordan lemma [19] it follows that

$$\lim_{R\to\infty} \int\limits_{ABC} e^{pt}\bar{f}(p)\,dp = 0 \text{ for } t > 0.$$

Thus, the integral (3.49) in the calculation of $f(t)$ for $t > 0$ can be replaced by an integral about a closed contour consisting of the straight line c and the arc ABC of a circle, the radius of which is $R \to \infty$. It is obvious that all the singular points of the function are inside of this contour. According to the Cauchy's theorem, the integral over a closed contour divided by $2\pi j$ is equal to the sum of the residues relative to the poles of the integrand inside the integration contour. Consequently, if the isolated points of the function $\bar{f}(p)$ are only isolated poles (of any order), then

$$f(t) = \frac{1}{2\pi j}\int_{c-j\infty}^{c+j\infty} e^{pt}\bar{f}(p)dp = \sum_{k=1}^{n} \operatorname{Re} s[e^{pt}\bar{f}(p)]_{p=p_k}, p > 0 \tag{3.54}$$

As is well known, the residue of a function $e^{pt}\bar{f}(p)$ with respect to the pole p_k of order s is calculated by the formula

$$\operatorname{Re} s[e^{pt}\bar{f}(p)]_{p=p_k} = \frac{1}{(s-1)!}\lim_{p\to p_k} \frac{d^{s-1}[(p-p_k)^s e^{pt}\bar{f}(p)]}{dp^{s-1}}. \tag{3.55}$$

This formula is also suitable for calculating the subtraction of a function with respect to a simple pole ($s = 1$), if we assume that $0! = 1$. Formulas (3.54) and (3.55) can be used to calculate the originals of the most frequently encountered transforms, the singular points of which are isolated poles. In particular, from these formulas the expressions for the second expansion theorem (Heaviside) are obtained for transforms that are a relation of two polynomials and any meromorphic function (the ratio of two integral transcendental functions, such as, for example, th p = sh p/ch p). If the integrand (actually a function $\bar{f}(p)$) has more complex singular points (for example, branch points) inside the contour chosen by the indicated method, then the computation becomes more complicated. In this case, the transform property $\bar{f}(p)$ and, in particular, the character of its singular points have an exceptional value in determining the original $f(t)$.

3.1.7 *Transfer Function*

The transfer function of the system is the relation of the transform of its output quantity to the transform of the input quantity, provided that the system is at rest at the initial instant of time.

This definition does not depend on the type of transform used. Denoting the transfer function by $W(p)$, we have

$$W(p) = \frac{\bar{y}(p)}{\bar{y}_0(p)} = \frac{\bar{y}_{LK}(p)}{\bar{y}_{0LK}(p)}. \tag{3.56}$$

For the class of linear systems under consideration, the equation of motion, in the most general case, has the form

$$\left(a\frac{d^n}{dt^n} + a_1\frac{d^{n-1}}{dt^{n-1}} + \cdots a_n\right)y(t) = \left(b_0\frac{d^m}{dt^m} + b_1\frac{d^{m-1}}{dt^{m-1}} + \cdots b_m\right)y_0(t). \tag{3.57}$$

Since at the initial moment the system is at rest, Eq. (3.57) in the operational form can be written as follows:

$$D(p)\bar{y}(p) = D_0(p)\bar{y}_0(p), \tag{3.58}$$

where

$$\left.\begin{aligned} D(p) &= a_0p^n + a_1p^{n-1} + \cdots + a_n, \\ D_0(p) &= b_0p^m + a_1p^{m-1} + \cdots + b_n, \end{aligned}\right\} \tag{3.59}$$

where $m \leq n$.

Obviously, the operational form (3.58) of Eq. (3.57) is also preserved for *LK*-transforms $y_{LK}(p) = = py(p)$ and $\bar{y}_{0LK}(p) = p\bar{y}_0(p)$.

From (3.58) we obtain

$$W(p) = \frac{D_0(p)}{D(p)}. \tag{3.60}$$

Thus, the transfer function $W(p)$ does not depend on the type of the input quantity, characterizing its own system. It follows from the comparison of Eqs. (3.56) and (3.57) that the denominator of the transfer function, equated to zero, gives the characteristic equation of the system $D(p) = 0$. The transfer function can be interpreted as the *LK*-transform of the transition function $h(t)$. Indeed, the transition function [14] is the output signal of the system $y(t) = h(t)$ obtained under the condition that at the initial moment the system is at rest and the input signal is a unit function, i.e. $y_0(t) = 1(t)$. In this case, according to (3.37) $\bar{y}_{0LK}(p) = 1_{LK}(p) = p1(p) = 1$, and using (3.56), we find

$$W(p) = \bar{h}_{LK}(p) = p\bar{h}(p). \tag{3.61}$$

Comparing (3.61) with (3.60), we obtain

$$h(t) \doteq \bar{h}(p) = \frac{W(p)}{p} = \frac{D_0(p)}{pD(p)}. \tag{3.62}$$

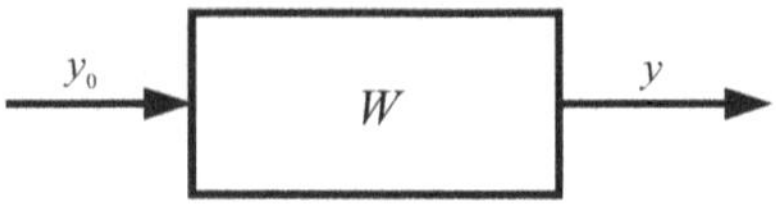

Fig. 3.5 Schematic representation of the directional unit

If at the initial moment the system is at rest and the input signal is of an arbitrary form $y_0(t) \doteq \bar{y}_0(p)$, then for the output signal we have

$$y(t) \doteq \bar{y}(p) = W(p)\bar{y}_0(p) = p\bar{h}(p)\bar{y}_0(p). \tag{3.63}$$

Finally, if the transient function of the system $h(t)$ is known, then for a preset input signal $y_0(t)$, the output signal $y(t)$ can be immediately obtained on the basis of Formulas (3.63) and (3.46) in the form of an integral

$$y(t) = \frac{d}{dt}\int_0^t h(t-\tau)y_0(\tau)\,d\tau. \tag{3.64}$$

It is to be recalled that Formulas (3.56)–(3.64) are applicable only under the condition that the system is at rest at the initial instant of time.

3.2 Structural Methods for Studying Mechanical Oscillatory Systems

The transfer functions considered above and defined by the expression (3.60) are a very convenient form of coupling between the transforms of the output and input quantities of a system of any complexity that is at rest at the initial instant of time. This coupling applies for an arbitrarily complex system belonging to the class of linear systems under consideration. When investigating a particular system whose equation of motion is known, the transfer function is easily found and, consequently, the transient function of the system $h(t)$ and the desired output signal $y(t)$ can be found by known rules [19].

The established coupling between the transforms of the output and input signals can also be used to explain the so-called structural research method, which has been

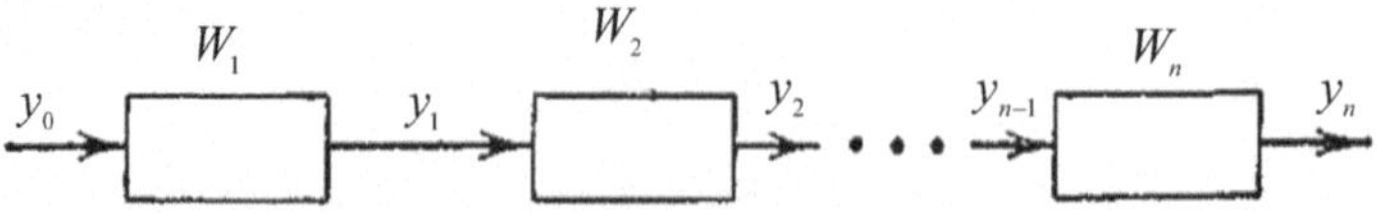

Fig. 3.6 Consecutive connection in automatic control systems

applied in the theory of oscillations and its applications to various problems of dynamics of machines. Of course, the consideration of the structural method cannot depend on the operational calculus, which is only a variant of mathematical apparatus of research. However, the application of the operational calculus simplifies the presentation.

With a structural approach to the consideration of systems, the concept of a unit of the directional effect plays an important role. A unit of the directional effect (which, for brevity sake, can be called a directional unit), in turn, can be a system of any complexity, the distinctive property of which lies in the fact that the output quantity of the unit $y(t)$ depends on the input quantity $y_0(t)$, but there is no reverse effect of the output upon the input. Any other unit connected to the output of the unit of the directional effect does not change the transfer function of the first unit. The unit can be of any physical nature, since from the point of view that is of interest to us, the system is classified according to the corresponding equation of motion connecting the output and input signals. Therefore, the unit of the directional effect will be represented as a rectangle indicating the input y_0 and output y signals and the transfer function W (Fig. 3.5). It is also characteristic for the unit that the ratio (3.60) takes place between the transforms of its output and input signals for the initial state of rest, which is the result of the algebraization of the equation of motion of the unit [13, 16].

A system of any complexity can be compared with a structural diagram consisting of simple units of directed effect, connected to each other in one way or another. The selection of individual units and the way they are connected is limited by a very general requirement: the relationship between the output and input quantities of the structural diagram must coincide with the equation of motion of the system. Therefore, the structural diagram, in a general sense, can be conditional on the choice of individual units and their connection methods, which opens up opportunities for mathematical modeling of physical systems and, conversely, for the physical modeling of mathematical equations.

3.2.1 Basic Ways of Connecting Units

Very diverse diagrams of connecting directional units can be reduced to three main methods: consecutive, parallel and rules of connection with feedback.

The consecutive connection of directional units is shown in Fig. 3.6.

In this case we have

$$\bar{y}_1(p) = W_1(p)\bar{y}_0(p), \bar{y}_2(p) = W_2(p)\bar{y}_1(p), \ldots, \bar{y}_n(p) = W_n(p)\bar{y}_{n-1}(p).$$

From these equations it follows that

$$\bar{y}(p) = \bar{y}_n(p) = W_1(p)W_2(p)\ldots W_n(p)\bar{y}_0(p) = W_{ser}(p)\bar{y}_0(p), \tag{3.65}$$

where the transfer function of n consecutively connected units is

$$W_{ser}(p) = W_1(p)W_2(p)\ldots W_n(p). \tag{3.66}$$

Such an approach is used in the automatic control theory, in which the typical elements, if we bear in mind their transfer functions, reflect the specific aspects of signal processing without establishing dimensional rules in the joining of elementary units [1, 21]. The situation is different in mechanical circuits and their analogs—electrical circuits, for which certain laws must be fulfilled. For example, for mechanical circuits, the principles of d'Alembert and Newton's laws are valid, while for electric circuits the Kirchhoff laws are valid. When considering linear mechanical oscillatory systems on the basis of structural approaches with respect to the typical elementary units used, a certain commonality of properties is proposed. It consists in the fact that all units (springs, dampers, motion transformation devices, etc. [15, 16]) have a signal corresponding to the displacement at the input, and to the effort at the output. The elastic element or spring corresponds to a directional unit with the transfer function $W = k$, i.e. there is displacement at the input, and force at the output, and the magnitude of the force is determined by the displacement and elasticity of the spring. For the damper, the transfer function is $W = bp$, where b is the coefficient of viscous damping, $p = \frac{d}{dt}$ is the differentiation symbol (p is also a complex variable: $p = j\omega; j = \sqrt{-1}$). When the input signal is in the form of the offset, at the output we get a resistance force proportional to the speed. If the typical unit is a motion transformation device [19], then the transfer function of the elementary unit has the form $W = ap^2$, where a is the reduced mass-and-inertia parameter of the unit; in this case we have an offset at the input, and at the output there is a force that depends on the acceleration parameters of the relative motion of the elements.

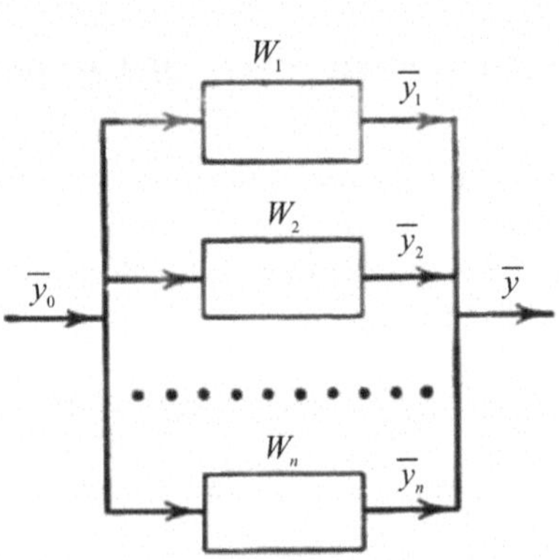

Fig. 3.7 The diagram of parallel unit connection

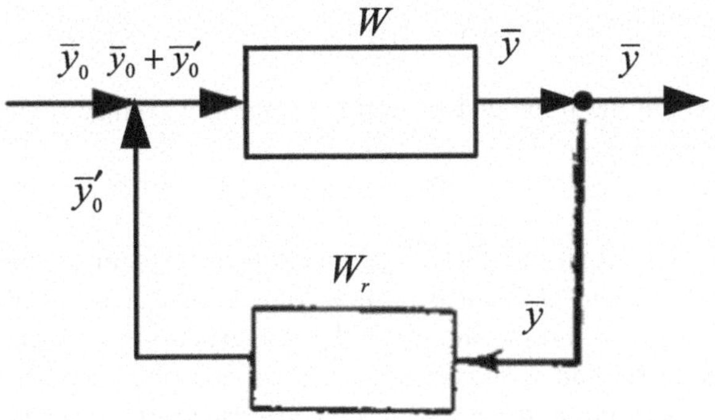

Fig. 3.8 The diagram of connection of units with feedback

With an extended set of typical elements of vibration protection systems, the integrating units of the first and second orders, as well as the unit of pure delay, are included in the number of elementary standard units [16]. Since in the case of consecutive connection of homogeneous typical elements arranged in a chain, the progressing or transmitted force from the initial point to the final one will be the same, the formula for determining the reduced or general stiffness of the elastic elements takes the following form for the two springs $W_1 = k_1$, $W_2 = k_2$:

$$W = \frac{k_1 k_2}{k_1 k_2}, \tag{3.67}$$

and for three springs $W_1 = k_1$, $W_2 = k_2$, $W_3 = k_3$, respectively

$$W = \frac{k_1 k_2 k_3}{k_1 k_2 + k_1 k_3 + k_2 k_3}. \tag{3.68}$$

The consecutive connection formulas can easily be generalized to any number of successive connections. At the same time, any other typical elements (dampers, motion transformation devices, etc.) can act as springs. After all, with the general property of these elements (the input is displacement, the output is force), all of them, in fact, can be regarded as some springs. In [16], blocks or compacts of such elements are called generalized springs, and their reduced stiffness at rest and in dynamics (with harmonic force impact) is determined from the transfer function. The peculiar features noted do not contradict the apparatus of the automatic control theory, since the construction basis for the structural diagrams are the differential equations of motion obtained. However, the peculiarities of the transformations must be taken into account in the development of structural approaches both in the theory of circuits and in the automatic control theory [22].

The parallel connection of the units is shown in Fig. 3.7.

The output quantities of the units are summed, and their sum forms the total output quantity of the system. Directly from Fig. 3.7 it can be seen that

$$\bar{y}(p) = \bar{y}_1(p) + \bar{y}_2(p) + \cdots + \bar{y}_n(p) = W_{par}(p)\bar{y}_0(p), \qquad (3.69)$$

where the transfer function n of parallel connected units

$$W_{par}(p) = W_1(p) + W_2(p) + \cdots + W_n(p). \qquad (3.70)$$

Note that the parallel connection of elements in mechanical oscillatory systems, taking into account the same dimensions of input and output signals, should not cause any doubt, although the construction of block structures, resulting in some reduced unit with reduced stiffness, may be necessary.

The connection of the elementary units (and also of the reduced ones) on the basis of feedback rules is shown in Fig. 3.8.

The output quantity of the first unit is fed to the input of the second unit (marked by the index r), the output quantity of which is fed to the input of the first unit in the sum with the input quantity of the system x_0. For the compound under consideration, we have

$$\bar{y}(p) = W(p)\left[\bar{y}_0(p) + \bar{y}_0'(p)\right] = W(p)\bar{y}_0(p) + W(p)W_r(p)\bar{y}(p), \qquad (3.71)$$

from which

$$\bar{y}(p) = \frac{W(p)}{1 - W_r(p)W(p)}\bar{y}_0(p) = W_3(p)\bar{y}_0(p), \qquad (3.72)$$

where

$$W_3(p) = \frac{W(p)}{1 - W_r(p)W(p)}. \qquad (3.73)$$

Diagrams with a specially applied feedback were widely used in control theory and its applications [16, 17]. The presence of feedback is typical for systems in which automatic control is implemented.

The feedback connection of units is often encountered in the development of structural diagrams of mathematical models. Please note one formal circumstance, which, however, leads to a significant practical conclusion. Let there be a system whose equation of motion in the transforms has the form

$$\bar{y}(p) = W(p)\bar{y}_0(p).$$

Obviously, the equality is not violated by adding the same expression to its left-hand and right-hand members, for example

$$\bar{y}(p) + W_r(p)W(p)\bar{y}(p) = W(p)\bar{y}_0(p) + W_r(p)W(p)\bar{y}(p), \qquad (3.74)$$

where $W_r(p)$ is some function of p. Then

$$\bar{y}(p) = W'(p)\left[\bar{y}_0(p) + \bar{y}'_0(p)\right], \tag{3.75}$$

where

$$W'(p) = \frac{W(p)}{1 + W_r(p)W(p)}; \bar{y}'_0(p) = W_r(p)\bar{y}(p). \tag{3.76}$$

It follows that any unit can be represented in the form of some other unit covered by the appropriate feedback. The possibility of such a representation which is, to a greater extent, arbitrary, is used in constructing structural diagrams of mathematical models. For example, if the unit has a transfer function $W = bp$, i.e. is a damper, to introduce this unit into an elastic unit, we introduce a feedback, which is defined as follows. We assume that

$$W = \frac{bp}{1 + W_1 bp}. \tag{3.77}$$

If we choose $W = k$, we get

$$W_1 = \frac{bp - k}{bpk}. \tag{3.78}$$

Thus, any unit (formally), using the appropriate transformations, can be turned into a unit of the required kind. Equivalent transformation of units and their connections are quite diverse. And we can assume the existence of a certain conventionality of the rules of their connection. Assuming that the transfer functions of the elementary units can be negative, then the operations of consecutive and parallel connections can be transformed one into another. In this case, the original mathematical model in the form of a system of differential equations of motion (mechanical oscillatory systems being considered) remains unchanged. It can be shown that the characteristic equation will be the same, from whatever part of the closed system the output quantity is taken. This situation is valid for any number of units.

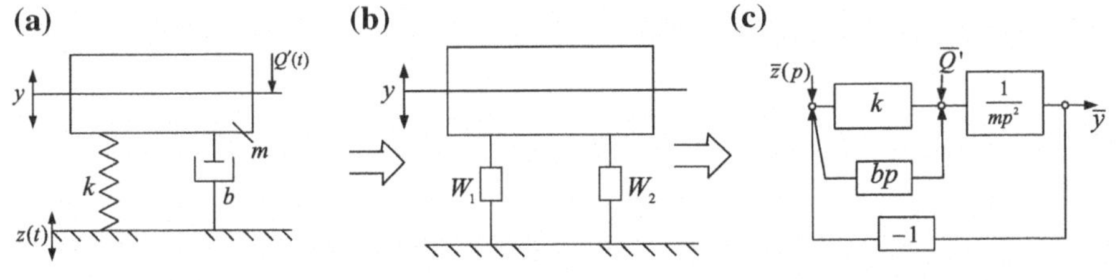

Fig. 3.9 The computational scheme (**a**, **b**) and structural (**c**) diagrams of a mechanical system with one degree of freedom

The denominator of the transfer function remains the same even in the case when the input signal is transferred to the input of any of the units of the closed system. If the transfer function of the open system has the form $W(p) = \frac{Q(p)}{P(p)}$, then the transfer function of the same system closed through the inverter (the unit with the transfer function equal to -1) is given by the expression

$$W_3(p) = \frac{W(p)}{1 + W(p)} = \frac{Q(p)}{Q(p) + P(p)}. \tag{3.79}$$

The characteristic equation of a closed system can be found by equating the denominator of the following expression to zero:

$$Q(p) + P(p) = 0 \tag{3.80}$$

This equation differs from the equation $P(p) = 0$ for an open system. It follows that processes in a closed system can take place differently than in an open system. Consider, for example, a mechanical system in the form of a solid body resting on the base through resiliently damping elements and moving along the coordinate y under the action of external perturbations $Q(t)$ and $z(t)$. In this case, $Q'(t)$ is of a force nature, and $z(t)$ is kinematic (Fig. 3.9).

The equation of motion of the system has the form

$$m\ddot{y} + b\dot{y} + ky = kz + b\dot{z} + Q, \tag{3.81}$$

where k is the spring stiffness, b is the coefficient of viscous friction, $p = j\omega$ ($j = \sqrt{-1}$) [17]. The transfer functions of the system have the form

$$W(p)_{z=0} = \frac{\bar{y}}{\bar{Q}} = \frac{1}{mp^2 + bp + k}; \; W'(p)_{Q=0} = \frac{\bar{y}}{\bar{z}} = \frac{k + bp}{mp^2 + bp + k}. \tag{3.82}$$

The transfer function of an open loop can be represented as

$$W''(p) = \frac{k + bp}{mp^2}. \tag{3.83}$$

Using (3.79), we find that $Q(p) = k + bp$, $P(p) = mp^2$, which for $Q'(t) = 0$ constitutes a match with (3.80). If we assume that $z(t) = 0$, and $Q'(t) \neq 0$, then it is necessary to bring the structural diagram in Fig. 3.9, c to the form at which the input action or signal $Q'(t)$ will be transferred to the application point of the input signal $z(t)$. In this case we obtain the same relationship between the open-loop and closed-loop transfer functions. It can be argued that any structural diagram can be transformed to a form in which a direct link is identified that has the open-loop transfer function for a single feedback. The transfer functions of open systems are widely used in the automatic control theory [1, 21].

3.2.2 Elementary Units of Structural Diagrams

An analysis of the equations of motion of linear systems allows us to conclude that any system can be compared with a dynamically equivalent system consisting of one or another connected elementary units that have some standard dynamic properties. This unit can be called elementary and can be denoted as a rectangle with the input y_0 and output y signals (see Fig. 3.5). At the heart of the structural approach in the control and constraint theory are the following provisions:

- any linear system can be represented in the form of elementary linear units connected to each other;
- every elementary unit is directional;
- there is a small number of different types of elementary units forming a complete system (that is, sufficient to leave a structural diagram of any system of a given class);
- the only basis for the classification of elementary units is their dynamic properties, i.e. the nature of the mathematical dependence of the output quantity on the input quantity, which, in particular, was reflected in [17].

Note that the structural approach also extends to parametric and nonlinear systems, naturally, under certain restrictions. For the output signals of the open-loop and closed-loop systems considered above, whose initial state is rest, the equation of motion in the transforms has the form

$$D(p)\bar{y}(p) = D_0(p)\bar{y}_0(p). \tag{3.84}$$

The left-hand member of (3.84) is a known polynomial $D(p)$, multiplied by the desired transform of the output signal, and the right-hand member is the known polynomial $D_0(p)$ multiplied by the known function $y_0(p)$, which is the transform of the input signal. Obviously, the equation remains unchanged if the true system is replaced by some artificial open-loop system having the same characteristic equation. The transform of the input signal $y_u(t)$ of this system must have the following form $\bar{y}_u(p) = D_0(p)\bar{y}_0(p)$. Then the transfer function of such an artificial system will take the form

$$W_u(p) = \frac{\bar{y}(p)}{\bar{y}_u(p)} = \frac{1}{D(p)}. \tag{3.85}$$

If the system in the initial state is not at rest, then on the right-hand member of Eq. (3.84) there will occur an additional known polynomial in p whose coefficients contain the given initial values. In this case, the entire right-hand member should be taken as the transform of the conditional input signal of the same artificial system with the transfer function (3.85).

Thus, the question of the dynamic correspondence between an artificial and a preset system will be solved if both systems have a transfer function (3.84). As it is known, the polynomial $D(p)$ of the nth degree can be represented in the form

$$D(p) = a_0(p - p_1)(p - p_2)\ldots(p - p_n), \tag{3.86}$$

where a_0 is the coefficient of p^n and $p_1, p_2, \ldots, p_n$ are the roots of the characteristic equation

$$D(p) = 0. \tag{3.87}$$

Depending on the properties of the roots of the characteristic Eq. (3.87), the individual factors in (3.86) can have different forms. If there is a zero root, there is a factor p; in the presence of real roots—factors of the form $p - a$; in the presence of complex-conjugate roots—the factors of the form $(p - b)^2 + \omega^2$ (the conjugate roots are denoted as $p_{12} = b \pm j\omega$). In general, the quantity of a can have both a negative and a positive value; the quantity of b is negative, positive and zero. If a root is multiple, then the corresponding factor will enter into (3.85) to a degree equal to the order of the multiplicity of the root. It follows from (3.85) that, in the general case, the transfer function of an artificial system must be a product the factors of which have a denominator of any of the factors listed above, and any constant for the numerator.

As it was established, the transfer function of consecutively connected units is equal to the product of the transfer functions of individual units. Consequently, a linear system with any characteristic equation in the form of a polynomial can be represented as a system consisting of consecutively connected units with transfer functions of only three kinds:

$$\frac{1}{p}, \frac{1}{p - a}, \frac{1}{(p - b)^2 + \omega^2}. \tag{3.88}$$

As noted above, such an approach is characteristic for the theory of control and communication, in which the problem the connections of elements is not based on the necessity of observing certain special restrictions. For example, such limitations exist in mechanics (the d'Alembert principle, Newton's laws, etc.), and also in the theory of electrical circuits (Kirchhoff's laws). Therefore, the choice of standard units in accordance with (3.88) assumes the corresponding kind of rules for the serial connection. This question, fundamental for mechanical oscillatory systems, was considered above, from which a different position in the choice and formation of a system of elementary units follows. Note that in the structural theory of vibration protection systems [16–18] or the structural theory of mechanical oscillatory systems, not all typical elementary units (3.88) are elementary, but consist of the simplest elementary units of other types. In particular, this refers to an oscillating unit. In the automatic control theory, a set of standard units is represented by a fairly wide set of standard units, overlapping the set significantly in accordance with (3.88). Details of the relationship between the concepts of elementary units in the automatic control theory and the structural theory of mechanical systems (in particular, vibration protection) are considered in [16, 17].

Despite the above-mentioned differences, quite natural and explainable between systems and approaches to the evaluation of their dynamic properties, it is possible to consider some common properties. In such a situation, each of the elementary units will have its own mathematical model, in the general case—the equation connecting the input and output. Developing such ideas, it is expedient to note the necessity of singling out specific problems connected with determining the conditions for the passage of signals along the system, i.e. problems of transformation of signals as such, in abstracting from the parameters of the physical nature (forces, displacements, velocities, etc.). Such approaches make sense in the problems of managing robotic devices, active vibration protection systems and special technical objects [23].

If the units with the indicated transfer functions (3.87) are physically feasible, then it is possible to construct real models of any linear systems of the class under consideration. Such models can naturally be called mathematical, since they, in general, may not reflect the essence of the physical processes occurring in the system being simulated. On the other hand, such a model can be constructed for the experimental solution of the differential equation. To fully simulate the initial equation of motion of the system (3.84), it is necessary to introduce a unit, representing the right-hand member of equation, into the structural diagram:

$$\bar{y}_0'(p) = D_0(p)\bar{y}_0(p). \tag{3.89}$$

For systems of the class under consideration, the expression corresponds to the representation of the sum of the derivatives of different orders (including zero) of the input signal. Consequently, there is a need to introduce units that multiply by a constant factor, differentiation and summation, which are characteristic of mechatronics.

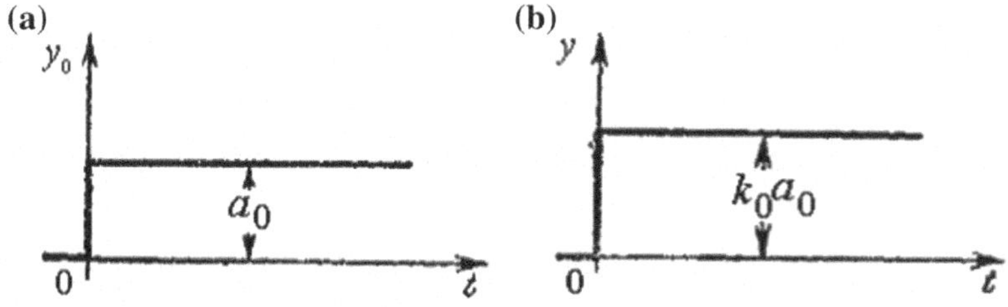

Fig. 3.10 The step function (**a**) and its reaction (**b**)

Considering in detail the listed types of units, which are usually called elementary, the transition function of each unit is determined, i.e. the motion of the unit is studied in standard conditions: the input signal is a unit function $1(t)$ under zero conditions.

1. *The amplifying unit*. For the amplifying unit, the dependence of the output signal on the input signal has the form

$$y(t) = k_0 \cdot y_0(t), \tag{3.90}$$

where k_0 is a constant value, called the unit gain coefficient. The gain coefficient k_0 can have any real value and any dimension, depending on the dimensions of y and y_0.

This simplest of the elementary units is sometimes called proportional, and also ideal, because it instantly reproduces the input quantity at the output without distortion. Output quantity may differ from input one only by scale or dimension.

If the input quantity has the nature of a step function, i.e. $y_0(t) = a_0 \cdot 1(t)$, where $a_0 = \text{const}$, then the output quantity will also have a step-like form, as shown in Fig. 3.10.

Obviously, the transition function of the amplifying unit $h(t) = k_0$, is equal to the constant value—the gain coefficient of the unit.

2. *The inertial unit.* The form of the coupling between the input y_0 and the output y values for the inertial unit (also called aperiodic) is expressed by equation

$$y + T\frac{dy}{dt} = ky_0. \tag{3.91}$$

The constant value k is called the gain coefficient of the inertial unit, T is the time constant ($T > 0$). The form of the general solution of (3.91) obviously depends on the form of the function $y_0(t)$. To determine the transient function of the inertial unit $h(t)$, a number of conditions are required:

Fig. 3.11 Transitional function of the inertial unit

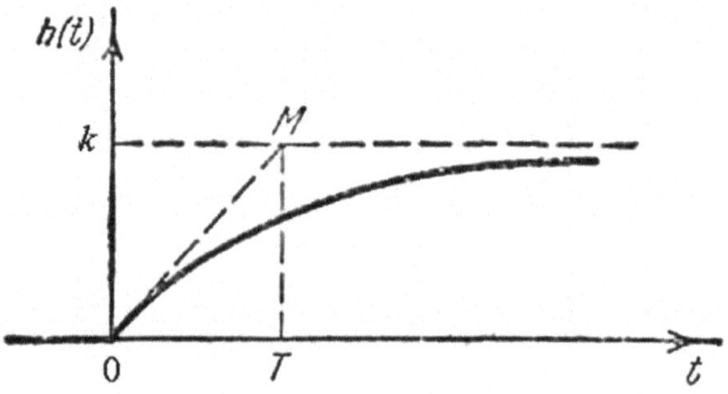

- a unit function is fed to the input of the unit at the instant $t = 0$

$$y_0(t) = \begin{cases} 1 & for\, t \geq 0, \\ 0 & for\, t < 0; \end{cases}$$

- it is considered that until the instant $t = 0$ the unit was at rest, i.e. the output is $y = 0$ for $t < 0$. In this case, the initial condition is $y(0) = 0$, because for an intermittent change of x, the quantity $\frac{dy}{dt}$ would an infinite value, and Eq. (3.91) would be violated.

The solution of Eq. (3.91) under the indicated conditions has the form

$$y(t) = h(t) = k\left(1 - e^{-\frac{t}{T}}\right). \tag{3.92}$$

The type of the transition function is shown in Fig. 3.11.

The straight line OM, which is tangent to the curve $h(t)$ at the point $t = 0$, intersects the asymptote $h(\infty) = k$ at the point M having the abscissa T. It follows from (3.91) that the greater the time constant T, the slower the transition process. On the other hand, as $T \rightarrow 0$, the inertial unit turns into a simple amplifying unit. If you use the operating method, then you need to determine the transfer function of the unit. Let's demonstrate this with the example of this unit, and in the future we will immediately apply the operating method. Equation (3.94) in the operational form with zero initial value $x(0) = 0$ will have the form

$$\bar{y}(p) + pT\,\bar{y}(p) = k\bar{y}_0(p).$$

Hence for the transfer function we obtain the expression

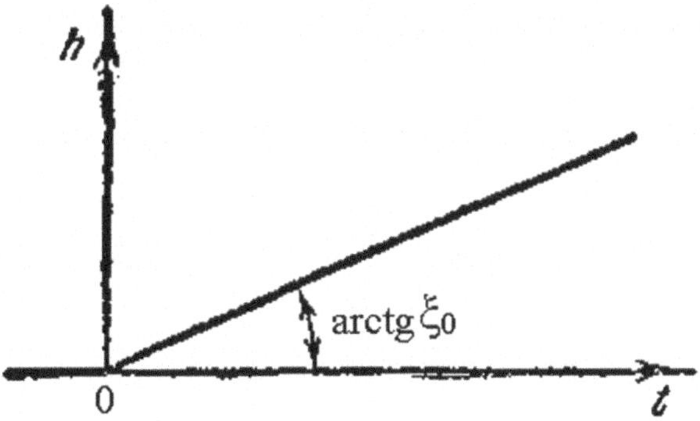

Fig. 3.12 Transitional function of the integrating unit

$$W(p) = \frac{\bar{y}(p)}{\bar{y}_0(p)} = \frac{k}{1+pT} = \frac{k}{T}\frac{1}{p-\alpha}, \tag{3.93}$$

where

$$\alpha = -\frac{1}{T} < 0.$$

Thus, according to the definition, the inertial unit is one of the elementary units (3.90) and corresponds to the real negative root of the characteristic equation of the linear system. Based on the known relations, we have

$$\bar{h}(p) = \frac{W(p)}{p} = \frac{k}{T}\frac{1}{p(p-\alpha)} \doteq h(t) = k\left(1 - e^{-\frac{t}{T}}\right), \tag{3.94}$$

which coincides with (3.92).

3. *Integrating unit*. The equation of this unit has the form

$$\frac{dy}{dt} = \xi_0 y_0, \tag{3.95}$$

where $y_0(0) = 0$.

The transfer function of the unit, obviously, will be

$$W(p) = \frac{\xi_0}{p}. \tag{3.96}$$

Fig. 3.13 The graph of the transition function of the differentiating unit

The integrating unit also belongs to the elementary units (3.88) and corresponds to the zero root of the characteristic equation of the system. The constant value ξ_0 is called the coefficient of amplification of the integrating unit. The transient function of the integrating unit is equal to

$$h(t) = \xi_0 t. \tag{3.97}$$

The form of this function is shown in Fig. 3.12.

A distinctive feature of the transition function of the integrating unit is that it does not have a steady-state (for $t \to \infty$) finite value. This property is, for example, the reason for the fundamental difference between astatic automatic control systems containing an integrating unit from static ACS that do not have this unit. Note that the inertial unit (3.95) during a certain period of time after the inclusion, namely, while $\frac{t}{T} \ll 1$, behaves almost like an integrating unit with a $\xi_0 = \frac{k}{T}$ gain coefficient, since $h(t) = k\left(1 - e^{-\frac{t}{T}}\right) \approx \frac{kt}{T}$ under this condition. The greater the time constant T, the longer the inertial unit retains the property of the integrating unit with a given degree of accuracy over a longer period of time. However, the larger T, the smaller the gain coefficient ξ_0.

4. *The differentiating unit.* This unit corresponds to equation

$$y = \mu \frac{dy_0}{dt} \tag{3.98}$$

and, consequently, the transfer function (for $x_0(0) = 0$)

$$W(p) = \mu p \tag{3.99}$$

For the transition function, we obtain

$$\bar{h}(p) = \mu \doteq h(t) = \mu\delta(t). \tag{3.100}$$

The constant μ is called the gain coefficient of the differentiating unit. The transition function $h(t)$ of this unit (Fig. 3.13) is zero for all $t \neq 0$ and has an infinite value for $t = 0$.

However, the integral of the transition function of the differentiating unit within unlimited time limits is finite; namely: $\int_{-\infty}^{\infty} h(t)\, dt = \mu$, since, according to the definition of the impulse function $\delta(t)$, there is an equality

$$\int_{-\infty}^{\infty} \delta(t)\, dt = 1.$$

The ideal integrating and differentiating units, characterized by Eqs. (3.95) and (3.98), respectively, can be approached by applying more complex electrical circuits and using electronic devices. In the ideal differentiating unit, the rule established earlier is violated: the order of the operator of the right-hand member of the equation of motion of the real system is not higher than the order of the operator of its left-hand member. However, the above diagrams of practical implementation of the differentiating unit, in fact, do not contradict this rule. As a result, in practice, differentiation is performed in principle approximately only. Nevertheless, in some cases, the differentiating unit introduced into the real diagram is considered theoretically as ideal, and then this rule can be violated.

5. *Oscillating unit*. The equation of the unit has the following form:

$$\frac{d^2y}{dt^2} + 2b_0\omega_0\frac{dy}{dt} + \omega_0^2 y = \omega_0^2 k y_0. \tag{3.101}$$

The initial state of this unit corresponds to the conditions $y(0) = 0$ and $\left(\frac{dy}{dt}\right)_{t=0} = 0$.

Here k is a constant called the unit gain coefficient, ω_0 is the resonant frequency, and b_0 is the damping constant, which is positive for the vibrational unit, but less than unity ($0 < b_0 < 1$). For $b_0 \geq 1$, the unit satisfying the Eq. (3.101) is aperiodic and, as will be shown below, can be represented as a chain of two consecutively connected inertial units. The transfer function of a oscillating unit has the form

$$W(p) = \frac{\omega_0^2 k}{p^2 + 2b_0\omega_0 p + \omega_0^2} = \frac{\omega_0^2 k}{(p - p_1)(p - p_2)}, \tag{3.102}$$

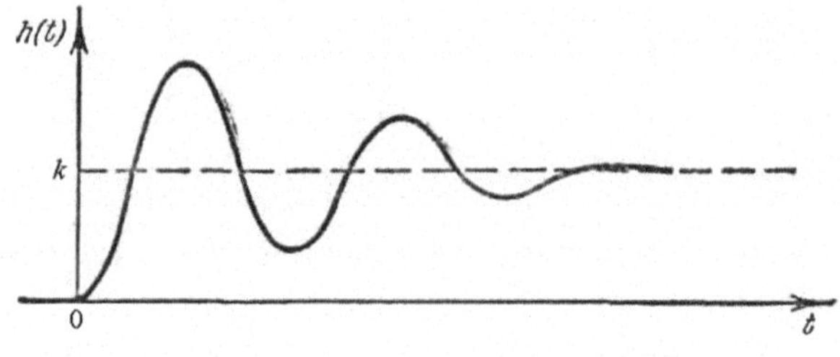

Fig. 3.14 The graph of the transient function of the oscillating unit

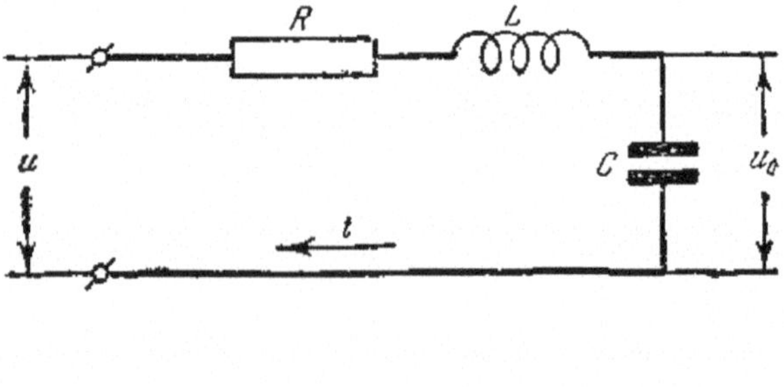

Fig. 3.15 The vibrational circuit of the electrical circuit

where p_1 and p_2 are the roots of the characteristic equation of the unit $p^2 + 2b_0\omega_0 + \omega_0^2 = 0$, equal to

$$p_{1,\,2} = \omega_0\left(-b_0 \pm j\sqrt{1 - b_0^2}\right). \tag{3.103}$$

For $b_0 > 1$, both roots will be real and negative. Consequently, the unit in question will be equivalent to two consecutively connected inertial units with different time constants. If $b_0 = 1$, we get a multiple negative real root and the considered unit will again be equivalent to two consecutively connected inertial unit, but with the same time constants.

On the basis of (3.103) we have

$$\bar{h}(p) = \frac{\omega_0^2 k}{p(p - p_1) \cdot (p - p_2)}.$$

To calculate the transition function, which is the original of the transform $\bar{h}(p)$, note that the latter has three simple poles at the points $p = 0$; $p = p_1$; $p = p_2$. Then

$$h(t) = \omega_0^2 k\left[\frac{1}{p_1 p_2} + \frac{e^{p_1 t}}{p_1(p_1 - p_2)} + \frac{e^{p_2 t}}{p_2(p_2 - p_1)}\right].$$

Substituting the expressions (3.100) for the roots p_1 and p_2 under the condition $0 < b_0 < 1$, after the corresponding transformations, we obtain

$$h(t) = k - Ce^{-\omega_0 b_0 t}\sin\left[\omega_0\sqrt{1 - b_0^2}t + \varphi\right], \tag{3.104}$$

where

$$C = \frac{k}{\sin\varphi};\ \mathrm{tg}\,\varphi = \frac{\sqrt{1 - b_0}}{b_0}. \tag{3.105}$$

The form of the transient function of the oscillating unit is shown in Fig. 3.14. The larger the damping constant b_0, the smaller the greatest deviation h from the steady-state value $k(\lim_{t\to\infty} h = k)$ and the faster the transient process decays. An example of an oscillation unit is a simple electric oscillating loop (Fig. 3.15), if the input value is the voltage u, and the output value is the voltage across the capacitor u_c.

Since the current in the loop is $i = C\frac{du_c}{dt}$, according to the Kirchhoff law, we get

$$LC\frac{d^2u_c}{dt^2} + RC\frac{du_c}{dt} + u_c = u.$$

Denoting $\omega_0 = \frac{1}{\sqrt{LC}}; b_0 = \frac{1}{2}R\sqrt{\frac{C}{L}} = \frac{R}{2\rho}$,
where $\rho = \sqrt{\frac{L}{C}}$ is the wave resistance of the loop, we reduce the equation to the standard form (3.101):

$$\frac{d^2u_c}{dt^2} + 2\omega_0 b_0\frac{du_c}{dt} + \omega_0^2 u_c = \omega_0^2 u.$$

If for $t < 0$ the loop was at rest, then $u_c(0) = 0$ and $i(0) = C\left(\frac{du_c}{dt}\right)_{t=0} = 0$.

The equation obtained shows that the gain coefficient of the considered oscillation unit is $k = 1$. It is known from the theory of electrical circuits [13, 14] that the oscillating loop becomes an aperiodic one, if $R \geq 2\sqrt{\frac{L}{C}}$, which coincides with the condition $b_0 \geq 1$.

6. *The unit of delay*. This unit is special, and its properties are displayed by the dependency

$$x(t) = x_0(t - \tau). \tag{3.106}$$

According to the delay theorem, for the transfer function of the unit, we have

$$W(p) = e^{-p\tau} \tag{3.107}$$

Equation (3.106) shows that the output quantity x exactly repeats the input value x_0, but with a time shift (with delay) to the interval τ. Thus, if a signal having the character of a single function (step function) is applied to the input of the unit, i.e. $y_0(t) = a_0(t)$, where a_0 = const, then the output is $y(t) = a_0 1(t - \tau)$. An example of the delay unit is an idealized pipeline, the pressure at which is transmitted at a finite rate in the form of an elastic wave propagating in a medium filling the tube [21].

7. *The summing unit.* This is the only kind of linear unit with several input quantities $y_1, y_2, \ldots, y_n$ and one output quantity x that is the sum of the input quantities

$$y = y_1 + y_2 + \cdots + y_n. \tag{3.108}$$

The simplest example of such a unit is an electrical circuit in which the input values are the currents $i_1, i_2, \ldots, i_n$ in the resistances $R_1, R_2, \ldots, R_n$, and the output value is the current i in the resistance R; obviously, $i = \sum_{k=1}^{n} i_k$. In practical diagrams, the input and output quantities are usually voltage. If we introduce the conductivities $G_i = \frac{1}{R_i}; G = \frac{1}{R}$, we obtain $Gu = \sum_{i=1}^{n} G_i(u_i - u)$, from which $u = \sum_{i=1}^{n} k_i u_i$,

where

$$k_i = \frac{G_i}{G + \sum_{i=1}^{n} G_i}.$$

3.2.3 Open-Loop and Closed-Loop Systems

Despite the fact that the general form of the transfer functions of both closed-loop and open-loop systems is the same (the ratio of polynomials), the dynamical properties of these systems can be very different. One of the most important problems of investigating closed-loop systems is to determine the stability of the system. When the characteristic polynomial is decomposed (3.87), it does not seem possible to compare the factors corresponding to the positive real roots of the characteristic equation (3.88) or to complex roots with a positive real part, with the corresponding elementary units. Nevertheless, the characteristic equation of a real system can have such roots if there are feedbacks in the system. In conclusion, it can be noted that the operator approach, developed on Laplace transforms, creates a methodological basis for solving various problems of control and constraint theory using the transfer function of the system. The reaction of the system to typical signals makes it possible to introduce an estimate of the dynamic properties of the system, without recourse to solving differential equations. Solved problems predetermine the peculiarities of structural approaches. In particular, this is reflected in the specifics of the formation of a set of elementary units, and in the implementation of the rules of structural transformations, which is associated with the formulation of certain conditions for transformations concerning electrical and mechanical circuits [15–17, 22]. At the same time, the specifics of structural approaches can be noted if the problem is related to the transformation of signals passing through the system. We assume that with while using dynamic analogies, physical features and specific problems of signal transmission and external influences, characteristic of real systems, should always be specified.

In the concept of electromechanical analogies, the introduction of a transformer into the electrical circuit and a solid body, which performs a plane motion, into the mechanical circuit can be considered processes with a similar basis. However, a solid body in a mechanical circuit is interpreted by a fairly complex system of lever linkages that occur as a result of the use of lever mechanisms. Considerations on the special aspects of lever linkages developed by the authors are based on the assumptions about the possibilities of separating the connection points of two elements having the form of two-terminal networks. The influence of the lever linkage type, which is implemented through lever mechanisms of the first and second types, on the properties of the mechanical circuit differs from mechanical circuits with solids from electric circuits with transformers, in particular, because of the possibility of change of the ratio sign in the lever mechanism.

It should be noted that the characteristic feature of mechanical circuits with included units in the form of a solid body is revealed in the studies, which consists in the fact that the lever mechanism introduces a "metric" into units' connection and ensures the emergence of parallel chains. This phenomenon is observed in electrical circuits, in which the introduction of a transformer leads to a violation of the galvanic integrity of the circuit.

References

1. Kim, P.D.: The theory of automatic control: in 2 vols. In: Linear Systems, vol. 1, p. 288. Fizmatlit Publ., Moscow (2003)
2. Feldbaum, A.A.: Fundamentals of the Theory of Optimal Automatic Systems, p. 624. GIFML Publ., Moscow (1966)
3. Karlov, N.V., Kirichenko, N.A.: Oscillations, Waves, Structures, p. 496. Fizmatlit Publ., Moscow (2003)
4. Kurosh, A.G.: The Course of Higher Algebra, p. 431. Nauka Publ., Moscow (1968)
5. Eliseev, S.V., Artyunin, A.I., Logunov, A.S., et al.: Dynamic Interactions of Machine Elements: Computational Schemes and Mathematical Models of Vibration States, p. 319. IrGUPS Publ., Irkutsk. Dep. in VINITI 08.11.13, No. 313-B 2013 (2013)
6. Tetelbaum, I.M., Shlykov, F.M.: Electrical Modeling of the Dynamics of Electric Drive Mechanisms, p. 192. Energia Publ., Moscow (1970)
7. Babakov, I.M.: Theory of Oscillations, p. 549. Nauka Publ., Moscow (1968)
8. Gardner, M.F., Burns, J.L.: Transients in linear systems (Transl. from English), 3rd edn, p. 551. Fizmatgiz Publ., Moscow (1961)
9. Feldbaum, A.A., Dudykin, A.D., Manovtsev, A.P., et al.: Theoretical Foundations of Communication and Control, p. 963. Fizmatgiz Publ., Moscow (1963)
10. Den-Hartog, J.P.: Mechanical Oscillations, p. 530. Publishing house of foreign literature, Moscow (1962)
11. Irodov, I.E.: Wave Processes. Basic Laws, p. 256. BINOM Publ., Moscow (2007)
12. Jie, F.S., Morse, I.E., Hinkle, R.T.: Mechanical Vibrations, p. 508. Mashinostroenie Publ., Moscow (1966)
13. Atabekov, G.I.: Linear Electrical Circuits, p. 592. Energia Publ., Moscow (1978)
14. Bakalov, V.P., Dmitrikov, V.F., Kruk, B.I.: Fundamentals of the Theory of Circuits. Textbook, p. 589. Radio and svyaz' Publ., Moscow (2000)
15. Druzhinsky, I.A.: Mechanical Chains, p. 240. Mashinostroenie Publ., Moscow (1977)
16. Eliseev, S.V., Reznik, YuN, Khomenko, A.P.: Mechatronic Approaches in the Dynamics of Mechanical Oscillatory Systems, p. 384. Nauka Publ., Novosibirsk (2011)
17. Eliseev, S.V., Reznik, YuN, Khomenko, A.P., et al.: Dynamic Synthesis in Generalized Problems of Vibration Protection and Vibration Isolation of Technical Objects, p. 523. ISU Publ., Irkutsk (2008)
18. Eliseev, S.V., Khomenko, A.P.: Dynamic Damping. Feedback Concept and Structural Methods of Mathematical Modeling, p. 357. Nauka Publ., Novosibirsk (2014)
19. Lurie, A.I.: Operational Calculus and Application in Engineering, p. 368. Nauka Publ., Moscow (1959)
20. Eliseev, S.V., Khomenko, A.P., Upyr, R.Yu.: Mechatronics of vibration protection systems with lever linkages. In: Modern Technologies. System Analysis. Modeling, vol. 3, no. 23, pp. 104–119. IrGUPS Publ., Irkutsk (2009)
21. Ivashchenko, N.N.: Automatic Adjustment. Theory and Elements of Systems, 4th edn, p. 736. Mashinostroenie Publ., Moscow (1978). Updated and revised

22. Eliseev, S.V., Moskovskikh, A.O., Bolshakov, R.S., Savchenko, A.A.: The possibilities of integrating the methods of the theory of circuits and the theory of automatic control in problems of the dynamics of machines [Electronic resource]. technomag.edu.ru: Science and education: electronic science.-tech, 6 edn. Access mode http://technomag.edu.ru/doc/378699.html (2012). Accessed 10.06.2012
23. Kolesnikov, V.I., Vorobiev, V.B., Shapovalov, V.V., et al.: Improving the Interaction of the Track and Rolling Stock, p. 365. Marshrut Publ., Moscow (2006)

Chapter 4
Construction of Mathematical Models of Mechanical Vibrational Systems: Additional Couplings and Equivalent Transformations

Determining the parameters of the dynamic interaction of elements and parts of machines and equipment is an important stage in solving the problems of ensuring the reliability of machines. In this respect, the estimation of the reduced parameters of systems at the level of mass-and-inertia, elastic, and dissipative characteristics is of great importance. Of particular importance is the identification of various dynamic effects that can arise when establishing non-conventional forms of coupling between individual types of motions of systems and their elements [1, 2]. Structural approaches, due to their visibility and the possibilities of obtaining integral estimates associated with the use of mathematical models in the form of structural diagrams and transfer functions, have certain advantages over other methods [3].

Within the scope of the structural theory of vibration protection systems [4, 5], the elements typical of mechanical oscillatory and vibration protection systems are typical for machine dynamics. Among them, besides mass-and-inertia elements, these are springs, dampers, motion transformation devices, etc. When considering small oscillations (in the representations of the linear theory), the initial mathematical models in the form of differential equations with constant coefficients can be transformed into equivalent models in the form of structural diagrams of automatic control systems (ACS), in which the object of control corresponds to the object of protection, and the standard units and constraints of the ACS correspond to the standard elements and constraints of mechanical vibrational systems. In the structural models of oscillatory systems, conventional springs are represented by standard units with the transfer functions of the amplifying element, in which the gain coefficient corresponds to the coefficient of stiffness of the elastic element. When forming the transfer function of the mechanical system as a whole, including for systems with several degrees of freedom, the elastic elements (springs) can enter the joints and form blocks made of springs.

The possibilities of some technological positions regarding the detection of algorithms for constructing and transforming structural circuits of vibration protection systems are considered. Apart from the elastic properties of dissipative

© Springer Nature Switzerland AG 2020
S. V. Eliseev and A. V. Eliseev, *Theory of Oscillations*, Studies in Systems, Decision and Control 252, https://doi.org/10.1007/978-3-030-31295-4_4

constraints, the possibility of taking into account the peculiar features of the influence of motion transformation devices is justified in this respect. Basically, the structure created by elastic elements, the object of protection (or control) and intermediate mass-and-inertia elements, predetermines simple procedures for consideration for dissipative and other factors. At the same time, it should be noted that the problems under discussion relate to linear problems.

4.1 Special Aspects of Transformation of Structural Mathematical Models

Evaluation of the interaction properties of the elements involves the use of appropriate transfer functions with the allocation of the input and output signals, followed by a "zeroing" of the complex variable p ($p = j\omega$). The basis for obtaining or determining the transfer function is a structural diagram that can be constructed from the initial mathematical model, i.e. from differential equations transformed with the use of Laplace transforms into a system of corresponding algebraic equations. In Fig. 4.1 shows the computational schemes and structural diagrams of

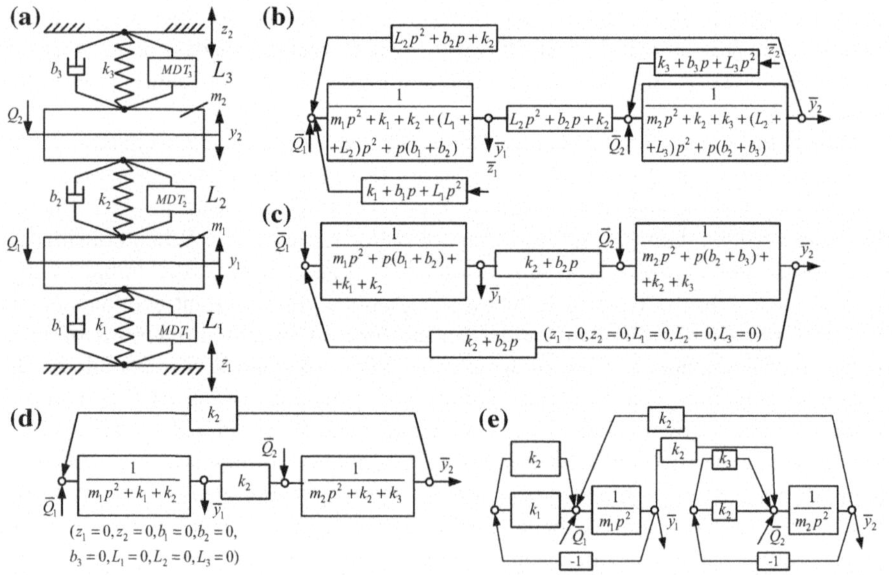

Fig. 4.1 Computational schemes and structural diagrams of a mechanical oscillatory system with additional elements. (**a** is the computational scheme with a complete set of typical elements; **b** is the structural diagram for $b1 \neq 0$, $b2 \neq 0$, $b3 \neq 0$, $L1 \neq 0$, $L2 \neq 0$, $L_3 \neq 0$; **c** is the structure diagram for $z_1 = 0$, $z_2 = 0$, $L_1 = 0$, $L_2 = 0$, $L_3 = 0$; **d** is the structural diagram for $b_1 = 0$, $b_2 = 0$, $b_3 = 0$, $L_1 = 0$, $L_2 = 0$, $L_3 = 0$; **e** is the detailed structural diagram with allocation of mass inertial elements at $L_1 = 0$, $L_2 = 0$, $L_3 = 0$, $b_1 = 0$, $b_2 = 0$, $b_3 = 0$, $z_1 = 0$, $z_2 = 0$.)

a mechanical oscillatory system with two degrees of freedom, containing dampers and motion transformation devices; variants of introducing additional constraints are shown. Note a characteristic feature of the change in the complexity of the structure upon transition from the diagrams, in Fig. 4.1b, to the diagrams in Fig. 4.1c, d. The additional constraints represent "input fragments" (what is meant here is dampers and motion transformation devices (MTD)), added to the spring available in the circuit. So, in the structural diagram (see Fig. 4.1b), the elements L_1p^2 and b_1p are added to the spring k_1; in their turn, L_2p^2 and b_2p are added to the spring k_2, and L_3p^2 and b_3p are added to the spring k_3, respectively. Thus, in view of the special aspects of the proposed method, not only the damping properties of the system, but also the inertial-type coupling, formed by motion transformation devices, including lever linkages, can be taken into account.

In what follows, it is assumed that the MTDs have reduced mass-and-inertia parameters that are referred to as L_1, L_2, L_3. Dissipative properties are taken into account within the concept of viscous friction (b is the coefficient of viscous friction).

In the case represented in Fig. 4.1a structural diagram reflecting the properties of the physical model in Fig. 4.1b, can be constructed by "parallel addition" of the corresponding standard elements. In this connection, it becomes possible to justify the method of taking into account the effect on the dynamic properties of the systems of effects of introducing not only elastic and damping elements, but also mass-and-inertia additional ones. In this case, the transfer functions of the systems can be determined on the assumption that only elastic couplings are initially used in the system: thus, in the beginning, the main contour is built, then a corresponding element is added in parallel to each elastic element determined by the stiffness coefficient k_i ($i = 1, \bar{n}$), having transfer functions b_ip and L_ip^2. That is, the consideration of the forces of additional interactions formed by dampers and MTDs can be performed on the basis of the previously defined transfer functions of a system containing only elastic elements. In this case, systems can have any number of degrees of freedom [6].

Note that the number of degrees of freedom is determined by the number of mass-and-inertia elements (reducible to independent material points) and the number of solid bodies (in the plane motion, it has two degrees of freedom, if we consider vertical motions). The object of protection in a mechanical oscillatory system is usually associated with a solid body in the form of a material point or a solid body with two degrees of freedom. The construction of the vibration protection system, and its structural diagram unfolds accordingly. Figure 4.1d represents a detailed structural diagram, on which two mass-and-inertia units m_1 and m_2 are allocated. In the structural diagram (see Fig. 4.1e), these elements are represented by two units with transfer functions $\frac{1}{m_1p^2}$ and $\frac{1}{m_2p^2}$ reflecting the properties of the object. For specific calculations, the object of protection can be m_1 or m_2 with subsequent structural transformations and feedback loops.

In this respect, although MTDs are mass-and-inertia elements, but their state does not require the introduction of separate coordinates; it is determined by the parameters of the relative motion of the mass-and-inertia elements (m_1 and m_2). For example, in the computational scheme (Fig. 4.1a), there are three devices for converting the motion—L_1, L_2 and L_3. The kinetic energy of the system (see Fig. 4.1a) is determined by expression

$$T = \frac{1}{2}m_1\dot{y}_1^2 + \frac{1}{2}m_2\dot{y}_2^2 + \frac{1}{2}L_1(\dot{y}_1 - \dot{z}_1)^2 + \frac{1}{2}L_2(\dot{y}_2 - \dot{y}_1)^2 + \frac{1}{2}L_3(\dot{y}_2 - \dot{z}_2)^2, \quad (4.1)$$

which predetermines, within the scope of the structural theory of vibration protection systems, the account of the dynamic properties of the MTDs as elements matched with elastic and mass-and-inertia units. In the structural diagram (see Fig. 4.1b), the partial system with mass-and-inertia element m_1p^2 has some generalized structure consisting of parallel units L_1p^2, L_2p^2, b_1p^2, b_2p^2, k_1 and k_2. Such a structure can be considered as some generalized spring with a transfer function

$$W_{gen}(p) = p^2(L_1 + L_2) + p(b_1 + b_2) + k_1 + k_2. \quad (4.2)$$

For $p = 0$, expression (4.2) is reduced to an ordinary spring with stiffness $k_1 + k_2$. As for expression (4.2), it can be regarded as the coefficient of dynamic stiffness of the generalized spring. In this case, the input signal of the unit with the transfer function (4.2) is the displacement (more precisely, the Laplace transform) of the corresponding coordinate, and the output signal is the force impact (more precisely, the Laplace transforms of some force factor). The conducted researches reveal details of structural transformations of the initial computational schemes in the form of mechanical oscillatory systems with one or several degrees of freedom. In particular, it was revealed that dynamic stiffness can be grouped into certain structures that could also be called quasi-springs, since they behave like ordinary springs in the transformations of structures. The difference between quasi-springs and ordinary ones is that their properties depend on the frequency of external influences.

The definition of dynamic responses is well algorithmized and makes it possible to reduce various vibration protection systems to fairly simple basic forms (or systems) with one and two degrees of freedom. The scientific novelty of the approach is that quasi-springs (or generalized springs), with simplifications, begin to take the form of units with transfer functions of a fractional-rational form. The order of the display details the introduction of additional elements in Fig. 4.2. In a system with one degree of freedom (Fig. 4.2a–e is a basic model, position I), the structural diagram has the form shown in Fig. 4.2a (position II). Taking into consideration friction forces in the computational scheme (introduction of a viscous friction damper is position III) corresponds to the occurrence on the structural diagram (position IV) of parallel additional bonds with the transfer functions bp. If we introduce into the system (Fig. 4.2a, position V) motion transformation devices with the transfer function Lp^2, where $L = \frac{J}{r_{cp}^2 tg^2\alpha}$ (J, r_{aver}) and α are parameters of the non-locking screw mechanism [5, 6], as a result, in the structural diagram (position VI) parallel to the elastic element k, the corresponding unit Lp^2 (position VI) will be introduced.

If a mechanical oscillatory system with two degrees of freedom is considered (Fig. 4.2b, position I), then its structural diagram can be presented in a detailed form (position II) with the identification of possible protection objects, as well as in a form that shows the presence of partial systems (position III), which has its own application in dynamic calculations. Transformations based on the use of partial systems are shown in Fig. 4.2c (positions I, II). It is shown that structural transformations are fairly simple (positions III and IV), which makes it possible not only

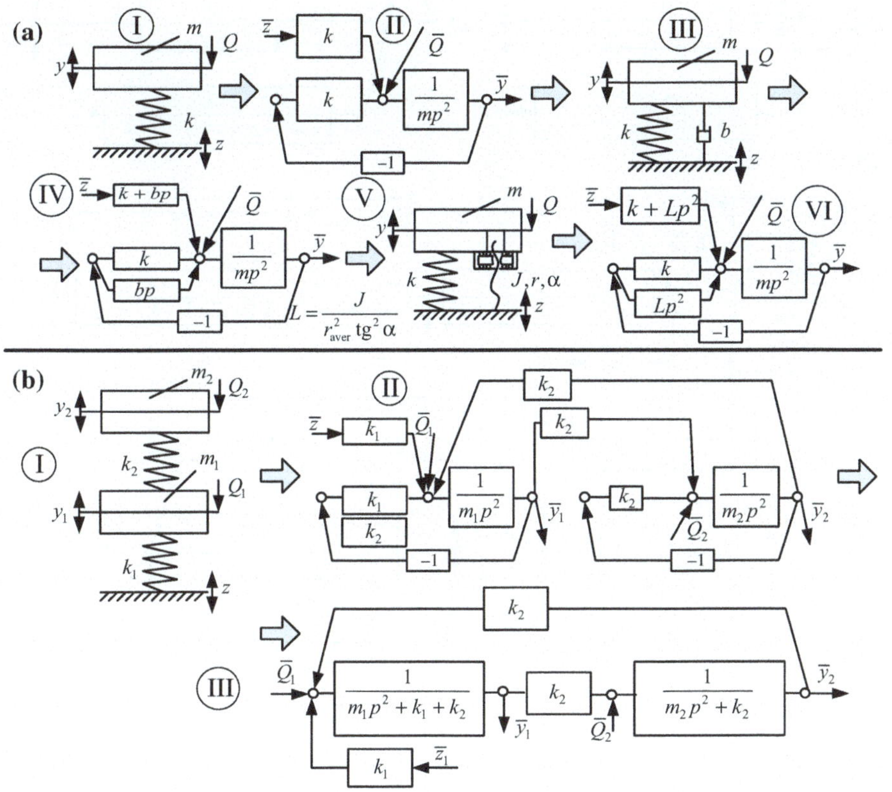

Fig. 4.2 The diagram of consecutive transformations of computational schemes and structural diagrams for the evaluation of interaction of the system elements. **a** is the computational scheme with one degree of freedom (positions I–IV are the introduction of a damper, V, VI are the introduction of MTD); **b** is the system with two degrees of freedom (positions I–III are formation of feedbacks with the object of protection m_2); **c** is the system with two degrees of freedom (positions I, II are the introduction of dampers in the first and second stages, III and IV are the formation of feedbacks for the partial system and the object of protection m_2); **d** is taking into account the features of the introduction of a damper and an MTD in one cascade

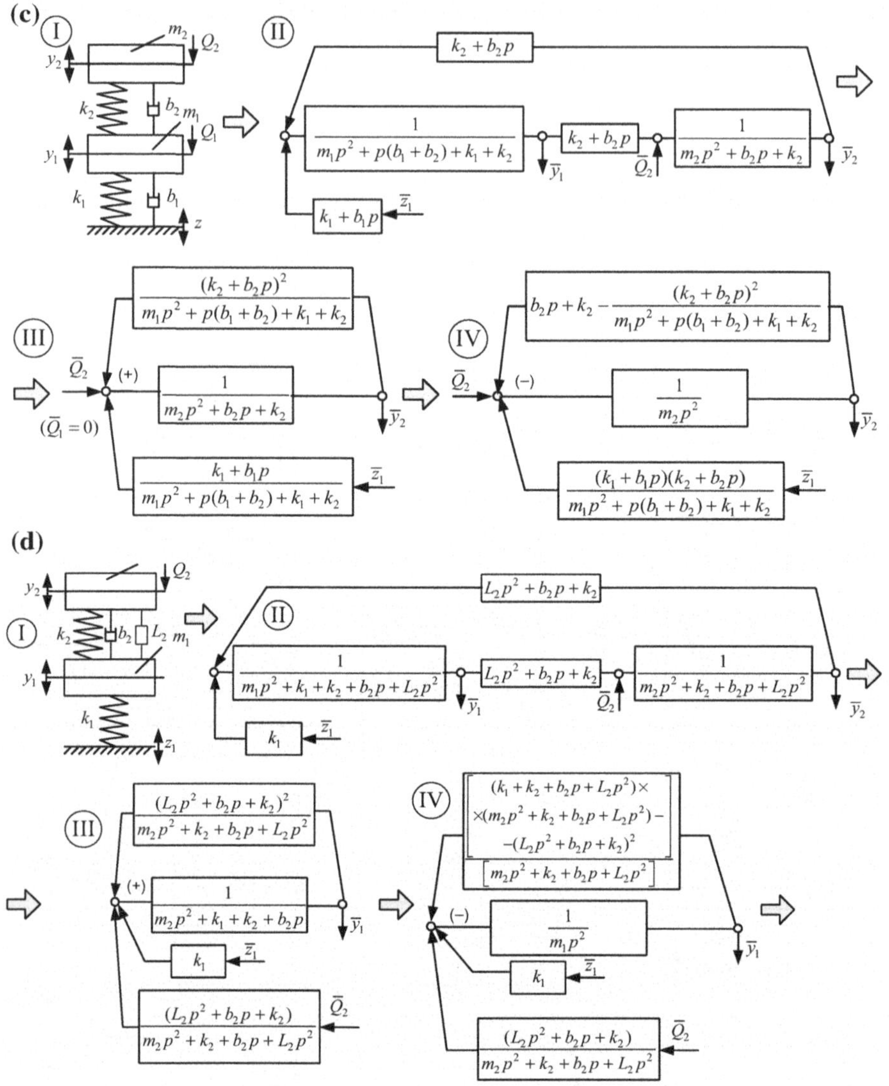

Fig. 4.2 (continued)

to form the appropriate feedback circuits, but also to solve problems of equivalent transformations of external influences that must be reduced to the corresponding coordinates of the state of systems.

In Fig. 4.2d into the system with two degrees of freedom (position I), a viscous friction damper b_2 and a motion transformation device are introduced into the upper stage. At positions II–IV, it is shown that the parameters b_2p and Lp^2 enter the structure of the transfer functions of the corresponding units. The influence of the resistance forces and the reduced mass-and-inertia parameters of the MTD can be taken into account in the usual manner using the frequency characteristics of methods of dynamic synthesis [5].

Thus, the technology of structural transformations in the construction of mathematical models of mechanical oscillatory systems as computational schemes of vibration protection systems consists in the implementation of several stages. When choosing a protection object, an integrating unit with a transfer function $\frac{1}{mp^2}$ can be defined, and the rest of the mechanical system is "convolved" into mechanical circuits that form additional feedbacks with respect to the protection object. The rules for converting mechanical circuits in a linear configuration are determined by the rules for consecutive and parallel connection of springs. Regarding the consideration of dissipative factors and the influence of dynamic relationships created by motion transformation devices, it should be noted that the simplicity of the approach that forms the method of structural transformations is based on the idea of the possibility of choosing and justifying the existence of typical elements that implement various functions. However, for each element, the input signal in the form of displacement is transformed by a standard element into the corresponding force factor.

4.2 Construction of Mathematical Models of Dynamic Interactions of Elements of a Generalized Form

The forces of resistance in oscillatory systems play an important role, given the physical features of their influence, however, in linear systems, resistance forces are often associated with the concepts of equivalent representations based on viscous resistance forces, when the frictional forces correspond to viscous friction. It is proposed to develop a method that allows assessing the possibilities brought by additional units and motion transformation devices, acting within the concept of building generalized couplings. These couplings can not only be elastic elements, which was considered in sufficient detail in [5, 7–9], but also to have physical forms of elastic elements with the properties of energy dissipation and motion transformation.

Taking viscous resistance into consideration. Dissipative elements, connected on the basis of the rules of parallel and consecutive combining of springs, give a general transfer function corresponding to a dissipative unit, or, as was determined in [5], the transfer function of a differentiating unit of the first kind. Questions related to the interconnection of several dissipative units and elastic elements, found a detailed reflection in [10, 11].

Let us consider some examples of first-order differentiating units being connected with each other. Figure 4.3 represents several options for a possible connection, where b_i ($i = \overline{1,3}$) corresponds to the coefficient of viscous resistance [12]. Since for $p = 0$ ($p = j\omega$ is a complex variable), the conditions for maintaining static

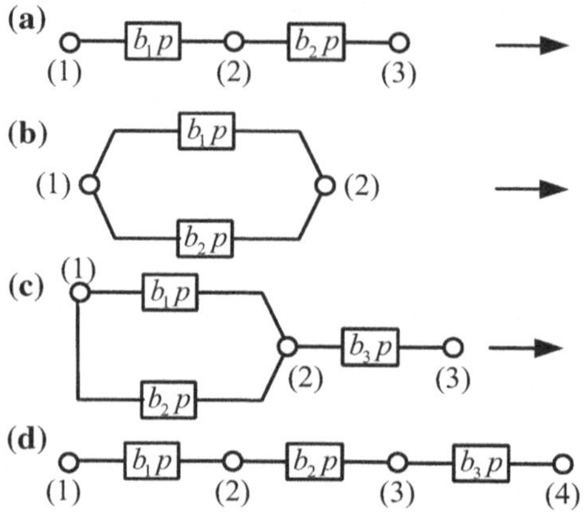

$$W_{13}(p) = \frac{b_1 b_2}{b_1 + b_2} p \tag{4.3}$$

$$W_{12}(p) = (b_1 + b_2) p \tag{4.4}$$

$$W_{13}(p) = \frac{(b_1 + b_2) b_3}{b_1 + b_2 + b_3} p \tag{4.5}$$

$$W_{14}(p) = \frac{b_1 b_2 b_3 p}{b_1 b_2 + b_2 b_3 + b_1 b_3} \tag{4.6}$$

Fig. 4.3 Examples of consecutive and parallel connections of dissipative elements. Connection **a–d** *is*: **a**—consecutive; **b**—parallel; **c**—cascaded; *r*—consecutive

equilibrium cannot always be ensured; therefore, compacts collected according to the rules of parallel and consecutive connection of springs often have elastic units in their composition. Actually, the parallel connection of the elastic and dissipative elements determined an appropriate place for the damper in the transport suspension theory [13–15]. Figure 4.3 shows several examples of mechanical circuits that can be assembled from typical elementary units in the form of elastic and dissipative elements that are connected consecutively, in parallel or in a combination of rules. Since the elements b_ip have the dimension of an elastic element, the union of only the dissipative elements gives a transfer function of the same form, although the parameters of the compound unit will vary, as in expressions (4.3)–(4.6) (see Fig. 4.3).

Expressions (4.3)–(4.6), shown in Fig. 4.3, represent the transfer functions of a mechanical circuit consisting of standard dissipative units or standard elements of differentiation of the first order. Such structural blocks have an displacement as an input signal, and at the output there is a resistance force, which depends on the b_i, parameter of viscous friction and the speed of dynamic interactions. If even one of the parameters of b_i (for example, expression (4.6)) is equal to zero, the output signal will have a zero value.

Introduction of elastic elements. In their turn, several variants of the connection of elastic and dissipative elements are presented in Fig. 4.4. The transfer functions of the mechanical circuits are determined by the expressions (4.7)–(4.10) (see Fig. 4.4). Note that with an appropriate choice of the location of the elastic element, the properties of the circuit as a whole can differ significantly when "zeroing" the coefficients of viscous friction.

The combination of units of different nature forms compacts (or blocks of units) with properties that distinguish them from elementary units. For example, in Fig. 4.4a the consecutive connection of the elastic and dissipative elements leads to

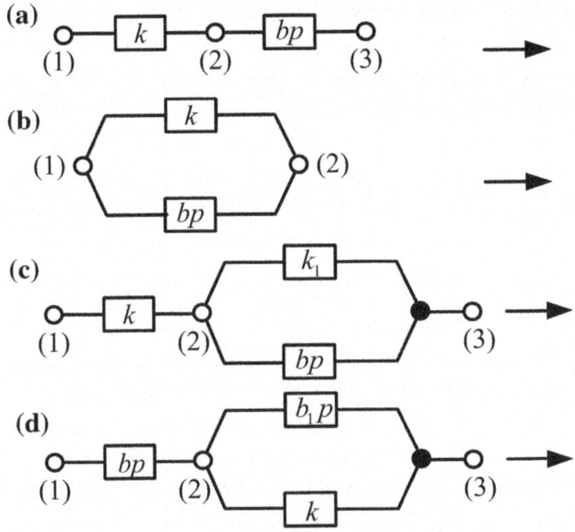

$$W_{13}(p)=\frac{kbp}{k+bp} \tag{4.7}$$

$$W_{12}(p)=k+bp \quad (4.8.)$$

$$W_{13}(p)=\frac{(k+bp)k}{k+k_1+bp} \tag{4.9}$$

$$W_{13}(p)=\frac{(k+b_1p)bp}{k+(b+b_1)p} \tag{4.10}$$

Fig. 4.4 Examples of compacts of elastic and dissipative elements. Connection **a–d** is: **a**—consecutive; **b**—parallel; **c**—a cascade with two elastic elements; **d**—a cascade with two dissipative elements

the formation of a forcing aperiodic unit; cascade connection according to the diagram in Fig. 4.4c—to the formation of a unit with a transfer function of an aperiodic unit of a general form. In its turn, the cascade connection according to the diagram shown in Fig. 4.4d, shows a compact of a more complex device, in which the transfer function is a fractional-rational expression with a numerator having the second order; the denominator of the transfer function of a compact is of the first order. The noted properties of compacts can be used in the development of non-traditional vibration protection systems.

Possibilities of simplifying representations. In the linear theory of oscillations of mechanical systems, the hypothesis is often used to account for the resistance forces, in which the possibility of an equivalent representation of resistance by viscous friction is substantiated [12]. A characteristic feature of the parameters of complex structures is their frequency dependence. With small coefficients of viscous friction and low frequencies, the transfer functions can acquire a simpler form, as well as the compacts. Considering the possibilities of "zeroing out" the transfer functions of dissipative elements, one can obtain a significant variety of intermediate structures that will possess different properties, including changes in the structure of the compact. In this respect, the dissipative elements, in which the resistance coefficients bp can vary depending on the control signals (including taking the values $b = 0$ or $b = \infty$), may prove promising for constructing active vibration protection systems.

The increase in the resistance forces to very large values leads to significant changes in the dynamic properties, which is clearly manifested in the influence of viscous friction on the form of the transient and amplitude-phase frequency characteristics. With a coefficient of resistance tending to a very large value, a "coupling" of two mass-and-inertia elements is possible. With elastic elements and motion transformation devices linked by second-order differentiating units, for b ∞, it is possible to "take out" these elements from the formulas that define the reduced parameters of compacts in the same way as the action of springs with stiffness that tends to very large values. Dissipative elements, in comparison with elastic ones, can be "nullified" at frequencies of external influences close to zero, as mass-and-inertia elements. That is, the dissipative units do not affect the static state.

Motion transformation devices. While enlarging a set of typical elements of vibration protection systems to the known elements reflecting the properties of springs and dissipative units, a typical element with the transfer function of a second-order differentiating unit can be added. The capabilities of physical implementations of such properties can be quite wide and presented in various constructive and technical solutions [16, 17]. Consider the examples shown in Fig. 4.5. Expressions (4.11)–(4.15) (see Fig. 4.5) are the transfer functions of the corresponding types of mechanical circuits.

Note that the introduction of MTD does not change the nature of the influence of the resistance forces: when $p \to 0$, the impact of the device Lp^2 is not manifested; the consecutive connection of bp and Lp^2 yields structure of a new element, the

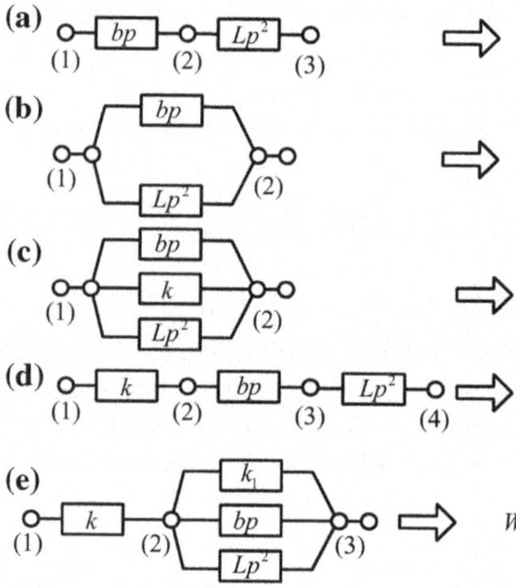

$$W_{13}(p) = \frac{bLp}{b + Lp^2} = \frac{bLp^2}{b + Lp} \tag{4.11}$$

$$W_{12}(p) = bp + Lp^2 \tag{4.12}$$

$$W_{12}(p) = k + bp + Lp^2 \tag{4.13}$$

$$W_{14}(p) = \frac{\dfrac{bpLp^2k}{bp + Lp^2}}{\dfrac{bpLp^2}{bp + Lp^2} + k} = \frac{bLp^2}{bpLp^2 + kLp^2 + kbp} \tag{4.14}$$

$$W_{13}(p) = \frac{k(k_1 + bp + Lp^2)}{k + k_1 + bp + Lp^2} = \frac{Lp^2k}{kb + Lp(bp + k)} \tag{4.15}$$

Fig. 4.5 Some types of connections of a dissipative unit with standard elements of mechanical oscillatory systems. **a** is the consecutive connection of the elements bp and Lp^2 (L is the reduced mass of the motion transformation unit); **b** is the parallel connection of the elements bp and Lp^2; **c** is the parallel connection of three elements; **d** is the consecutive connection of three elements; **e** is the combined connection

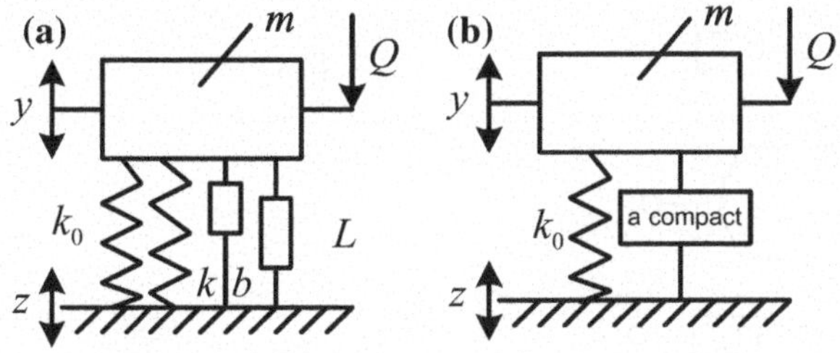

Fig. 4.6 The computational scheme of a vibration protection system with an extended set of elements. **a** the base spring (k_0) that works in parallel with the typical elements k, bp, Lp^2; **b** the base spring (k_0) works in parallel with the compact ($k + bp + Lp^2$)

properties of which require separate consideration. With parallel connection, it is also necessary to further study the properties of new elements; in addition, compacts can be introduced into the usual computational scheme of the vibration protection system (Fig. 4.6).

In more detail, the dynamic properties of vibration protection systems with elements k, bp, Lp^2 included in the compact are considered in [18]. A compact can be called a formation or a structure of several interconnected standard elements. Note that with the use of the extended set of elements ($k + bp + Lp^2$), the type of external influence becomes very important. In this case, Fig. 4.6 shows: Q as a power perturbation, and z as kinematic. In such cases, the system transfer functions will reflect the influence of the inertial forces of moving space under the kinematic perturbation, which is introduced by the unit Lp^2.

Some peculiarities of taking into account the dynamic properties of elements with the transfer function Lp^2. Let us consider a number of examples of connecting motion transformation devices with various elements of oscillatory structures (Fig. 4.7).

Carrying out a comparative analysis of the connection diagrams, it can be noted that the connection of the monotypic units gives structures of the same type as the original ones (Fig. 4.7a, b). A similar result can also be obtained by combining elastic and dissipative units. However, heterogeneous connections can bring a wide variety of transfer functions of the constructed compacts. Thus, the consecutive connection of the elastic unit with the device Lp^2 leads to a structure (see Fig. 4.7c), which has resonant properties. The frequency of resonant oscillations is determined as follows:

$$\omega_{nat}^2 = \frac{k}{L}. \tag{4.23}$$

In the circuitry in Fig. 4.7d, the introduction of a dissipative units, taking into consideration elastic-inertial constraints, makes it possible to obtain the transfer function of an oscillating unit of the forcing type, in which the numerator's order is greater by one than the order of the denominator. The transfer function of a compact has the same form according to the diagram shown in Fig. 4.7f. The transfer

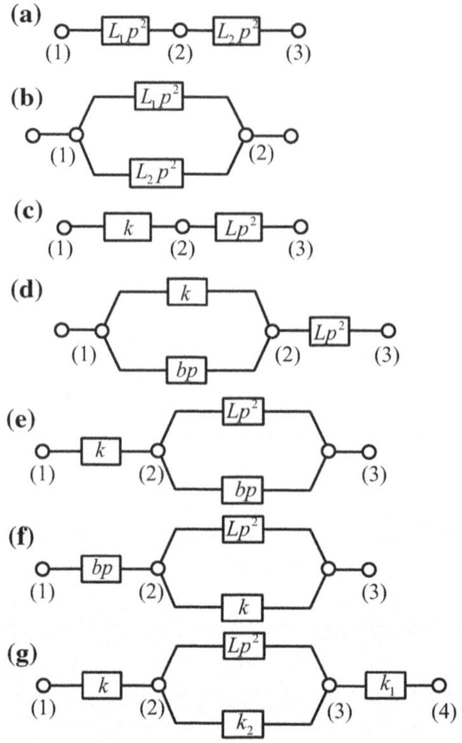

$$W_{13}(p) = \frac{L_1 L_2}{L_1 + L_2} p^2 \tag{4.16}$$

$$W_{12}(p) = p^2 (L_1 + L_2) \tag{4.17}$$

$$W_{13}(p) = \frac{kLp^2}{k + Lp^2} \tag{4.18}$$

$$W_{13}(p) = \frac{(k + bp)Lp^2}{k + bp + Lp^2} \tag{4.19}$$

$$W_{13}(p) = \frac{(Lp^2 + bp)k}{Lp^2 + k + bp} \tag{4.20}$$

$$W_{13}(p) = \frac{bp(Lp^2 + k)}{Lp^2 + k + bp} \tag{4.21}$$

$$W_{14}(p) = \frac{kk_1Lp^2 + kk_1k_2}{Lp^2(k + k_1) + kk_1 + k_1k_2 + k_2k} \tag{4.22}$$

Fig. 4.7 Connection diagrams of elementary units. **a** is the consecutive connection of two monotypical elements; **b** is the parallel connection; **c** is the consecutive connection of the elastic element k and the device Lp^2; **d** is the combination of the cascade compound ($k + bp$) and Lp^2; **e** is the combination of the cascade connection (Lp^2, bp) with the spring k; **f** is the combination of the cascade connection (Lp^2, k) with the unit bp; **g** is the combination of the cascade connection (Lp^2, k) in the consecutive connection with two springs k_1 and k_2

function, according to the diagram in Fig. 4.7e is analogous to the transfer function of the compact in Fig. 4.7c, but differs by the presence of a dissipative element. In Fig. 4.7, we consider a structure consisting of four units, which is displayed by a transfer function of the form

$$W(p) = \frac{\frac{kk_1}{k+k_1}(Lp^2 + k_2)}{\frac{kk_1}{k+k_1} + Lp^2 + k_2}, \tag{4.24}$$

from which it follows that when

$$\omega_{dyn}^2 = \frac{k_2}{L}, \tag{4.25}$$

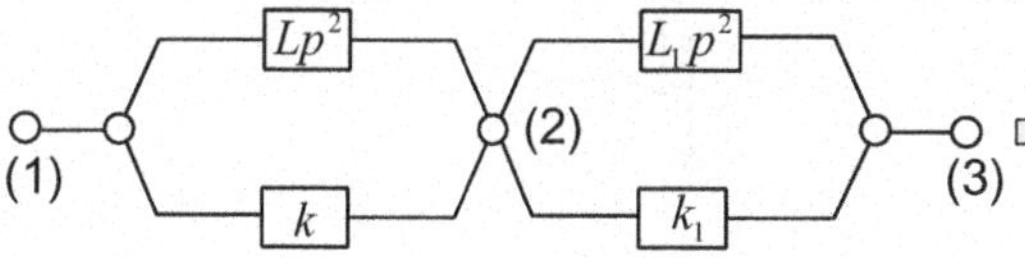

$$W_{13}(p) = \frac{(k + Lp^2)(k_1 + L_1 p^2)}{(L + L_1)p^2 + k + k_1}$$

Fig. 4.8 The diagram of a mechanical circuit of two blocks or cascades

then the mode of "zeroing" the numerator is possible, and when

$$\omega_{nat}^2 = \frac{kk_1 + k_2k_1 + k_2k}{L}, \tag{4.26}$$

the resonance properties of the circuit are manifested.

A more complex structure of the compact is shown in Fig. 4.8, which shows that the compact consists of two consecutively connected blocks. This kind of problem refers to the so-called cascade vibration protection.

This kind of mechanical circuit (or compact) has two frequencies: $\omega_1^2 = \frac{k}{L}$ and $\omega_2^2 = \frac{k_1}{L_1}$, at which the numerator of the transfer function of the chain will be zero. In turn, at a frequency

$$\omega_{nat}^2 = \frac{k + k_1}{L + L_1} \tag{4.27}$$

The denominator of the transfer function is also zeroed. An increase in the number of cascades makes it possible to implement the idea of constructing a mechanical filter for periodic signals, which was considered, in particular, in [19]. The introduction of lever linkages makes it possible to change the elastic properties of a compact. Consider the circuit in Fig. 4.9.

The transfer function of the system in Fig. 4.9 can be written as follows

$$W(p) = \frac{\bar{y}}{\bar{Q}} = \frac{1}{mp^2 + Lp^2 + k + i^2(k_1 + L_1p^2)} = \frac{1}{p^2(m + L + L_1) + k + k_1 i^2}, \tag{4.28}$$

Fig. 4.9 The mechanical system with elastic-inertial and lever linkages

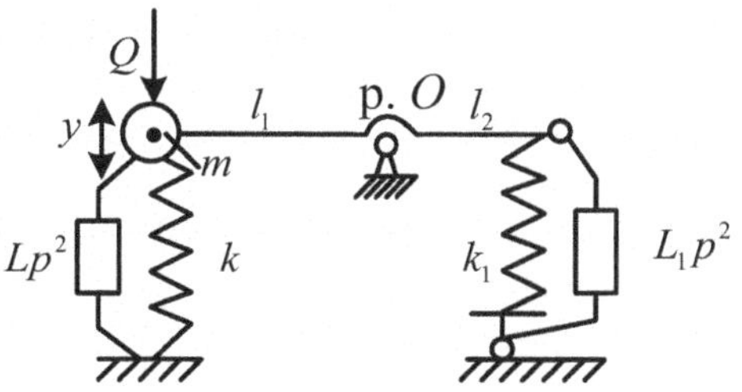

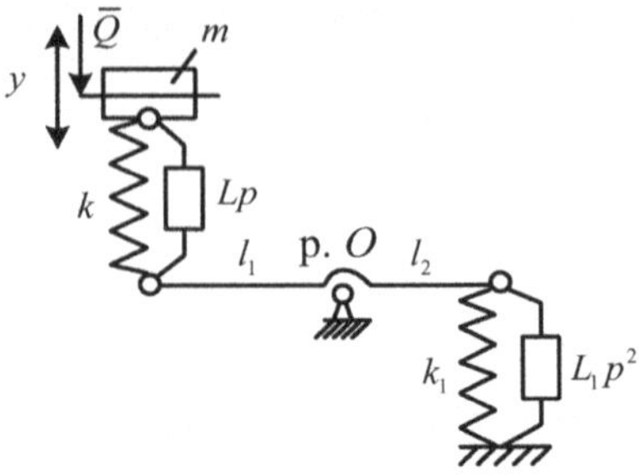

Fig. 4.10 The mechanical system with two cascades

which reveals the physical meaning of the dynamic reduced stiffness of the system. If we assume that the configuration of the basic model has elements m and k in its structure, then the reduced dynamic stiffness of the system is defined as follows:

$$k_{red.d} = \left|k_1 i^2 - \omega^2(L + L_1 i^2)\right|. \tag{4.29}$$

In the static mode, $k_{\text{red}} = k_1 i^2$, which coincides with the previously obtained results. For a static load $Q = \text{const}$, a motion transformation device or a second-order differentiating unit, like the dissipative unit, does not affect the elastic properties of the system in the static state. Consider an example with a mechanical system, as shown in Fig. 4.10.

As a result of the corresponding transformations, the transfer function of the system takes the form

$$W(p) = \frac{\bar{y}}{\bar{Q}} = \frac{1}{mp^2 + \frac{(Lp^2+k)(L_1p^2+k_1)i^2}{(Lp^2+k)+(L_1p^2+k_1)i^2}} =$$
$$= \frac{p^2(L+L_1 i^2)+k+k_1 i^2}{p^4(m(L+L_1 i^2)+i^2 LL_1)+p^2(m(k+k_1 i^2)+Lk_1 i^2+kL_1 i^2)+kk_1 i^2}. \tag{4.30}$$

If we assume that $L = 0$, then

$$W(p) = \frac{\bar{y}}{\bar{Q}} = \frac{p^2 L_1 i^2 + k + k_1 i^2}{p^4(mL_1 i^2 +) + p^2(mk + k_1 i^2 m + kL_1 i^2) + kk_1 i^2} \tag{4.31}$$

In the system that is structured that way, one dynamic absorbing mode and two resonance modes are possible, which does not differ in principle from usual situations and offers extra possibilities for adjusting systems due to transfer ratios. Standard elementary units of mechanical oscillatory systems and vibration protection systems, in particular, within the scope of structural approaches, can be divided into at least two groups. The first is elastic elements, the properties of which do not depend on the frequency for harmonic motions. The second group includes elements whose transfer functions contain explicitly the complex variables

$p(p = j\omega)$, which makes the properties of the elements dependent on the frequency of the emerging motions.

At the same time, typical elements of the extended set in the formation of structural groups and blocks (the proposed name is compacts) are implemented according to the rules of parallel and consecutive connection of springs, which is typical for the theory of mechanical circuits. The proposed method for constructing the transfer functions of mechanical circuits can be transferred to the determination of the transfer functions of vibration protection systems as a whole, taking into account the characteristics of the external perturbation. Introduction to the structure of the transfer function compacts of typical elements that depend on frequency makes it possible to estimate the changes in the properties of the system during the transition from static to dynamic state, which is ensured by "zeroing" of the corresponding parameters of standard units. Such approaches offer opportunities for new constructive and technical solutions in the development of controlled vibration protection systems.

4.3 Mathematical Models of Vibration Protection Systems: Taking into Consideration the Special Aspects of Motion of the Protection Object

In the applied theory of vibration protection systems, the simplest basic models are usually represented in the form of a mechanical oscillatory system with one degree of freedom; while the object performs rectilinear small vertical oscillations. The object can be connected to the bearing surface by means of connecting elements in the form of springs and dampers. The bearing surface for the protection object can be represented by zones of contacts I and II (Fig. 4.11), and the protection object itself can have guides providing vertical motion (or motion along one y coordinate).

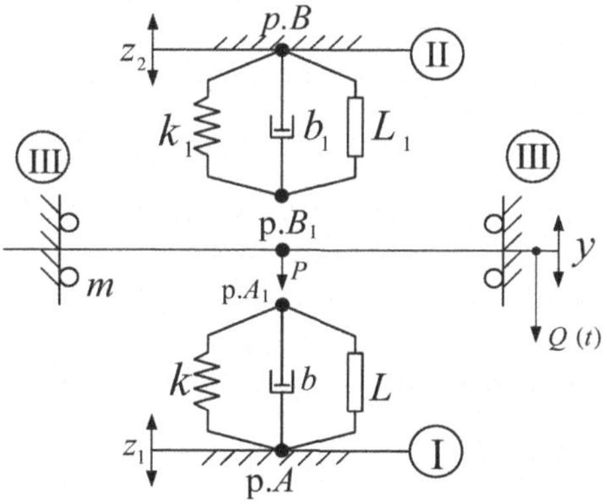

Fig. 4.11 The basic computational scheme of the vibration protection system with one degree of freedom

In the structure of the computational scheme (see Fig. 4.11), the elements have stiffness k and k_1; the interactions of the dissipative nature are reduced to equivalent viscous friction dampers with coefficients b and b_1. In addition, motion transformation devices are introduced into the system in Fig. 4.11. These motion transformation devices may have various structural and technical forms, and their properties are reflected by the parameters L and L_1, depending on the acceleration arising from the relative motion of the elements. The guides III in Fig. 4.11 are shown due to the emphasis of the fact that in the motion of an object only the displacement along the coordinate y is selected specifically and supported by certain means. In many cases it is assumed that the motion along the y coordinate is dominant, and the remaining forms of motion are considered negligibly small in this case. As external forces exerted on the object of protection, the force of weight P, the external force $Q(t)$ and the motions of the bearing surfaces I and II, denoted respectively by $z_1(t)$ and $z_2(t)$, are taken into account. The kinematic perturbations $z_2(t)$ and $z_2(t)$ can be reduced to equivalent force actions applied to the protection object m. Points A, A_1, B and B_1 denote contact points in which static and dynamic responses of constraints can be determined.

4.3.1 Description of System Features

The mathematical model of the linear system (see Fig. 4.11) can be represented (for the general case) by equations at the preliminary determinating the expressions for kinetic, potential energy, and also for the energy dissipation function:

$$T = \frac{1}{2} m(\dot{y})^2 + \frac{1}{2} L \cdot (\dot{y} - \dot{z})^2 + \frac{1}{2} L_1 \cdot (\dot{y} - \dot{z}_1)^2; \tag{4.32}$$

$$\Pi = \frac{1}{2} k \cdot (y - z)^2 + \frac{1}{2} k_1 \cdot (y - z_1)^2; \tag{4.33}$$

$$\Phi = \frac{1}{2} b \cdot (\dot{y} - \dot{z})^2 + \frac{1}{2} b_1 \cdot (\dot{y}_1 - \dot{z}_1)^2. \tag{4.34}$$

Applying the usual approaches connected with the use of Laplace transforms, we find that:

$$\begin{aligned} & [(m + L + L_1)p^2 + (b + b_1)p + k + k_1] \cdot \bar{y} \\ & = \left(mp^2 + bp + k\right) \cdot \bar{z} + \left(mp^2 + b_1 p + k_1\right) \cdot \bar{z}_1 + \bar{P} + \bar{Q}, \end{aligned} \tag{4.35}$$

where $p = j\omega$ ($j = \sqrt{-1}$) is a complex variable; $\bar{y}$, $\bar{P}$, $\bar{Q}$ are symbolic Laplace representation of functions [20].

If we assume that small motions y occur with respect to the position of static equilibrium, then under zero initial conditions and harmonic forms of external influences $\bar{Q}$, $\bar{z}_1$ and $\bar{z}_2$, the transfer functions of the system can be obtained:

$$\underset{\substack{Q=0\\ z_1=0}}{W(p)} = \frac{\bar{y}}{\bar{z}} = \frac{Lp^2 + bp + k}{(m + L + L_1)p^2 + (b + b_1)p + k_1 + k}; \quad (4.36)$$

$$\underset{\substack{Q=0\\ z=0}}{W(p)} = \frac{\bar{y}}{\bar{z}_1} = \frac{L_1p^2 + b_1p + k_1}{(m + L + L_1)p^2 + (b + b_1)p + k_1 + k}; \quad (4.37)$$

$$\underset{\substack{z_1=0\\ z=0}}{W(p)} = \frac{\bar{y}}{\bar{Q}} = \frac{1}{(m + L + L_1)p^2 + (b + b_1)p + k_1 + k}. \quad (4.38)$$

The denominator (4.36)–(4.38) is common to all transfer functions and is called the characteristic equation:

$$A_0 = (m + L + L_1)p^2 + (b + b_1)p + k_1 + k. \quad (4.38')$$

From (4.38′), the frequency of the natural oscillations can be determined, and with the help of (4.36)–(4.38) the corresponding frequency characteristics can be constructed and the dynamic properties can be evaluated through responses (or reactions) to typical external influences.

The transfer functions of the system (4.36)–(4.38) can also be determined using the structural diagram (Fig. 4.12), which is equivalent to Eq. (4.35), in the sense that the structural diagram in the symbol of the concepts of automatic control theory [21] is a graphical or structural analogue of the linear differential Eq. (4.35).

The structural diagram in Fig. 4.12 reflects the dynamic interactions that arise between the security object (mass inertia element m) and bearing surfaces whose motion is predetermined and forms kinematic perturbations. In particular, in Fig. 4.12c it is shown that the kinematic perturbations are transformed into the force impacts applied at the point 1. At the same time, in the contact of the protection object with a fixed surface III, which provides a vertical motion (in this case), interactions can also occur, determined by the corresponding parameters of the units k_2, b_2 and L_2. Such units can be shown conventionally in Fig. 4.13a (the computational scheme), and also on the structural diagram shown in Fig. 4.13b.

The comparison of the structural diagrams in Figs. 4.12 and 4.13 shows that there is a certain difference in the reflection of the properties formed by the relative and absolute motion of the units. For example, the parameters k_2, b_2, L_2 are related only to the parameters of the absolute motion of the protection object. In the expressions determining the transfer functions (4.36)–(4.38), the numerators of the

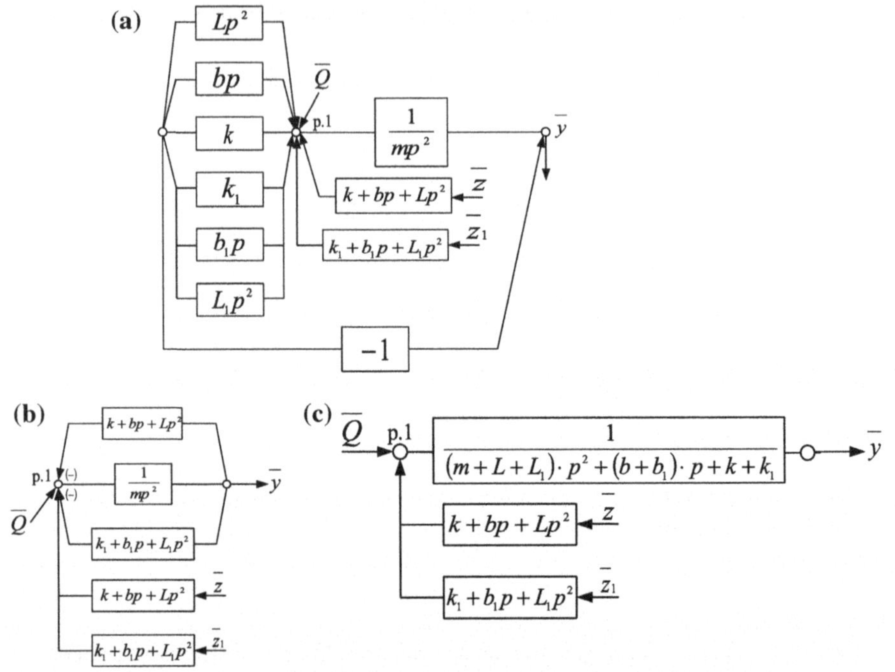

Fig. 4.12 Variants of construction of structural diagrams for the basic computational scheme in Fig. 4.11. **a** is the detailed structural diagram with the allocation of a single feedback; **b** is the structural diagram with allocation of feedbacks in the form of elastic elements with reduced stiffness and the object of protection as an integrating unit of the second order; *c* is the generalized scheme

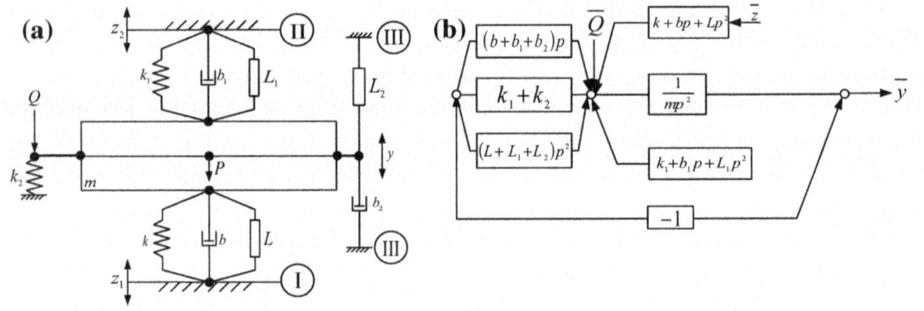

Fig. 4.13 Diagrams representing the interaction of the protection object with the bearing surface III (Fig. 4.11): **a** is the computational scheme; **b** is the structural diagram of the system

fractional-rational expressions do not change, but the characteristic Eq. (4.38′) takes the following form:

$$A_0' = (m + L + L_1 + L_2)p^2 + (b + b_1 + b_2)p + k + k_1 + k_2 = 0. \qquad (4.39)$$

In this connection, peculiar features of the estimation of the dynamic properties of the system arise under the influence of various external perturbations. It must be taken into account that the estimates made are rather arbitrary because of the idealization of the real properties of the elementary units and the assumptions made regarding the peculiar features of motion.

The structural diagrams considered in Figs. 4.12 and 4.13 are equivalent and can be obtained by conventional structural transformations. The point 1 in all versions of structural diagrams is symbolic in the sense that all forces corresponding to the concepts of the d'Alembert principle converge in it. Essentially, the sum of all the forces converging at the point 1 (see Fig. 4.12) corresponds to writing the Eq. (4.35) in symbolic form.

4.3.2 Taking into Consideration Spatial Metrics

The above remarks are related to the object of protection of a solid body in the form of a material point. However, a solid body can have finite dimensions, as shown in Fig. 4.14. Such situations arise when a solid body consists of several parts, and the problem of vibration protection implies the possibility of contact failure with the emergence of gaps and subsequent collisions of elements.

To ensure contact and avoid gaps in the systems (Fig. 4.13), preload conditions are created, for example, by pre-compression of springs or by special additional

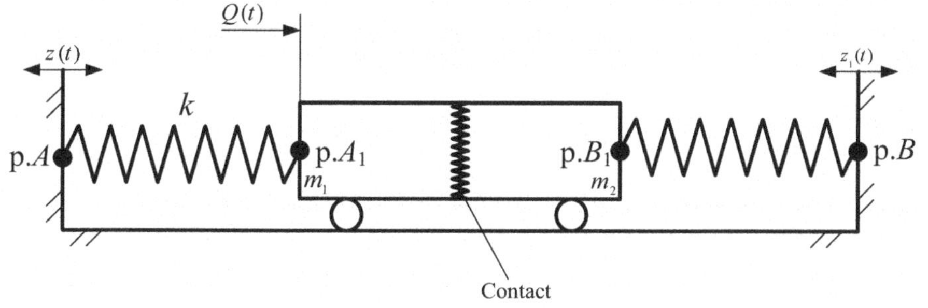

Fig. 4.14 The computational scheme of a vibration protection system with a protection object of two parts ($m = m_1 + m_2$)

influences. Some peculiar features of the occurrence of systems of this kind, having so-called unilateral or one-sided constraints.

4.3.3 Mathematical Model of the Rotational System

Finiteness of the size of the object of protection can be manifested in the calculation schemes that reflect not translational (rectilinear) oscillatory motions, but rotational oscillatory motions, where the object in the form of a material point of mass m is connected with the center of rotation by a weightless absolutely rigid rod (Fig. 4.15).

The mathematical model for the computational scheme in Fig. 4.15 can be obtained on the basis of the usual approaches under the assumption that the angular motion φ and the corresponding vertical motions along the y coordinate (in the fixed coordinate system) are small. In the computational scheme, interaction with the bearing surface I, which moves according to the known law $z(t)$, is considered. The external force is applied to the object of protection (mass m); the interaction of an element of mass m with the fixed basis (k_2, b_2, L_2) is taken into account similarly to the method considered above. The equations of motion relative to the static equilibrium position (for $z_1 = 0$) can be written in the coordinates y and φ, respectively:

$$\left[(m+L)p^2+bp+k\right]\cdot\bar{y}=\left(Lp^2+bp+k\right)\cdot\bar{z}+\overline{Q}, \tag{4.40}$$

$$\left[(J+Ll^2)p^2+bl^2p+kl^2\right]\cdot\bar{\varphi}=l^2\cdot\left(Lp^2+bp+k\right)\cdot\bar{z}+\overline{M}\cdot\bar{\varphi}. \tag{4.41}$$

where $J = ml^2$ is the moment of inertia of the object of protection relative to the fixed point O; l is the length of the rod; $y = \varphi \cdot l$; $M = \overline{Q} \cdot l$.

Comparison of (4.40) and (4.41) shows the relationship between the equation in terms of a constant coefficient l2. If we take as the basis the equation in the form (4.10), then the mathematical model of the system with the calculated circuit in

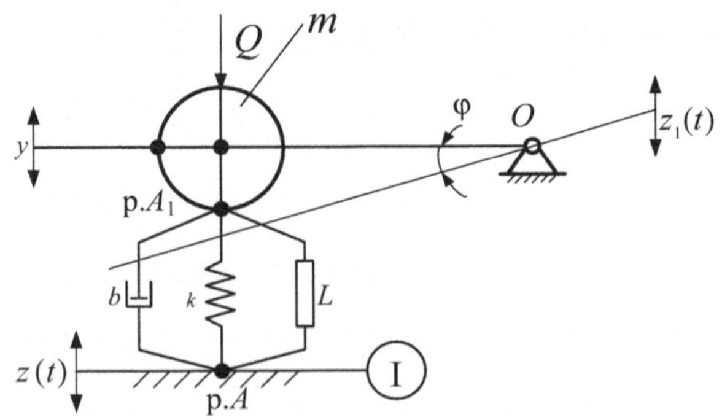

Fig. 4.15 Vibration protection system with an object of mass m, performing rotational oscillations relative to a fixed point

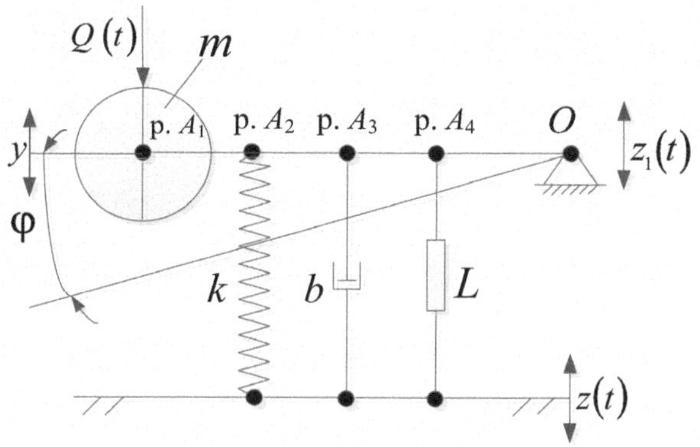

Fig. 4.16 The computational scheme of the system with separate points of attachment of elementary units

Fig. 4.15 does not fundamentally differ from Eq. (4.35). That is, in a certain sense, the calculated schemes of two types (Figs. 4.11a and 4.15) can be considered equivalent. At the same time, there are principal differences in systems: the object of protection in Fig. 4.11a is a material point (and in this case does not have a spatial "metric"). As for the rotational type system (Fig. 4.15), the protection object is represented here by a material point of mass m, connected by a weightless rigid rod of length l with a fixed center of rotation at point O. In this case, the system has a spatial metric, which predetermines the emergence of a number of features consisting in the fact that the attachment points of elementary units may not coincide with the material point m, but lie along the length of the rod.

A rod of the length l in Fig. 4.15 can be regarded, in a sense, as a certain lever of the first kind, on which the attachment points of typical elements are parallel (see Fig. 4.16). The mathematical model of the system is constructed for $Q = 0$; $z_1 = 0$; $z \neq 0$.

For further calculations, we denote that $OA_1 = l$; $OA_2 = l_1$; $OA_3 = l_2$; $OA_4 = l_3$. We write the expressions for the kinetic and potential energy, as well as for the energy dissipation function:

$$T = \frac{1}{2}m(\dot{y})^2 + \frac{1}{2}L(\dot{\varphi}l_3 - \dot{z})^2; \tag{4.42}$$

$$\Pi = \frac{1}{2}k \cdot (\varphi l - z)^2; \tag{4.43}$$

$$\Phi = \frac{1}{2}b \cdot (\varphi l - z)^2. \tag{4.44}$$

After a series of transformations, we obtain the equation of motion with the coordinate $\varphi(y = \varphi \cdot l)$:

$$\left[\left(ml^2 + L \cdot l_3^2\right)p^2 + \left(bl_2^2\right)p + kl_1^2\right] \cdot \bar{\varphi} = \left(Ll_3^2 p^2 + bl_2^2 p + kl_1^2\right) \cdot \bar{z}. \tag{4.45}$$

The transfer function of the system in accordance with (4.45) takes the following form:

$$\underset{\substack{Q=0\\z_1=0}}{W(p)} = \frac{\bar{\varphi}}{\bar{z}} = \frac{Ll_3 p^2 + bl_2 p + kl_1}{\left(ml^2 + Ll_3^2\right)p^2 + bl_2^2 p + kl_1^2}. \tag{4.46}$$

From the expression (4.46), using the relation $y = \varphi \cdot l$, we can pass to the transfer function:

$$\underset{\substack{Q=0\\z_1=0}}{W'(p)} = \frac{\bar{y}}{\bar{z}} = \frac{Li_3 p^2 + bi_2 p + ki_1}{\left(m + Li_3^2\right)p^2 + bi_2^2 p + ki_1^2}, \tag{4.47}$$

where $i_3 = \frac{l_3}{l}$; $i_2 = \frac{l_2}{l}$; $i_1 = \frac{l_1}{l}$ are gear ratios of the lever linkages.

It should be noted that in this case the gear ratios have a positive sign, since in the first-order levers there is no change in directions during the transfer of motion. If we assume that $z_1 \neq 0$ and $z = 0$ and $Q = 0$, then the expressions (4.42)–(4.44) take the corresponding form:

$$T = \frac{1}{2}m(\varphi \cdot l + \dot{z}_1)^2 + \frac{1}{2}L(\varphi \cdot l_3); \tag{4.48}$$

$$\Phi = \frac{1}{2}b(\varphi \cdot l_2)^2; \tag{4.49}$$

$$\Pi = \frac{1}{2}k \cdot (\dot{\varphi} l_1)^2. \tag{4.50}$$

The equation of motion of the system is as follows:

$$\left[\left(ml^2 + Ll_3^2\right)p^2 + b \cdot l_2^2 + k \cdot l_1^2\right] \cdot \bar{\varphi} = \left[ml \cdot p^2 + bl_2 p + kl_1\right] \cdot \bar{z}_1. \tag{4.51}$$

The transfer function of the system for $z_1 \neq 0$ ($z = 0$; $Q = 0$) can be determined from (4.51):

$$W(p) = \frac{\bar{y}}{\bar{z}_1} = \frac{mp^2}{\left(m + Li_3^2\right)p^2 + bi_2^2 p + ki_1^2}. \tag{4.52}$$

It follows from (4.52) that the vibrations of the point of rotation O change the transfer function.

We write (4.51) not in the coordinates of φ, but in the coordinates of y, assuming that $y = \varphi l + z_1$. In this case, (4.48)–(4.50) take the following form:

$$T = \frac{1}{2}m(\dot{y})^2 + \frac{1}{2}L\left(\frac{\dot{y} - \dot{z}_1}{l}\right)^2 \cdot l_0^2; \tag{4.53}$$

$$\Phi = \frac{1}{2}b \cdot \left(\frac{\dot{y} - \dot{z}_1}{l}\right)^2 \cdot l_2^2; \tag{4.54}$$

$$\Pi = \frac{1}{2}k \cdot \left(\frac{\dot{y} - \dot{z}_1}{l}\right)^2 \cdot l_1^2. \tag{4.55}$$

Then Eq. (4.51) will be as follows:

$$\bar{y} \cdot \left[\left(m + L\frac{l_3}{l^2}\right)p^2 + p\left(b\frac{l_2^2}{l^2}\right) + k\left(\frac{l_1^2}{l^2}\right)\right] = \bar{z}_1 \cdot \left[\left(L\frac{l_3}{l^2}\right)p^2 + p\left(b\frac{l_2^2}{l^2}\right) + k\left(\frac{l_1^2}{l^2}\right)\right]. \tag{4.56}$$

The transfer function of the system in the coordinate of y will have a different form than that presented in (4.52):

$$W_y(p) = \frac{\bar{y}}{\bar{z}_1} = \frac{Li_3^2p^2 + b_2i_2^2p + ki_1^2}{(m + Li_3^2)p^2 + bi_2^2p + ki_1^2}. \tag{4.57}$$

Let us note that the denominators (4.52) and (4.57) coincide, but the numerators of the transfer functions will be different. A more general case is the simultaneous action of external perturbations, when $z_1 \neq 0$, $z \neq 0$ ($Q = 0$). In this case, two external actions can result in a single input signal. It is assumed that $z = z_1$, although there may be situations when there are certain relations between z and z_1. The features of such interactions can be attributed to the known forms of control in automatic control systems for force effects.

We write expressions (4.48)–(4.50), assuming that $z \neq 0$; $z_1 \neq 0 (Q = 0)$, and we obtain

$$T = \frac{1}{2}m \cdot (\varphi \cdot l + \dot{z}_1)^2 + \frac{1}{2}L \cdot (\dot{\varphi}l_3 - \dot{z})^2 \tag{4.58}$$

$$\Phi = \frac{1}{2}b \cdot (\dot{\varphi}l_2 - \dot{z})^2; \tag{4.59}$$

$$\Pi = \frac{1}{2}k \cdot (\dot{\varphi} \cdot l - \dot{z})^2. \tag{4.60}$$

The equation of motion takes the form

$$\left[(ml^2 + Ll_3^2)p^2 + bl_2^2 p + kl_1^2\right] \cdot \bar{\varphi} - mlp^2 \cdot \bar{z}_1 + \left[Ll_3 p^2 \bar{z} + bl_2 p + kl_1\right] \cdot \bar{z}. \quad (4.61)$$

In accordance with Eq. (4.61), the system is under the influence of two external factors: and. To build the response of the system to typical external input effects, the superposition principle can be used. If there is reason to believe that, where α can be determined from the condition $-\infty < \alpha < \infty$ and is a real number, then (4.61) can be reduced to the form:

$$\underset{\bar{z}=\alpha\bar{z}_1}{W(p)} = \frac{\bar{\varphi}}{\bar{z}_1} = \frac{(\alpha Ll_3 - ml)p^2 + \alpha bl_2 p + \alpha kl_1}{(ml^2 + Ll_3^2)p^2 + bl_2^2 p + kl_1^2}. \quad (4.62)$$

Note that the introduction of a coupling between $\bar{z}$ and $\bar{z}_1$ affects only the numerator of the transfer function (4.61), but not the denominator. Such features of dynamic state control in the theory of automatic control are associated with the perturbation-stimulated control principle.

Similarly, (4.51) can be transformed into an equation using the coordinate y. In this case, the transfer function will also differ from (4.52).

4.3.4 The Variant of a Force Perturbation

When taking into account the action of the force perturbation, when $Q \neq 0$, and $z_1 = 0$, $z = 0$, we obtain

$$\underset{\substack{\bar{z}=0\\ \bar{z}_1=0}}{W_y(p)} = \frac{\bar{y}}{\bar{Q}} = \frac{1}{(m + Li_3^2)p^2 + bi_2^2 p + ki_1^2}. \quad (4.63)$$

In the expressions (4.47), (4.57) and (4.63), the denominators of the transfer functions are the same, however, the numerators depend on the type and nature of the action of the external forces. Note that in systems with rotational motion external influences, for example, external force factors, can have application points not only on the object of protection, but also on a weightless rod. With that, transfer functions will change in a certain way, which lays the groundwork for the appropriate adjustment of vibration protection means.

We also note that the rotational system shown in Fig. 4.15 can have a weightless rod that has a point of rotation. This point of rotation creates lever linkages that are characteristic for levers of the second kind. Let us consider the computational scheme of a vibration protection system (Fig. 4.17) of a rotary type with the point of rotation ensuring the effect of different signs.

1. We assume that $Q = 0$, $z_1 = 0$, and $z \neq 0$ и $z_2 \neq 0$. Let the motion of the protection object be determined by the coordinate φ. Then the expressions for

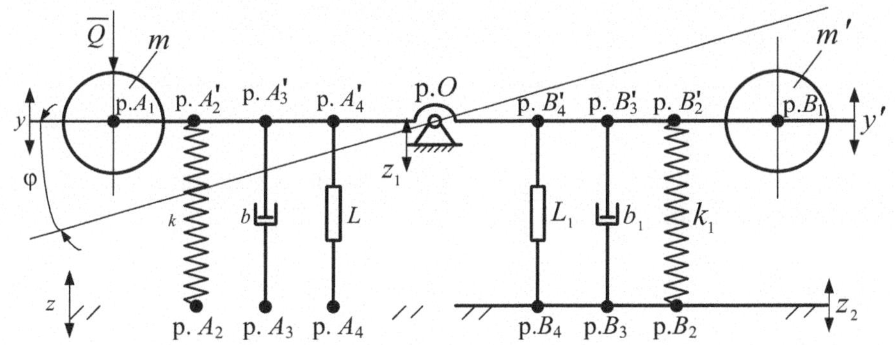

Fig. 4.17 The computational scheme of a vibration protection system with separation of points of attachment of standard elements on a weightless rod with an intermediate center of rotation

the kinetic and potential energy, as well as for the dissipation energy, have the following form:

$$T = \frac{1}{2}m\cdot(\dot{\varphi}l)^2 + \frac{1}{2}L(\dot{\varphi}l_3 - \dot{z})^2 + \frac{1}{2}m'\cdot(\dot{\varphi}l')^2 + \frac{1}{2}L_1\cdot[-\dot{\varphi}(l_3') - \dot{z}_2]^2; \tag{4.64}$$

$$\Phi = \frac{1}{2}b_1\left(-\dot{\varphi}l_2' - \dot{z}_2\right)^2 + \frac{1}{2}b(\dot{\varphi}l_2 - \dot{z}_2)^2; \tag{4.65}$$

$$\Pi = \frac{1}{2}k(\dot{\varphi}l_1 - \dot{z})^2 + \frac{1}{2}k_1\left(-\dot{\varphi}l_1' - \dot{z}_2\right)^2, \tag{4.66}$$

where $OB_1 = l'$; $OB_2' = l_1'$; $OB_3' = l_2'$; $OB_4' = l_3'$ are the parameters of the attachment points (see Fig. 4.17) of the elements m', L_1, b_1 and k_1, respectively.

The equation of motion of the system in Fig. 4.17 for the case Q = 0, z1 = 0, z ≠ 0, z2 ≠ 0 has the form

$$\begin{aligned}\bar{\varphi}\cdot&\left[p^2\left(ml^2 + m'(l')^2 + Ll_3^2 + L_1\cdot(l_3')^2\right) + p\left(bl_2^2 + b_1(l_2')^2\right) + k(l_1^2) + k_1(l_1')^2\right] = \\ &= \bar{z}\cdot\left[Ll_3p^2 + pbl_2 + kl_1\right] + \bar{z}_2\cdot\left(-L_1l_3'p^2 - pb_1l_2' - k_1l_1'\right).\end{aligned} \tag{4.67}$$

If we assume that $\bar{z} = \bar{z}_2$, then (4.67) is transformed to the form

$$\bar{\varphi}\cdot\left[p^2\left(ml^2+m'(l')^2+Ll_3^2+L_1\cdot(l_3')^2\right)+p\left(bl_2^2+b_1(l_2')^2\right)+k(l_1^2)+k_1(l_1')^2\right]$$
$$=\bar{z}\cdot\left\{p^2\cdot\left[Ll_3p^2-L_1l_3'\right]+p\cdot\left(bl_2-b_1l_2'\right)+kl_1-k_1l_1'\right\}. \tag{4.68}$$

The transfer function of the system in the coordinates φ and under external perturbation takes the form

$$W'_{\varphi}(p)=\frac{\bar{\varphi}}{\bar{z}}$$
$$=\frac{p^2\cdot\left[Ll_3-L(l_3')\right]+p\cdot\left(bl_2-b_1l_2'\right)+kl_1-k_1l_1'}{p^2\cdot\left[ml^2+m'(l')^2+Ll_3^2+L_1(l_3')^2\right]+p\left[bl_2^2+b_1(l_2')^2\right]+kl_1^2+k_1(l_1')^2} \tag{4.69}$$

The expression (4.69) can be reduced to the coordinate $\bar{y}$, then

$$W'_y(p)=\frac{\bar{y}}{\bar{z}}$$
$$=\frac{p^2\cdot\left[Li_3-L(i_3')\right]+p\cdot\left(bi_2-b_1i_2'\right)+ki_1-k_1i_1'}{p^2\cdot\left[m+m'(i')^2+Li_3^2+L_1(i_3')^2\right]+p\left[bi_2^2-b_1(i_2')^2\right]+ki_1^2+k_1(i_1')^2}, \tag{4.70}$$

where $i'=\frac{l'}{l}$; $i_1'=\frac{l_1'}{l}$; $i_2'=\frac{l_2'}{l}$; $i_3'=\frac{l_3'}{l}$.

Expressions (4.69) and (4.70) for transfer functions differ from each other by the constant factor l, since $\varphi = y/l$. However, this situation changes if $z_1 \neq 0$. In this case

$$\varphi=\frac{y-z_1}{l}. \tag{4.71}$$

Expressions (4.64)–(4.66), respectively, take the form

$$T=\frac{1}{2}m(\dot{\varphi}l+\dot{z}_1)^2+\frac{1}{2}m'\cdot(-\dot{\varphi}l'+\dot{z}_1)^2+\frac{1}{2}(\dot{\varphi}l_3+\dot{z}_1-\dot{z})^2$$
$$+\frac{1}{2}L_1\cdot(-\dot{\varphi}l_3'+\dot{z}_1-\dot{z}_2)^2; \tag{4.72}$$

$$\Phi=\frac{1}{2}b\cdot(\dot{\varphi}l_2+\dot{z}_1-\dot{z})^2+\frac{1}{2}b_1\cdot(-\dot{\varphi}l_2'+\dot{z}_1-\dot{z}_2)^2; \tag{4.73}$$

$$\Pi=\frac{1}{2}k(\varphi l+z_1-z)^2+\frac{1}{2}k_1(-\varphi l'+z_1-z_2)^2. \tag{4.74}$$

The equation of motion of the system in coordinates and taking into account the compatible input action ($z = 0$; $z_1 \neq 0$; $z_2 \neq 0$) has the following form:

$$\begin{aligned}
&\bar{\varphi} \cdot \left\{p^2 \cdot \left[ml^2 + m'(l')^2 + Ll_3^2 + L_1\left(l_3'\right)^2\right] + p \cdot \left[bl_2^2 + b_1\left(l_2'\right)^2\right] + kl_1^2 + k_1\left(l_1'\right)^2\right\} \\
&= \bar{z}_1 \cdot \left[p^2 \cdot \left(-ml + m'l' - Ll_3 + L_1 l_3'\right) + p \cdot \left(-bl_2 + b_1 l_2'\right) - kl_1 + kl_1'\right] \\
&\quad + \bar{z} \cdot \left[p^2 Ll_3 + pbl_2 + kl_1\right] + \bar{z}_2 \cdot \left[-L_1\left(l_3'\right)^2 p^2 - p \cdot \left(b_1 l_2'\right) - k_1 l_1'\right].
\end{aligned} \tag{4.75}$$

The response to the total effect of kinematic perturbations is possible on the basis of the application of the superposition principle. Suppose that $z = z_1 = z_2$, then, on the basis of (4.75), the transfer function of the system takes the form

$$\underset{\substack{z=z_1=z_2 \\ Q=0}}{W_{\varphi}(p)} = \frac{\bar{\varphi}}{\bar{z}} = \frac{p^2 \cdot (m'l' - ml)}{p^2 \cdot \left[ml^2 + m'(l')^2 + Ll_3^2 + L_1(l_3')^2\right] + p \cdot \left[bl_2^2 + b_1(l_2')^2\right] - kl_1^2 + k_1(l_3')^2}. \tag{4.76}$$

In the simplified form (4.76) it will look like this

$$\underset{\substack{z=z_1=z_2 \\ Q=0}}{W(p)} = \frac{\bar{\varphi}}{\bar{z}} = \frac{p^2 \cdot (m'l_1' - ml)}{p^2 \cdot \left[ml^2 + m'(l')^2 + Ll_3^2 + L(l_3')^2\right] + p \cdot \left[bl_2^2 + b_1(l_2')^2 + kl_1^2 + k_1(l_1')^2\right]}. \tag{4.77}$$

In the y coordinate system, the expressions (4.72)–(4.74) will have the form

$$\begin{aligned}
T = &\frac{1}{2} m \cdot (\dot{y})^2 + m' \cdot [(-\frac{\dot{y} + \dot{z}_1}{l}) \cdot l' + z_1]^2 + \frac{1}{2} L \cdot [\frac{(\dot{y} - \dot{z}_1)}{l} l_3 + \dot{z}_1 + \dot{z}]^2 \\
&+ \frac{1}{2} L_1 \cdot [-\frac{(\dot{y} - \dot{z}_1)}{l} l_3' + \dot{z}_1 - \dot{z}_2]^2.
\end{aligned} \tag{4.78}$$

$$\Phi = \frac{1}{2} b \cdot [-\frac{(\dot{y} - \dot{z}_1)}{l} l_2 + \dot{z}_1 - \dot{z}]^2 + \frac{1}{2} b_1 \cdot [-\frac{(\dot{y} - \dot{z}_1)}{l} \cdot l_2' + \dot{z}_1 - \dot{z}_2]^2. \tag{4.79}$$

$$\Pi = \frac{1}{2} k \cdot [\frac{(y - z_1)}{l} \cdot l_1 + z_1 - z]^2 + \frac{1}{2} k_1 [-\frac{(y - z)}{l} \cdot l_1' + z_1 - z_2]^2. \tag{4.80}$$

We write the equations of the system using the coordinate y:

$$\begin{aligned} &\bar{y} \cdot \left\{p^2 \cdot \left[m + m'(i')^2 + Li_3^2 + L_1\left(i_3'\right)^2\right] + p\left[bi_2^2 + b_1\left(i_2'\right)^2\right] + ki_1^2 + k_1\left(i_1'\right)^2\right\} \\ &= \bar{z}_1 \cdot \left\{ \begin{array}{c} \left[-m' \cdot i' \cdot (1 - i') - L_3 i_3 \cdot (1 - i_3) + L_1 i_3' \cdot (1 + i_3')\right] \cdot p^2 \\ +p \cdot \left[-bi_2 \cdot (1 - i_2) + b_1 \cdot i_2' \cdot (1 + i_2')\right] + k_1 i_1' \cdot (1 + i_1') - ki_1(1 - i_1) \end{array} \right\} \\ &+ \bar{z} \cdot \left[Li_3 p^2 + bi_2 + ki_1\right] + \bar{z}_2 \cdot \left[-L_1 i_3 p^2 - b_1 \cdot i_2' p - k_1 \cdot i_1\right]. \end{aligned} \tag{4.81}$$

Using (4.81), we obtain an expression for the transfer function, assuming that z = z1 = z2 (Q = 0):

$$\begin{aligned} \underset{\substack{z=z_1=z_2 \\ Q=0}}{W(p)} = \frac{\bar{y}}{\bar{z}} = &\frac{-m'i' \cdot (1-i') \cdot p^2 + p^2 \cdot \left[L_1 i_3' \cdot \left(1 + i_3'\right) - Li_3 \cdot (1 - i_3) + Li_3 - L_1 i_3\right] +}{\left[m + m'(i')^2 + Li_3^2 + L_1\left(i_3'\right)^2\right]p^2 + p \cdot \left(bi_2^2 + b_1\left(i_2'\right)^2\right) + ki_1^2 + k\left(i_1'\right)^2} \\ &\frac{+p \cdot \left(-bi_2 \cdot (1 - i_2) + b_1 i_2' \cdot \left(1 + i_2'\right) + bi_2 - b_1 i_2'\right) + k_1 i_1' \cdot \left(1 + i_1'\right) - ki_1 \cdot (1 - i_1) + ki_1 - k_1' i'}{+p \cdot \left(bi_2^2 + b_1\left(i_2'\right)^2\right) + ki_1^2 + k\left(i_1'\right)^2} \end{aligned} \tag{4.82}$$

Let us make some simplifications and we will obtain:

$$\underset{\substack{z=z_1=z_2 \\ Q=0}}{W(p)} = \frac{\bar{y}}{\bar{z}} = \frac{\left\{\left[-m'i' \cdot (1 - i') \cdot p^2 + L_1 i_3' + Li_3\right] \cdot p^2 + \right.}{\left\{\left[m + m'(i')^2 + Li_3^2 + L_1\left(i_3'\right)^2\right]p^2 + \right.} \cdots \frac{\left. +p \cdot \left[b_1 i_2^2 + b_1\left(i_2'\right)^2\right] + ki_1^2 + k_1\left(i_1'\right)^2\right\}}{\left. +p \cdot \left(bi_2^2 + b_1\left(i_2'\right)^2\right) + ki_1^2 + k\left(i_1'\right)^2\right\}}. \tag{4.83}$$

4.3.5 *Comparative Analysis of the Possibilities of Vibration Protection Systems of Two Types*

Basic models of vibration protection systems represented in the form of mechanical vibrational structures with one degree of freedom in Figs. 4.11 and 4.17 are the most common.

The following characteristics are common to the models of translational and rotational type:

1. The minimum complexity configuration consists of a mass-and-inertia and elastic elements.
2. The object of protection in the form of a solid body with translational motion in an idealized representation is regarded as a material point.
3. The elastic element connects a material point (or protection object) with a spring, which, in turn, is connected to the bearing surfaces.

4. There can be several bearing surfaces. When considering the straight-line movements, stiffnesses of the springs are grouped according to the rules of parallel addition.
5. When an object is subjected to several elastic elements at different angles, the total spring stiffness can be considered as a reduced characteristic. In this case, the number of elements and the grade angles of the lines of the elastic forces are taken into account.
6. Connections of elastic elements with the object of protection and bearing surfaces are considered as holding or bilateral constraints.
7. The mass-and-inertia element (object of protection) can be considered in the mode of interaction with a fixed basis (surface III in Fig. 4.11). Such interaction reflects the concept of the object of protection as a unit, making forward motion in contact with a stationary bar of the so-called initial mechanism. In this case, the kinematic pair of class V (translational) is considered. This is due to the fact that the object of protection, being in the general case a solid body (with the corresponding level of abstract representation), implements only motion with respect to one coordinate.
8. External forces can have the form of force factors. In this case, they are applied directly to the protection object. When the kinematic perturbation is known, the motion of the bearing surface is known. The kinematic perturbation can be considered as a separate effect or "reduced" to an equivalent force [1].
9. To describe the motion of a vibration protection system, a coordinate system associated with a fixed basis can be used; coordinate system in relative motion. The bearing surface, as already noted, can have a known law of motion. Then the relative coordinate is the sum of the motion of the base and the object of protection.
10. The mathematical model of the basic vibration protection system may include dissipative elements. Then it is necessary to take into account the features of their location. In the general case, dissipative elements (we mean the elements of viscous friction or damper) can be introduced as a parallel complement to elastic elements using an extended set of type elements, which is reflected in the structural theory of vibration protection systems; the introduction and accounting of additional units is carried out in a similar way.

Some generalizations of the concepts of basic models. In the framework of the traditional approach to mathematical models of vibration protection systems of progressive motion with one degree of freedom, the systems of interaction of elements in the structure do not assume the existence of mechanisms that transform the motion. Although the object of protection, as already mentioned, can be regarded as the initial unit of the mechanism. It must be pointed out that the notions of mechanical systems will change if the units of the mechanism as a mechanical circuit include not only solid, but also elastic bodies. We also note that the

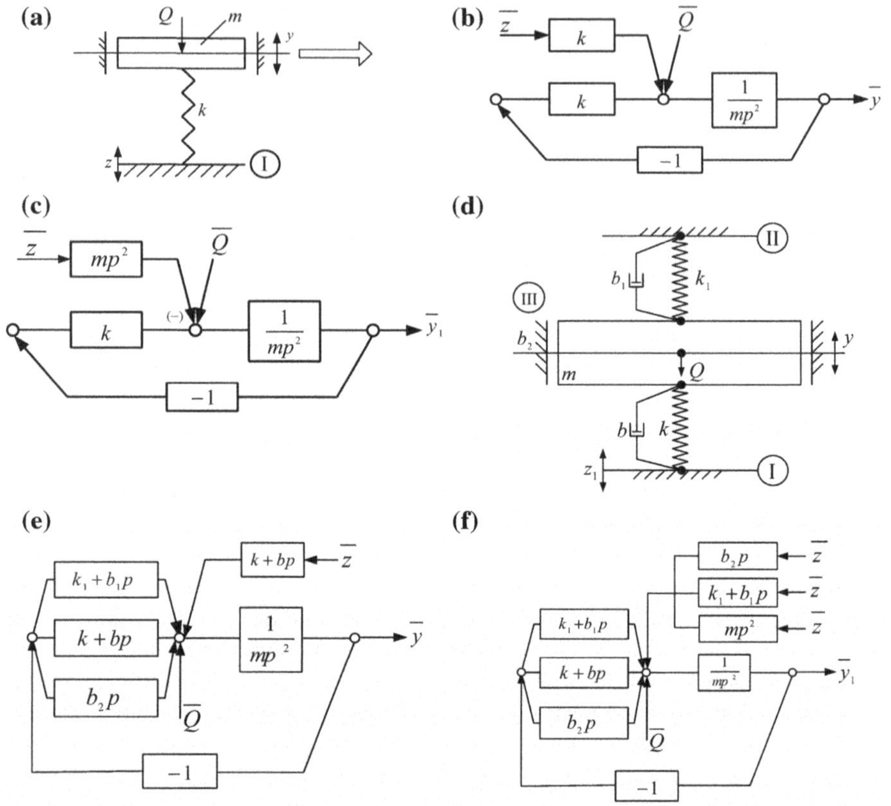

Fig. 4.18 Variants of complicating the basic model of the vibration protection system with one degree of freedom. **a** basic computational scheme; **b** structural diagram of the basic system (fixed coordinate system); **c** structural diagram of the system in relative coordinates ($y_1 = y - z$); **d** the computational scheme of the basic system with additional elastic and dissipative units; **d** structural diagram of the system in absolute coordinates (y); **e** structural diagram of the system in relative coordinates ($y_1 = y - z$)

considered elements of models of vibration protection systems are abstract representations of real objects possessing integral properties, therefore the separation into elementary standard units is conditional.

In Fig. 4.18, there are variants of complicating the basic computational scheme of a vibration protection system of a translational type with one degree of freedom.

Analysis of the structural diagrams (see Figs. 4.18b, c, and also Figs. 4.18d, e) shows that taking into account the features of the dynamic interactions in the vibrational system is associated with the occurrence among the traditional elastic and dissipative elements of links with the transfer function of the differential unit of the second order. This unit, like the spring (k) and the damper (bp), has an input signal in the form of an offset, and an output signal in the form of a force. Thus, the structural model of the basic computational scheme has three typical elements (k, bp

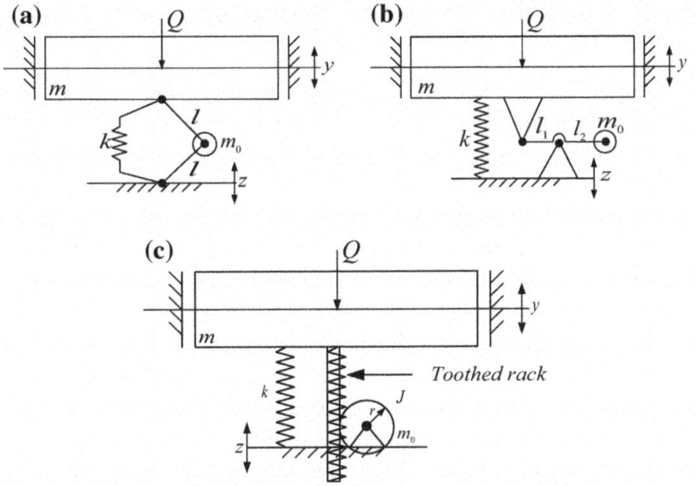

Fig. 4.19 Computational schemes of the basic vibration protection system with additional constraints (or standard elements): **a** articulation linkage; **b** the lever of the first kind; **c** gearing

and mp^2) having a displacement as the input signal, and an effort as the output. In essence, the above-mentioned elements are peculiar kind of "springs"; it is possible to correlate such forms of generality of dynamic properties with the properties of generalized springs. Similar justifications are associated with the methodological basis of the structural theory of vibration protection systems [5].

The object of protection, as follows from structural models, is represented by an integrating unit of the second order. An effort is applied to the input of this unit, and the output signal in this case is a displacement. Such a structural interpretation and functional differences between the typical elements of the structural scheme of the initial mechanical oscillatory system are explained by the features of the analytical tools of the automatic control theory. As for the properties of mass-and-inertial elements, they can take two forms: when the mass-and-inertia element acts as an object of protection (or control), or when it operates as a generalized spring [18].

The search for ways of expanding the set of typical elements leads to the introduction of additional constraints implemented by mechanisms, which is shown in Fig. 4.19.

The curves in Fig. 4.19 Computational schemes were considered in works [1, 5], in which it was shown that, in the mechanized form, the properties of additional constraints (or mechanisms) are interpreted by a standard unit implementing the functions of a differentiating unit of the second order. Further complication of additional constraints introduced into the basic scheme gives structures that have transfer functions in the form of fractional-rational expressions, which allows, even in simple models, displaying complex dynamic interactions of real technical objects with a spatial metric.

4.3.6 Basic Vibration Protection System of Rotary Type

If in the system of translational motion of the object of protection the mass-and-inertia element can be represented as a material point, then in the system of rotational motion the object of protection possesses a moment of inertia, i.e. has a "spatial" metric.

For a solid body that performs motions in a plane, one can distinguish (in contrast to a system of translational type) two characteristic points. Firstly, there is a fixed point (see Fig. 4.15), around which the body oscillates. Secondly, the solid body also has a barycentre. These points do not always coincide, which forms certain features of the motion. The idealized motion of a solid body with a fixed point of rotation can be represented as a material point attached to one end of a weightless rigid rod. The second end of this rod (see Fig. 4.15) is attached at a fixed point. Such a coupling forms a kinematic rotational pair of class V, whereas in the case of translational motion of this kind, the coupling between the object and the column is in the form of a translational kinematic pair of class V. The motion of the system can be described using the angular coordinate φ, and also with the help of the coordinates of the rectilinear vertical motion y and y_1, as was discussed earlier.

In Fig. 4.17, it is possible to isolate the bearing surfaces with which elastic or other standard elements contact. A peculiar feature of the rotational type of motion is the capability of forming an additional kinematic perturbation through the motion of the point of rotation of a solid body. Since a characteristic feature of the system is the presence of a weightless rigid rod or a lever having spatial dimensions, in such a system, the coordinates of the points of attachment of standard elements acquire a significant value. Such features form dynamic interactions that depend on geometric parameters. An important role is played by the form of the lever, which predetermines the location of the elements not only from one side, but also from both sides of the point of rotation, which provides for the possibility of mutual compensation when summing the force factors. The transmission functions of systems in angular and linear coordinates will also be different.

The most important circumstance is that a rotary-type vibration protection system (or mechanical oscillatory system), as a characteristic feature, implies the existence of a lever or a lever mechanism. Taking into consideration such factors does not deny the analogy between the two types of vibration protection systems, but focuses attention on the existence of peculiar aspects. In any case, a complete analogy should assume the possibility of manifesting such effects in systems of translational type. In addition, the difference between rotational type systems is in taking into account the relative location of the barycentre and the center of rotation. This results in the fact that a solid body in a plane motion relative to a fixed point can be represented as two material points located on a weightless rigid rod with the center of rotation between the selected material points.

Thus, the problems of the reducibility of systems of one kind to another (translational motion to rotational, and vice versa) require attention to the features of systems, the manifestation of which is associated with the manifestations of the lever linkages inherent in this type of motions.

4.4 Determination of Parameters and Characteristics of the State of the Object of Protection with Support of Rotation

In Sect. 4.3, when considering the basic computational scheme of the vibration protection system, it was assumed that the barycentre and the center of rotation coincided, which made it possible to obtain the necessary transfer functions both in force and in kinematic perturbations. At the same time, the barycentre and the center of rotation of the solid body may not coincide, which in general results in the necessity to take into account a number of peculiaritites in the dynamic interactions of the elements of the system.

4.4.1 Motion of the Object of Protection According to the Scheme of the Lever of the Second Kind

Consider a system consisting of a lever mechanism of the second kind, as shown in Fig. 4.20.

Let us write the expressions for the kinetic and potential energy when moving relative to the static equilibrium position, assuming that the body performs rotational oscillations with respect to the point O. ($l_0 = OO_1$), $Q = Q_0 \sin\omega t$;

$$T = \frac{1}{2}\left(J + Ml_0^2\right) \cdot (\dot{\varphi})^2; \tag{4.84}$$

$$\Pi = \frac{1}{2}k_1 \cdot (\varphi l_1)^2 + \frac{1}{2}k_2 \cdot (\varphi l_2)^2. \tag{4.85}$$

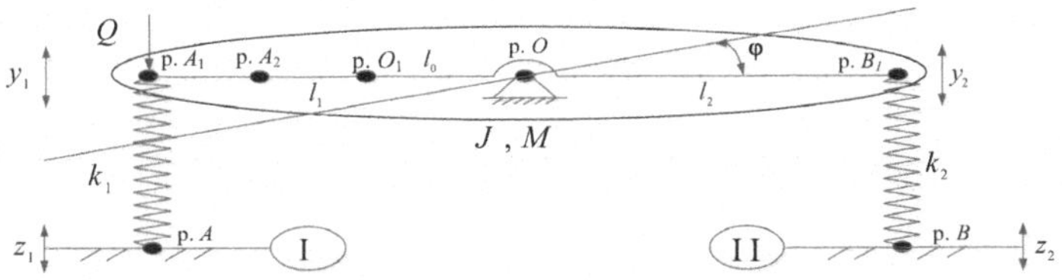

Fig. 4.20 The schematic diagram of a rotational type system with one degree of freedom

The equation of motion of the system with respect to the coordinate φ has the form

$$\left(J + Ml_0^2\right) \cdot \ddot{\varphi} + \left(k_1 l_1^2 + k_2 l_2^2\right) \cdot \varphi = l_1 Q_0 \sin \omega t. \tag{4.86}$$

Using Laplace transforms, and with zero initial conditions, we find the transfer function of the system with an "input signal" $M_{dv} = l_0 Q_0 \sin \omega t$ and an "output signal" φ:

$$W(p) = \frac{\bar{\varphi}}{\bar{M}_{dv}} = \frac{l_1^2}{\left[(J + Ml_0^2)p^2 + k_1 l_1^2 + k_2 l_2^2\right]}. \tag{4.87}$$

For fixed bases I and II ($z_1 = 0$; $z_2 = 0$), we can write that $y_1 = \varphi \cdot l_1; y_2 = \varphi \cdot l_2$, then from (4.87) one can find:

$$W_1(p) = \frac{\bar{y}_1}{\bar{Q}} = \frac{l_1^2}{\left[(J + Ml_0^2)p^2 + k_1 l_1^2 + k_2 l_2^2\right]}. \tag{4.88}$$

Assuming that the lever is a weightless rigid rod, we obtain

$$W_1'(p) = \frac{\bar{y}_1}{\bar{Q}} = \frac{1}{k_1 + k_2 i^2}, \tag{4.89}$$

where $i = \frac{l_2}{l_1}$ is the transfer ratios of the lever of the second kind.

With respect to the concentrated force Q, the lever system under the imposed restraints can be regarded as two parallel springs k_1 and $k_2 i^2$. In this case, it is assumed that the force Q is applied at point A_1 (see Fig. 4.20). If the force is applied at another point – A_2 (see Fig. 4.20), where $l_{00} = OA_2$, then the transfer function (4.88) takes the form

$$W_1''(p) = \frac{\bar{y}_1}{\bar{Q}_{p.A_2}} = \frac{l_1 \cdot l_{00}}{(J + Ml_0^2)p^2 + k_1 l_1^2 + k_2 l_2^2}. \tag{4.90}$$

Under the same restraints as in the original case, we obtain

$$W_1'''(p) = \frac{\bar{y}_1}{\bar{Q}_{p.A_2}} = \frac{i_0}{k_1^2 + k_2 i^2}, \tag{4.91}$$

where $i_0 = \frac{l_{00}}{l_1}$ characterizes the place of application of the concentrated force Q.

Is should be noted that the reduced stiffness of the system, determined by the relation and the displacement, can be found by inversion of expression (4.91), then:

$$k_{red\,A_2} = \frac{k_1}{i_0} + \frac{k_2 i^2}{i_0}. \tag{4.92}$$

It follows from (4.92) that the smaller $i_{0,}$ the greater the reduced stiffness, and $i_0 = 0$, $k_{red\,A_2} \rightarrow \infty$. Physically, such a situation is quite understandable. We also note that the reduced stiffness from expressions (4.88) and (4.90) can be obtained under the condition $p = 0$ ($p = j\omega$ is a complex variable). In physical terms, this corresponds to the transformation of the periodic force Q into a constant.

Thus, the reduced stiffness, determined from (4.88) and (4.90) under the condition that $p = 0$, makes it possible to evaluate the elastic properties of the system under static force. It is necessary to note one more characteristic circumstance: the expression (4.88) is used to determine the reduced stiffness in the case of applying the force Q at point A_1. If the force Q is applied at point A_2. In the first case, $i_0 = 1$, in the second case, $0 \leq i_0 \leq 1$. If i0 becomes such that $i_0 > 1$, then the reduced stiffness $k_{red\,A_2}$ determined by the expression (4.92) will decrease. Physically, this is quite understandable, since the moment of the force will increase, and consequently the displacement of the point A_1 also increases at the same time. Note that in the case under consideration an attempt is made to separate the point of application of the concentrated force and the point to which the elastic element joins.

That is, for $i_0 = 1$, the stiffness determines the deformation of an elastic element with stiffness k_1. If $i_0 < 1$, then the force Q is applied to the point A_2, and the deformation of the spring is determined by the motion of p. A_1 (or the y coordinate). For $i_0 > 1$, the point A_2 of the force application moves beyond the length of the lever arm l1 (see Fig. 4.20), although the deformation of the elastic element in A_1 is taken into account. Similar features were considered in Sect. 4.3 when assessing the possibility of distributing the attachment points of typical elements (springs, dampers, etc.) over the shoulders of levers of the first and second kinds.

4.4.2 *Static Reactions in a System with Objects in Angular Vibrations*

Let us consider a solid body (see Fig. 4.20), which has a fixed point of rotation and is supported by the elastic constraints k_1 and k_2. If we assume that the gravitational force is at the point of rotation O, then in order to maintain the equilibrium state in the absence of an external perturbation Q, it is necessary to consider two variants of structural design (Fig. 4.21).

We assume that for $l_0 = 0$, the springs k_1 and k_2 are not deformed, and the lever position is parallel to the surfaces *I* and *II*. In the initial position, the static reactions R_A, R_{A_1} and R_B, R_{B_1} will be equal to zero: the reaction at the point $O - R_0 = Mg$. If the lever is deflected (see Fig. 4.21a) through angle φ, the following reactions occur at points A and A_1 : $|R_A| = |R_{A_1}| = \varphi \cdot l_1 \cdot k_1$. The reactions will be equal in

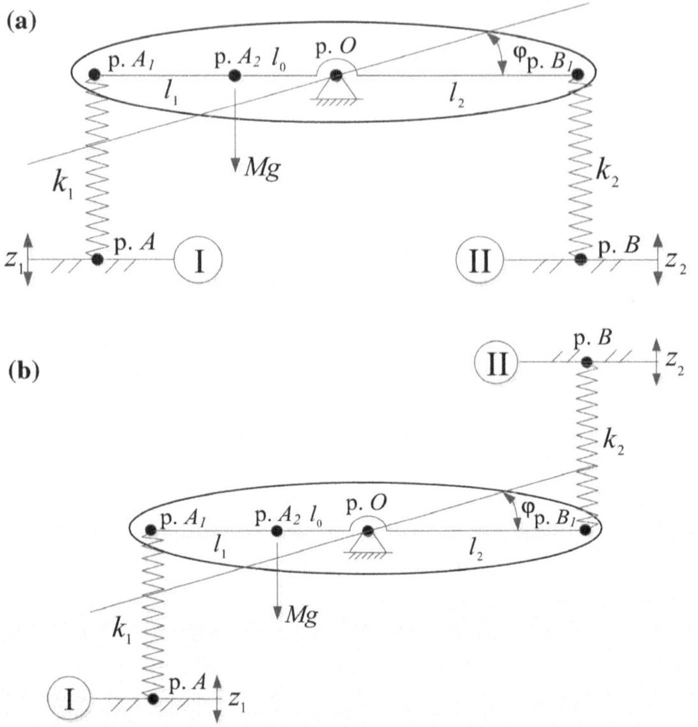

Fig. 4.21 Variants of computational schemes of vibration protection systems of rotational type. **a** asynchronous interaction of elements; **b** synchronous impact

magnitude, but oppositely directed, as shown in Fig. 4.22a. In turn, simultaneously at points B and B_1 there will also be reactions. The reactions of R_A and R_B, and also R_{A_1} and R_{B_1} will have opposite directions.

In the absence of preliminary deformation of the springs k_1 and k_2, in order to create "tension", the elastic element k_1 will compress and provide appropriate responses at points A and A_1. Simultaneously, the elastic element k_2 will stretch, and reactions B and B_1 will occur. In such a situation, it is necessary to have unilateral constraints at points B and B_1, otherwise contact at the points B and B_1 may be lost. A similar situation arises (see Fig. 4.22a), if the deviation of the angle φ (the swing of the object) is implemented in the other direction, i.e. clockwise. Typical for the loading option of the supports (see Fig. 4.22a) is that the supports will always operate in different modes: one will be compressed and the other will be stretched, i.e. will work asynchronously.

In turn, with this arrangement of the elastic elements (see Fig. 4.22b), when the solid body is rotated clockwise, the elastic elements k_1 and k_2 will be simultaneously compressed; when turning clockwise, they will be simultaneously stretched, which can be called a synchronous effect. In any cases of loading the supports (see

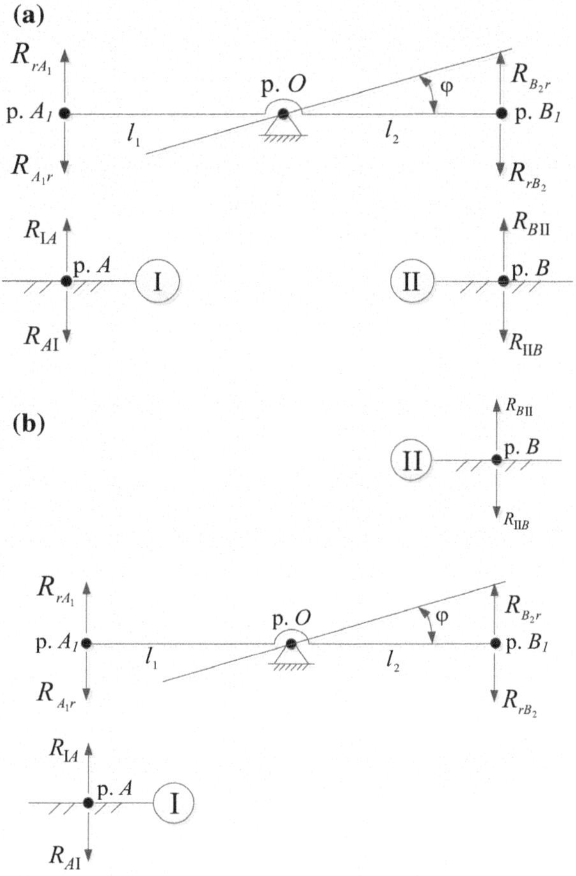

Fig. 4.22 Variants of the location of the bearing surfaces and the corresponding reactions between the elastic elements. **a** k_1 and the bearing surface I—R_{AI} and R_{IA}, as well as k_1 and solid body—R_{A1r} and R_{rA1}; **b** k_2 and bearing surface II—R_{BII} and R_{IIB}; k_2 and solid body—R_{B1} and R_{rB1}

Fig. 4.22), in the absence of preliminary deformation of the springs, reaction at the points of contact of the elastic element with both the bearing surface (I and II) and with the solid body, will be alternating, and when the lever is in parallel position, the reaction values will be zero.

We assume that the elastic element k_1 (see Fig. 4.22a) has a sufficient length for preliminary compression by an amount of Δ_1. If such a compressed spring is introduced into the span AA_1 and attempted to maintain the lever position parallel to the bearing surfaces *I* and *II*, the spring k_2 must be deformed by compression by an amount of Δ_2, while maintaining the static equilibrium, the following condition will be met:

$$\Delta_1 \cdot k_1 \cdot l_1 = \Delta_2 \cdot k_2 \cdot l_2 \tag{4.93}$$

or

$$\Delta_1 = \frac{l_2}{l_1} \cdot \frac{k_2}{k_1} \cdot \Delta_2 = \frac{k_2}{k_1} \cdot i \cdot \Delta_2, \tag{4.94}$$

where $i = \frac{l_2}{l_1}$ is the ratio of the lever of the second kind.

Pre-compression results in the occurrence of static compression reactions; these reactions will be of the same sign:

$$R_A = R_I = k_1 \cdot \Delta_1; \tag{4.95}$$

$$R_B = R_{II} = k_2 \cdot \Delta_2 \cdot \frac{k_2}{k_1} \cdot l. \tag{4.96}$$

If the barycentre of the solid body (see Fig. 4.22a) is displaced by the value of l_0, then the force of weight $P = Mg$ will lead to a change in the reactions at points A_1 and B.

Given the static nature of the loading, it can be found that applying force P at point A_2 will result in a rotation of the lever from the initial (parallel) position by an angle φ:

$$\varphi = \frac{Mg \cdot l_0}{k_1 l_1^2 + k_2 l_2^2}. \tag{4.97}$$

The reaction at point A, i.e. on the bearing surface I, is determined:

$$R_A = \varphi \cdot l_1 \cdot k_1 + \Delta_1 \cdot k_1 = k_1 \cdot (\varphi l_1 + \Delta_1). \tag{4.98}$$

In its turn,

$$R_B = \Delta_2 k_2 - \varphi l_2 \cdot k_2 = k_2 \cdot (\Delta_2 - \varphi l_2). \tag{4.99}$$

It follows from (4.99) that when $R_B = 0$, the critical case, defined by

$$\Delta_2 = \frac{Mg \cdot l_0 \cdot l_2}{k_1 l_1^2 + k_2 l_2^2} = \frac{Mg \cdot i_0}{k_1 + k_2 i^2}. \tag{4.100}$$

Knowing Δ_2, we can find the critical value l_0, assuming that

$$i_0 = \frac{l_0}{l_1} = \frac{\Delta_2 \cdot (k_1 + k_2 i^2)}{Mg}, \tag{4.101}$$

$$l_0 = \frac{\Delta_2 \cdot l_1 \cdot (k_1 + k_2 i^2)}{Mg}. \tag{4.102}$$

Note that the fulfillment of condition (4.102) or R_B being equal to zero in the presence of a two-sided (bilateral) constraint brings about a situation where the contact between the elastic element and the supporting surface *II* is preserved, and with further changes in the parameters, the reaction at points B and B_1 will change signs and the spring k_2 will expand. If the constraint between k_2 and the bearing surface II is one-sided (unilateral), the system loses its static stability.

For the arrangement of the elastic elements k_1 and k_2 in accordance with the scheme in Fig. 4.22b the consideration of the weight forces (Mg) through the definition of the angle φ leads to the following reaction values:

$$R_A = \Delta_1 \cdot k_1 + \varphi \cdot l_1 \cdot k_1 = k_1 \cdot (\Delta_1 + \varphi l_1) = k_1 \cdot (\Delta_1 + \frac{Mgl_0 \cdot l_1}{k_1 l_1^2 + k_2 l_2^2})$$
$$= \frac{k_1 \cdot [\Delta_1 (k_1 + k_2 i^2 + Mgl_0)]}{k_1 + k_2 i^2}; \qquad (4.103)$$

$$R_A = \frac{Mgl_0 \cdot l_2}{k_1 l_1^2 + k_2 l_2^2} - \Delta_2 k_2. \qquad (4.104)$$

Assuming that $R_B = 0$, we find that the conditions defined by (4.104) coincide with the conditions (4.102) with all the conclusions from the given relations that the use of bilateral constraints is necessary.

4.4.3 *Movement of the Object According to the First-Order Lever Scheme*

When using levers of the first kind, as shown in Fig. 4.23, the procedure for calculating static reactions when weight forces are taken into account will remain the same.

In Fig. 4.23a the following designations are accepted: $l_0 = A_2O$, $l_1 = A_1O$, $l_2 = A_2O$. The computational scheme (see Fig. 4.23a) can also be considered in two

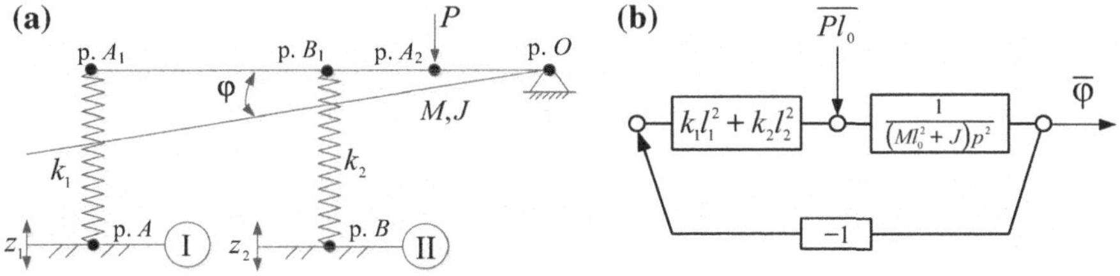

Fig. 4.23 Computational scheme **a** and structural diagram **b** of the vibration protection system with lever linkages (levers of the first kind)

versions: synchronous deformation and asynchronous one. In the latter case, the surface II in Fig. 4.23, and should be rotated by 180° along the axis of the elastic element k_2.

In addition to the weight forces in the calculation schemes shown in Figs. 4.20, 4.21, 4.22 and 4.23, constant forces that can coincide with the force of weight or have an opposite direction can be taken into account. In this case, an essential value is the application of a constant force with respect to a fixed point of rotation. With a suitable arrangement of a constant force and a correct choice of its direction, the mismatch of the point of rotation with the barycentre can be compensated, i.e. it is necessary to fulfill the following conditions:

$$Mgl_0 + Q' \cdot l' = 0$$

where Q' is a constant force; l' is the distance to the center of rotation.

Such an approach to adjusting the properties of a mechanical oscillatory system with one degree of freedom coincides with the methods of static balancing known in engineering science (in particular, in the theory of mechanisms and machines) [22].

Figure 4.24 shows the location of the additional constant force Q, which can be used to compensate for the mismatch of the barycentre and the center of rotation of the object in the form of a solid body.

In Fig. 4.24a the constant force Q' is directed in the direction opposite to the force of the weight $P = Mg$, while the sum of the moments of the forces Q′ and P relative to point O must be zero. In the structural diagram, force factors will be applied to one point, but have different signs. In Fig. 4.24. The force Q_1 is on the other arm of the lever and has the same direction as the force of the weight P. The balancing can be made in this case also by an additional load, but this will change the moment of inertia of the solid body. When passing to the other arm, the force Q' may have the opposite direction with respect to P.

After determining the static reactions at points A_1, B_1 (see Figs. 4.20 and 4.23), one can find a reaction at point O, using the static equation:

$$\overline{R}_{A_1} + \overline{R}_{B_1} + \overline{P} + \overline{R}_0 = 0, \tag{4.105}$$

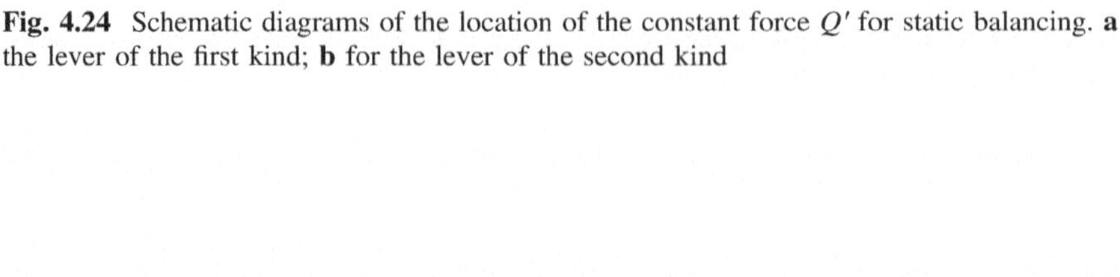

Fig. 4.24 Schematic diagrams of the location of the constant force Q' for static balancing. **a** for the lever of the first kind; **b** for the lever of the second kind

i.e., the reaction at point $O(R_O)$ is defined as the result of geometric summation of the reactions of constraints and external forces (forces of weight and constant forces). It is assumed that all the force factors are parallel vectors, which makes it possible to pass to algebraic sums, assuming that the possible displacements of the protection object are small.

When the time-dependent force perturbation is acting on the object of protection, the object comes into motion relative to the static equilibrium position and is formed at the points of contact of the connector elements with the bearing surfaces, as well as at the points of contact of the standard elements with the object of protection and the contact of the latter with the rack (dynamic responses). The definition of dynamic responses will be considered below. The sum of dynamic and static reactions provides an idea of the overall reaction. Since a overall reaction can take zero values, such situations with unilateral constraints can result in contact failure and the emergence of gaps with subsequent collisions of contact surfaces.

In any case, the knowledge of the overall reaction in the contacts of the elements of mechanical systems makes it possible to produce the necessary force calculations to determine the parameters of springs, dampers and other elements.

4.5 On the Relationship Between the Problem of Vibration Protection and Static Balancing

As was shown above, the overall reaction at the points of support and contacts of the elements of mechanical oscillatory systems consists of static and dynamic components. Static reactions are formed under the influence of the weight forces of the mass-and-inertia units, and also under the action of constant forces. The latter can be applied not only to the mass-and-inertia unit of the basic model, but also to the elements of the mechanical oscillatory system. This idea of the effect of static loads is associated with the expansion of the notion of a set of typical standard units of vibration protection systems.

The problems of vibration protection and vibration isolation, with their specific content, are determined by the specificity of the structural and technical forms of the protection objects and the conditions of interaction with the vibrational environment. At the same time, the separation of force and kinematic perturbations is rather conditional and can be reduced to generalized concepts about the effect of external perturbations on technical objects. The system of electromechanical analogies, developed in the theory of electrical circuits, also creates certain opportunities for expanding the understanding of mechanical oscillatory systems within the scope of the structural theory of vibration protection systems.

In the general case, in the computational scheme of a vibration protection system, it is always possible to distinguish three main parts: the source of the perturbation, the object of protection, the actual vibration protection device (or that part of the mechanical oscillatory system that performs the function of reducing the level of impacts).

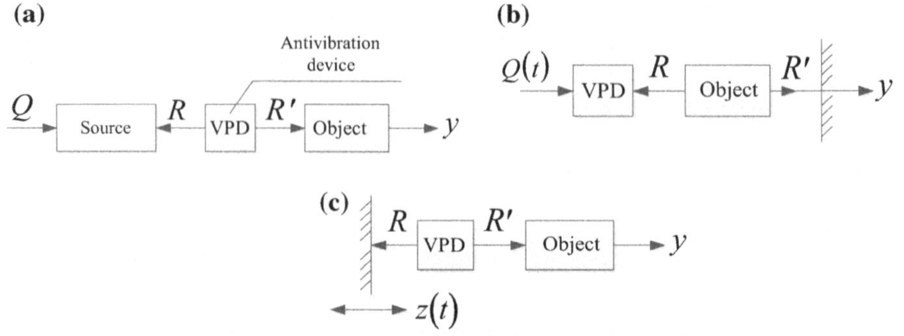

Fig. 4.25 The schematic diagram of the vibration protection system with one degree of freedom. **a** is a general case; **b** is the force perturbation; **c** is the kinematic perturbation

In the simplest cases, when the computational scheme is described by one or more coordinates, the source and the object can be considered solid bodies. For further consideration, translational motions along some y axis are used (Fig. 4.25), although rotational motions with the possibilities of obtaining the same results can be chosen. At the same time, rotational motions possess certain characteristic features.Fig. 4.25.

4.5.1 Peculiarities of Formation of External Perturbations

Vibration protection device (VPD) in Fig. 4.25 is located between the perturbation source and the protection object. In most cases, the mass of one of the bodies in the system—the source or the object—significantly exceeds the mass of the other body—the object or source, respectively. Then the motion of the body of the "larger" mass can be considered independent of the motion of the body of the "smaller" mass. If, in particular, the "large" mass has an object, then it is usually considered immobile, the motion of the system is caused in this case by external forces applied to the source, representing the force excitation $Q = Q(t)$ (see Fig. 4.25b). If the "large" mass has a source, then the law of its motion z = z (t) can be considered given; this motion plays the role of kinematic excitation of the system (more precisely, of the object, see Fig. 4.25c). In both cases, the body of the "greater" mass is called the carrying body, or the base, and the body of the "small" mass is called weightless. The diagram shown in Fig. 4.25b, is usually applied when considering the protection of buildings, structures, ceilings or foundations from the dynamic influences generated by machines with unbalanced moving parts or other vibroactive equipment installed on them. The diagram shown in Fig. 4.25c, are used in the problems of vibration protection of instruments, apparatuses, precise mechanisms and machine tools, i.e. equipment that is sensitive to vibrations and mounted on oscillating bases or on moving objects. It can be assumed that the

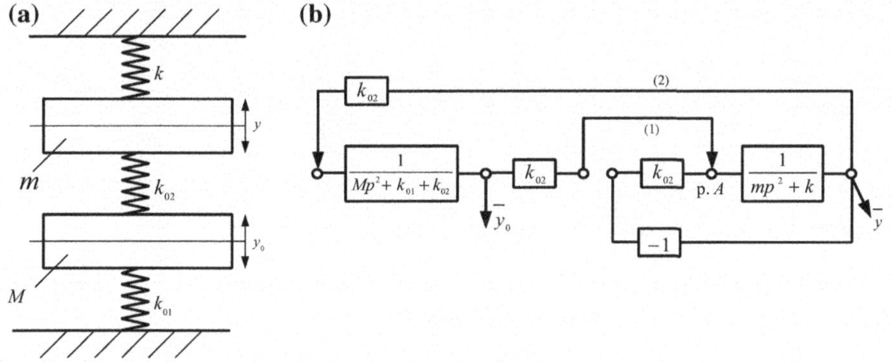

Fig. 4.26 The diagram of formation of force perturbation in the vibration protection system. **a** is the computational scheme taking into account the existence of an energy source; **b** is the structural diagram of a vibration protection system with a source of force perturbation

source of the perturbation is a form of interaction with a certain system that has incomparably large energy possibilities. In Fig. 4.26 it is assumed that the partial system M, k_{01}, k_{02} reflects the properties of the power perturbation source.

Since in the calculation of the action of vibrations the force disturbance is considered independent of the load, the source of the force action can be represented in the form of contacts with the oscillatory system (M, k_{01}, k_{02}) performing undamped free harmonic oscillations. In this case, the condition $M \gg m$ is fulfilled, which allows us to consider the action of the partial system (M, k_{01}, k_{02}) only via the constraint (1), which forms the force $Q = k_{02}\, y_0$, applied to the object by the mass m. For $M \gg m$, the feedback response via the constraint (2) can be considered negligible.

Important for subsequent studies is the fact that the force perturbation is implemented as a force applied to the object (see A, Fig. 4.26b). In addition, the existence of this type of perturbation is associated with the recognition of the asymmetry of the action of the constraints (1) and (2) between the partial systems (M, k_{01}, k_{02}) and (m, k_{01}, k). With kinematic perturbation, the diagram for the formation of the effects occurs according to the same principle (see Fig. 4.21b), but a kinematic perturbation (see Fig. 4.26a) will be applied at point B, as shown in Fig. 4.27.

In contrast to the force perturbation (see Fig. 4.26b), the displacement $\bar{y}_0$ acts through the element with stiffness k_{02}. In this case, it is also assumed that the constraint (2) does not affect the motion of the partial system (M, k_{01}, k_{02}). Comparison of the two forms of external influences in Fig. 4.26b and 4.27 shows the relativity of considerations on the separation of external perturbations into force and kinematic ones. In fact, they can be reduced to each other. The latter, in the end, served as the basis for generalized approaches.

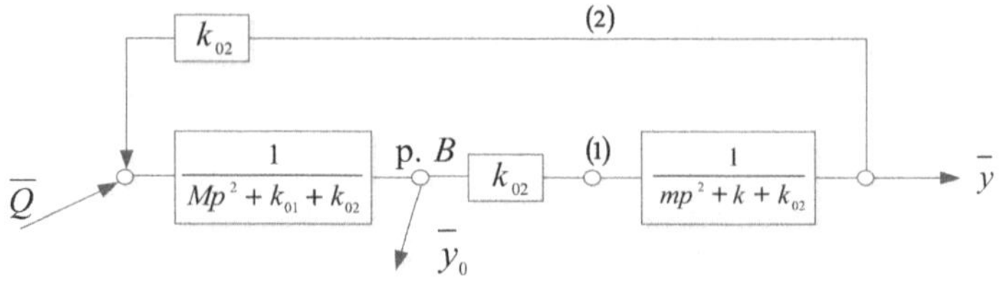

Fig. 4.27 Schematic diagram of the formation of kinematic impact

4.5.2 Properties of System Elements

The vibration isolation device is the most important part of the vibration protection system. Its purpose is to create such a motion mode, initiated by given perturbations, in which the object protection objective is implemented. In many cases, this can be achieved when using an inertia-free vibration isolation device, which for the diagrams shown in Fig. 4.20, can represent a uniaxial vibration isolator. For such a vibration isolator, the reactions R and R' coincide in magnitude ($R = R'$). Thus, in the simplest case, the reaction R can be considered proportional to the displacement and the displacement speed of the object:

$$R = ky + b\dot{y}. \tag{4.106}$$

In this case it is assumed that the vibration isolator has a simple structure in which the motion of the protection object forms the dynamic state of the vibration isolator. The computational scheme of the vibration protection system (VPS) has the form as in Fig. 4.28.

Dependence (4.105) describes the linear characteristic of a simple inertial-free vibration isolator; the coefficients k and b are called respectively the stiffness and the damping coefficient. For $b = 0$ (4.106) describes the characteristic of a linear elastic element (spring); at $k = 0$, the characteristic of a linear viscous damper. Thus, the model of a vibration isolator with the characteristic (4.106) can be represented as a parallel connection of a spring and a damper (see Fig. 4.28). The stiffness coefficient k of the vibration isolator with a linear characteristic (4.106) determines the natural frequency of the system:

Fig. 4.28 The computational scheme of a vibration protection system with parameters m, k, b with one degree of freedom

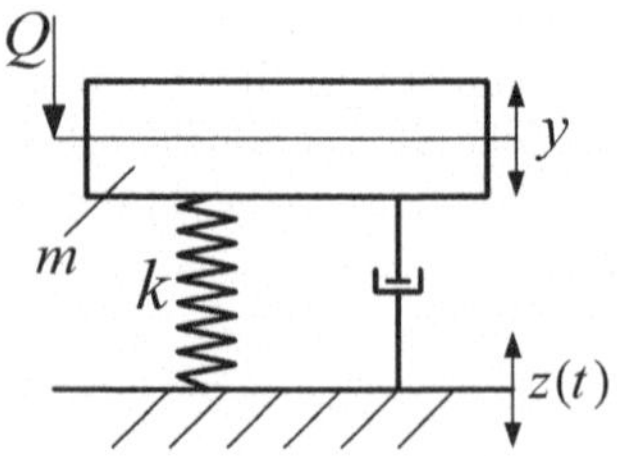

$$\omega_0 = \sqrt{\frac{k}{m}}, \tag{4.107}$$

as well as static deformation y_{ot} of the vibration (sinking) of the vibration isolator. The damping properties of the system presented in Fig. 4.23, are characterized by a damping factor:

$$n = b/2m \tag{4.108}$$

and relative damping:

$$\nu = n/\omega_0. \tag{4.109}$$

At $\nu = 1$, critical damping is implemented in the system [19]. The effect of damping in the general case has various forms and significantly changes the properties of vibration protection systems.

The basic model of a vibration protection system can be extended by adding elements that implement more complex functions of transforming motions and force interactions. Figure 4.29 shows how the initial model (see Fig. 4.28) can be transformed if the technical object is sufficiently complex and, with its simplifications, it is necessary to take into account a number of structural and technical features.

4.5.3 *The Problem of Vibration Protection: Possible Generalizations*

In the most general form, the problem of vibration protection can be represented as an installation between the object and the bearing surface of a vibration protection device (VPD) (see Fig. 4.29a). As external influences, a perturbing force is applied to the object with mass m (force Q), as well as a perturbation from the side of the bearing surface or base (kinematic perturbation $z(t)$). At the contact points of the VPD with the protection object (point A_1) and the supporting surface, dynamic responses occur. In addition to the dynamic at points A, A_1, static reactions are also generated, caused by constant forces, including the forces of gravity of the object (and elements of the VPD). In the general case, the reactions at points A and A_1 are not equal to each other. However, in the simplest situations, when VPDs consist only of elastic and dissipative elements, they can be identical. External perturbations can also be applied to VPD elements.

In its simplest form, the VPD can be represented (in a set or one by one) by standard elementary units in the form of springs, dampers and motion transformation devices that have simple transfer functions of the amplifying unit, as well as differentiating units of the first and second orders (see Fig. 4.29b). In Fig. 4.29c–e

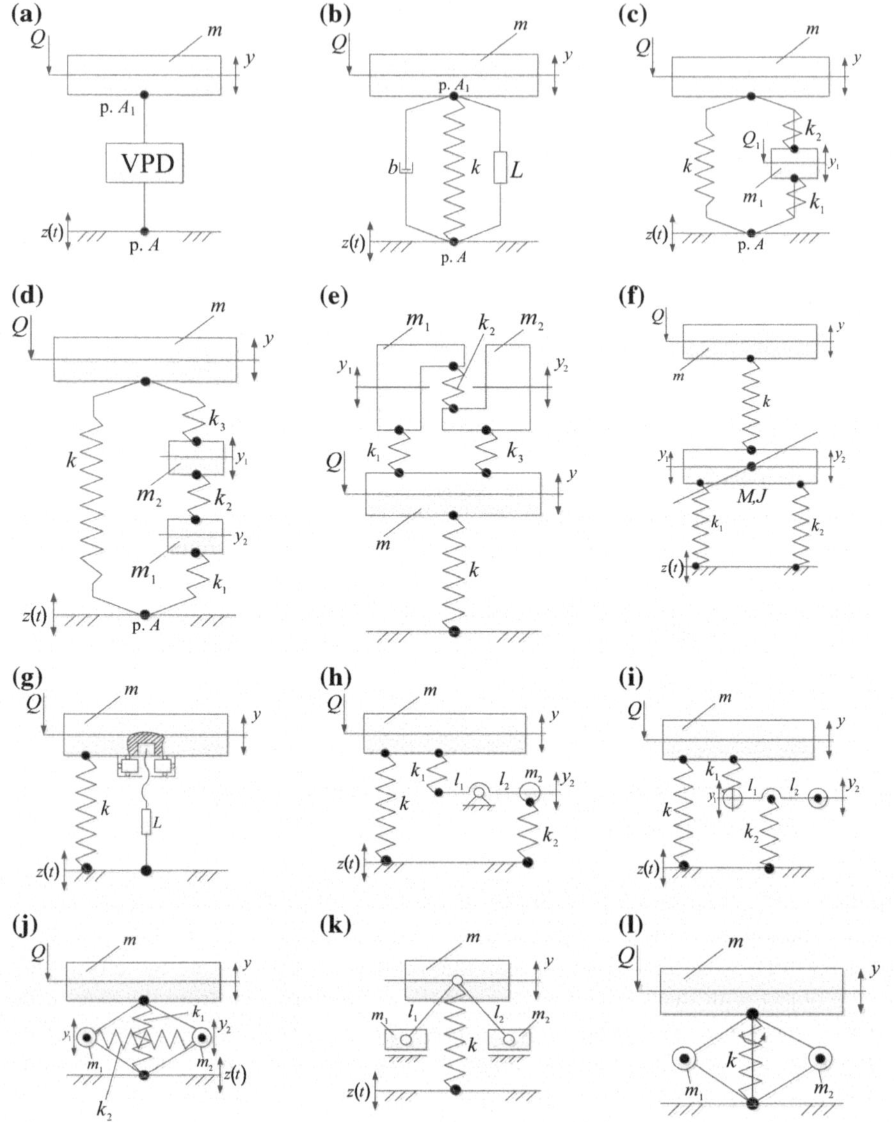

Fig. 4.29 Variants of structural and technical design of vibration protection devices in systems of translational rectilinear motion of a protection object with one degree of freedom (variants of the VPDs are described in the text)

it was shown that the vibration protection device can be a mechanical circuit or a more complex mechanical vibrational structure. In this case, the VPD can have internal sources of perturbation. It should be noted that here we consider linear systems that perform small oscillations. The VPDs in Fig. 4.29c–e can be convoluted into generalized springs with transfer functions of fractional-rational form.

For example, non-locking screw mechanisms (see Fig. 4.29l), lever mechanisms with a fixed (see Fig. 4.29h) and a movable (see Fig. 4.29i) support points, and also lever linkage mechanisms of various types (see Fig. 4.29j–l) represent a special class of VPD. Consideration of the mechanisms in the structure of the vibration protection system is fully justified for many reasons, including the fact that, in a real structure, the standard elements are mechanisms (shock absorbers, vibration absorbers, springs, dampers, torsion bars, etc.).

The examples given reflect only a small part of the possible solutions. It should be noted that VPDs differ just as greatly in rotary-type systems in which toothed, cam and lever mechanisms are widely used.

The representation of the object of protection in the form of a solid body with one degree of freedom is reflective of, as a rule, preliminary assignments of problems. Many real objects have two or more degrees of freedom, which was the subject of attention in the review chapter of the monograph. It should be noted that the vibration protection of objects with several degrees of freedom, which is typical, in particular, for transport dynamics, is a complex problem, requiring the search and development of adequate approaches and methods for estimating the dynamic state.

4.5.4 Vibration Protection Efficiency: Efficiency Coefficients for Harmonic Oscillations

The efficiency of vibration protection is understood as the degree of implementation of vibration protection by the vibration protection device. For a force harmonic excitation, $Q(t) = Q_0 \sin \omega t$, where Q_0 and ω are the amplitude and frequency of the driving force, respectively. The purpose of protection can be to reduce the amplitude R_0 of the force transferred to a fixed object:

$$R_0 = \frac{Q_0\sqrt{\omega_0^4 + 4n^2\omega^2}}{\sqrt{\left(\omega_0^2 - \omega^2\right)^2 + 4n^2\omega^2}}, \tag{4.110}$$

which follows from the kinetostatics equation for the element m for $z(t) = 0$:

$$m\ddot{y} + b\dot{y} + ky = Q. \tag{4.111}$$

Using Laplace transforms, we find that

$$\overline{R}_0 = \bar{y} \cdot (bp + k). \tag{4.112}$$

In turn, the decrease in the amplitude of the steady-state forced oscillations of the object under the action of the force is defined as follows:

$$y_0 = \frac{Q_0}{m\sqrt{\left(\omega_0^2 - \omega^2\right)^2 + 4n^2\omega^2}}. \tag{4.113}$$

The transfer function of the system with the input signal and output is as follows:

$$W_R(p) = \frac{\overline{R}_0}{\bar{Q}} = \frac{bp + k}{mp^2 + bp + k}, \tag{4.114}$$

as

$$\bar{y} = \frac{\bar{Q}}{mp^2 + bp + k}. \tag{4.115}$$

It should be noted that the constraint reaction $\overline{R}_0$ is determined from (4.89):

$$\overline{R}_0 = \bar{y} \cdot (bp + k) = \overline{Q} - mp^2\bar{y}. \tag{4.116}$$

Assuming that $\bar{y}$ is determined from (4.115), we obtain:

$$\overline{R}_0 = \overline{Q} - \frac{mp^2\overline{Q}}{mp^2 + bp + k} = \frac{(bp + k)}{mp^2 + bp + k}\overline{Q}$$

or

$$W_R(p) = \frac{\overline{R}_0}{\bar{Q}} = \frac{bp + k}{mp^2 + bp + k},$$

which coincides with the expression (4.114). That is, in the vibration protection system with one degree of freedom, the transfer function W_R (p) determines the ratio of the magnitude of the coupling reaction $\overline{R}_0$ of the applied harmonic force $\overline{Q}$ applied to the object of protection. The question is whether we can determine W_R (p) directly from the structural diagram shown in Fig. 4.30, which corresponds to the computational scheme in Fig. 4.28.

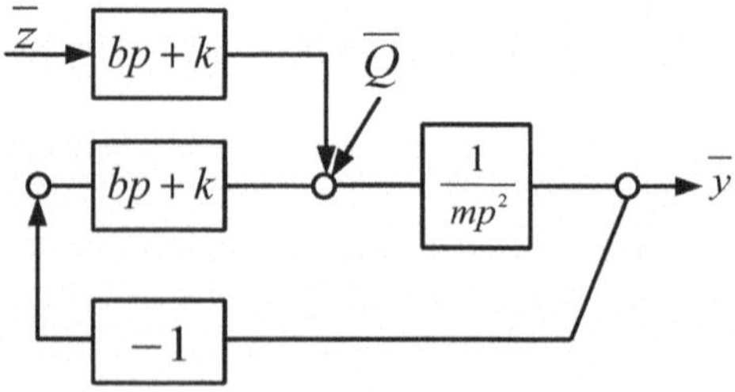

Fig. 4.30 The structural diagram of the vibration protection system with one degree of freedom

Note that the structural diagram (see Fig. 4.30) is a graphic analogue of the mathematical model of the VPD in the form of a differential equation, which can be obtained by known methods [5].

The kinematic perturbation $z(t)$ characterizes the motion of the base and is transmitted to the protection object, which can be described by means of a transfer function:

$$W(p)_z = \frac{\bar{y}}{\bar{z}} = \frac{bp + k}{mp^2 + bp + k}. \tag{4.117}$$

The physical meaning of expression (4.117) is that it corresponds to the ratio of the amplitude of the forced oscillations of the object to the amplitude of the oscillations of the base. Expressions (4.114) and (4.117) are the same, which indicates the possibility of finding W_R (p) through $W(p)_z$, using the structural scheme shown in Fig. 4.30. In this case one can keep in mind the possibility of the equivalence of the action of two external perturbations, since it is true that $Q = kz$. Note that for $z(t) = z_0 \sin \omega t$ and $Q(t) = 0$, the purpose of the vibration protection can be to reduce the amplitude of the absolute acceleration (overload) of the object:

$$y_0 = \frac{z_0\omega^2\sqrt{\omega_0^2 + 4n^2\omega^2}}{\sqrt{\left(\omega_0^2 - \omega^2\right)^2 + 4n^2\omega^2}}, \tag{4.118}$$

and in the decrease in the amplitude of the oscillations of the object relative to the base, i.e. in the coordinate system $y' = y - z$:

$$y' = \frac{z_0\omega^2}{\sqrt{\left(\omega_0^2 - \omega^2\right)^2 + 4n^2\omega^2}}. \tag{4.119}$$

Thus, for each of the vibration protection problems, dimensionless efficiency coefficients can be selected. Such coefficients are determined by the corresponding transfer functions of the initial structural model (see Fig. 4.30).

With force perturbation, we get:

$$k_R = |W_R(p)| = |W_z(p)| = k_y\frac{ky_0}{Q}; k_z = |W_z(p)| = \left|\frac{\bar{y}}{\bar{z}}\right|; k_R = k_z; W_Q(p) = \frac{\bar{y}}{\bar{Q}} = \frac{1}{mp^2 + bp + k}. \tag{4.120}$$

In the case of kinematic effects, the following relationships are considered:

$$k_R = \frac{k_{\Pi ep}}{\omega^2 z_0}; k_{y'} = \frac{y_0'}{z_0}. \tag{4.121}$$

In the technical literature [19] k_R is called the coefficient of vibration isolation, and k_y is the coefficient of dynamism. If we introduce the dimensionless parameter $\gamma = \frac{\omega}{\omega_0}$ and $\nu = \frac{n}{\omega_0}$ $\left(2n = \frac{b}{m}\right)$, then

$$k_R = \sqrt{\frac{1+4\nu^2\gamma^2}{(1-\gamma^2)^2+4\nu^2\gamma^2}}; k_y = \frac{1}{\sqrt{(1-\gamma^2)^2+4\nu^2\gamma^2}}; k_{y'} = \frac{\gamma^2}{\sqrt{(1-\gamma^2)^2+4\nu^2\gamma^2}}. \tag{4.122}$$

When calculating the simplest vibration protection systems, the efficiency condition according to the criteria k_R, k_y, $k_{y'}$ is formed as inequations:

$$k_R \leq 1; k_y \leq 1; k_{y'} \leq 1. \tag{4.123}$$

Since these coefficients depend on frequency, one can consider the effectiveness of vibration protection at a given frequency γ or in a given frequency range $\gamma_1 \leq \gamma \leq \gamma_2$. Note that the effectiveness of vibration protection by the criterion $k_R \leq 1$ is ensured at any level of damping in the frequency range

$$\gamma \geq \sqrt{2}. \tag{4.124}$$

For any γ from the range (4.124), the higher the efficiency, the weaker the damping; the best efficiency is provided by an ideally elastic vibration isolator ($\nu = 0$).

In its turn, the vibration protection efficiency by the criterion $k_y \leq 1$ is also ensured in the range (4.124) for any values of ν. When vibration protection is effective in the entire frequency range $0 < \gamma < \infty$; for $\nu < \frac{1}{\sqrt{2}}$, the efficiency taking place within the range

$$\gamma > \sqrt{2(1-2\nu^2)}. \tag{4.125}$$

For a fixed value of γ, the efficiency increases with increasing damping. Vibration protection, according to the criterion $k_{y'} \leq 1$, is effective in the whole frequency range, if $\nu > \frac{1}{\sqrt{2}}$, and at $\nu < \frac{1}{\sqrt{2}}$ – within in the range

$$0 < \gamma < \frac{1}{\sqrt{2(1-2\nu^2)}}. \tag{4.126}$$

The degree of efficiency at a fixed frequency γ increases with increasing damping; in the worst case (for $\nu = 0$), the efficiency range corresponds to the band:

$$0 < \gamma < \frac{1}{\sqrt{2}}. \tag{4.127}$$

The dependences $k_R = k_R(\gamma, \nu)$; $k_y = k_y(\gamma, \nu)$; $k_{y'} = k_{y'}(\gamma, \nu)$ for fixed ν can be considered as the amplitude-frequency characteristics of the given system with respect to the corresponding input and output. The different frequency characteristics of the vibration protection system, respectively, are determined:

$$\varepsilon = arctg\frac{2\nu\gamma}{1-\gamma^2}; \tag{4.128}$$

$$\eta = arctg\frac{2\nu\gamma^3}{1-\gamma^2(1-4\gamma^2)}, \tag{4.129}$$

and represent the dependence of the phase shifts of the processes $y(t)$, $y'(t)$ (4.128) and $R(t)$, $\ddot{y}(t)$ (4.129) with respect to the perturbations $Q(t)$ and $z(t)$.

As shown in Fig. 4.30, the vibration isolation device is often performed in the form of joining several elements forming a complex *vibration isolator*. Under certain conditions, the reaction R of such a compound can be approximated by the dependence (4.106), where δ is the deformation of the compound as a whole. Then the complex vibration isolator under consideration is equivalent (in the sense of affecting the source and the object) to a simple one, the characteristics $d_э$ and $b_э$ of which can be called *equivalent coefficients* of stiffness and damping coefficients.

Thus, for the development of structural approaches in the theory of vibration protection and vibration isolation of machines, equipment, instruments and equipment, it is essential to take into account the features of the motion of the protection object. In the simplest cases, the object of protection can be considered as a material point or a system of material points performing rectilinear translational motion. If the security object carries out angular oscillatory motions around a certain point, it is necessary to take into account the dimensions of the protection object. In this case, a solid body is considered, on which the attachment points of the elements of the vibration protection system are positioned. The method of constructing mathematical models of mechanical oscillatory systems is based on taking into account the spatial location in the vibration protection system of standard elementary units. This makes it possible to take into account the effects that arise when elastic, dissipative and mass-and-inertia elements operate within a certain spatial structure. The geometry of the location of the attachment points of elements is of particular importance for the protection of objects making angular oscillatory motions. Studies show that there are significant features of the dynamic properties in relation to the basic models of vibration protection systems of translational and rotational types of motion. It can be noted that the mathematical models of systems that perform angular vibrational motion should be built taking into account the emerging lever linkages. Such linkages can be implemented with the help of levers of the first and second kinds.

There are analogies between the basic computational schemes of vibration protection systems, but the problem of adequacy of models requires taking into account a number of specific details. In the rotational-oscillating system, static and dynamic equilibration modes can arise. There are also quite definite differences in the estimates of the effect of external perturbations of the force and kinematic kinds.

In systems with several degrees of freedom, the basic vibration protection systems can be divided into several groups: systems with elements that perform rectilinear translational motion, systems of rotational type, and combined systems. Each of the distinguished classes of systems has its own peculiarities, which must be taken into account when determining static and dynamic responses in the contacts of the elements of systems between themselves and with bearing surfaces.

The approach developed in the problems of dynamic synthesis of vibration protection systems makes it possible to use the ideas about the possibilities of new types of additional constraints created by mechanical circuits in general and mechanisms in particular. The latter is due to the fact that, in many cases, the elastic-dissipative and mass-and-inertia constraints in real vibration protection systems take the form of special devices and mechanisms.

4.6 Features of Transformation of Mechanical Circuits on the Basis of Introduction of Intermediate Devices into the Compounds

The mechanical oscillatory system is often used as a computational scheme, designed to solve a specific problem; first of all, the prospect of obtaining a mathematical model is evaluated. In many cases, the mathematical model can be based on known methods of constructing differential equations of motion or the dynamic state of the system. The subsequent linearization and simplification result in a certain balance during which the basic model is fixated. In the solution of the problems of vibration protection and vibration isolation, as was shown in [23, 24], structural methods based on the application of the theory of circuits and the approaches developed in the theory of automatic control can be used. Possible integration of approaches involves the selection of a protection object, which, in the automatic control theory, coincides with understanding of the existence of the control object.

If we assume that in a vibration protection system one can distinguish a protection object in respect of which the technology of searching and selecting methods and means for controlling the dynamic state is implemented, then the formation of a model containing the protection object and also the circuits forming the system becomes a natural extension of the modeling process. Basically, in this sense, some new generality or model is synthesized, in which the constituent circuits, which can be of a different nature, are closed to the object of control (protection). As a result, we obtain an integral model that allows us to approach the evaluation of the dynamic state from the standpoint of the automatic control theory and use its analytical tools to the fullest extent [25].

On the other hand, the surrounding communication object can be considered as fragments of mechanical circuits and be transformed into some structures having transfer functions. The physical meaning of structures is manifested in the fact that they act as additional feedbacks in relation to the object of protection. In such circuits, the forces acting on the object are formed. The parameters of the circuits become the reduced stiffness of the generalized springs, which, on the whole, results in a substantial simplification of the initial structural diagram of the system, but the rules of transformations in the structure of the theory of circuits, and the automatic control theory have their own specifics and the applicability limits.

4.6.1 Transformations When Connecting Elements in a Mechanical Circuit

In the theory of mechanical circuits, the rules of consecutive and parallel connection of elements are used for transformations. If the mechanical circuit (in its simplest form) consists of the elements S_1 and S_2, which are connected by one node (point 2 in Fig. 4.31), then the equivalent dual unit will be a two-terminal network, the parameters of which are determined by the formula

$$S_3 = \frac{S_1 S_2}{S_1 + S_2}. \tag{4.130}$$

Consecutive connections are characteristic for mechanical and electrical circuits. It is necessary to make sure that the units S_1 and S_2 coincide in the parameters "input-output" for each units. For example: in Fig. 4.31 two elastic elements are used; in both springs the input is displacement, and the output is the force; i.e., this will be a consecutive connection of the springs.

The connection of elements through a solid body in the form of a material point. We will assume that the connection of the two elements occurs in such a way that point 2 represents some "assembly", as shown in Fig. 4.32. In this case, when $m = 0$, it is easy to return to the original version. If we assume that $m \neq 0$, then the

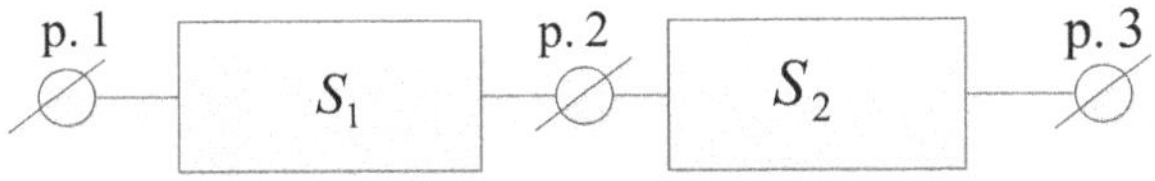

Fig. 4.31 The schematic diagram of the consecutive connection of elements

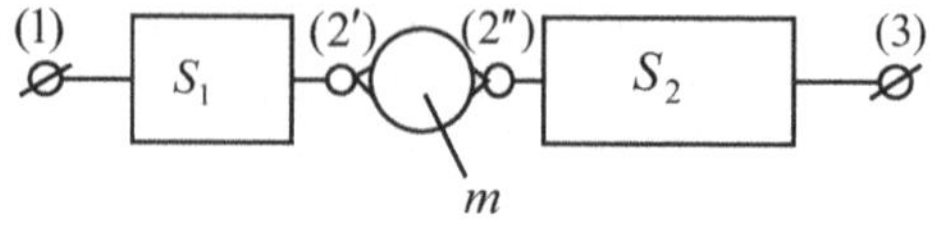

Fig. 4.32 The schematic diagram of the connection of two elements S_1 and S_2 through point 2 having a mass m

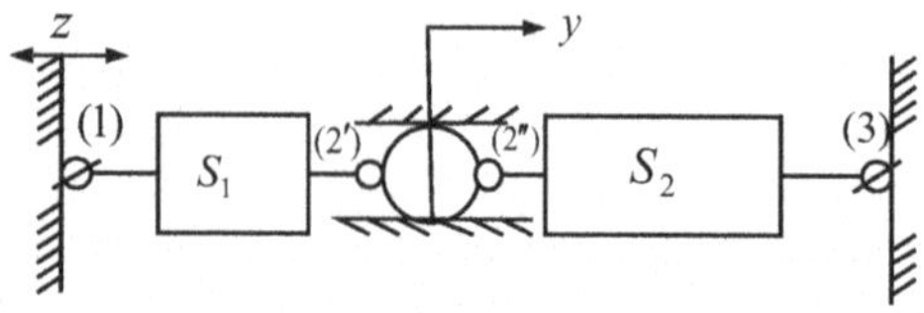

Fig. 4.33 The computational scheme of the consecutive connection of the elements S_1 and S_2 through the material point (m)

system in Fig. 4.32 is transformed into a mechanical oscillatory system with one degree of freedom (Fig. 4.33).

Since the material point has no dimensions, it can be assumed that a solid body in the form of a material point with mass m will bring some dynamic effects, but the line of action of the forces of the elastic elements remains in the form of a straight line. The difference between the schemes in Figs. 4.31 and 4.33 is that the coupling between points 1 and 3 will depend on the frequency when the displacement is applied to p. 1 in the form z_1, which corresponds to the kinematic disturbance of the chain (in this case, we can use the concept of a branch for which the simplest form is two links connected in series).

Definition of the reduced parameters. Let us use some techniques of the structural theory of vibration protection systems to determine the necessary data [5], for which we construct the structural diagram in Fig. 4.33, where the input is the kinematic perturbation z, and the output is the force acting on the mass m (in this case, it is the dynamic response Q_p on the fixed stand, on which the spring k_2 rests (Fig. 4.34a)).

If you convert the structural diagram in Fig. 4.34a, taking the equivalence of the force and kinematic perturbations, we obtain the reduced scheme (Fig. 4.34b), which, in turn, can be simplified (Fig. 4.34c), and find the transfer function:

$$W_3(p) = \frac{\bar{Q}_p}{\bar{Z}} = \frac{k_1(mp^2 + k_2)}{mp^2 + k_1 + k_2}, \tag{4.131}$$

where Q_p is the dynamic response at point (3); $p = j\omega$. If we assume that $p = 0$, i.e. that the system is considered in statics, then

$$W_3(p)_{p=0} = \frac{k_1 k_2}{k_1 + k_2}, \tag{4.132}$$

which completely coincides with the expression (4.130). The physical meaning of (4.132) is that the introduction of a solid body in the form of a material point

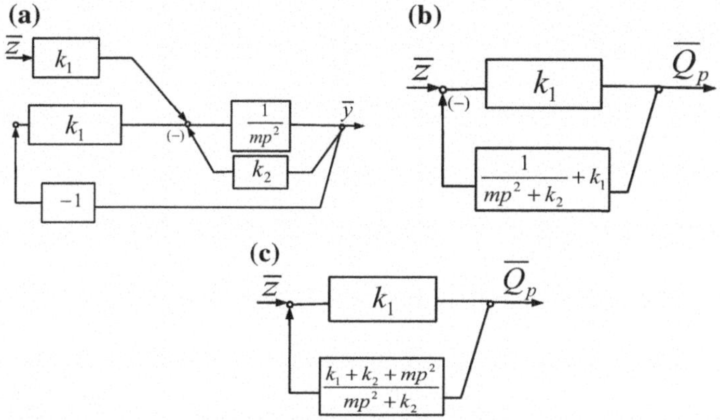

Fig. 4.34 The structural diagram of the system corresponding to the scheme in Fig. 4.33: **a** is the general view; **b** is reduced elements; **c** is the simplified reduced scheme

between elements S_1 and S_2 does not change anything, but only in statics. At the same time, with the introduction of a solid body, if we mean studying the possibilities of combining two typical elements (in a consecutive connection), pursuing some goals of the development of the structure in its topological sense, nothing unusual happens in the statics, as the material point has only mass, but does not have dimensions.

The mass of the intermediate element plays a role in the course of the dynamic processes. Module $W_3(p)$ is the dynamic stiffness of the series connection of two springs. For $p = 0$, the dynamic stiffness (we denote it by k_{dyn}) coincides with the static stiffness $k_0 = \frac{k_1 k_2}{k_1 + k_2}$, but the situation changes depending on the frequency of the dynamic process, which is shown in Fig. 4.35.

On the graph, $k_{dyn}(\omega)$ can be found at the point of intersection with the line k_0, at frequencies ω_1 and ω_2 – $k_{dyn} = k_0$; i.e. at two frequencies the dynamic system has properties, as in statics. Similarly, one can find the frequencies u at the intersection of $k_{dyn}(\omega)$ with the line k_0. However, the above is valid only for $m = 1$ (taken for preliminary estimates).

In fact, the dynamic stiffness depends on two parameters: the frequency ω and the mass value m. Thus, for $m \neq 0$, $m \neq 1$

$$W_3(p) = \frac{k_1 p^2 + \frac{k_1 k_2}{m}}{p^2 + \frac{k_1}{m} + \frac{k_2}{m}}. \tag{4.133}$$

If $m \to \infty$, then $|W_3(p)| \to k_1$. Physically, this means that $m \to \infty$, the element S_2 or the spring (in our case) with the stiffness k_2 are blocked (the stiffness k_2 becomes large, and $k_0 = k_1$).

In intermediate cases, the change in m at each frequency gives a family of graphs. Analysis shows that the introduction of a solid body in the form of a

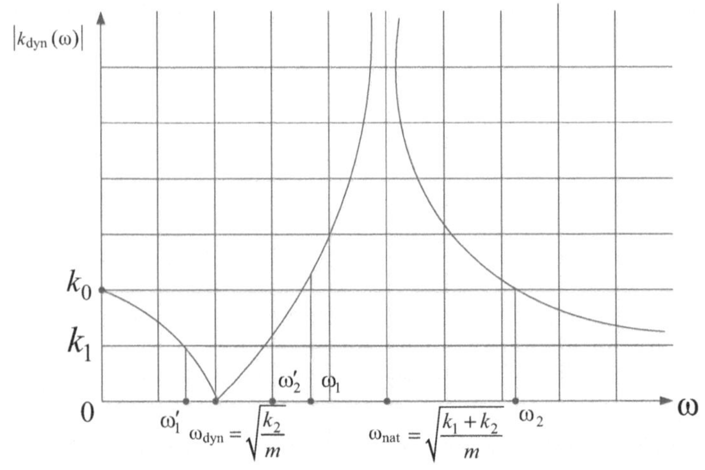

Fig. 4.35 The dependency of the dynamic stiffness of the consecutive connection of springs k_1 and k_2 from frequency

material point as a new element designed to expand the possibilities of the very rule of a consecutive connection, has a number of limitations. This provides an idea of the complex dependence of the properties of the branch on the combination of the parameters of the elements. It should be noted that the basic qualities in this case (because we are talking about an ordinary spring, if $m = 0$ and $p = 0$), occur at some frequencies ω_1 and ω_2. In addition, the connection can behave like an ordinary spring at frequencies and, however, instead of k_{sp} (or k_0 for $p = 0$), we will have k_1. It should be also noted that such stiffness is taken by the branch for $m \to \infty$. That is, the process of consecutive connection, when dynamic factors are taken into account, gives an idea of the properties of a branch that has a stiffness less than in statics, which will be observed after ω_2. Among the fundamentally new properties, one can attribute the presence of a mode with ω_{dyn}. At this frequency, when the springs k_1 and k_2 are consecutively connected through mass m, dynamic zero stiffness will be observed. In turn, at ω_{nat}, the dynamic stiffness becomes infinitely large and the branch is "locked".

Solid body in the form of a rod. A more general case with respect to a material point is a loaded solid rod, which has a mass and moment of inertia. Such an intermediate element in the serial connection S_1 and S_2 can be used if the problems are solved using a system of forces that will not be in one row but act in a plane (for example, creating a system of parallel forces). In this case, the topology of the system, more precisely, the topological scheme of the system, changes if by this we mean some configurations of equilibrium of the system of forces on the plane with the use of moments of forces and angular displacements.

In the general case, a solid body in the plane performs complex motions, consisting of translational motion along the trajectory of the center of mass and

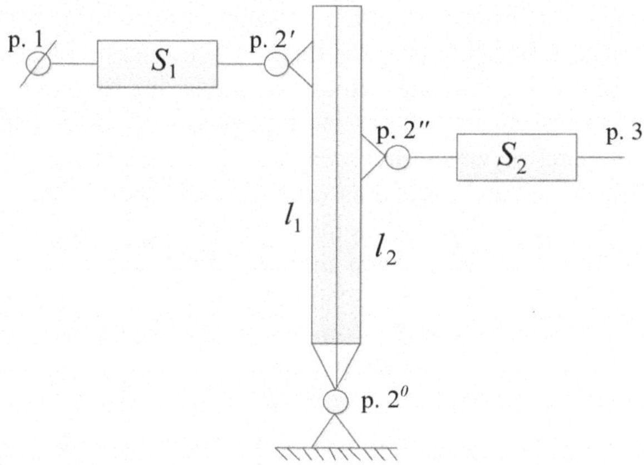

Fig. 4.36 The schematic diagram of the consecutive connection through a weighty rod

rotational motion around the center of mass. In our case, it can be assumed that the translational motion will be rectilinear. If a solid body changes its position in the form of a weighty rod, participating in a plane motion, it is always possible to find the instantaneous center of velocities, and the transition of the body from one position to another takes place due to one turn [26]. Such representations assume the possibility of inserting a rod as some connecting device between the elements S_1 and S_2 in a consecutive connection, as shown in Fig. 4.36, where point 2^0 is the instantaneous center of velocities or the instantaneous center of rotation in view of the smallness of the displacements themselves.

The rod in Fig. 4.36 can be considered as a lever to which the elements S_1 and S_2 join at points 2′ and 2″. Because of the peculiarities of the lever, a relation arises between S_1 and S_2, determined by the position of the points 2′and 2″ with respect to the center of rotation—that is, to the point 2^0. The position of point 2^0 depends on several reasons and can be different. In the case shown in Fig. 4.36, the lever of the second kind is considered [27]. However, if point 2^0 moves and holds a position between the points 2′ and 2″, then this will be a lever of the first kind. The mentioned levers have their own peculiarities. The identification of the fulcrum with the instantaneous velocity center seems to be justification, taking into account, as mentioned above, a number of limitations, since the ultimate goal of the research is to study the effect of the intermediate device in the form of a weightless lever when connected in series. If the physical implementation of the lever interaction is related to the installation of the support points of the lever in the instantaneous velocity center, this situation will greatly simplify the process of studying the dynamic properties of the connection. The authors develop the idea of "taking apart the points of connection of dual elements" [28], but in these studies, a slightly different approach is proposed. Attempts to justify, in a simpler form, the theory of constructing mechanical filters were made in [25]. With representations of the lever as weightless rod, there is a restriction of representation about a set of dynamic properties of mechanical circuits in general and filters in particular. In the theory of electrical circuits [16, 29], an ideal transformer is considered as a device that introduces effects analogous to effects of a lever in mechanical circuits. In particular, it is noted in [16, 28] that the introduction of a transformer, in a certain sense, "breaks" the galvanic circuit; with this, the topology of the circuit system changes (it can be understood as the emergence of some spatial forms of coupling). From our point of view, the essence of the problem lies in the fact that in the transition from mechanical systems (and circuits in particular) consisting of material points connected by typical elements (a spring, a damper) to mechanical systems containing solids in a planar motion, and also their joints, it is necessary to introduce the lever linkages and relationships into the techniques of constructing mathematical models and in the structure of the models themselves, as well as lever mechanisms of various types. Practically, one has to face these problems in the dynamics of power transmissions, study of computational schemes and mathematical models of reducers, differentials, speed boxes, etc. Actually, through the dynamics of planar interaction, a methodology is developed for studying elastic oscillations in systems with spatial solid bodies.

4.6.2 Some Suggestions on Understanding the Role of the "Floating Support"

Before we examine the features of the lever with supports, let us consider a weightless lever, whose fulcrum is not defined and is floating. Such an approach can be found, for example, in [25]. Note that if the weightless lever is implemented in the coupling of S_1 and S_2, in which the motion S_1 and S_2 is determined by the construction means, the lever can occupy an arbitrary position only if one of the connecting elements has a kinematic pair of class IV, which ensures the rotation of the lever relative to the connection point of the element S_1, and another kinematic pair, creating the possibility of sliding the lever, along its axis. If this does not happen, the lever will jam the motion. When considering small motions of the lever and elements in non-degenerate situations, when the conditions ($l_1 = 0$ or $l_2 = 0$) are satisfied, the rod (the lever without the fulcrum) can transmit the motion, forming a random configuration of interaction between the two boundary cases. In the first one, the rod having the ability to slip along one of the kinematic pairs is perpendicular to the lines of motion S_1 and S_2, and in the second the rod is in a position close to the parallel motion S_1 and S_2. In these cases, interaction with a weightless rod with a floating support point can be considered unimplementable.

The coupling of elastic elements through a lever of the second kind. Let us consider a lever scheme with a lever of the second kind, shown in Fig. 4.37. We assume that J is the moment of inertia of the lever relative to point 2^0. Let us find the kinetic energy of the system

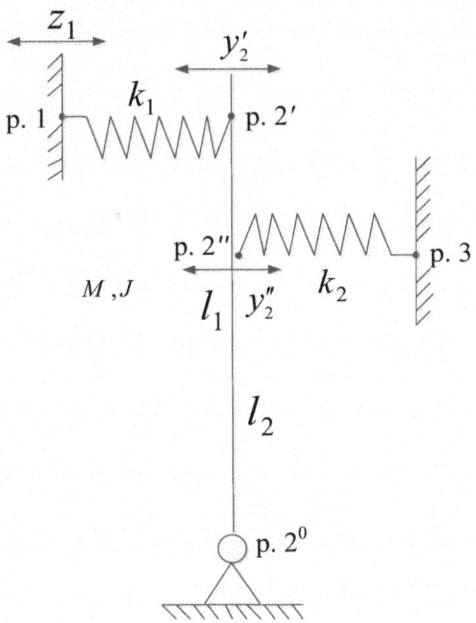

Fig. 4.37 The schematic diagram of the connection of elastic elements with a lever of the second kind

$$T = \frac{1}{2} J\dot{\phi}^2. \tag{4.134}$$

The potential energy of the system takes the following form

$$\Pi = \frac{1}{2} k_1 (y_2' - z_1)^2 + k_2 (y_2'')^2. \tag{4.135}$$

If φ is the angle of a small rotation of the lever, then $y_2' = \phi l_1$, and $y_2'' = \phi l_2$, then

$$\Pi = \frac{1}{2} k_1 (\phi l_1 - z_1)^2 + k_2 (\phi l_2)^2. \tag{4.136}$$

We write the equation of motion for the system shown in Fig. 4.37:

$$J\ddot{\phi} + k_1 l_1^2 \phi - k_1 l_1 z_1 + k_2 l_1^2 \phi = 0$$

or

$$J\phi'' + \phi(k_1 l_1^2 + k_2 l_1^2) = k_1 l_1 z_1. \tag{4.137}$$

The structural diagram of the system has the form shown in Fig. 4.38.

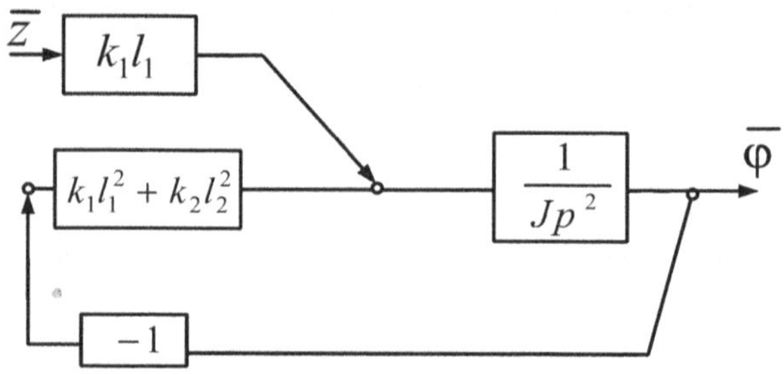

Fig. 4.38 The structural diagram of the system shown in Fig. 4.37, in the coordinates of φ

Let us construct a mathematical model of the system, using the relation $\phi = \frac{y_2''}{l_2}$, then the kinetic and potential energy of the system will take the following form:

$$T = \frac{1}{2}J\left(\frac{\dot{y}_2''}{l_2}\right)^2;\ \Pi = \frac{1}{2}(y_2' - z_1)^2 k_1 + \frac{1}{2}k_2(y_2'')^2.$$

We assume that $y_2' = y_2''\frac{l_1}{l_2}; i = \frac{l_1}{l_2}$, and write the expression for the potential energy:

$$\Pi = \frac{1}{2}k_1(y_2''i - z_1)^2 + \frac{1}{2}k_2(y_2'')^2..$$

Equation (4.137) is transformed to the form

$$m_0\ddot{y}_2'' + y_2''(k_1 i^2 + k_2) = k_1 iz,\ \left(\frac{J}{l_2^2} = m_0\right). \tag{4.138}$$

To find the dynamic response at the point of contact between the element k_2 and the rack, we construct a structural diagram (Fig. 4.39).

To determine the reduced stiffness, it is necessary to find the dynamic response Qp, which occurs in volume 3, which can be done using the circuit in Fig. 4.39b. Let us find the transfer function of the system:

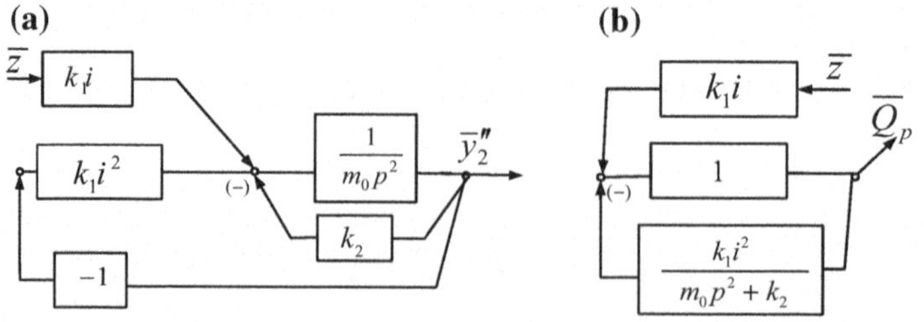

Fig. 4.39 The structural diagram of the system shown in Fig. 4.38. **a** is the general scheme; **b** is the scheme to determine the dynamic response at point 3—$\bar{Q}_p$

$$W(p) = \frac{\bar{Q}_p}{\bar{z}} = \frac{k_1 i(m_0 p^2 + k_2)}{m_0 p^2 + k_1 i^2 + k_2}. \tag{4.138'}$$

The physical meaning of $|W(p)|$ is that the expression module (4.138′) is the reduced dynamic stiffness of the generalized intermediate spring (k_{dyn}). For p = 0, the dynamic stiffness becomes static one, denoted as k_0.

To determine the interaction condition in statics, we take $m_0 = \frac{J}{l_2^2} = 0$ and obtain (for p = 0):

$$W_3(p) = \frac{\bar{Q}_p}{\bar{z}} = \frac{k_1 i k_2}{k_1 i^2 + k_2}. \tag{4.139}$$

It should be noted that, in this case, there is some kind of a consecutive connection. For i = 1, this connection coincides with the expression (4.131), which is consistent for the case m = 0, with which one can fully agree. For m = 0, there is a correspondence in our assumptions to a weightless rod; we note that, for i = 1, there is also a coincidence with the expression (4.139). If $i \neq 1$, then the static stiffness

$$k_0' = \frac{k_1 i k_2}{k_1 i^2 + k_2} \tag{4.140}$$

depends on the gear ratio i. If i = 0, then $k_0' = 0$; if i = 1, then $k_0' = k_0 = \frac{k_1 k_2}{k_1 + k_2}$, which corresponds to the case considered above, when S_1 and S_2 are connected to each other for m = 0, p = 0. The value of i determines the boundary or the measure of the mechanical circuit change in the topological sense. To "emphasize" the effect, one can introduce such a notion as the difference R of reduced stiffness in the presence of a difference in position of the points 2′, 2″:

$$\Delta = l_2 - l_1, \Delta/l_2 = 1 - i. \tag{4.141}$$

We write in this case that

$$\begin{aligned} R &= \frac{k_1 i k_2}{k_1 i^2 + k_2} - \frac{k_1 k_2}{k_1 + k_2} = \frac{k_1^2 k_2 i - k_1^2 k_2 i^2}{(k_1 i^2 + k_2)(k_1 + k_2)} \\ &= \frac{(k_1 k_2)[i^2(k_2 - k_1) + k_1 i - k_2]}{(k_1 + k_2)(k_1 i^2 + k_2)} = \frac{k_1 k_2 (k_1 i + k_2 i - k_1 i^2 - k_2)}{(k_1 + k_2)(k_1 i^2 + k_2)}. \end{aligned} \tag{4.142}$$

The physical meaning of R is that this index is determined through $i = \frac{l_1}{l_2}$, i.e. characterizes the measure of the change in the position of the line of action of forces in the plane (conditionally, this measure reflects the emergence of two-dimensionality in the location of the action of forces). If i = 0, then the intermediate rod (the lever in which l = 0) does not participate in the transmission of the action, and the circuit decomposes. Then, in comparison with the variant without a rod, one can obtain $R = \frac{-k_1 k_2}{k_1 + k_2}$, which is to be expected. For $l = 1$, we get that $R = 0$, that is, the presence of a rod does not change the situation in comparison with its absence. If $i \neq 0$ and $i \neq 1$, then it makes sense to return to the expression (4.142), from which it is possible to find and estimate topological changes for other values of i (excluding the case $i = 0$, $i = 1$ and $i = \infty$). Let us note that the case $i = \infty$ corresponds to $l_2 = 0$, which is analogous to the case $l_1 = 0$ in the physical essence. We reduce (4.142) to the following form:

$$R = \frac{(k_1 k_2)[i^2(k_1 + k_2) - k_1 i^2 - k_2]}{(k_1 + k_2)(k_1 i^2 + k_2)}. \tag{4.143}$$

From the numerator (4.143) we can isolate the equation with respect to i:

$$i^2 - \frac{(k_1 + k_2)i}{k_1} + \frac{k_2}{k_1} = 0, \tag{4.144}$$

solving which, one can find

$$i_{1,2} = \frac{k_1 + k_2}{2k_1} \pm \frac{k_1 - k_2}{2k_1}. \tag{4.145}$$

Then $i_1 = 1$, and $i_2 = \frac{k_2}{k_1}$, whence follows that R can take the value $R = 0$ not only for $l = 1$, but also for $i = \frac{k_2}{k_1}$.

If we exclude the value $i = 0$, $i = 1$ and $i = \infty$, $i = \frac{k_2}{k_1}$ from the consideration, then we can construct the graph of $R(i)$, taking for example $k_1 = 2k$, $k_2 = k_1$, then

$$R = \frac{2k^2[i^2 3k - 2ki^2 - k]}{3k(2ki^2 + k)} = \frac{2}{3} \cdot \frac{2k(-2i^2 + 3i - i)}{2i^2 + 1}. \tag{4.146}$$

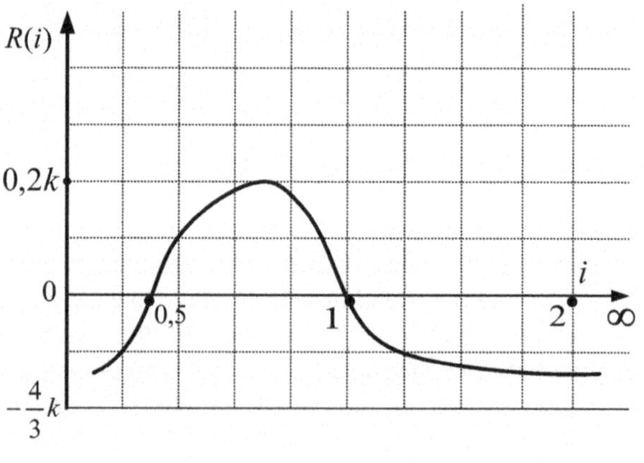

Fig. 4.40 The plot of the dependence $R(i)$

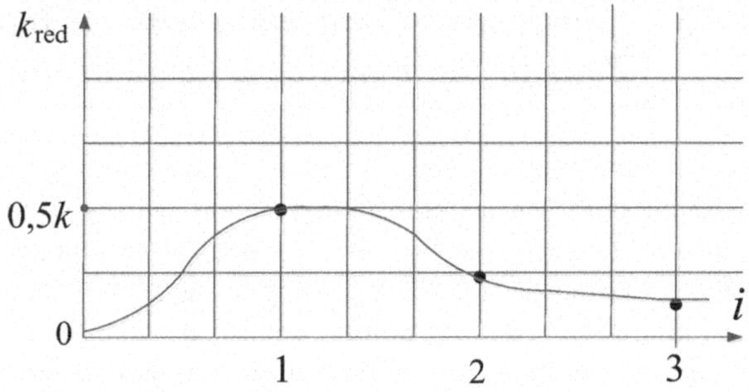

Fig. 4.41 The plot of the dependence of the reduced stiffness on i

Figure 4.40 shows the graph of the dependence $R(i)$, from which (with the elimination of points $i = 0$ and $i = \infty$), that R for $i = 0.5$ and $i = 1$ takes the zero value, and as i increases, the value of R tends to some limit $R = \frac{4}{3}k$ in the negative region.

But if $i = 1$, then for $J = 0$ (a weightless lever), such a function as the separation of interactions along parallel lines or the formation of some interaction space is not implemented.

The properties of the reduced stiffness given the reduced ratio of the lever. Let us consider an example of the dependence of k_{red}, when

$$k_{red} = \frac{ik^2}{i^2k + k} = \frac{ki}{i^2 + 1}. \quad (4.147)$$

From Fig. 4.41 it follows that if i is increased, then the difference between l_1 and l_2 increases. In this case $y_2'' = y_2' \frac{1}{i}$, i.e. the displacement becomes smaller. For $i = \infty$, the point 2″ will be connected with the immobile center, and nothing will be transferred to the element k_2, which can be interpreted as converting it into an element with infinitely high stiffness, then $k_{red} = \frac{k_1 i k_2}{k_1 i^2 + k_2} = 0$, which is typical for the considered fixing conditions.

In turn, if $i = 0$, then this is possible for $l_1 = 0$, when the element K_1 abuts against a fixed point (the center of rotation of the lever or the instantaneous center of velocities). In this case, $k_{\text{red}} = 0$, and the element S_2 moves freely. The circuit breaks, and the reduced stiffness is zero. A similar situation occurs if $i = \infty$ and, for example, $l_2 = 0$: the element S_2 rests on a fixed point, and S_1 starts moving freely. The special cases under consideration are implemented only for $J = 0$, $m_0 = 0$, and the frequency of the dynamic process is not taken into account. The gear ratio in the range $0 < i < \infty$ is a kind of regulator of reduced stiffness, as it was shown in Fig. 4.41 for $k_1 = k = k_2$. If this condition is violated (for example, $k_1 = 10\,k$, $k_2 = k$, then $k_{\text{red}} \approx 0{,}9\,k$), i.e. k_{red} varies from small values to practical coincidence with the stiffness of one of the elements, but does not exceed it.

4.6.3 Consecutive Connection with an Intermediate Solid Body in the Form of a Lever of the First Kind

We assume that in this case there is an analogy in the dynamic processes of a material point on an elastic suspension with a system of rotational type. Let us consider the case (see Fig. 4.41), when the instantaneous velocity center is located between points 2′ and 2″, according to the scheme corresponding to the lever of the first kind (Fig. 4.42) As before, we assume that the lever has the moment of inertia J, and z is the kinematic perturbation. Let us find the expression for the kinetic and potential energy:

$$T = \frac{1}{2}J\dot{\phi}^2; \Pi = \frac{1}{2}k_1(C_2' - z)^2 + \frac{1}{2}k_2(y_2'')^2,$$

where a number of relations are satisfied

$$\phi = \frac{y_2'}{l_1};\quad \phi = \frac{y_2''}{l_2};\quad y_2' = \phi\frac{l_1}{i};\quad y_2'' = y_2' i;\quad i = \frac{l_2}{l_1}.$$

However, here there is a significant difference from the previous case, since the velocities $\dot{y}_2''$ and $\dot{y}_2'$ have different signs. In the general case, it can be assumed that $i = -\frac{l_2}{l_1}$ (and this has its own peculiarities and is taken into account in dynamic interactions, if necessary).

Let

$$\Pi = \frac{1}{2}k_1\left(\frac{y_2''}{i} - z\right)^2 + \frac{1}{2}k_2(-y_2'')^2 = \frac{1}{2}k_1\left(\frac{\phi l_2}{i} - z\right)^2 + \frac{1}{2}k_2(-\phi l_2)^2. \quad (4.148)$$

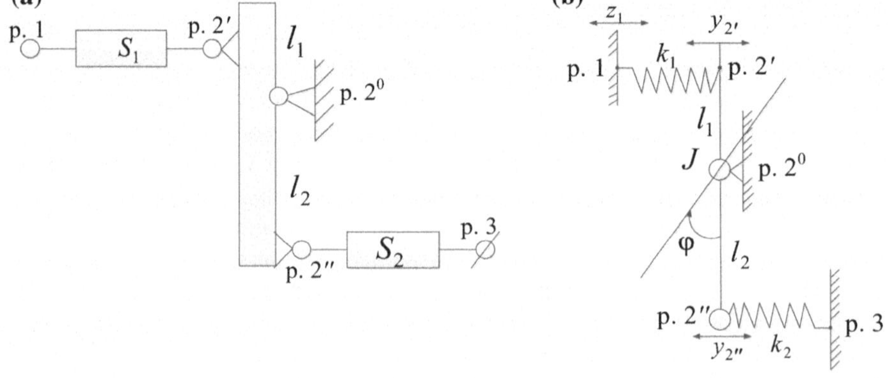

Fig. 4.42 The schematic diagram of the serial connection of elements S_1 and S_2 according to the first-order lever scheme: **a** is the general scheme; **b** is elastic elements

To derive the equation of motion, we take into account that:

$$y_2 = \phi l_1; \quad y_2'' = \phi l_2; \quad y_2' = \frac{l_1}{l_2} y_2''; \quad i = \frac{l_1}{l_2},$$

then

$$\Pi = \frac{1}{2} k_1 (\phi l_2 i - z)^2 + \frac{1}{2} k_2 (\phi l_2)^2,$$

whence

$$J\ddot{\phi} + \phi(k_1 l_2^2 i^2 + k_2 l_2^2) = z k_1 l_2 i. \tag{4.149}$$

Note that $\varphi = \frac{y_2''}{l_2}$, and $m_0 = \frac{J}{l_2^2}$, then (4.149) is transformed:

$$m_0 \ddot{y}_2'' + \ddot{y}_2 (k_1 i^2 + k_2) = z k_1 i. \tag{4.150}$$

Structural diagrams of the considered system are presented in Fig. 4.43.

The transfer function of the system, obtained on the basis of the scheme (Fig. 4.43a) has the following form:

$$W(p) = \frac{k_1 i (m_0 p^2 + k_2)}{m_0 p^2 + k_2 + k_1 i^2}. \tag{4.151}$$

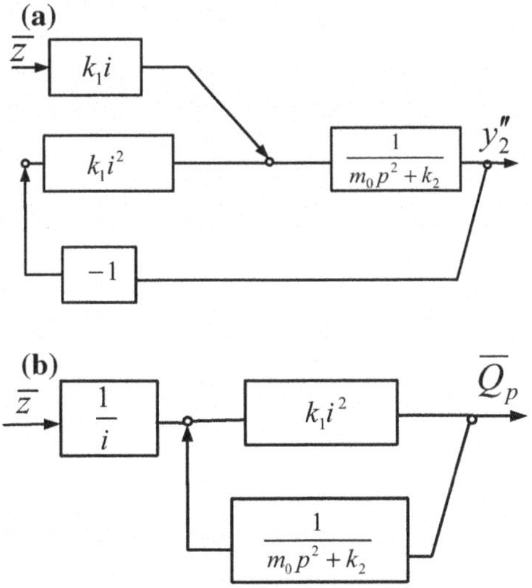

Fig. 4.43 The structural diagram of the system. *a* is the coordinate y_2''; *b* is to determine the dynamic response $\bar{Q}_p$

Since the expression (4.151) coincides with the structure of the expression (4.138′), then studies of possible situations will have a character similar to that of a second-kind lever. Let us consider particular cases of motion of an intermediate solid body.

1. **Progressive rectilinear motion**. In this case, the instantaneous velocity center is at infinity.

Therefore, the displacement transfer coefficient, previously designated as i, will be 1. For $i = 1$, no changes in the nature of the interaction of the elements occur.

But if the mass of a solid (or a lever) is taken into account, then this case is reduced to the previously considered interaction through a material point.

2. **A solid body rotates about a fixed point, and there is no translational motion**. In this case, the problem is reduced to the previously considered cases of a weighty (or weightless) lever of the first and second kind.

4.6.4 Some Connection Properties

1. In the electrical and mechanical circuits connecting elements (typical elementary stars) during the implementation of consecutive connections, some intermediate devices can be used. Ideal voltage transformers, for example, can act as such devices in electrical circuits. Such transformers result in the disruption of the galvanic circuit, since the connection between the voltages of the transformer windings is carried out through an electromagnetic field. In transformers, this is provided by means of cores. The same occurs in mechanical circuits, in particular, in the consecutive connection of elements of a typical extended set of units. The physical form of such a device may be a weightless (or weighty) rod having a point of rotation. The use of such devices in the implementation of consecutive connections in respect to mechanical circuits makes it possible to change the metric of the system, i.e. to form a system of forces at least in the plane. It should be also noted that the idea of introducing a transformer has been implemented in the effects of using the electromagnetic field in various branches of technology.
2. In the mechanical circuits of the device for intermediate coupling with small mass-and-inertia parameters (weightless rods), kinematic functions are performed, which allows us to change the given parameters of the mechanical circuits. The latter can be extended to other combinations of consecutive connections from different types of homogeneous elements.
3. Lever linkages make it possible to introduce a generalized procedure for determining the parameters of the consecutive connection of two elements. In the general form of providing kinematic interaction, the control parameter is the gear ratio of the lever; for $i = 1$ the generalized procedure reduces to the usual one.

4. The proposed approach can be considered as a methodological basis for taking into account in dynamic calculations the properties of real assemblies, equipment and transmissions of various machines, since any kind of mobile couplings in structural implementations are often quite complex, multi-unit mechanisms and devices.
5. The presence in the developed mechanical systems of units in the form of solid bodies with a coupling makes it promising to study the particularities of the linkages discussed above, but in the presence of a fulcrum belonging to the moving unit.

4.6.5 Lever as a Combination of Two Elements: Dynamic Aspects

Let's consider the lever scheme of a technological machine in the form of a chain mechanical system with two degrees of freedom, as shown in Fig. 4.44.

The following designations are used in Fig. 4.44: m_1 and m_2 are the mass of the elements of the chain system; $k_1 - k_4$ are the stiffnesses of the elastic elements; z_1 and z_2 are kinematic external actions; point 1 is the point of connection of the elements k_2 and k_3; points 1′ and 1″ are the ends of the lever with lengths l_1 and l_2, that is, point 1 in Fig. 4.44a is the point of connection of the two elements k_1 and k_2; point 1 in Fig. 4.44b–e is the lever support; in Fig. 4.44f is the center of mass of the system. The input lever in Fig. 4.44b–e is weightless, in Fig. 4.44f—lever has mass m and moment of inertia J. To describe the state of the system, the coordinates $y_1 - y_4$ in the fixed reference frame are used; y and φ (in Fig. 4.44f) are the coordinates of the lever as a solid body.

4.6.6 Comparative Evaluation of the Dynamic Properties of Systems

The problem consists in evaluating the possible forms of joining elements with parameters k_2 and k_3 (see Fig. 4.44a) if the junction point 1 can be separated by the lever (l_1 and l_2) at points 1′ and 1″. The first variant involves considering the support point (item 1) of the lever connected to the fixed stand, as shown in Figs. 4.44b, c. To simplify the subsequent calculations, we assume that the lever also has a moment of inertia J and is able to rotate through an angle φ. In what follows, after the derivation of the equations, it will be assumed that $J = 0$. Thus, the initial computational scheme will have three degrees of freedom y_1, y_2 and φ, where φ is uniquely determined by y_3 and y_4, using the lever linkages.

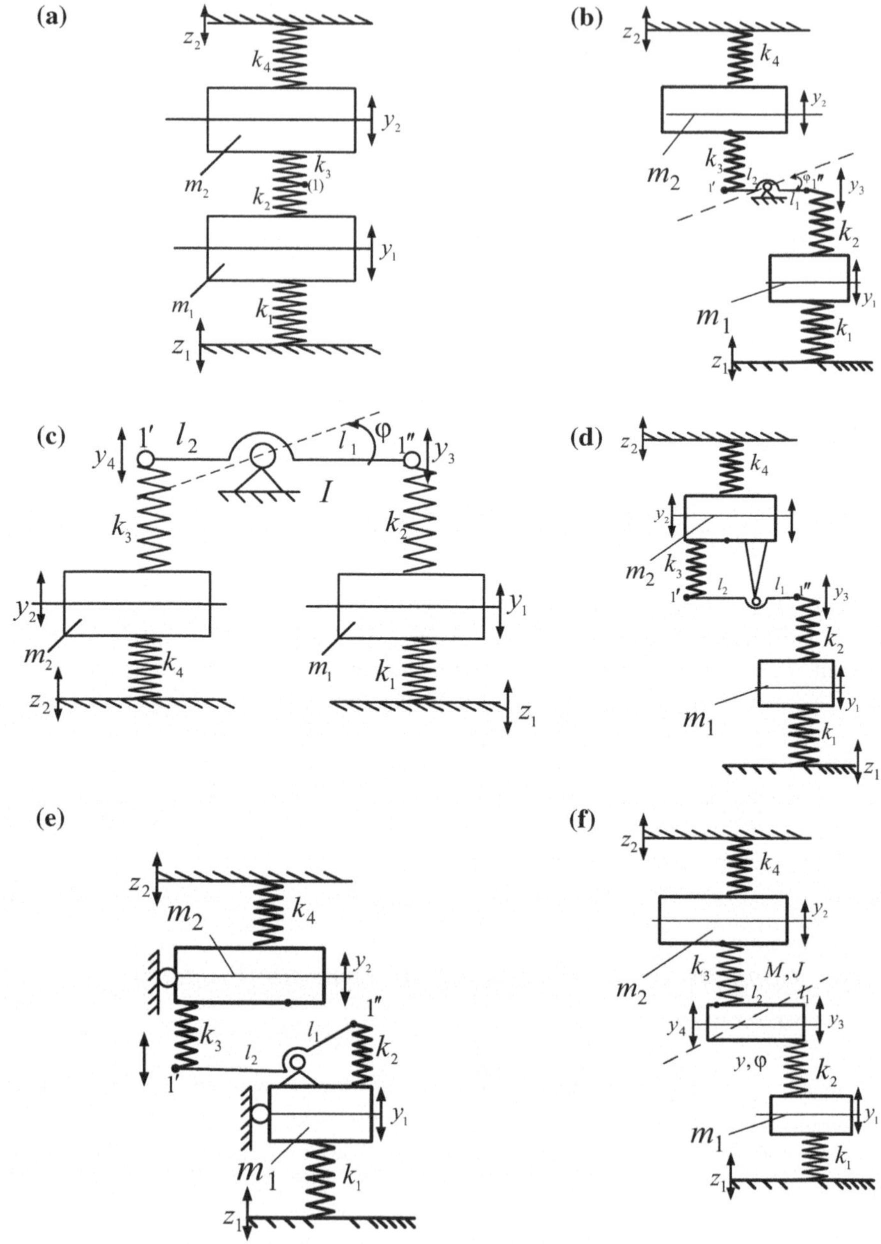

Fig. 4.44 Variants of the location of the contact point of two typical elements k_2 and k_3 when using lever ties. **a** is the initial computational scheme; **b** is the interaction through a lever with a fixed point; **c** is the interaction with a lever that has a symmetrical loading scheme; **d** the lever has a support on element m_2; **e** the lever has a support on an element m_1; *f*—lever has mass M and moment of inertia J

I. As a basic model, let us consider the computational scheme in Fig. 4.44a, where the elements k_2 and k_3 are connected in series, which can be reflected through the reduced stiffness:

$$k_{red} = \frac{k_2 k_3}{k_2 + k_3}.$$

The expression for the kinematic and potential energy of the basic model (see Fig. 4.44a) has the following form:

$$\mathrm{T} = \frac{1}{2} m_1 \dot{y}_1^2 + \frac{1}{2} m_2 \dot{y}_2^2; \tag{4.152}$$

$$\Pi = \frac{1}{2} k_1 (y_1 - z_1)^2 + \frac{1}{2} k_{red} (y_2 - y_1)^2 + \frac{1}{2} k_4 (y_2 - z_2)^2. \tag{4.153}$$

Let us write the system of differential equations of motion:

$$m_1 \ddot{y}_1 + y_1 (k_1 + k_{red}) - y_2 k_{red} = k_1 z_1; \tag{4.154}$$

$$m_2 \ddot{y}_2 + y_2 (k_{red} + k_4) - y_1 k_{red} = k_4 z_2. \tag{4.155}$$

The structural diagram of the dynamically equivalent automatic control system is shown in Fig. 4.45

The system under consideration, as follows from Fig. 4.45, has two partial systems with partial oscillation frequencies:

$$\omega_1^2 = \frac{k_1 + k_{red}}{m_1}; \tag{4.156}$$

$$\omega_2^2 = \frac{k_2 + k_{red}}{m_2} \tag{4.157}$$

Note that the structural diagram is a graphic analog of Eqs. (4.154), (4.155) in the region of Laplace transforms ($p = j\omega$; $j = \sqrt{-1}$). In the system (see Fig. 4.45), the couplings between the partial structures are elastic. In the case under consideration, the resistance forces are considered small. In the system, there are two "input signals" (kinematic perturbation $\bar{z}_1$ and $\bar{z}_2$), which allows us to find a number of transfer functions. We assume that in all these cases the perturbation has one frequency. If, $\bar{z}_1 \neq 0$, and $\bar{z}_2 = 0$, then the transfer function from z_1 with respect to y_1 and y_2 takes the following form:

$$\bar{W}_1 = \frac{\bar{y}_1}{\bar{z}_1} = \frac{k_1 (m_2 p^2 + k_2 + k_{red})}{(m_1 p^2 + k_1 + k_{red})(m_2 p^2 + k_2 + k_{red}) - k_{red}^2}; \tag{4.158}$$

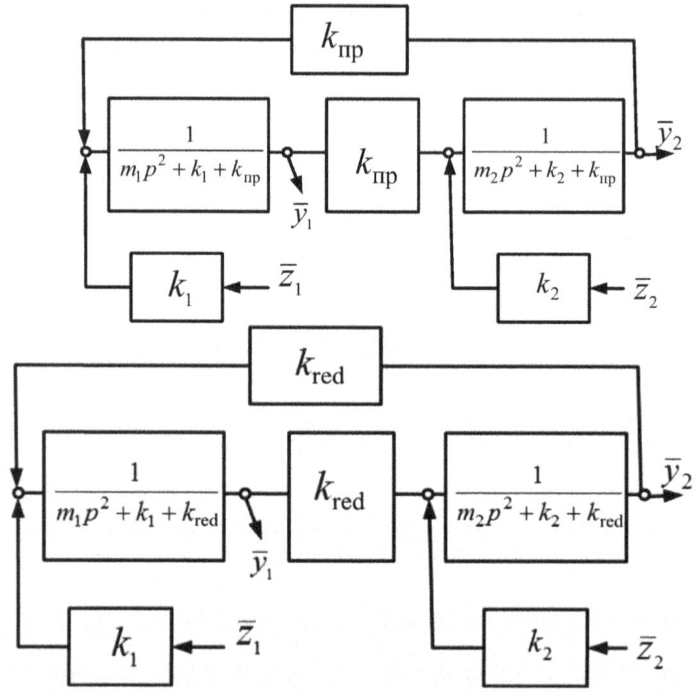

Fig. 4.45 The structural diagram of the system corresponding to Fig. 4.44a

$$\bar{W}_2 = \frac{\bar{y}_2}{\bar{z}_1} = \frac{k_1 k_{red}}{(m_1p^2 + k_1 + k_{red})(m_2p^2 + k_2 + k_{red}) - k_{red}^2}, \tag{4.159}$$

Similarly, other necessary transfer functions can be found.

II. Let us find the expression for the kinetic and potential energy for the system in Fig. 4.44b.

$$T = \frac{1}{2}m_1\dot{y}_1^2 + \frac{1}{2}m_2\dot{y}_2^2 + \frac{1}{2}J\dot{\phi}^2, \tag{4.160}$$

$$\Pi = \frac{1}{2}k_1(y_1 - z_1)^2 + k_2(y_3 - y_1)^2 + \frac{1}{2}k_3(y_4 - y_2)^2 + \frac{1}{2}k_4(y_2 - z_2)^2. \tag{4.161}$$

In the case under consideration

$$\phi = \frac{y_3}{l_1} \text{ or } \phi = \frac{y_4}{l_2},$$

Then, taking into account the type of the lever linkage, it can be assumed that

$$y_3 = \frac{l_1}{l_2} y_4 \text{ or } y_3 = -iy_4, \tag{4.162}$$

where $i = \frac{l_1}{l_2}$ is the gear ratio of the lever. Note that with a lever of a different type, one must take into account the sign i or the relation (4.141) in the directions of the velocities of the lever points.

Then the expressions (4.160) and (4.161) can be transformed to this form:

$$T = \frac{1}{2} m_1 \dot{y}_1^2 + \frac{1}{2} m_2 \dot{y}_2^2 + \frac{y_4^2}{l_2^2}; \tag{4.163}$$

$$\Pi = \frac{1}{2} k_1 (y_1 - z_1)^2 + \frac{1}{2} k_2 (-y_1 - iy_4)^2 + \frac{1}{2} k_3 (y_4 - y_2)^2 + \frac{1}{2} k_4 (y_2 - z_2)^2. \tag{4.164}$$

We assume that $\frac{J}{l_2^2} = m_0$, then the system of differential equations of motion will be as follows:

$$m_1 \ddot{y}_1 + y_1 (k_1 + k_2) + y_2 i y_4 = k_1 z_1; \tag{4.165}$$

$$m_2 \ddot{y}_2 + y_2 (k_3 + k_4) - y_4 k_3 = k_2 z_2; \tag{4.166}$$

$$m_0 \ddot{y}_4 + y_4 (k_2 i^2 + k_3) + y_1 k_2 i - k_3 y_2 = 0. \tag{4.167}$$

The structural diagram of the system in Fig. 4.44b in the coordinates y_1, y_2 and y_4 is shown in Fig. 4.46.

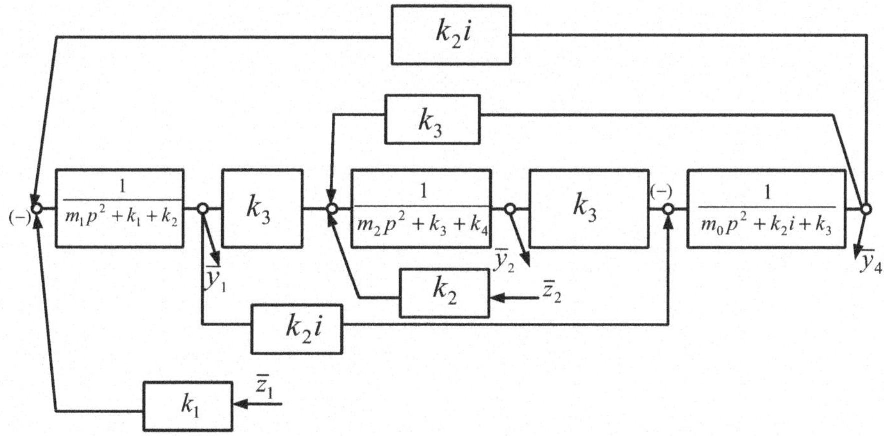

Fig. 4.46 The structural diagram of the system corresponding to the circuit in Fig. 4.44b

From Fig. 4.44b it follows that the introduction of the mass-and-inertia lever added, in comparison with the scheme in Fig. 4.44a, an additional degree of freedom. In this case, the cross-couplings of the system are elastic. There will be three partial systems and their frequencies:

$$\omega_1^2 = \frac{k_1 + k_2}{m_1}; \tag{4.168}$$

$$\omega_2^2 = \frac{k_3 + k_4}{m_2}; \tag{4.169}$$

$$\omega_3^2 = \frac{k_2 i}{m_0}. \tag{4.170}$$

To take into account the effects of introducing a weightless lever (rod), we assume that $m_0 = 0$. From (4.146), we find that

$$y_4 = -y_1 \frac{k_2 i}{k_2 i^2 + k_3} + y_2 \frac{k_3}{k_2 i^2 + k_3}. \tag{4.171}$$

We assume that

$$a = \frac{k_2 i}{k_2 i^2 + k_3}; \quad b = \frac{k_3}{k_2 i^2 + k_3}, \tag{4.172}$$

then

$$y_4 = -y_1 a + y_2 b. \tag{4.173}$$

Substituting (4.173) into (4.165) and (4.166), we obtain the following system of equations

$$m_1 \ddot{y}_1 + y_1 (k_1 + k_2 - k_2 a i) + k_2 i b y_2 = k_1 z_1; \tag{4.174}$$

$$m_2 \ddot{y}_2 + y_2 (k_3 + k_4 - k_3 b) + k_3 a y_1 = k_2 z_2. \tag{4.175}$$

With allowance for the contractions, these equations take the form

$$m_1 \ddot{y}_1 + y_1 \left(\frac{k_1 k_2 i^2 + k_1 k_3 + k_2 k_3}{k_2 i^2 + k_3} \right) + y_2 \frac{k_2 k_3 i}{k_2 i^2 + k_3} = k_1 z_1; \tag{4.176}$$

$$m_2 \ddot{y}_2 + y_2 \left(\frac{k_2 k_3 i^2 + k_4 k_2 i^2 + k_4 k_3}{k_2 i^2 + k_3} \right) + y_1 \frac{k_2 k_3 i}{k_2 i^2 + k_3} = k_2 z_2. \tag{4.177}$$

For $i = 1$ (4.176) and (4.177) reduce to a system of equations that differs from the system of Eqs. (4.144), (4.145) in that the signs before the terms $k_{red} y_1$ in

(4.144) and $k_{red}\, y_2$ in (4.145) are not minuses, but pluses. This means that the introduction of a lever of the first kind at $i = 1$ produces a change in the sign of the coupling between the partial systems. If in the initial system (without the introduction of a first-order lever) such a coupling was positive (4.145), then taking into account the introduction of the linkage of this type, the type of cross-coupling between partial systems changes, it becomes negative. This results in a change in the denominator of the transfer function (4.148), (4.149), i.e. to a change in the characteristic equation of the system. As for the choice of i, for which the system of Eqs. (4.176) (4.177) would coincide with the system of Eqs. (4.144), (4.145), does not exist.

If $i = 0$, this corresponds to $l_1 = 0$, then (4.176), (4.177) take the following form:

$$m_1\ddot{y}_1 + y_1(k_1 + k_2) = k_1 z_1;. \tag{4.178}$$

$$m_2\ddot{y}_2 + y_2 k_4 = k_2 z_2. \tag{4.179}$$

The presence of a support for the lever, in this case, "breaks down" the original system. It turns into a system of two oscillating loops, which do not interact with each other. The remaining part of the lever $l_2 \neq 0$ for $l_1 = 0$ has no effect on motion. However, if $J \neq 0$, then the loop m_2, k_2 forms, together with the fragment of the lever $l_2 \neq 0$, a system with two degrees of freedom, but the interaction with the loop m_1, k_1, k_2 is destroyed. If $i = \infty$, then $l_2 = 0$; in this case, the situation is symmetric with respect to the situation considered above.

III. If the coupling between the elements k_1 and k_2 is implemented using a lever of the second kind whose center of rotation coincides with one of the ends of the lever, then the system of equations of motion (4.176), (4.177) will be as follows:

$$m_1\ddot{y}_1 + y_1\left(\frac{k_1k_2i^2 + k_1k_3 + k_2k_3}{k_2i^2 + k_3}\right) - y_2\frac{k_2k_3i}{k_2i^2 + k_3} = k_1z_1; \tag{4.180}$$

$$m_2\ddot{y}_2 + y_2\left(\frac{k_2k_3i^2 + k_4k_2i^2 + k_4k_3}{k_2i^2 + k_3}\right) - y_1\frac{k_2k_3i}{k_2i^2 + k_3} = k_2z_2. \tag{4.181}$$

With such a transition it is necessary to take into account that expression (4.162) takes the form

$$y_3 = iy_4, \tag{4.182}$$

which results in the significant changes in the properties of the system.

If i does not take extremal values, then for $i < 1$ and $i > 1$ one can observe a change in the dynamic properties of the system. We compare two systems of equations of motion (4.144), (4.145) and (4.180), (4.181):

$$\left.\begin{array}{r}m_1\ddot{y}_1 + y_1(k_1 + k_{red}) - y_2 k_{red} = k_1 z_1;\\ m_2\ddot{y}_2 + y_2(k_{red} + k_4) - y_1 k_{red} = k_4 z_2;\\ m_1\ddot{y}_1 + y_1\left(\frac{k_1 k_2 i^2 + k_1 k_3 + k_2 k_3}{k_2 i^2 + k_3}\right) - y_2 \frac{k_2 k_3 i}{k_2 i^2 + k_3} = k_1 z_1;\\ m_2\ddot{y}_2 + y_2\left(\frac{k_2 k_3 i^2 + k_4 k_2 i^2 + k_4 k_3}{k_2 i^2 + k_3}\right) - y_1 \frac{k_2 k_3 i}{k_2 i^2 + k_3} = k_2 z_2.\end{array}\right\} \tag{4.183}$$

From the system of Eqs. (4.183)

$$k_2 + k_3 i = k_2 i^2 + k_3, \tag{4.184}$$

whence

$$i^2 k_2 - i(k_2 + k_3) + k_3 = 0. \tag{4.185}$$

The solution (4.185) with respect to i has the following form:

$$i = \frac{k_2 + k_3}{2k_2} \pm \frac{(k_2 - k_3)}{2k_2}. \tag{4.186}$$

Thus, we can find the values of i for which the two systems of Eqs. (4.144), (4.145) and (4.180), (4.181) are the same:

$$i_1 = 1; i_2 = \frac{k_3}{k_2}. \tag{4.187}$$

A similar result was obtained in Sect. 4.6 (by the expression (4.145)).

Consideration of the calculation scheme shown in Fig. 4.44c gives results analogous to those obtained for the scheme in Fig. 4.44b. However, the use of the scheme shown in Fig. 4.44c gives a clearer picture of dynamic interactions.

4.6.7 The Case of a Support on a Movable Element

Consider a system whose computational scheme (see Fig. 4.14d) assumes the existence of a support on the mass m_2. In this case, we consider a lever of the first kind, having a moment of inertia J with respect to the support point and mass m. The barycentre of the lever coincides with the fulcrum. Since there is a joint at point 1, the scheme can be transformed according to the scheme shown in Fig. 4.44d (Fig. 4.47).

Note that during the transformation, the fulcrum of the first-kind lever (point 1) is divided by a spring with a stiffness that connects the lever and the mass m2. In further calculations, this spring will be removed under the assumption. We introduce the variable $y - y_2 = y_0$, then we write the expression for the kinetic and potential energy, using the coordinates:

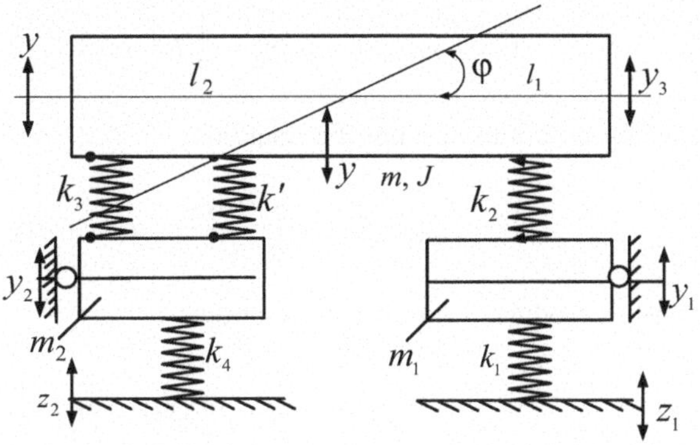

Fig. 4.47 The transformed computational scheme corresponding to the scheme in Fig. 4.44

$$T = \frac{1}{2}J\dot{\phi}^2 + \frac{1}{2}m\dot{y}^2 + \frac{1}{2}m_1\dot{y}_1^2 + \frac{1}{2}m_2\dot{y}_2^2; \tag{4.188}$$

$$\Pi = \frac{1}{2}k_1(y_1 - z_1)^2 + \frac{1}{2}k_2(y_3 - y_1)^2 + \frac{1}{2}k_4(y_2 - z_2)^2 + \frac{1}{2}k_3(y_4 - y_2)^2 + \frac{1}{2}k_0(y_0)^2. \tag{4.189}$$

For calculations, we introduce a series of relations:

$$y = ay_3 + by_4, \quad a = \frac{l_1}{l_1 + l_2}; \quad b = \frac{l_2}{l_1 + l_2}; \quad \phi = c(y_4 - y_3); \quad c = \frac{1}{l_1 + l_2}. \tag{4.190}$$

We transform (4.188), (4.189), using the relations (4.190), and we obtain

$$T = \frac{1}{2}m_1\dot{y}_1^2 + \frac{1}{2}m_2\dot{y}_2^2 + \frac{1}{2}Jc^2(\dot{y}_4 - \dot{y}_3)^2 + \frac{1}{2}m(a\dot{y}_3 + b\dot{y}_4)^2; \tag{4.191}$$

$$\Pi = \frac{1}{2}k_1(y_1 - z_1)^2 + \frac{1}{2}k_2(y_3 - y_1)^2 + \frac{1}{2}k_4(y_2 - z_2)^2 + \frac{1}{2}k_3(y_4 - y_2)^2 + \frac{1}{2}k_0(y_0)^2,$$

where $y_0 = y - y_2$. Then assuming that $ay_3 + by_4 - y_2 = y_0$, we write the potential energy in the following form:

$$\Pi = \frac{1}{2}k_1(y_1 - z_1)^2 + \frac{1}{2}k_2(y_3 - y_1)^2 + \frac{1}{2}k_4(y_2 - z_2)^2 + \frac{1}{2}k_3(y_4 - y_2)^2 + \frac{1}{2}k_0(ay_3 + by_4 - y_2)^2. \tag{4.192}$$

We note that $\dot{y} = y_0 + y_2 = ay_3 + by_4$, then in the expression for the kinetic energy

$$T = \frac{1}{2} m_1 \dot{y}_1^2 + \frac{1}{2} m_2 \dot{y}_2^2 + \frac{1}{2} Jc^2 (\dot{y}_4 - \dot{y}_3)^2 + \frac{1}{2} m (\dot{y}_0 + \dot{y}_2)^2$$

one can take into account the relations

$$y_3 = \frac{y_0 + y_2 - by_4}{a} = a_1 y_0 + a_1 y_2 - b_1 y_4.$$

We introduce a number of notations:

$$y_3 = a_1 y_0 + a_1 y_2 - b_1 y_4; \quad a_1 = \frac{1}{a}; \quad b_1 = \frac{b}{a}.$$

We write the kinetic energy in this form:

$$T = \frac{1}{2} m_1 \dot{y}_1^2 + \frac{1}{2} m_2 \dot{y}_2^2 + \frac{1}{2} Jc^2 (\dot{y}_4 - a_1 \dot{y}_0 + a_1 \dot{y}_2 - b_1 \dot{y}_4)^2 + \frac{1}{2} m (a a_1 \dot{y}_0 + a a_1 \dot{y}_2 - b_1 a \dot{y}_4 + b \dot{y}_4)^2 \tag{4.193}$$

or

$$T = \frac{1}{2} m_1 \dot{y}_1^2 + \frac{1}{2} m_2 \dot{y}_2^2 + \frac{1}{2} Jc^2 (b_2 y_4 - a_1 y_0 - a_1 y_2)^2 + \frac{1}{2} m (\dot{y}_0 + \dot{y}_2)^2, \tag{4.193$'$}$$

where coordinates (y_1, y_2, y_4, y_0) are present. The potential energy of the system in this case takes the form

$$\begin{aligned} \Pi = & \frac{1}{2} k_1 (y_1 - z_1)^2 + \frac{1}{2} k_2 (a_1 y_0 + a_1 y_2 - b_1 y_4 - y_1)^2 + \frac{1}{2} k_4 (y_2 - z_2)^2 \\ & + \frac{1}{2} k_3 (y_4 - y_2)^2 + \frac{1}{2} k_0 (y_0)^2. \end{aligned} \tag{4.194}$$

We make a number of auxiliary calculations and write the system of equations of motion:

$$m_1 \ddot{y}_1 + y_1 (k_1 + k_2) - y_2 (-k_2 a_1) + y_4 (k_2 b_1) + y_0 (-k_2 a_1) = k_1 z_1; \tag{4.195}$$

$$\begin{aligned} & \ddot{y}_2 (m_2 + Jc^2 a_1^2 + m) + y_2 (k_2 a_1^2 + k_4 + k_3) + y_1 (-k_2 a_1) + \ddot{y}_4 (-Jc^2 a_1 b_2) \\ & + y_4 (-k_2 a_1 b_1 - k_3) + \ddot{y}_0 (Jc^2 a_1^2 + m) + y_0 (k_2 a_1^2) = k_4 z_2; \end{aligned} \tag{4.196}$$

$$\begin{aligned} & y_1 (k_2 b_1) + \ddot{y}_2 (-Jc^2 a_1 b_2) + y_2 (-k_2 a_1 b_1 - k_3) + \ddot{y}_4 (Jc^2 b_2^2) \\ & + y_4 (k_2 b_1^2 + k_3) + \ddot{y}_0 (-Jc^2 a_1 b_2) + y_0 (-k_2 a_1 b_1) = 0; \end{aligned} \tag{4.197}$$

$$\begin{aligned} & \ddot{y}_2 (Jc^2 a_1^2) + y_2 (k_2 a_1^2) + y_1 (-k_2 a_1) + \ddot{y}_4 (-Jc^2 a_1 b_2) \\ & + y_4 (-k_2 a_1 b_1 + k_3) + \ddot{y}_0 (Jc^2 a_1^2) + y_0 (k_0 + k_2 a_1^2) = 0. \end{aligned} \tag{4.198}$$

Table 4.1 The coefficients of the equations of motion (4.195), (4.196)

a_{11} (y_1)	a_{12} (y_2)	a_{13} (y_4)	a_{14} (y_0)
$m_1p^2 + k_1 + k_2$	$-k_2a_1$	k_2b_1	$-k_2a_1$
a_{21}	a_{22}	a_{23}	a_{24}
$-k_2a_1$	$(m_2 + m + Jc^2a_1^2)p^2 +$ $+ k_2a_1^2 + k_3 + k_4$	$-Jc^2a_1b_2p^2 - k_3 - k_2a_1b_1$	$(m + Jc^2a_1^2)p^2 + k_2a_1^2$
a_{31}	a_{32}	a_{33}	a_{34}
k_2b_1	$-Jc^2a_1b_2 - k_2a_1b_1 - k_3$	$Jc^2b_2p^2 + k_2b_1^2 + k_3$	$-Jc^2a_1b_2 - k_2a_1b_1$
a_{41}	a_{42}	a_{43}	a_{44}
$-k_2a_1$	$-Jc^2a_1b_2 - k_2a_1b_1 - k_3$	$Jc^2b_2p^2 + k_2b_1^2 + k_3$	$-Jc^2a_1b_2 - k_2a_1b_1$
Q_1	Q_2	Q_3	Q_4
k_1z_1	k_4z_2	0	0

Note Q1–Q4 are generalized forces

The coefficients of the equations of motion in coordinates (y_1, y_2, y_4, y_0) are shown in Table 4.1.

To proceed to the coordinate system $(y_1 - y_3)$, we use the method described in [6], and produce "zeroing" of y_0, we obtain a system of equations in coordinates (y_1, y_2, y_4). Assume that $J = 0$ and $m = 0$, then we first obtain equations

$$\begin{cases} m_1\ddot{y}_1 + y_1(k_1 + k_2) - y_2k_2a_1 + y_4k_2b_1 = k_1z_1; \\ m_2\ddot{y}_2 + y_2(k_2a_1^2 + k_4 + k_3) - y_1k_2a_1 - y_4(k_2a_1b_1 + k_3) = k_4z_2; \\ y_1k_2b_1 - y_2(k_2a_1b_1 + k_3) + y_4(k_2b_1^2 + k_3) = 0. \end{cases} \quad (4.199)$$

The third equation from system (4.199) can be written in the following form

$$y_1a_3 - a_4y_2 + a_5y_4 = 0, \quad (4.200)$$

where $a_3 = k_2b_1; \quad a_4 = k_2a_1b_1 + k_3; \quad a_5 = k_2b_1^2 + k_3$, which is used to obtain the coordinates:

$$y_4 = \frac{a_4y_2 - a_3y_1}{a_5} = b_3y_2 - b_4y_1, \quad (4.201)$$

where, in turn, $b_3 = \frac{a_4}{a_5}$; $b_4 = \frac{a_3}{a_5}$.

Let us carry out the substitution (4.201) into the system of Eqs. (4.199) and obtain

$$m_1\ddot{y}_1 + y_1(k_1 + k_2 - k_2b_1b_4) - y_2(k_2a_1 - k_2b_1b_3) = k_1z_1; \quad (4.202)$$

$$m_2\ddot{y}_2 + y_2(k_2a_1^2 + k_4 + k_3 - k_2a_1b_1b_3 - k_3b_3) - y_1(k_2a_1 - k_2a_1b_1b_4 - k_2b_1b_4) = k_4z_2. \quad (4.203)$$

For further application, we shall formulate a list of relations

$$\begin{gathered} a=\frac{l_1}{l_1+l_2};b=\frac{l_2}{l_1+l_2};c=\frac{1}{l_1+l_2};a_1=\frac{1}{a};b_1=\frac{b}{a};b_2=1+b_1=\frac{a+b}{a}; \\ a_3=k_2b_1;a_4=k_2a_1b_1+k_3;a_5=k_2b_1^2+k_3;b_3=\frac{a_4}{a_5};b_4=\frac{a_3}{a_5}. \end{gathered} \tag{4.204}$$

Using (4.204) and taking the values of l_1 and l_2 equal to zero simultaneously and alternately, we can simplify the generalized computational schemes in Fig. 4.44b–e, which are eventually reduced to the basic or initial scheme shown in Fig. 4.44a.

We assume that $i=\frac{l_2}{l_1}$, then the relations (4.204) can be transformed as follows:

$$\begin{gathered} a=\frac{l_1}{l_1+l_2}=1+i;b=\frac{l_2}{l_1+l_2}=\frac{i}{1+i};b_1=i;b_2=1+i;a_3=k_2i; \\ a_4=k_2(i+1)i+k_3;a_5=k_2i^2+k_3;b_3=\frac{k_2(i+1)i+k_3}{k_2i^2+k_3};b_4=\frac{k_2i}{k_2i^2+k_3}. \end{gathered} \tag{4.205}$$

The simplifications lead to comparable results.

4.6.8 *The Case of Joining Two Elements with an Intermediate Link in the Form of a Lever of the Second Kind*

An option is considered for the computational scheme in Fig. 4.44d, when a lever of the second kind is used. In this case, the interaction scheme of the elements will be different, as shown in Fig. 4.48.

The computational scheme (Fig. 4.48) takes the following notation: the segment (1)(1″) corresponds to l_2, the segment (1)(1′) corresponds to l_1, the lever has the moment of inertia J relative to the support (p. 1), φ is the deflection angle of the lever from the mass element m_2; y_3 is the coordinate of the spring attachment point k_3 − p. (1′), $i=\frac{l_2}{l_1}$ The expression for the kinetic and potential energy has the following form:

$$T=\frac{1}{2}m_1\dot{y}_1^2+\frac{1}{2}m_2\dot{y}_2^2+\frac{1}{2}m_0\dot{y}_3^2; \tag{4.206}$$

$$\Pi=\frac{1}{2}k_1(y_1-z_1)^2+\frac{1}{2}k_2(y_3'-y_1)^2+\frac{1}{2}k_3(y_3-y_2)^2+\frac{1}{2}k_4(y_4-z_2)^2. \tag{4.207}$$

We introduce a number of relationships: the angle of rotation φ in the relative motion of the lever $\phi=\frac{y_2-y_4}{l_2}$; $y_4'=y_2-y_4$; $y_3'=\frac{y_4'}{i}$; $y_4'=\phi l_2$; y_4', y_3' are the coordinates of the points of attachment of the springs in relative motion; we express

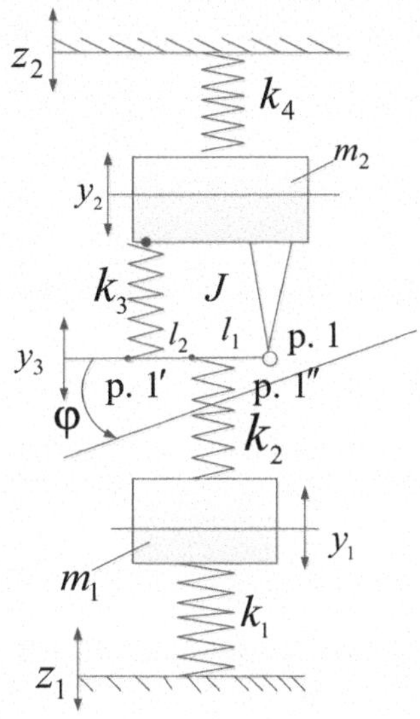

Fig. 4.48 The computational scheme of the system in Fig. 4.44d, but with a lever of the second kind

y_4' through y_3': $y_4' = y_3' i$. Then the absolute velocity of the reduced mass $\left(m_0 = \frac{J}{l_2^2}\right)$ is determined $\dot{y}_4 = \dot{y}_2 + \dot{y}_4'$. The displacement of the attachment point of the spring k_2 takes the form $y_3 = y_2 + y_3' = y_2 + y_4'$. We write (4.206), (4.207) in the following form:

$$T = \frac{1}{2} m_1 \dot{y}_1^2 + \frac{1}{2} m_2 \dot{y}_2^2 + \frac{1}{2} m_0 \dot{y}_4^2 = \frac{1}{2} m_1 \dot{y}_1^2 + \frac{1}{2} m_2 \dot{y}_2^2 + \frac{1}{2} m_0 (\dot{y}_2 - \dot{y}_4')^2; \quad (4.208)$$

$$\Pi = \frac{1}{2} k_1 (y_1 - z_1)^2 + \frac{1}{2} k_2 (y_2 - y_3' - y_1)^2 + \frac{1}{2} k_3 (y_4')^2 + \frac{1}{2} k_4 (y_2 - z_2)^2. \quad (4.209)$$

We assume that $y_4' = y_2 - y_1$; $y_3' = \frac{y_4'}{i} = \frac{y_2 - y_4}{i}$. Let us we make a series of computations and obtain a system of equations

$$m_1 \ddot{y}_1 + y_1 (k_1 + k_2) + y_2 (-k_2) + \frac{k_2 y_4'}{i} = k_1 z_1; \quad (4.210)$$

$$\ddot{y}_2 (m_0 + m_2) + y_2 (k_2 + k_4) - y_1 k_2 - y_4' k_2 - \ddot{y}' m_0 = k_2 z_2; \quad (4.211)$$

$$\ddot{y}_4' m_0 + y_4' \left(k_3 + \frac{k_2}{i}\right) + y_1 \frac{k_2}{i} - \ddot{y}_2'' m_0 - \frac{k_2}{i} y_2'' = 0. \quad (4.212)$$

If $m_0 = 0$ $\left(m_0 = \frac{J}{l_2^2}\right)$, then the system of equations is as follows:

$$m_1\ddot{y}_1 + y_1(k_1 + k_2) - y_2k_2 + k_2y_4' = k_1z_1; \tag{4.213}$$

$$m_2\ddot{y}_2 + y_2(k_2 + k_4) - y_1k_2 - \frac{y_4'k_2}{i} = k_2z_2; \tag{4.214}$$

$$k_3y_4' + \frac{y_4'k_2}{i} - \frac{y_2k_2}{i} + \frac{y_1k_2}{i} = 0. \tag{4.215}$$

From (4.215) we find the value of the coordinate:

$$y_4' = y_2\frac{k_2i}{k_3i^2 + k_2} - \frac{k_2i}{k_3i^2 + k_2}y_1. \tag{4.216}$$

After substituting (4.216) into (4.213) and (4.214), we obtain

$$m_1\ddot{y}_1 + y_1\left(k_1 + k_2 - \frac{k_2^2i}{k_3i^2 + k_2}\right) - y_2\left(k_2 - \frac{k_2^2i}{k_3i^2 + k_2}\right) = k_1z_1; \tag{4.217}$$

$$m_2\ddot{y}_2 + y_2\left(k_2 + k_4 - \frac{k_2^2i}{k_3i^2 + k_2}\right) - y_1\left(k_2 - \frac{k_2^2i}{k_3i^2 + k_2}\right) = k_2z_2, \tag{4.218}$$

or

$$m_1\ddot{y}_1 + y_1\left[\frac{(k_1 + k_2)k_3i^2 + k_1k_2}{k_3i^2 + k_2}\right] - y_2\left[\frac{k_2k_3i^2}{k_3i^2 + k_2}\right] = k_1z_1; \tag{4.219}$$

$$m_2\ddot{y}_2 + y_2\left[\frac{(k_2 + k_4)k_3i^2 + k_2k_4}{k_3i^2 + k_2}\right] - y_1\left[\frac{-k_3i^2k_2}{k_3i^2 + k_2}\right] = k_2z_2. \tag{4.220}$$

If we assume that $l_2 = 0$, then (4.219), (4.220) are transformed into a system of equations:

$$m_1\ddot{y}_1 + y_1k_1 = k_1z_1; \tag{4.221}$$

$$m_2\ddot{y}_2 + y_2k_4 = k_2z_2. \tag{4.222}$$

For $i = 1$ $(l_2 = l_1)$ we obtain respectively

$$m_1\ddot{y}_1 + y_1\left(\frac{k_1k_3 + k_2k_3 + k_1k_2}{k_2 + k_3}\right) - y_2\frac{k_3k_2}{k_3 + k_2} = k_1z_1; \tag{4.223}$$

$$m_2\ddot{y}_2 + \frac{k_2k_3 + k_3k_4 + k_2k_4}{k_3 + k_2} - y_2\frac{k_3k_2}{k_2 + k_3} = k_2z_2. \tag{4.224}$$

We transform (4.223), (4.224) to the form

$$m_1\ddot{y}_1 + y_1(k_1 + k_{\text{red}}) - y_2 k_{red} = k_1 z_1; \qquad (4.225)$$

$$m_2\ddot{y}_2 + (k_4 + k_{red}) - y_2 k_{red} = k_2 z_2, \qquad (4.226)$$

where $k_{red} = \frac{k_3 k_2}{k_3 + k_2}$.

We note that the system of Eqs. (4.225), (4.226) corresponds to the configuration of the computational scheme in Fig. 4.44a. Such a situation can only be for $i = 1$. If $l_1 = 0$, then $i \to \infty$, in this case the elastic element k_3 "leaves" the interaction.

When considering the computational scheme in Fig. 4.48 note that for $l_1 = 0$, the spring k_2 connects m_1 and m_2 directly, and the elastic element k_3 "exits" from the interaction, since $l_2 \neq 0$. The system in this case completely coincides with the chain system of the usual kind. A similar situation occurs for $i = 1$. If $l_2 = 0$ for $l_1 = 0$, then the situation remains the same ($i = 1$). If $l_2 = 0$, and $l_1 \neq 0$, then the system splits into two blocks $(m_1 p^2 + k_1)$ and $(m_2 p^2 + k_4)$, and the coupling of these blocks occurs only at $l_1 = 0$.

4.6.9 The Version of the Unsupported Lever

Consider a system of general form when the lever has no support (Fig. 4.49).

Expressions for the kinetic and potential energy have the following form:

$$T = \frac{1}{2}m_1\dot{y}_1^2 + \frac{1}{2}m_2\dot{y}_2^2 + \frac{1}{2}\dot{y}^2 + \frac{1}{2}J\dot{\phi}^2; \qquad (4.227)$$

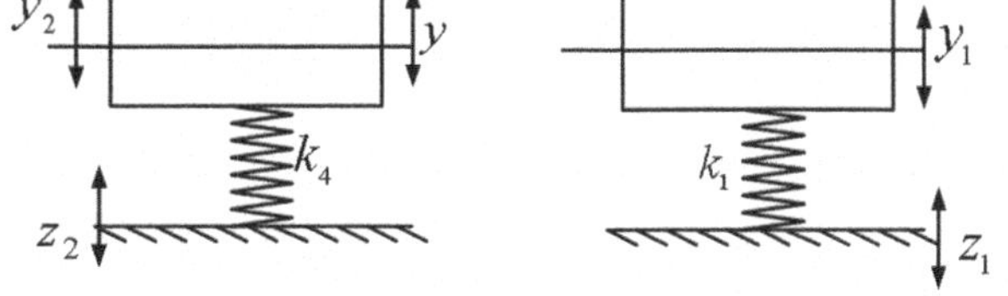

Fig. 4.49 The computational scheme of a system with two degrees of freedom with a lever without a fixed point of support of the lever

$$\Pi = \frac{1}{2}k_1(y_1 - z_1)^2 + \frac{1}{2}k_2(y_3 - y_1)^2 + \frac{1}{2}k_3(y_4 - y_2)^2 + \frac{1}{2}k_4(y_2 - z_2)^2. \quad (4.228)$$

We introduce a number of relations:

$$\begin{aligned} &y_3 = y + l_2\phi; \quad y_4 = y - l_2\phi; \quad \phi = \frac{y_3 - y_4}{l_1 + l_2}; \quad y = ay_4 + by_3; \\ &a = \frac{l_1}{l_1 + l_2}; \quad b = \frac{l_2}{l_1 + l_2}; \quad c = \frac{1}{l_1 + l_2}. \end{aligned} \quad (4.229)$$

We transform (4.227), (4.228) to the form

$$T = \frac{1}{2}m_1\dot{y}_1^2 + \frac{1}{2}m_2\dot{y}_2^2 + \frac{1}{2}(a\dot{y}_4 + b\dot{y}_3)^2 + \frac{1}{2}Jc^2(\dot{y}_3 - \dot{y}_4)^2; \quad (4.230)$$

$$\Pi = \frac{1}{2}k_1(y_1 - z_1)^2 + \frac{1}{2}k_2(y_3 - y_1)^2 + \frac{1}{2}k_3(y_4 - y_2)^2 + \frac{1}{2}k_4(y_2 - z_2)^2. \quad (4.231)$$

We make a series of intermediate calculations and write the system of equations of motion in coordinates $(y_1, y_2, y_4, y_0) \rightarrow (y_1, y_2, y_3, y_4)$:

$$m_1\ddot{y}_1 + y_1(k_1 + k_2) + y_3(-k_2) = k_1z_1; \quad (4.232)$$

$$m_2\ddot{y}_2 + y_2(k_3 + k_4) - y_4k_3 = k_4z_2; \quad (4.233)$$

$$\ddot{y}_3(Mb^2 + Jc^2) + y_3(k_2) - \ddot{y}_4(Jc^2 - Mab) + y_1(-k_2) = 0; \quad (4.234)$$

$$\ddot{y}_4(Ma^2 + Jc^2) + y_4k_3 - \ddot{y}_3(Jc^2 - Mab) - k_3y_2 = 0. \quad (4.234')$$

We assume that $M = 0$, $J = 0$, i.e. the lever in this case is weightless, but there is a turning point in it, then

$$m_1\ddot{y}_1 + y_1(k_1 + k_2) - y_3k_2 = k_1z_1; \quad (4.235)$$

$$m_2\ddot{y}_2 + y_2(k_2 + k_4) - y_4k_3 = k_4z_2; \quad (4.236)$$

$$y_3k_2 - y_1k_2 = 0; y_4k_3 - y_2k_3 = 0, \quad (4.237)$$

from where $y_4 = y_2$.

After the substitutions in $y_3 = y_1$ (4.235), (4.236) we obtain

$$m_1\ddot{y}_1 + y_1(k_1 + k_2) - y_1k_2 = k_1z_1; \quad (4.238)$$

$$m_2\ddot{y}_2 + y_2k_3 + k_4y_2 - y_2k_3 = k_4z_2. \quad (4.239)$$

In this case the system (4.232)–(4.234) reduces to the form:

$$m_1\ddot{y}_1 + y_1k_1 = k_1z_1; \tag{4.240}$$

$$m_2\ddot{y}_2 + y_2k_4 = k_4z_2. \tag{4.241}$$

If the lever does not have mass-and-inertia properties, then it does not play any role in the circuit in Fig. 4.44f; the system of equations splits into two independent equations—(4.240) and (4.241).

We use the Laplace transforms and write down

$$\left.\begin{array}{c}\bar{y}_3(A) - \bar{y}_4(B) - \bar{y}_1k_2 = 0,\\ \bar{y}_3B + y_4C - k_3y_2 = 0.\end{array}\right\} \tag{4.242}$$

If J and M are not equal to zero, then it follows from the system of equations that when

$$A = (Mb^2 + Jc^2)p^2 + k_2; \quad B = (Mab - Jc^2)p^2; \quad C = (Ma^2 + Jc^2)p^2 + k_3$$

it is possible to transform the system of Eqs. (4.232)–(4.234). Using (4.242), we find that

$$y_4 = \frac{y_2Ak_3 + y_1k_2B}{AC - B^2}; \tag{4.243}$$

$$y_3 = \frac{y_1k_2C + y_2k_3B}{AC - B^2}. \tag{4.244}$$

Let $y_4 = k_2a_1y_1 + k_3b_1y_2$, where $a_1 = \frac{B}{AC-B^2}$; $b_1 = \frac{A}{AC-B^2}$; in turn, $y_3 = a_2k_2y_1 + b_2k_3y_2$ where i.e. $a_1 = \frac{B}{AC-B^2}$; $b_1 = \frac{A}{AC-B^2}$; that is, $b_2 = a_1$.

We make the substitution of (4.242), (4.243) into (4.232) and (4.233), we obtain:

$$\bar{y}_1(m_1p^2 + k_1 + k_2) - k_2(a_2k_2y_1 + b_2k_3y_2) = k_1z_1; \tag{4.245}$$

$$\bar{y}_2(m_2p^2 + k_3 + k_4) - k_3(a_1k_2y_1 + b_2k_3y_2) = k_2z_2 \tag{4.246}$$

or

$$\bar{y}_1(m_1p^2 + k_1 + k_2 - a_2k_2^2) - y_2k_2k_3b_2 = k_1\bar{z}_1; \tag{4.247}$$

$$\bar{y}_2(m_2p^2 + k_3 + k_4 - b_1k_3^2) - y_1k_3k_2a_1y_1 = k_2z_2. \tag{4.248}$$

Since $a_1 = b_2$, then in the system of Eqs. (4.247), (4.248) there is some symmetry. Let us consider

$$k_2 - a_2 k_2^2 = k_2 - \frac{k_2^2 C}{AC - B^2} = \frac{ACk_2 - k_2 B^2 - k_2^2 C}{AC - B^2};$$

$$k_2 - a_2 k_2^2 = \frac{\left[(Mb^2 + Jc^2)p^2 + k_2\right]\left[(Ma^2 + Jc^2)p^2 + k_3\right]k_2 -}{AC - B^2}$$
$$\frac{-k_2\left[(Mab - Jc^2)p^2\right]^2 - k_2\left[(Ma^2 + Jc^2)p^2 + k_3\right]}{AC - B^2}.$$

If $p = 0$, then $k_2 - a_2 k_2^2 = \frac{k_2^2 k_3 - k_2^2 k_3}{k_2 k_3} = 0$. If $p^2 \to \infty$, then

$$k_2 - a_2 k_2^2 = \frac{k_2(Mb^2 + Jc^2)(Ma^2 + Jc^2) - (Mab - Jc^2)^2 k_2}{(Mb^2 + Jc^2)(Ma^2 + Jc^2) - (Mab - Jc^2)^2} = k_2.$$

A similar result will be found for $k_3 - k_3^2 b_1$.

That is, for $p \to \infty$ and $M \neq 0$, $J \neq 0$ the system splits into two independent blocks (the motions for y_1 and y_2 are not connected). If $p = 0$, then there is no interaction effect, as for $p \to \infty$, but the reasons for this are different. If $p \neq 0$ and $p \neq \infty$, then we have an intermediate case. With regard to the evaluation of the influence of the intermediate body between m_1 and m_2 through point 1, there are the following peculiarities of introduction of a weighty lever with a floating support.

In the static case, the connection between m_1 and m_2 for $p = 0$ was ensured through an elastic element with reduced stiffness determined by the expression (4.151) $k_{red} = \frac{k_2 k_3}{k_2 + k_3}$. When a lever is introduced for $p = 0$, we must take into account the coefficients in the cross-coupling $a_1 = b_2$. For $p = 0$ they are zero, since $b_2 = \frac{B}{AC - B^2}$, and $B = (Mab - Jc^2)p^2$. Let us estimate the values of the coefficients a_2 and b_1: $a_2 = \frac{C}{AC - B^2} = \frac{k_3}{k_2 k_3} = \frac{1}{k_2}, b_1 = \frac{1}{k_3}$. If we substitute these values into Eqs. (4.247), (4.248), then these equations break up into autonomous ones; in this case, the connections between the elements m_1 and m_2 through the elastic elements k_2 and k_3 are not carried out. If we take into account the effect of forces at different frequencies, then there is one symmetry relation that makes the value B = 0 and, consequently, b_1 and a_2 too. In this case, at any frequency, the elements m_1 and m_2 do not interact.

General assessment of possible dynamic properties. In the general case, the system, the calculation scheme of which is shown in Fig. 4.44e has four degrees of freedom, taking into account the weighty lever. Therefore, we can expect that the interaction between m_1 and m_2 will undergo significant changes 4 times at resonant frequencies. The value of the two frequencies, associated with the lever, is determined from the characteristic equation

$$AC - B^2 = 0.$$

When resonance frequencies are reached, the coefficients b_1, b_2, a_2 become infinite. If we take each of the coefficients a_2 and b_1, then they take, at certain frequencies, zero values, determined by the following formulas:

$$\omega_{1dyn}^2 = \frac{k_2}{b^2 + Jc^2} \quad \omega_{2dyn}^2 = \frac{k_3}{a^2 + Jc^2}.$$

At these frequencies, the system of elements m_1 and m_2, initially having a chain structure and an cross-coupling of the elastic type, behaves like a solid on elastic supports $k_1 + k_2$ and $k_3 + k_4$ with a cross-coupling of the inertial type. A characteristic feature of the connection of two elements according to the scheme shown in Fig. 4.44a, with the introduction of a weighty lever with a floating support is an expansion of the range of dynamic properties. In this case, the parameters of the lever have significant values, since the dynamic stiffness of the interaction gives variants of variations at different frequencies.

References

1. Eliseev, S.V., Reznik, Y.N., Khomenko, A.P.: Mechatronic Approaches in the Dynamics of Mechanical Oscillatory Systems, p. 384. Nauka Publisher, Novosibirsk (2011)
2. Khomenko, A.P., Eliseev, S.V., Ermoshenko, YuV: System analysis and Mathematical Modeling in Mechatronics of Vibration Protection Systems, p. 288. IrGUPS Publisher, Irkutsk (2012)
3. Belokobylsky, S.V., Eliseev, S.V., Kashuba, V.B.: Applied Problems of the Structural Theory of Vibration Protection Systems, p. 363. Polytekhnika Publisher, St.Petersburg (2013)
4. Belokobylsky, S.V., Eliseev, S.V., Sitov, I.S.: Dynamics of Mechanical Systems. Lever Linkages and Inertial-Elastic Constraints, p. 319. Polytekhnika Publisher, St. Petersburg (2013)
5. Eliseev, S.V., Reznik, YuN, Khomenko, A.P., et al.: Dynamic Synthesis in Generalized Problems of Vibration Protection and Vibration Isolation of Technical Objects, p. 523. ISU Publisher, Irkutsk (2008)
6. Eliseev, S.V., Volkov, L.N., Kukharenko, V.P.: Dynamics of Mechanical Systems with Additional Constraints, p. 214. Nauka Publisher, Novosibirsk (1990)
7. Eliseev, S.V., Khomenko, A.P.: Dynamic Damping. Feedback Concept and Structural Methods of Mathematical Modeling, p. 357. Nauka Publisher, Novosibirsk (2014)
8. Eliseev, S.V., Khomenko, A.P., Barsukov, S.V.: Modern Problems of the Dynamics of Machines. Protection Against Vibrations and Shocks, 165p. IrGUPS Publisher, Irkutsk (Dep. in VINITI 03/21/11, No. 135-B 2011), (2011)
9. Khomenko, A.P., Eliseev, S.V., Artyunin, A.I., Eliseev, A.V., Bol'shakov, R.S., Kaimov, E. V.: Elastic Elements in Mechanical Systems. Structural Interpretations, 460p. IrGUPS Publisher, Irkutsk. (Dep. in VINITI 08.15.13, No. 230-B 2013) (2013)
10. Logunov, A.S.: Dynamics of pneumatic elements and motion transformation devices in the systems of vibration protection of objects: author's abstract of Ph.D. (Engineering) thesis, 20p. IrGUPS Publisher, Irkutsk (2009)
11. Fomina, I.V.: Development of methods for constructing mathematical models with joints of units: author's abstract of Ph.D. (Engineering) thesis, 18p. IrGUPS Publisher, Irkutsk (2011)
12. Dembaremdiker, A.D.: Shock Absorbers for Vehicles, p. 199. Mashinostroenie Publisher, Moscow (1985)

13. Kogan, A.Y.: Track Dynamics and its Interaction with Rolling Stock, p. 326. Transport Publisher, Moscow (1997)
14. Timoshenko, S.P., Gere, J.: Mechanics of Materials, p. 672. Lan' Publisher, St. Petersburg (2002)
15. Khachaturov, A.A., Afanasyev, V.L., Vasilyev, V.S. et al.: Dynamics of the road—bus—car—driver system. In: Khachaturov, A.A. (ed.). 536p. Mashinostroenie Publisher, Moscow (1976)
16. Atabekov, G.I.: Linear Electrical Circuits, p. 592. Energia Publisher, Moscow (1978)
17. Eliseev, S.V.: Structural Theory of Vibration Protection Systems, p. 224. Nauka Publisher, Novosibirsk (1978)
18. Harris, C.M., Shock, A.G.: Vibration Handbook, 877p. Mc Graw-Hill, N.–Y (2002)
19. Frolov K.V. (ed.).: Vibrations in technology: ref.: in 6 vols. Protection Against Vibrations and Shocks, vol. 6, 586p. Mashinostroenie Publisher, Moscow (1983)
20. Lurie, A.I.: Operational Calculus and Application in Engineering, p. 368. Nauka Publisher, Moscow (1959)
21. Ivashchenko, N.N.: Automatic Adjustment. Theory and Elements of Systems, 4th edn, 736p. Mashinostroenie Publisher, Moscow (1978) (updated and revised)
22. Shchepetilnikov, V.A.: The Balancing of Mechanisms, p. 256. Mashinostroenie Publisher, Moscow (1982)
23. Vargunin V.N., Gusarov V.N., Ivanov B.G. et al.: Designing and calculation of articulation devices and assemblies. In: Mulyukin, O.P. (ed.). SamGAPS Publisher, Samara, 86p (2006)
24. Weits, V.L., Kochura, A.E., Martynenko, A.M.: Dynamic Calculations of Drive Machines, p. 352. Mashinostroenie Publisher, Leningrad (1971)
25. Druzhinsky, I.A.: Mechanical Chains, p. 240. Mashinostroenie Publisher, Moscow (1977)
26. Loitsyansky, L.G., Lurie, A.I.: The Course of Theoretical Mechanics: in 2 vols. In: Dynamics, vol. 2, p. 640. Nauka Publisher, Moscow (1983)
27. Kraynev, A.F.: Reference Dictionary on Mechanisms, 2nd edn, p. 560. Mashinostroenie Publisher, Moscow (1987). updated and revised
28. Gardner, M.F., Burns, J.L.: Transients in Linear Systems: Transl. from English, 3rd edn, p. 551. Fizmatgiz Publisher, Moscow (1961)
29. Bakalov, V.P., Dmitrikov, V.F., Kruk, B.I.: Fundamentals of the Theory of Circuits. Textbook, p. 589. Radio and svyaz' Publisher, Moscow (2000)

Chapter 5
Joints in the Dynamics of Mechanical Oscillatory Systems

The features of the motion of mechanical oscillatory systems are determined by the presence of various constraints their structure and refer to several scientific directions, including analytical mechanics, the theory of mechanisms and machines, the theory of oscillations, etc. The following definition of the concept of a "constraint" is given in [1]: "Adaptations that implement the relationships between the quantities determining the position and speed of the points of the system are called constraints". That is, physically, constraints are in the form of some real limitations; first of all, such are the elements of mechanical oscillatory systems. The simplest and most important class are positional relationships: they perform dependencies between the coordinates of the points of the system and are analytically expressed by the relations

$$f_i(y_1, y_2, y_3, \ldots y_{3N}; t) = 0, \quad (i = \overline{1, r}), \tag{5.1}$$

called coupling equations (where r is the number of coupling equations).

In this case $r \leq 3N$, and the sign of equality corresponds to the ratio of the system according to a predetermined law of motion. In this sense, the resulting equations of motion of mechanical oscillatory systems in the form of Laplace transforms are constraint equations.

The Cartesian coordinates of all points in the system can be expressed in terms of quantities $q_1 \ldots q_n$ called generalized coordinates, and in time t. If the links are stationary, then we can construct a set of generalized coordinates so that the time t does not enter into the original equations, that is, it is necessary to exclude t.

Note that "the object of analysis in analytic mechanics is a material point, a system of a finite number of free points (in celestial mechanics) or material points subject to couplings, one or more bodies connected with each other"—exactly in this form the notion of joints solid bodies. The concept of "constraint", considered above, is given taking into account the concept of interactions between elements of mechanical systems. In the physical sense, constraints in systems, if mechanisms and devices are considered in which the constructive and engineering

© Springer Nature Switzerland AG 2020

S. V. Eliseev and A. V. Eliseev, *Theory of Oscillations*, Studies in Systems, Decision and Control 252, https://doi.org/10.1007/978-3-030-31295-4_5

implementation has features, are defined as kinematic pairs of different classes. Such approaches are used in the theory of mechanisms and machines [2]. Further development of the analytical apparatus for describing the kinematic relations of the parameters of articulated solids was obtained in [3]. In robotics, "joint" is used as a generalized concept, which allows unifying various forms of interaction between elements of contacting solids. In this respect, one could note the possibilities of "virtual" joints, by which some forms of self-organization of motion in oscillatory systems can be understood, when the bodies move in phase or the regime of dynamic absorbing of oscillations is observed in the system. That is, the use of the concept of "joint" does not contradict the concepts of "constraint" or "kinematic pair" of a particular kind. It can be assumed that, in the presence of a greater number of contacts between bodies, the notion of "joint" becomes more adequate in the ideas of possible constructive and engineering links. This situation is possible, for example, in vibration protection systems with lever linkages [4, 5]. A number of theoretical questions were considered in [6, 7], dealing with the technologies of constructing mathematical models with joints of units.

Forms of dynamic interactions. Generalizing the idea of the basic properties of dynamical systems, we note that the introduction of additional constraints is implemented through the addition of some elements to others. It is important to take into account the features and design forms of the couplings, because they often determine the dynamic capabilities of the systems. For example, it can be noted that the introduction of a dynamic absorber as an additional device corresponds to the introduction of so-called additional constraints into the mechanical oscillatory system [8]. The structural features of the additional constraints can be seen in the fact that the latter should have fixing points, i.e. take the form of dual elements, and provide the possibility of creating, in a dynamic sense, some structures or blocks from them, which is determined by the joints. In the simplest forms, the joint is interpreted as a kinematic pair. The latter allows us, in case of necessity, to consider in mechanical vibrational systems the joint of the elements of the systems or the joint of the units, as well as the joint of solids, often taking the form of joint between rods, or between rods and solids. Accounting for joints is necessary for dynamic calculations, since their presence changes the dynamics of the system in a certain way. In real constructions, a joint can have elastic and dissipative properties, which can be expressed through additional degrees of freedom of mutual motion. However, making the stiffness of the elastic joints high, we can proceed to the introduction or formation of the joint as an element limiting the freedom of motion. Accounting for joints is often associated with the consideration in the mechanical systems of lever linkages, mechanisms and constraints, which introduces a certain specificity in the assessment of the spectrum of the dynamic capabilities of mechanical oscillatory systems in general and vibration protection systems in particular.

Consideration of constructive and engineering forms of the implementation of the functions of elastic, dissipative and elastic-dissipative units in the technical systems shows that the so-called elementary units are permanently complex in their device, contain interconnections of solids and can be called hinges. In turn, the spheres are kinematic pairs, which in the theory of mechanisms and machines are represented in several classifications. The most widely used are kinematic pairs of class V (rotational hinges, translational pairs), class IV (gears, cam mechanisms) and pairs of class III (spherical hinges).

Some definitions. The joint of solids is characteristic of machines, since the latter consist of mechanisms, and those in turn are mechanical circuits consisting of solids connected by kinematic pairs. In theoretical mechanics, the analytical tools are developed that allow one to solve the problems of statics, kinematics, and dynamics of joints, for example, when considering the motion of physical pendulums with one and two degrees of freedom, bifilar suspensions, horizontal magnets of seismic instruments, etc. [9]. At the same time, in the structure of mechanical oscillatory systems there is a certain specificity, since such systems consist of solid bodies or material points connected by springs and dampers, while, as a rule, there is no focus on its details and the physical forms of the compound itself. At the same time, the form of the compound itself is important. The reliable operation of machines and mechanisms in most cases is provided by retaining holonomic constraints. If the constraints are unilateral, then the dynamics of the interaction of the connected bodies will have a specific character.

The joint localizes the place of dynamic interaction, which requires the development of a detailed method of constructing mathematical models that allow one to determine certain parameters of mechanical systems. In an idealized form, the joint introduces into the system certain kinematic constraints. So, the joint provides equal speeds to two points that belong to different bodies. At this point, there occur dynamic responses. The location of the joint changes the given values of the mass-and-inertia, elastic and other characteristics of the system. In real conditions, frictional forces arise in the joints. If the joints have elastic properties, then account for such features of the details associated with the increase in the complexity of mathematical models, with an increase in the number of considered independent degrees of freedom of motion. The transition to an idealized system can be accomplished by increasing the stiffness of the joint, which in the passage to the limit leads to a simplification of the mathematical model [10].

Dynamic responses in joints, peculiarities of forces arising in joints, self-braking conditions, influence of backlashes, self-oscillations, etc. are considered in special sections of the theory of mechanisms and machines. In Fig. 5.1 shows the computational schemes reflecting the different types of joints in oscillatory motions.

The above computational schemes give an idea of the physical features of motion, the mutual arrangement of jointed bodies. The study of the dynamic properties of systems with joints is naturally related to the study of the influence of the locations of the points of joint. If such points are located near the center of gravity and the masses of the attached bodies differ sufficiently strongly, then relatively small displacements of the center can be assumed. However, the situation changes if the displacements substantially change the inertia-and-mass characteristics.

One of the first works on the dynamics of vibrational motions in mechanisms was [11], in which the specificity of the motion of mechanisms with allowance for

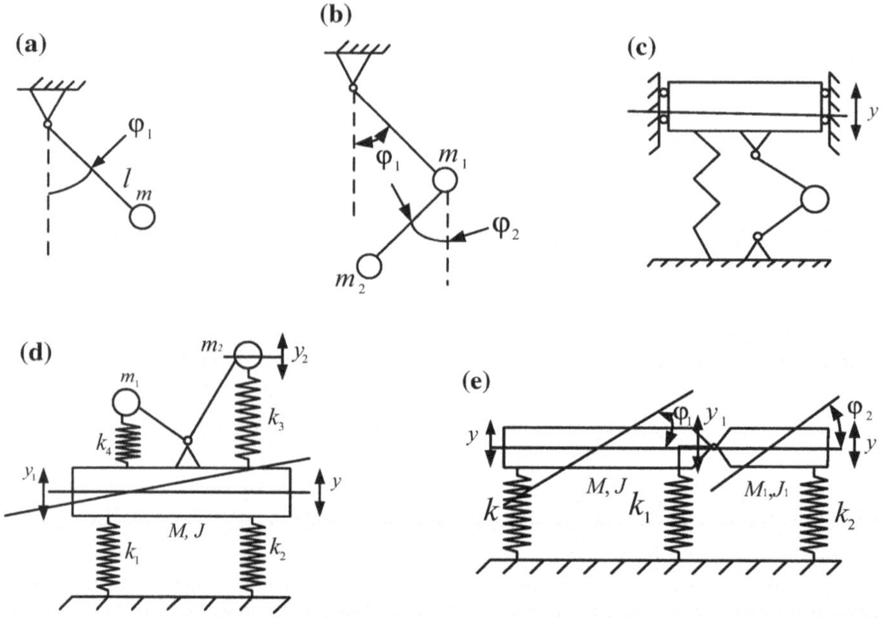

Fig. 5.1 Computational schemes of mechanical oscillatory systems with joints: **a** the rod with mass; **b** the double pendulum; **c** the system with a motion transformation device; **d** the L-shaped dynamic absorber of oscillations; **e** the suspension scheme

various kinematic pairs or the type of joint of solids is considered. A number of specific problems have been solved with respect to gear mechanisms used in power transmissions and drives [12], screw non-locking [13] and articulated mechanisms [14]. A large role of joints is played in various technical systems with quasi-zero stiffness [15], since the latter consist of mechanisms, and they, in turn, are mechanical chains of solids connected by kinematic pairs. In the theory of mechanisms and machines, an analytical apparatus has been developed that makes it possible to solve the problems of statics, kinematics, and dynamics of joints by determining reactions in hinges. At the same time, the kind of coupling itself is important. The reliable operation of machines and mechanisms is in most cases provided by bilateral holonomic constraints. If the constraints are unilateral, then the dynamics of interaction of the connected bodies will be specific [16].

In the framework of the structural theory of mechanical oscillatory systems, in particular, vibration protection systems, the elementary unit is a device having two points by which it is fixed in the structure of the system. It is from such positions that sequential and parallel connections of units of the extended set of typical elements are considered in [3, 17–19]. In systems with several degrees of freedom, solid bodies can be connected with each other by kinematic pairs of a certain type: rotational and translational pairs of class V, spherical hinges are most often used.

The presence of joints of the elements opens the possibility for generalization of the notions of possible forms of constraints in vibration protection systems, which allows us to proceed to representations on generalized dynamic constraints that take in various constructive versions different forms:

- Elementary units and their interconnections at the block level;
- Elementary units of the extended set of typical elements of mechanical oscillatory systems and their interconnection in structures on the basis of the rules of sequential and parallel connection of elements;
- Mechanical vibrating circuits;
- Mechanisms of various types (lever, screw, toothed, etc.);
- Active systems for the formation of controlled forces based on servo drives.

In general, focusing on joints, one way or another, is closely related to lever interactions, which is especially characteristic of the dynamics of transport systems.

5.1 Interaction of Solids with Joints of Rotational Type

The joints implemented by kinematic pairs of rotation of class V provide the possibility of solid bodies (or units) in the structure of a mechanical oscillatory system of reciprocating rotational motions relative to each other or relative to a stationary base. A solid body attached to a protection object can act as a dynamic vibration absorber. In the theory and practice of vibration protection and vibration isolation pendular and lever devices are known. A solid body can be included into the system with one or more connection points. Some examples of possible joints are shown in Fig. 5.2 [20].

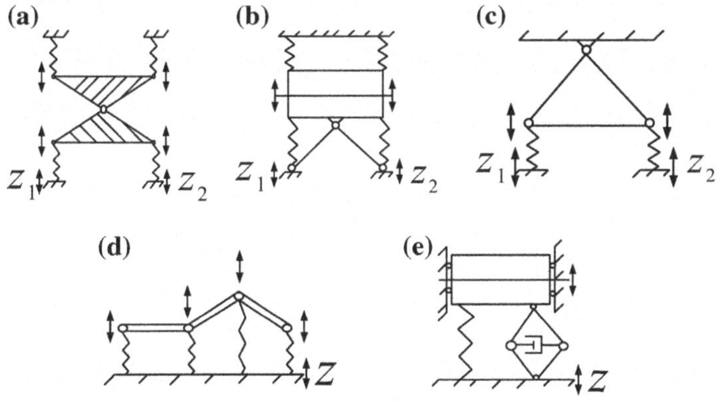

Fig. 5.2 Schematic diagrams of mechanical oscillatory systems with joints: **a** the system with three degrees of freedom with one joint; **b** the L-shaped (pendulum) absorber of vibrations (one joint); **c** the system with one degree of freedom (one joint with a movable base); **d** the multiply connected mechanical chain (two joints); **e** the system with joints in the lever mechanism

Various methods are used to study the dynamics of systems with articulated solids. In particular, in the place of the supposed joint of the rotational type, it is possible to introduce a generalized coordinate characterizing the relative displacement, and an elastic (or other) constraint after the construction of the mathematical model to make a very large in magnitude (elasticity, damping, inertial interaction). Then two coordinates characterizing the relative motion can in the limit "merge" into one, and the system "lose" one degree of freedom of motion. A prerequisite for this approach could be considered the consideration of cascade vibration protection systems, as well as the problem of vibration protection and vibration isolation, which takes into account the local elasticity of the place of fixing the vibration isolator or shock absorber.

Since the unified form of differential equations can use any system of generalized coordinates, including the coordinates of relative displacements, then the question arises about transformations of matrices based on permissible transformation rules [21, 22]. By introducing new variables that can potentially become null, you can exclude the corresponding column and matrix, which in the physical sense means the introduction of the joint. The number of degrees of freedom decreases. Further investigation of the system is carried out in the usual ways, but in a simplified scheme. Physically, this means that the relative motion between certain points is limited by the parameters of the connecting unit (for example, the stiffness in the coupling is several orders of magnitude higher than in other compounds). Then the system begins to oscillate as a system with a smaller number of degrees of freedom, which is quite common in engineering practice. Actually, the above representations are based approaches to the selection, justification and simplification of computational schemes for vibration protection systems of objects.

5.1.1 The Description of a Computational Scheme

Let us consider a mechanical oscillatory system with four degrees of freedom, shown in Fig. 5.3. Such a scheme can reflect, for example, the interaction of the transported spring-loaded weight in the body of the car and generally has four degrees of freedom.

It is of interest to create a technology for constructing models that would take into account the influence on the dynamic properties of the attachment features of elastic elements with stiffness k_3 and k_0 (points A_2 and A_1, and also p. B). Possible variants of transformation of oscillatory systems into systems with joints are shown in Fig. 5.4, which shows the possible points of joints that turn into joints. Thus, for $k_0 \to \infty$ the points B_1 and B_2 (Fig. 5.3a) can form a joint; and also A_1 and A_2 for $k_0' \to \infty$, C_1 and C_2 for $k_2 \to \infty$. In the case when $k_0' \to \infty$, $k_0 \to \infty$ and $k_2 \to \infty$, one can obtain a scheme of the known dynamic vibration absorber [19]. Introducing the coordinates of the relative displacement for the circuit in Fig. 5.4a: $y_A = y_{A1} - y_{A2}$ for $y_A \to \infty$, one can obtain a circuit, as in Fig. 5.4b, and so on.

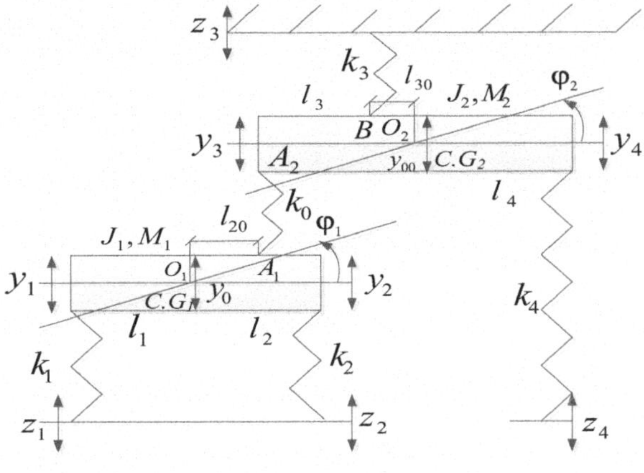

Fig. 5.3 Scheme of interaction of systems with four degrees of freedom

Thus, choosing points of joint, it is possible to obtain a sufficiently large number of variants of schemes, among which one can find the computational schemes of many known, for example, vibration protection systems.

5.1.2 *Joints in a Beam System with Two Degrees of Freedom*

Consider a beam system with two degrees of freedom, paying attention to the possibility of introducing joints at certain points by means of their "displacement". In the computational scheme shown in Fig. 5.5, this possibility is represented for the cases of the coincidence of the points A_1 and A_2, B_1 and B_2, when the system loses one degree of freedom, but the joint in the form of a kinematic pair of class V or a rotational hinge enables the solid body to perform oscillating and reciprocating motions, A and B. In the future, the possibilities of joints not only in the system of coordinates y_1 and y_2, but also in other coordinate systems will be considered. We note that, in addition to the joints at the points A and B, it is also possible to consider the constraints of motion with respect to the coordinates of pp. A_2 and B_2 at the same time, which can be determined by the condition $y_2 - y_1 = 0$. In this case, the system (see Fig. 5.5) with one degree of freedom and performs vertical translational motion on an elastic element with stiffness $k_1 + k_2$.

For the scheme shown in Fig. 5.5, we write the equations of kinetic and potential energy

$$\mathrm{T} = \frac{1}{2}M\dot{y}_0^2 + \frac{1}{2}J\dot{\varphi}^2; \tag{5.2}$$

$$\Pi = \frac{1}{2}k_1(y_1 - z_1)^2 + \frac{1}{2}k_2(y_2 - z_2)^2. \tag{5.3}$$

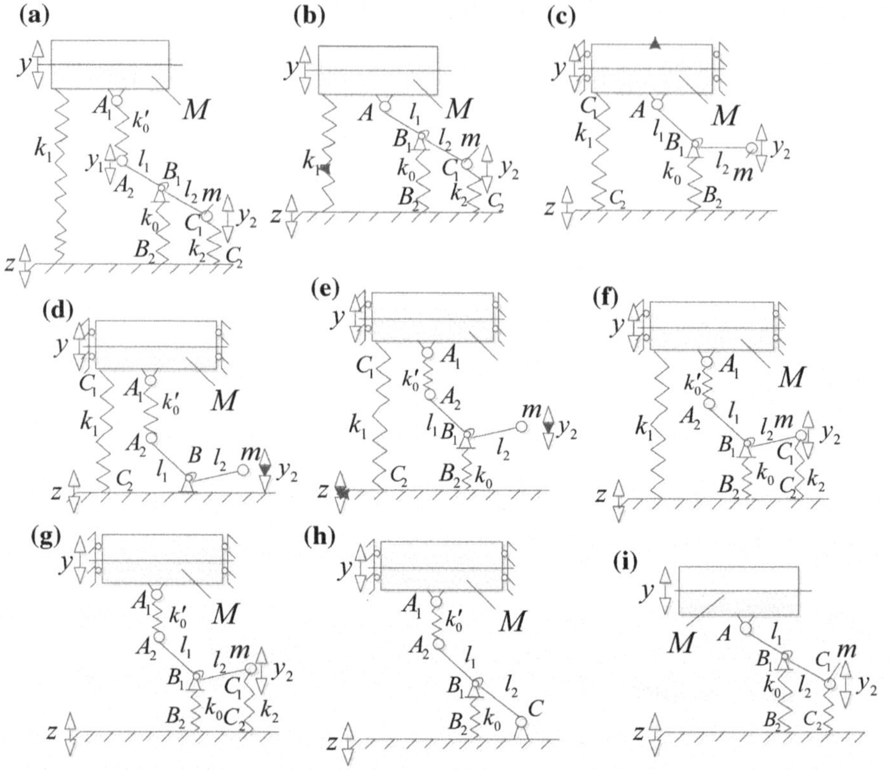

Fig. 5.4 Schematic diagrams of mechanical oscillatory systems, in which joints can arise for $k_0' \to \infty$, $k_0 \to \infty$ and $k_2 \to \infty$

Fig. 5.5 The computational scheme of a system having two elastic supports and moving in the system of coordinates y_1, y_2

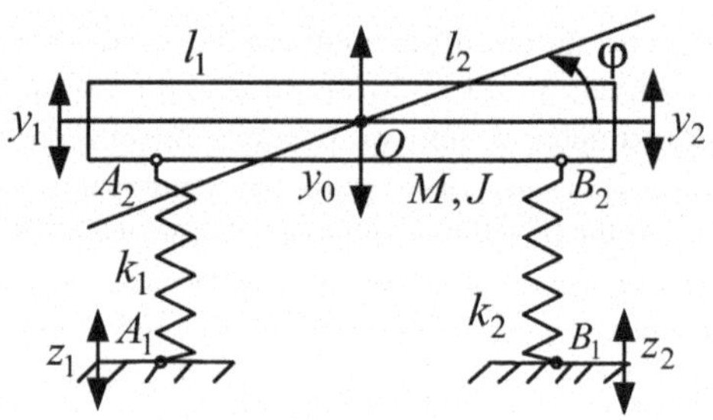

where y_1, y_2 are the coordinates of the points A_1 and B_2 in the conditionally fixed (absolute) coordinate system; y_0 is the coordinate of the center of gravity; φ is the angle of rotation with respect to the center of gravity (p. O); J is the moment of inertia with respect to the center of gravity (p. O); M is the mass of the beam; respectively, $l_1 = A_2O$; $l_2 = B_2O$

$$a = \frac{l_2}{l_1 + l_2}; \quad b = \frac{l_1}{l_1 + l_2}; \quad d = \frac{1}{l_1 + l_2}; \tag{5.4}$$
$$y_1 = y_0 - l_1\varphi; \quad y_2 = y_0 + l_2\varphi.$$

5.1.3 Features of the Choice of Coordinate Systems

1. Coordinates y_0, φ. Using the approaches described in [18], we write the differential equations of motion for the system shown in Fig. 5.4 in the system of coordinates y_0 and φ:

$$\ddot{y}_0 M + y_0 k_1 + k_2 y_0 - k_1 l_1 \varphi + k_2 l_2 \varphi = k_1 z_1 + k_2 z_2, \tag{5.5}$$

$$J\ddot{\varphi} + k_1 l_1^2 \varphi + k_2 l_2^2 \varphi - k_1 l_1 y_0 + k_2 l_2 y_0 = k_2 l_2 z_2 - k_1 l_1 z_1. \tag{5.6}$$

Expressions (5.5) and (5.6) define the matrix form of the notation in

$$A = \begin{vmatrix} M & 0 \\ 0 & J \end{vmatrix}, \quad B = \begin{vmatrix} k_1 + k_2 & k_2 l_2 - k_1 l_1 \\ k_2 l_2 - k_1 l_1 & k_1 l_1^2 + k_2 l_2^2 \end{vmatrix}, \quad C = \begin{vmatrix} k_1 z_1 + k_2 z_2 \\ -k_1 l_1 z_1 + k_2 l_2 z_2 \end{vmatrix}. \tag{5.7}$$

Let's construct a structural diagram of the dynamically equivalent automatic control system (Fig. 5.6).

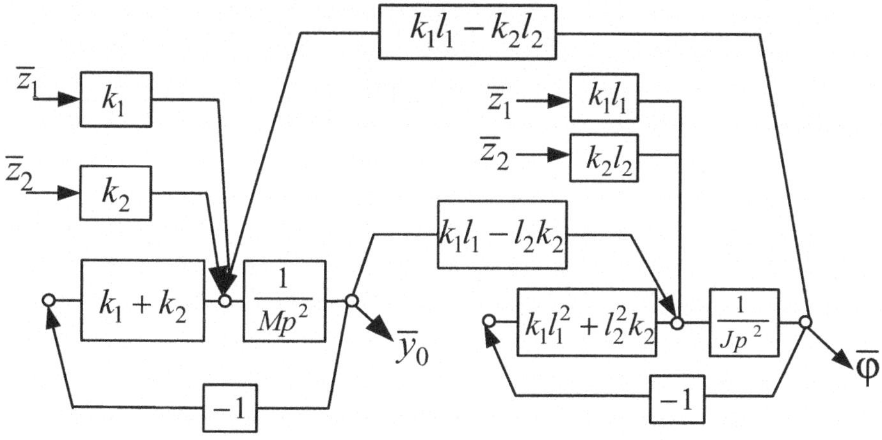

Fig. 5.6 The structural diagram of the dynamically equivalent ACS for the computational scheme, shown in Fig. 5.5, in the system of coordinates y_0 and φ

This system has a kinematic perturbation (z_1 and z_2), which can lead, in certain situations, to the emergence of dynamic absorbing modes. The system is characterized by elastic cross-couplings.

2. The coordinates y_1, y_2. For the structural diagram shown in Fig. 5.8, we write the equations of motion in the system of coordinates y_1 and y_2:

$$\ddot{y}_1(Ma^2 + Jd^2) + \ddot{y}_2(Mab - Jd^2) + k_1 y_1 = k_1 z_1; \tag{5.8}$$

$$\ddot{y}_2(Mb^2 + Jd^2) + \ddot{y}_1(Mab - Jd^2) + k_2 y_2 = k_2 z_2. \tag{5.9}$$

Moreover, the matrix structure (5.8) and (5.9) has the form

$$A = \begin{vmatrix} Ma^2 + Jd^2 & Mab - Jd^2 \\ Mab - Jd^2 & Mb^2 + Jd^2 \end{vmatrix}, \quad B = \begin{vmatrix} k_1 & 0 \\ 0 & k_2 \end{vmatrix}, \quad C = \begin{vmatrix} k_1 z_1 \\ k_2 z_2 \end{vmatrix}. \tag{5.10}$$

The structural diagram of the dynamically equivalent automatic control system for the computational scheme shown in Fig. 5.5, in the coordinates y_1 and y_2 takes the form, as shown in Fig. 5.7.

For the system under consideration, the nature of the cross couplings changes—they become inertial. Along with this, external influences also change, which now act specifically on the partial systems (see Fig. 5.7).

3. The coordinates y_1, φ. For the computational scheme shown in Fig. 5.5, we write the equations of motion in the system of coordinates y_1 and φ:

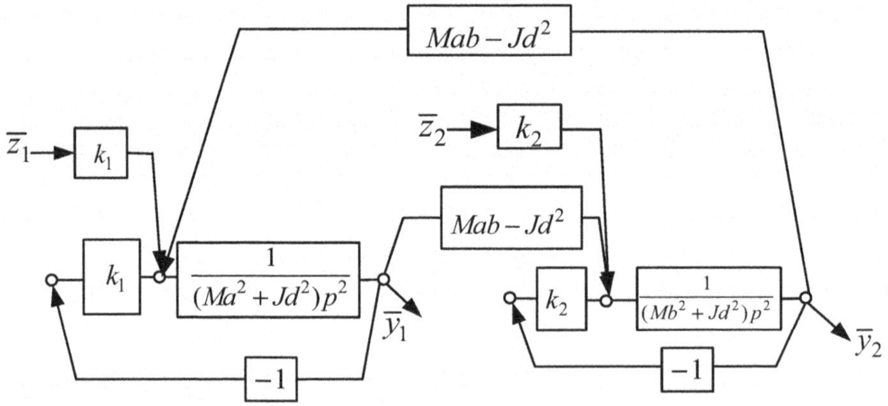

Fig. 5.7 The structural diagram of the dynamically equivalent ACS for the computational scheme, shown in Fig. 5.5, in the system of coordinates y_1 and y_2

$$\ddot{y}_1 M + M l_2 \ddot{\varphi} + k_1 y_1 + k_2 y_1 + k_2 (l_1 + l_2)\varphi = k_1 z_1 + k_2 z_2, \tag{5.11}$$

$$\ddot{y}_1 M l_2 + (J + M l_2^2)\ddot{\varphi} + k_2 (l_1 + l_2) y_1 + k_2 (l_1 + l_2)^2 \varphi = k_2 (l_1 + l_2) z_2. \tag{5.12}$$

For the expressions (5.11) and (5.12), the matrix structure has the form

$$A = \begin{vmatrix} M & M l_2 \\ M l_2 & J + M l_2^2 \end{vmatrix}, \quad B = \begin{vmatrix} k_1 + k_2 & k_2 (l_1 + l_2) \\ k_2 (l_1 + l_2) & k_2 (l_1 + l_2)^2 \end{vmatrix}, \quad C = \begin{vmatrix} k_1 z_1 + k_2 z_2 \\ k_2 (l_1 + l_2) z_2 \end{vmatrix}. \tag{5.13}$$

Structural diagram of the dynamically equivalent automatic control system (ACS) for the computational scheme shown in Fig. 5.5, in the coordinates y_1 and φ takes the form, as shown in Fig. 5.8. The peculiarity of this structural diagram is that here the cross-couplings acquire an elastic-inertial character and can be "zeroed" at certain frequencies, and the external disturbance acts only on one input.

4. The coordinates y_2, φ. For the computational scheme shown in Fig. 5.5, we write the equations of motion in the system of coordinates y_2 and φ:

$$\ddot{y}_2 M + M l_1 \ddot{\varphi} + k_1 y_2 + k_2 y_2 + k_1 (l_1 + l_2)\varphi = k_1 z_1 + k_2 z_2; \tag{5.14}$$

$$\ddot{y}_2 M l_1 + (J + M l_1^2)\ddot{\varphi} + k_1 (l_1 + l_2) y_2 + k_2 (l_1 + l_2)^2 \varphi = k_1 (l_1 + l_2) z_2. \tag{5.14'}$$

For the expression (5.14), the matrix structure has the form

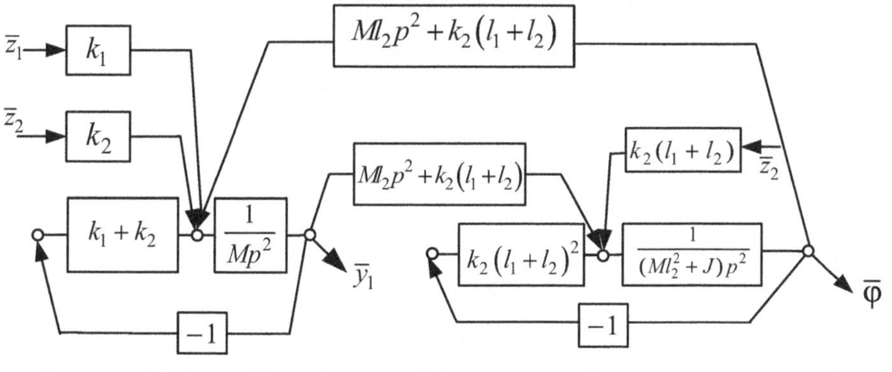

Fig. 5.8 The structural diagram of the dynamically equivalent ACS for the computational scheme, shown in Fig. 5.5, in the system of coordinates y_1 and φ

$$A = \begin{vmatrix} M & Ml_1 \\ Ml_1 & J + Ml_1^2 \end{vmatrix}, \quad B = \begin{vmatrix} k_1 + k_2 & k_1(l_1 + l_2) \\ k_1(l_1 + l_2) & k_1(l_1 + l_2)^2 \end{vmatrix}, \quad C = \begin{vmatrix} k_1 z_1 + k_2 z_2 \\ k_1(l_1 + l_2) z_2 \end{vmatrix}, \tag{5.15}$$

$$[A][\ddot{y}] + B[y] = C.$$

The structural diagram of the dynamically equivalent ACS for the computational scheme shown in Fig. 5.5, in the coordinates y_2 and φ takes the form, as shown in Fig. 5.9.

We note that changes in the generalized coordinates lead to a change in the transfer functions of the cross-couplings, which is due to the change in the frequencies in the dynamic absorbing modes.

5. Coordinates $\underline{y_0}$, $\underline{y_1}$. For the computational scheme shown in Fig. 5.5, we write the equations of motion in the system of coordinates y_0 and y_1:

$$\left(Ml_2^2 + J\right)\ddot{y}_0 - J\ddot{y}_1 + k_2(l_1 + l_2)^2 y_0 - k_2 l_1 (l_1 + l_2) y_1 = k_2 l_2 (l_1 + l_2) z_2, \tag{5.16}$$

$$-J\ddot{y}_0 + J\ddot{y}_1 + \left(k_1 l_1^2 + k_2 l_2^2\right)^2 y_1 - k_2 l_1 (l_1 + l_2) y_0 = k_1 l_2^2 z_1 - k_2 l_1^2 z_2. \tag{5.17}$$

In this case, for the expressions (5.16), (5.17) the matrix structure has the form

$$A = \begin{vmatrix} Ml_2^2 + J & J \\ J & J \end{vmatrix}, \quad B = \begin{vmatrix} k_2(l_1 + l_2)^2 & k_2 l_1 (l_1 + l_2) \\ k_2 l_1 (l_1 + l_2) & k_1 l_1^2 + k_2 l_2^2 \end{vmatrix}, \quad C = \begin{vmatrix} k_2 l_2 (l_1 + l_2) z_2 \\ k_1 l_2^2 z_1 + k_2 l_1^2 z_2 \end{vmatrix}. \tag{5.18}$$

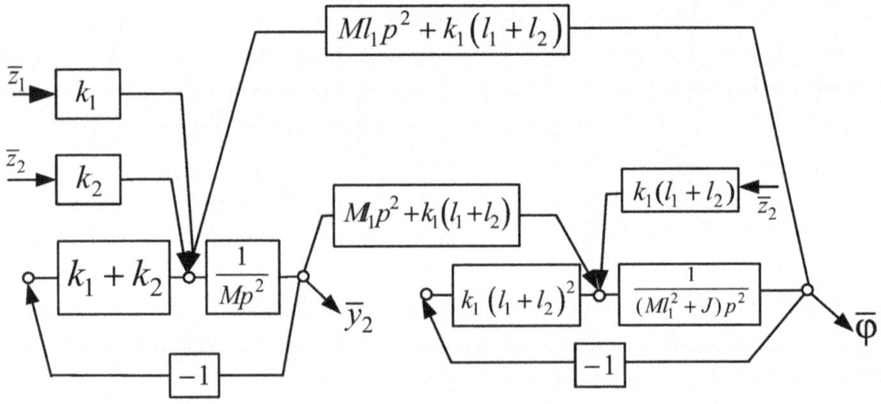

Fig. 5.9 The structural diagram of the dynamically equivalent ACS for the computational scheme, shown in Fig. 5.5, in the system of coordinates y_2 and φ

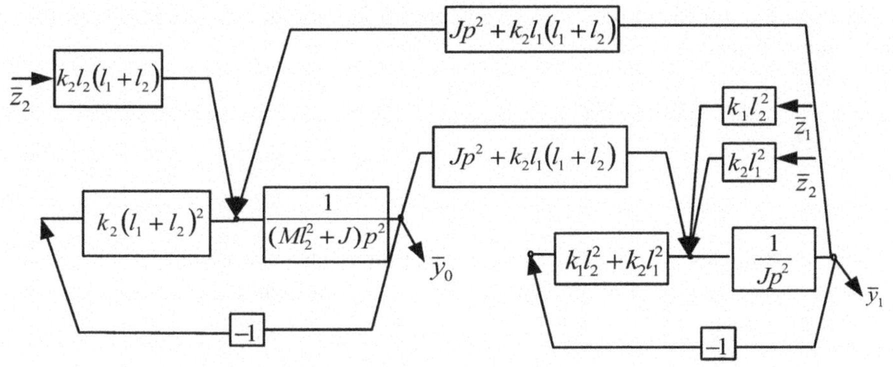

Fig. 5.10 The structural diagram of the dynamically equivalent ACS for the computational scheme, shown in Fig. 5.5, in the system of coordinates y_0 and y_1

The structural diagram of the dynamically equivalent ACS for the computational scheme shown in Fig. 5.5, in the coordinates y_0 and y_1 takes the form, as shown in Fig. 5.10, which reflects changes in both cross-couplings and in the parameters of partial systems.

6. The coordinates y_{01}, y_2. We introduce the system of coordinates y_{01} and y_2. To do this, we perform the following transformations: $y_1 - z_1 = y_{01}$, since $y_1 = y_0 - l_1\varphi$, respectively, $y_0 = (z_1 + y_0)\cdot a + y_2 b$; $y_0 = az_1 + y_{01}a + y_2 b$; $\varphi = d(y_2 - y_1) = dy_2 - dz_1 - dy_{01}$. We write the expressions (5.2) and (5.3) for the kinetic and potential energy taking into account the system of coordinates y_{01} and y_2:

$$\mathrm{T} = \mathrm{T}_1 + \mathrm{T}_2, \tag{5.19}$$

where $\mathrm{T}_1 = \frac{1}{2}M(a\dot{z}_1 + \dot{y}_{01}a + \dot{y}_2 b)^2$; $\mathrm{T}_2 = \frac{1}{2}J\dot{\varphi}^2 = \frac{1}{2}Jd^2(\dot{y}_2 - \dot{z}_1 - \dot{y}_{01})^2$.

We transform the expression (5.3) to the form

$$\Pi = \frac{1}{2}k_1 y_{01}^2 + \frac{1}{2}k_2(y_2 - z_2)^2. \tag{5.20}$$

For the computational scheme shown in Fig. 5.5, we write the equations of motion in the system of coordinates y_{01} and y_2:

$$(Ma^2 + Jd^2)\ddot{y}_{01} + y_{01}k_1 + \ddot{y}_2(Mab - Jd^2) = \ddot{z}_1(-Ma^2 - Jd^2); \tag{5.21}$$

$$(Mb - Jd^2)\ddot{y}_{01} + \ddot{y}_2(Mb + Jd^2) + y_2 k_2 = k_2 z_2 + \ddot{z}_1(-Mab + Jd^2). \tag{5.22}$$

In this case, for the expressions (5.21) and (5.22), the matrix structure will have the following form

$$A = \begin{vmatrix} Ma^2 + Jd^2 & Mab - Jd^2 \\ Mb - Jd^2 & Mb + Jd_2 \end{vmatrix}, \quad B = \begin{vmatrix} k_1 & 0 \\ 0 & k_2 \end{vmatrix}, \quad C = \begin{vmatrix} (-Ma^2 + Jd^2)z_1 \\ (-Mab - Jd^2)z_1 + k_2 z_2 \end{vmatrix}. \tag{5.23}$$

7. <u>The coordinates y_{01}, y_1</u>. For the computational scheme shown in Fig. 5.5, we write the equations of motion in the system of coordinates y_{01} and y_1, using the above actions:

$$\left(Mb^2 + Jd^2\right)\ddot{y}_{02} + y_{02}k_2 + \ddot{y}_1\left(Mab - Jd^2\right) = \ddot{z}_2\left(-Mb^2 + Jd^2\right); \tag{5.24}$$

$$\left(Mab - Jd^2\right)\ddot{y}_{02} + \ddot{y}_1\left(Ma^2 + Jd^2\right) + y_1 k_2 = \ddot{z}_2\left(-Mba - Jd^2\right) + k_1 z_1. \tag{5.25}$$

In this case, for the expressions (5.24) and (5.25), the matrix structure has the form

$$A = \begin{vmatrix} Mb^2 + Jd^2 & Mab - Jd^2 \\ Mab - Jd^2 & Ma^2 + Jd^2 \end{vmatrix}, \quad B = \begin{vmatrix} k_2 & 0 \\ 0 & k_1 \end{vmatrix}, \quad C = \begin{vmatrix} (-Mb^2 + Jd^2)z_2 \\ k_1 z_1 + (-Mab - Jd^2)z_2 \end{vmatrix}. \tag{5.26}$$

5.1.4 Comparative Analysis

Table 5.1 gives the coefficients of the equations considered, unified in different coordinate systems. Variants of introduction of joints, presented in Table 5.1, are shown in Fig. 5.11.

Computational schemes of a particular kind can be obtained by excluding a column and a row in the corresponding matrix associated with the system of coordinates: (1) y_0, φ—Fig. 5.11a, b; (2) y_1, y_2—Fig. 5.11c, d; (3) $y_{01} = 0$, $y_2 \neq 0$ ($y_{01} = y_1 - z_1$)—Fig. 5.11e; (4) $y_{01} = 0$, $y_2 \neq 0$ ($y_{01} = y_1 - z_1$)—Fig. 5.11f; (5) $y'_0 = 0, y_1 \neq 0, y_2 \neq 0 (y'_0 = y_0 - z)$—Fig. 5.11g; (6) $y'_A = (y_A - z) = 0$, $y_1 \neq 0$, $y_2 \neq 0$—Fig. 5.11h—p. A is between the center of gravity and the left elastic support; (7) $y'_B = (y_B - z) = 0, y_1 \neq 0, y_2 \neq 0$—Fig. 5.11i—p. B is outside the left elastic support. In each of the cases considered, each version of the joint has its own mathematical model. Note that in schemes in which both, it is necessary to take into account the relationship between the coordinates determined by the lever linkages. A special case is the choice as a joint of the points A and B, which either lie between the attachment points of the elastic elements, or go beyond this space,

Table 5.1 Coefficients of the equations of motion in different coordinate systems

The system of coordinates y_0 and φ			*The system of coordinates y_1 and y_2*		
The coefficients of equations (5.4) and (5.5)		b_i	The coefficients of equations (5.7) and (5.8)		b_i
a_{11}	a_{12}	b_1	a_{11}	a_{12}	b_1
$Mp^2+k_1+k_2$	$-k_1l_1+k_2l_2$	$k_1z_1+k_2z_2$	$(Ma^2+Jd^2)p^2+k_1$	$(Mab-Jd^2)p^2$	k_1z_1
a_{21}	a_{22}	b_2	a_{21}	a_{22}	b_2
$-k_1l_1+k_2l_2$	$Jp^2+k_1l_1^2+k_2l_2^2$	$-k_1l_1z_1+k_2l_2z_2$	$(Mab-Jd^2)p^2$	$(Mb+Jd^2)p^2+k_2$	k_2z_2
The system of coordinates y_1 and φ			*The system of coordinates y_2 and φ*		
The coefficients of the equations (5.10) and (5.11)		b_i	The coefficients of the equations (5.13) and (5.13′)		b_i
a_{11}	a_{12}	b_1	a_{11}	a_{12}	b_1
$Mp^2+k_1+k_2$	$Ml_2p^2+k_2(l_1+l_2)$	$k_1z_1+k_2z_2$	$Mp^2+k_1+k_2$	$Ml_1p^2+k_1(l_1+l_2)$	$k_1z_1+k_2z_2$
a_{21}	a_{22}	b_2	a_{21}	a_{22}	b_2
$Ml_2p^2+k_2(l_1+l_2)$	$(J+Ml_2^2)p^2+k_2(l_1+l_2)^2$	$k_2(l_1+l_2)$	$Ml_1p^2+k_1(l_1+l_2)$	$(J+Ml_1^2)p^2+k_1(l_1+l_2)^2$	$k_1(l_1+l_2)$
The system of coordinates y_0 and y_1			*The system of coordinates y_{01} and y_2 ($y_{01}=y_1-z_1$)*		
The coefficients of the equations (5.15) and (5.16)		b_i	The coefficients of the equations (5.20) and (5.21)		b_i
a_{11}	a_{12}	b_1	a_{11}	a_{12}	b_1
$(Ml_2^2+J)p^2+k_2(l_1+l_2)^2$	$-Jp^2-k_2(l_1+l_2)\,l_1$	$k_2l_2(l_1+l_2)z_2$	$(Ma^2+Jd^2)p^2+k_1$	$(Mab-Jd^2)p^2$	$(-Ma^2-Jd^2)p^2z_1$
a_{21}	a_{22}	b_2	a_{21}	a_{22}	b_2
$-Jp^2-k_2(l_1+l_2)l_1$	$Jp^2+k_1l_1^2+k_2l_2^2$	$k_1l_2^2z_1+k_2l_1^2z_2$	$(Mab-Jd^2)p^2$	$(Mb^2+Jd^2)p^2+k_2$	$(-Mab+Jd^2)p^2z_1+k_2z_2$
The system of coordinates y_{01} and y_1 ($y_{02}=y_2-z_2$)					
The coefficients of the equations (5.4) and (5.5)				b_i	
a_{11}		a_{12}		b_1	
$(Mb^2+Jd^2)p^2+k_2$		$(Mab-Jd^2)p^2$		$(-Mb^2+Jd^2)z_2$	
a_{21}		a_{22}		b_2	
$(Mab-Jd^2)p^2$		$(Ma^2+Jd^2)p^2+k_1$		$(-Mab-Jd^2)p^2z_2+k_1z_1$	

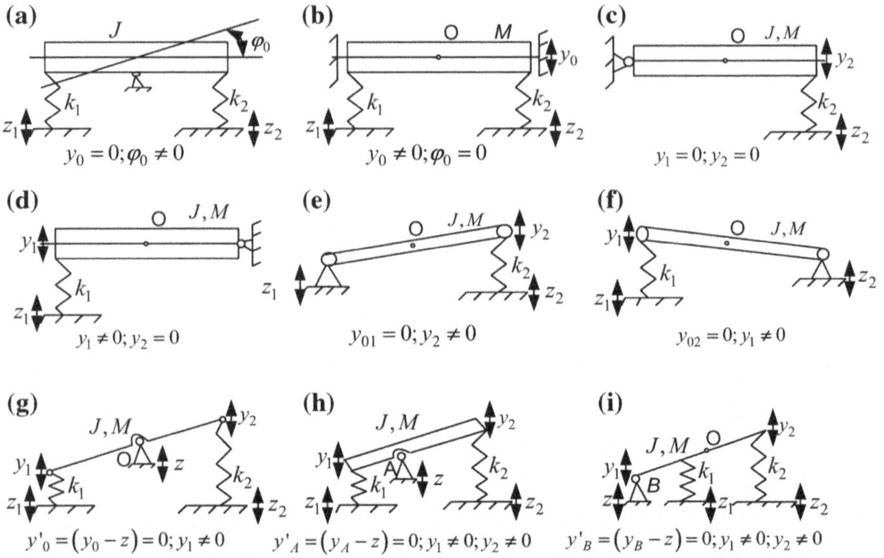

Fig. 5.11 Variants of introduction of a joint in a system with two degrees of freedom (see Fig. 5.5). The options are explained in the text

which requires taking into account the features of the coordinates in the mechanical system, which could be called observation points. In this situation, the observation point is considered as a point of possible joint.

5.1.5 Features of Dynamic Properties

Using the coefficient matrices for the systems presented in Table 5.2, it is possible to get mathematical models for any particular case by excluding columns and rows. Consider, for example, the problem of construction of a mathematical model for the computational scheme in Fig. 5.12, which corresponds to the operation of the system in Fig. 5.5 in the coordinates y_{01} and y_2. Note that $y_{01} = y_1 - z_1$. We use the relations $y_0 = ay_1 - by_2$ and $\varphi = d(y_2 - y_1)$, then

$$y_0 = [a(y_{01} + z_1) + by_2] = az_1 + y_{01}a + y_2 b. \tag{5.27}$$

Substituting (5.27) into the expressions (5.2) and (5.3), we obtain

$$\mathrm{T} = \frac{1}{2}M(a\dot{z}_1 + \dot{y}_{01}a + \dot{y}_2 b)^2 + \frac{1}{2}Jd^2(\dot{y}_2 - \dot{y}_{01} - \dot{z}_1); \tag{5.28}$$

Table 5.2 The frequency properties of the regimes for a system with different systems of coordinates of VPS, obtained through the introduction of joint (Fig. 5.5)

Coordinate system	Natural oscillations frequency	Frequency of dynamic absorbing	"Locking" at high frequencies	Notes (joint type)
y_0 and φ	$\omega_{\text{nat}}^2 = \frac{k_1 l_1^2 + k_2 l_2^2}{J}$	–	–	Joint is located at point O (hinge)
y_1 and y_2	$\omega_{\text{nat}}^2 = \frac{k_2}{Mb^2 + Jd^2}$	–	–	Joint $y_1 = 0$
	$\omega_{\text{nat}}^2 = \frac{k_1}{Ma^2 + Jd^2}$	–	–	Joint $y_1 = 0$
y_1 and φ	$\omega_{\text{nat}}^2 = \frac{k_2 (l_1 + l_2)^2}{J + Ml_2^2}$	–	–	Joint $y_1 = 0$
y_2 and φ		–	–	Joint $y_1 = 0$
y_0 and y_1	$\omega_{nat}^2 = \frac{k_1 l_1^2 + k_2 l_2^2}{Jd^2}$	–	–	Joint $y_0 = 0$
	$\omega_{\text{nat}}^2 = \frac{k_2 (l_1 + l_2)^2}{Ml_2^2 + Jd^2}$	–	–	Joint $y_1 = 0$
y_0 and y_2	$\omega_{\text{nat}}^2 = \frac{k_1 l_1^2 + k_2 l_2^2}{Jd^2}$	–	–	Joint $y_0 = 0$
	$\omega_{\text{nat}}^2 = \frac{k_1 (l_1 + l_2)^2}{Ml_1^2 + Jd^2}$	–	–	Joint $y_2 = 0$
y_{01} and y_2	$\omega_{\text{nat}}^2 = \frac{k_2}{Mb^2 + Jd^2}$		$\frac{Jd^2 - Mab}{Mb^2 + Jd^2}$	Joint $y_{01} = 0$ $z_1 = z_2 = z$ $y_{01} = y_1 - z_1$
	$\omega_{\text{nat}}^2 = \frac{k_1}{Ma^2 + Jd^2}$	–	$\frac{Jd^2 - Mab}{Mb^2 + Jd^2}$	Joint $y_0 = 0$ $z_1 = z_2 = z$

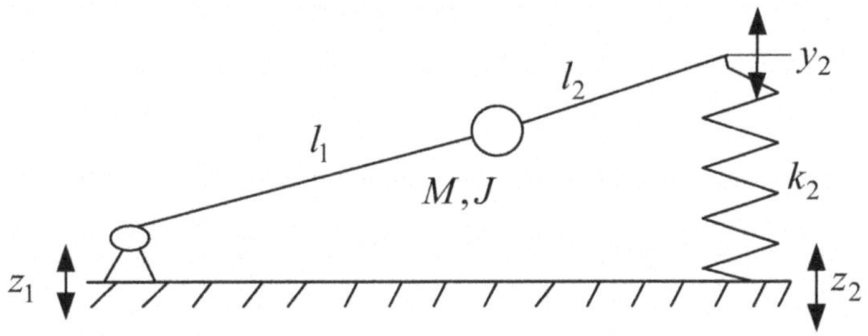

Fig. 5.12 The computational scheme of the VPS in the system of coordinates y_{01} and y_2

$$\Pi = \frac{1}{2} k_1 y_{01}^2 + \frac{1}{2} k_2 (y_2 - z_2)^2. \tag{5.29}$$

The system of equations of motion in the case under consideration takes the following form:

$$(Ma^2 + Jd^2)\ddot{y}_{01} + y_{01} k_{01} + \ddot{y}_1 (Mab - Jd^2) = \ddot{z}_2 (-Ma^2 - Jd^2); \tag{5.30}$$

$$(Mab - Jd^2)\ddot{y}_{01} + \ddot{y}_2 (Mb^2 + Jd^2) + y_2 k_2 = \ddot{z}_2 (-Mba - Jd^2) + k_2 z_2, \tag{5.31}$$

which coincides with Eqs. (5.21) and (5.22).

In the system in Fig. 5.12 the frequency of natural oscillations is determined by the formula

$$\omega_{\text{nat}}^2 = \frac{k_2}{Mb^2 + Jd^2}. \tag{5.32}$$

If we assume that $z_1 = z_2 = z$, in the system in Fig. 5.12 dynamic absorbing mode is possible at frequency

$$\omega_{\text{dyn}}^2 = \frac{k_2}{Jd^2 - Mab}. \tag{5.33}$$

The transfer function of the system has the form

$$W(p) = \frac{y_2(p)}{z(p)} = \frac{(Jd^2 - Mab)p^2 + k_2}{(Mb^2 + Jd^2)p^2 + k_2}. \tag{5.34}$$

At high frequencies, the system is locked and

$$\left| W(p) \right|_{p \to \infty} = \frac{Jd^2 - Mab}{(Mb^2 + Jd^2)}. \tag{5.35}$$

The amplitude-frequency characteristic of the system is represented in accordance with (5.35) and has the form as in Fig. 5.13.

According to the variants of the introduction of joints, Table 5.2 provides information on possible modes of operation (frequency aspect).

The list of versions for introducing joints can be supplemented with system of coordinates y'_{01} and y_1, where ($y'_{01} = y_0 - z_1$ and y_2); y'_{01} and y_2, where ($y'_{01} = y_0 - z_2$ and y_2), etc.

The introduction of joints in various versions on the basis of simplification of the initial computational scheme (see Fig. 5.4) allows forming and systematizing the class of mathematical models obtained by a certain technique from beam type systems. At the same time, any model from this class can be obtained autonomously, but the technique of constructing differential equations in each such case

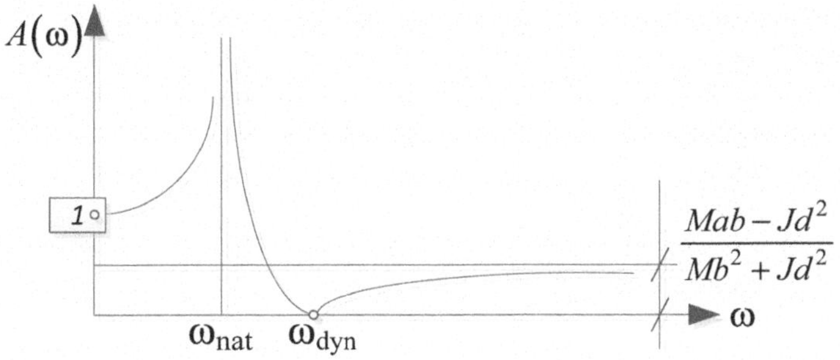

Fig. 5.13 The amplitude-frequency characteristic of the system, the computational scheme of which is shown in Fig. 5.12

will require the consideration of a number of specific details. The relationship itself between the coordinates y_1 and y_2 (and others) reflects the joints created by the virtual masses: $Ma^2 + Jd^2$, $Mb^2 + Jd^2$, which emerge in the transformations and are reduced mass-and-inertia parameters with respect to the solid body in the form of a beam. We also note that when virtual masses of two elements are connected in a mechanical system, a need for the lever arises. In turn, the place of fastening of the elastic elements is spaced on the beam, which also forms the lever linkages.

5.2 To the Question of the Possibility of Virtual Joints in Mechanical Oscillatory Systems

In a number of works that deal with the development of the theory of vibration protection systems, the joints of solid bodies are considered as a form of interconnecting units, if they have the form of solids, through kinematic pairs of one or another class. The joint can result in the fact that one body loses one or more degrees of freedom in relative motion.

For example, a solid body in the form of a beam, if its motion is considered with respect to the base (or reference point), has two degrees of freedom. The beam rests on elastic supports, i.e. the system consists of a certain set of typical units. If one of the ends of the beam is inserted into the joint (for example, of the hinged type), then the beam will obtain one instead of two degrees of freedom, which allows the beams to reciprocate relative to the joint (or hinge). If we do the same way with respect to the other support, then the motion of the solid body becomes impossible with respect to the base, but it can move together with the base.

In a system with one degree of freedom (for example, mass on an elastic element) the situation is simplified. If a joint arises between the solid and the base, the solid body "sticks" to the surface of the base (movable or immovable). In these cases, it can be shown that the presence of the condition that the relative motion parameters (the difference in coordinates or velocities) be zero is not yet sufficient for the joint to occur. Constraints of a unilateral nature are also needed.

Consider a system with two degrees of freedom (Fig. 5.14), whose motion is described in coordinates y_1 and y_2. The system has two elements of mass m_1 and m_2, as well as elastic elements k_1, k_2, k_3 and k_4. The perturbation of the system is of a kinematic nature (z_1, z_2, z_4).

We choose the coordinate system of motion u. Then the expressions for the kinetic and potential energy take the form

$$\mathrm{T} = \frac{1}{2} m_1 \dot{y}_1^2 + \frac{1}{2} m_2 \dot{y}_2^2; \tag{5.36}$$

$$\Pi = \frac{1}{2} k_1 (y_1 - z_1)^2 + \frac{1}{2} k_2 (y_2 - z_2)^2 + \frac{1}{2} k_3 (y_2 - y_1)^2 + \frac{1}{2} k_4 (y_2 - z_4)^2. \tag{5.37}$$

Using the Lagrange formalism, we formulate a system of equations of motion

$$\ddot{y}_1 (m_1) + y_1 (k_1 + k_3) + \ddot{y}_2 (0) + y_2 (-k_3) = k_1 z_1; \tag{5.38}$$

$$\ddot{y}_1 (0) + y_1 (-k_3) + \ddot{y}_2 (m_2) + y_2 (k_2 + k_3 + k_4) = k_2 z_2 + k_2 z_4. \tag{5.39}$$

Table 5.3 presents the coefficients of the equations of motion (5.38) and (5.39).

Generalized forces along the coordinates y_1 and y_2 with respect to the coordinates and are respectively

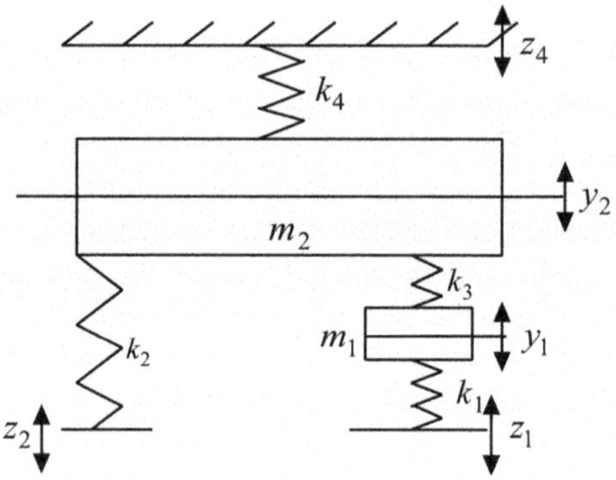

Fig. 5.14 Computational scheme of a VPS with kinematic perturbation

Table 5.3 The values of the coefficients of the equations of motion of the system in the coordinates y_1 and y_2

a_{11}	a_{12}
$m_1 p^2 + k_1 + k_3$	$-k_3$
a_{21}	a_{22}
$-k_3$	$m_2 p^2 + k_2 + k_3 + k_4$

$$Q_1 = k_1 z_1, \quad Q_2 = k_2 z_2 + k_2 z_4. \tag{5.40}$$

To introduce the joint between the elements of the system m_1 and m_2, the system of coordinates y_1 and y_0 is used, where

$$y_0 = y_2 - y_1. \tag{5.41}$$

Then the system of equations (5.38) and (5.39) can be transformed to the form:

$$\ddot{y}_1(m_1 + m_2) + y_1(k_1 + k_2 + k_4) + \ddot{y}_0(m_2) + y_0(k_2 + k_4) = k_1 z_1 + k_2 z_2 + k_4 z_4; \tag{5.42}$$

$$\ddot{y}_1(m_2) + y_1(k_2 + k_4) + \ddot{y}_0(m_2) + y_2(k_2 + k_3 + k_4) = k_2 z_2 + k_4 z_4. \tag{5.43}$$

Table 5.4 presents the coefficients of the equations of motion (5.42) and (5.43). Generalized forces with respect to the coordinates y_1 and y_0 are respectively

$$Q_1 = k_1 z_1 + k_2 z_2 + k_4 z_4; \quad Q_2 = k_2 z_2 + k_4 z_4. \tag{5.44}$$

Assuming that $y_0 \to \infty$ for $k_2 \to \infty$, we write the equation of motion with respect to y:

$$\ddot{y}_1(m_1 + m_2) + k_1 + k_2 + k_4 = k_1 z_1 + k_2 z_2 + k_4 z_4. \tag{5.45}$$

Let $z_2 = z_1 = z = z_4$, then the transfer function of the system takes the following form:

$$W(p) = \frac{\bar{y}_1}{\bar{z}} = \frac{k_2 + k_4 + k_1}{(m_2 + m_1)p^2 + k_2 + k_4 + k_1}. \tag{5.46}$$

From the analysis of (5.46) it is clear that, when "joining" m_1 and m_2, the initial system having two degrees of freedom turns into a system with one degree of freedom. In this system on frequency

$$\omega_{1nat}^2 = \frac{k_1 + k_2 + k_4}{m_1 + m_2} \tag{5.47}$$

resonance is possible: at $p = 0$, $|W(p) = 1|$; at $\omega \to \infty |W(p)| \to 0$. The properties of the system in the presence of an joint as a unilateral constraints will be

Table 5.4 The values of the coefficients of the equations of motion of the system in the coordinates y_1 and y_0

a_{11}	a_{12}
$(m_1 + m_2)p^2 + k_1 + k_2 + k_4$	$m_2 p^2 + k_2 + k_4$
a_{21}	a_{22}
$m_2 p^2 + k_2 + k_4$	$m_2 p^2 + k_2 + k_3 + k_4$

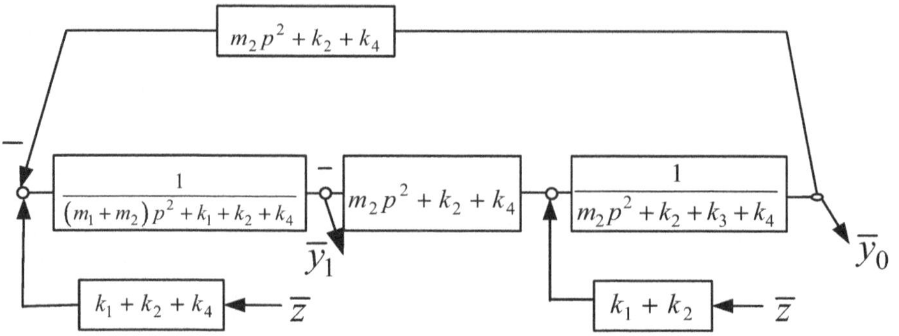

Fig. 5.15 Structural diagram of the VPS in coordinates y_1 and y_0

correspondingly manifested over the entire frequency range. Using the data of Table 5.4, we construct a structural diagram (Fig. 5.15) of the automatic control system (ACS), dynamically equivalent to the initial system (see Fig. 5.14).

When constructing the structural diagram, it was assumed that $z_2 = z_1 = z = z_4$. From the analysis of the scheme it follows that at a frequency

$$\omega_1^2 = \frac{k_1 + k_4}{m_2} \tag{5.48}$$

the connections between the motion m_1 and m_2 are "zeroed", which corresponds to the regime of independent motion along the coordinates y_1 and y_0. The amplitudes of the oscillations y_1 and y_0 can be determined through the transfer functions of the system when the frequency, determined from (5.46), is substituted into them.

If we consider the mechanical part k_1, k_3 and m_1 as elements of a generalized spring, then at the frequency

$$\omega_2^2 = \frac{k_1 + k_3}{m_1} \tag{5.49}$$

the stiffness of the generalized spring tends to infinity and the possible relative motion y_0 is blocked.

This circumstance can be considered as the emergence of the possibility of joint of two bodies in a virtual version. Dynamically, this means that at the frequency determined by Formula (5.49), a joint may occur. This joint has external signs: $y_0 = 0$ and $k_3 \to \infty$ (since resonance properties arise in the generalized spring). The peculiarity of this mode is that it occurs only on one frequency. When the frequency changes, the joint breaks up, since the constraint is unilateral.

An interesting situation is when, in the presence of some solid body having a mass m_2, two channels of transmission of influences from the external medium (oscillations of the basis), an elastic element k_1 and a mechanical circuit m_1, k_1 and k_3, and, possibly at a certain frequency $\omega_2^2 = \frac{k_1 + k_3}{m_1}$, to provide "sticking" of the mass m_1 to the mass m_2. The vibrations of the base, as it were, cause the intermediate elements to “pull” to a certain center. In justifying the behavior of the system, it was assumed that it belonged to the class of linear systems, and the influence of the resistance forces was not taken into account.

Thus, joints in mechanical oscillatory systems play an important role, since they reflect the peculiarities of possible couplings of units. The forms of joints are diverse and most often represent kinematic pairs of various types. The most common are kinematic pairs of class V, however, in complex mechanical systems, kinematic pairs of classes IV and III can be used. The introduction of joints in mechanical systems alters the number of degrees of freedom of motion of protection objects, as well as the structure of systems, and thereby forms a system of transfer of dynamic influences corresponding in form and effect.

Accounting for joints of solid bodies and joints brought by the joints can significantly change the dynamic properties of mechanical oscillatory systems, create additional modes of dynamic absorbing, provide blocking or locking of individual branches of the system at certain frequencies of external influence. Mathematical models of systems with joints can be constructed from more complex structures, by reducing the number of degrees of freedom, which is implemented by certain actions over the coefficient matrix of the unified system of equations of motion. Although the initial system of equations is more complex, its use as a basis for simplification is justified by the choice of various variants of joints with the corresponding expansion of the spectrum of the dynamic properties of systems as a whole. The starting point in the method of constructing mathematical models is the selection of a coordinate system that allows for certain coordinates the possibility of “zeroing” coordinates, which actually reflects the effect of the joint.

5.3 Mechanical Oscillating Systems with Translational Motions. Possible Forms of Link Joints

In the problems of vibration protection and vibration isolation of equipment and machines, especially at the preliminary stage of estimating dynamic properties, it becomes necessary to simplify the initial computational scheme [23]. In recent years, a number of papers have appeared that deal with the simplification of mechanical systems based on the rules for transforming structural diagrams of the dynamically equivalent automatic control systems based on the concepts of generalized springs (or quasi-springs): for example, [24]. The methods of formation of joints also have certain possibilities in this direction. Studies show that joints in translational motion can be implemented in systems; most often in situations like this, interacting elements with a very rigid elastic coupling are combined into one block. This approach not only simplifies the original system, but also reduces the number of degrees of freedom of motion, while leaving the possibility of assessing the legitimacy of the simplifications.

5.3.1 System Features

In the proposed section, the possibilities of choosing schemes for simplifying a mechanical system with three degrees of freedom are considered (Fig. 5.16). Expressions for kinetic and potential energy of motion can be represented in the following form:

$$\mathrm{T} = \frac{1}{2}m_1\dot{y}_1^2 + \frac{1}{2}m_2\dot{y}_2^2 + \frac{1}{2}m_3\dot{y}_3^2; \tag{5.50}$$

$$\begin{aligned} \Pi = &\frac{1}{2}k_1(y_1 - z_1)^2 + \frac{1}{2}k_2(y_2 - y_1)^2 + \frac{1}{2}k_3(y_3 - y_2)^2 \\ &+ \frac{1}{2}k_4(y_3 - z_2)^2 + \frac{1}{2}k_5(y_3 - y_1)^2 + \frac{1}{2}(y_2 - z_2)^2, \end{aligned} \tag{5.51}$$

where k_1, k_2, k_3, k_4, k_5, k_6 are the corresponding stiffness coefficients of springs connecting masses m_1–m_3, each of which can be a protection object. In the system under consideration (see Fig. 5.16), there are kinematic perturbations z_1 and z_2. With a given arrangement of the elastic elements, the system can not be referred to nonplanar systems. Using the usual techniques, we obtain a system of differential equations of motion in the system of coordinates y_1, y_2, y_3. In this case, the equations of motion of the system (see Fig. 5.16) take the form

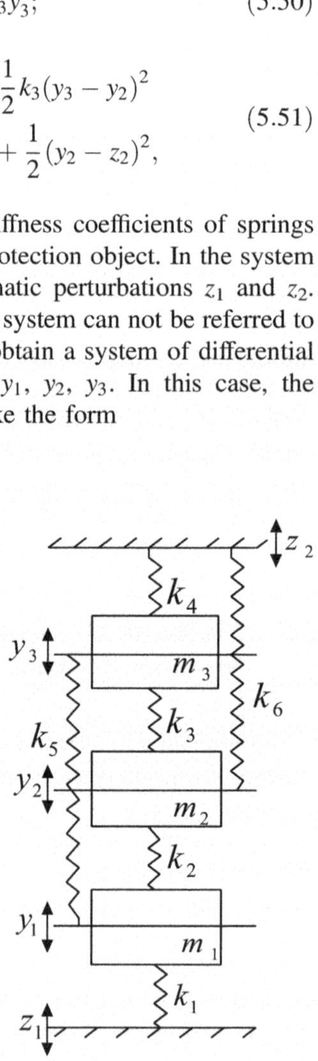

Fig. 5.16 The computational scheme of a vibration protection system with three degrees of freedom

$$
\begin{aligned}
&m_1\ddot{y}_1 + y_1(k_1 + k_2 + k_3) - k_2y_2 - k_3y_3 = k_1z_1;\\
&m_2\ddot{y}_2 + y_2(k_2 + k_3 + k_6) - k_2y_1 - k_3y_3 = k_6z_2;\\
&m_3\ddot{y}_3 + y_3(k_3 + k_4 + k_5) - k_3y_2 - k_3y_1 = k_4z_2.
\end{aligned}
\tag{5.52}
$$

We denote the right-hand sides of Eq. (5.52)

$$
b_1 = k_1z_1; \quad b_2 = k_6z_2; \quad b_3 = k_4z_2. \tag{5.53}
$$

Assuming that the properties of the joint of the masses m_1 and m_2 are related to another coordinate system, we introduce the relation

$$
y_0 = y_2 - y_1. \tag{5.54}
$$

and proceed to the system y_0, y_1, y_3. In this case, the expressions (5.50) and (5.51) are transformed to the form

$$
\mathrm{T} = \frac{1}{2}m_1\dot{y}_1^2 + \frac{1}{2}m_2(\dot{y}_0 + \dot{y}_1)^2 + \frac{1}{2}m_3\dot{y}_3^2; \tag{5.55}
$$

$$
\begin{aligned}
\Pi = {}& \frac{1}{2}k_1(y_1 - z_1)^2 + \frac{1}{2}k_2(y_0 + y_1 - y_1)^2 + \frac{1}{2}k_3(y_3 - y_0 - y_1)^2\\
& + \frac{1}{2}k_4(y_3 - z_2)^2 + \frac{1}{2}k_5(y_3 - y_1)^2 + \frac{1}{2}k_6(y_0 + y_1 - z_2)^2.
\end{aligned}
\tag{5.56}
$$

Using (5.53) and (5.54), we can transform the system of equations of motion to the coordinates y_0, y_1, y_3, which allows us to obtain

$$
\begin{aligned}
&(m_1 + m_2)\ddot{y}_1 + y_1(k_1 + k_3 + k_5 + k_6) + m_2\ddot{y}_0\\
&\quad + (k_3 + k_6)y_0 + y_3(-k_3 - k_5) = k_1z_1 + k_6z_2;\\
&m_2\ddot{y}_2 + y_1(k_3 + k_6) + m_2y_0 + y_0(k_2 + k_3 + k_6) + y_3(-k_3) = k_6z_2;\\
&m_3\ddot{y}_3 + y_3(k_3 + k_4 + k_5) + y_1(-k_3 - k_5) + y_0(-k_3) = k_4z_2.
\end{aligned}
\tag{5.57}
$$

Table 5.5 shows the coefficients of the equations (5.57) in unified form.

Table 5.5 The values of the coefficients of the equations system (5.57) in coordinates y_0, y_1, y_3

a_{11}	a_{12}	a_{13}
$(m_1 + m_2)p^2 + k_1 + k_3 + k_5 + k_6$	$m_2p^2 + k_3 + k_6$	$-k_3 - k_5$
a_{21}	a_{22}	a_{23}
$m_2p^2 + k_3 + k_6$	$m_2p^2 + k_2 + k_3 + k_6$	$-k_3$
a_{31}	a_{32}	a_{33}
$-k_3 - k_5$	$-k_3$	$m_3p^2 + k_3 + k_4 + k_5$

For a system of coordinates y_0, y_1, y_3, the generalized forces have the form

$$b_1 = k_1 z_1 + k_6 z_2; \quad b_2 = k_6 z_2; \quad b_3 = k_4 z_2. \tag{5.58}$$

When moving from one coordinate system to another, the generalized forces are usually determined through the corresponding equality of work on virtual displacements in two comparable coordinate systems. In this case, when the perturbation has a kinematic nature, the generalized forces are obtained simultaneously in the derivation of the equations.

When comparing the mathematical models reflected by Eqs. (5.52) and (5.57), it can be noted that the diagonal term a_{11} contains the sum of the masses $m_1 + m_2$. The cross-couplings a_{12} and a_{21} have also changed (see Table 5.5), which, in contrast to Eq. (5.52), acquire not elastic but an inertial-elastic character. In particular, with the frequency of external action

$$\omega_1^2 = \frac{k_3 + k_6}{m_2} \tag{5.59}$$

the decoupling of motions between partial systems a_{11} and a_{22} (see Table 5.5) is possible.

5.3.2 *Features of Different Coordinate Systems*

To introduce a system of coordinates of the form y_1, y_2, y_{00} it is assumed that

$$y_{00} = y_3 - y_2. \tag{5.60}$$

In these coordinates, the expressions for the kinetic and potential energy of the system (5.50) and (5.51) can be written in the following form:

$$\mathrm{T} = \frac{1}{2} m_1 \dot{y}_1^2 + \frac{1}{2} m_2 \ddot{y}_2^2 + \frac{1}{2} m_3 (\dot{y}_{00} + \dot{y}_2)^2; \tag{5.61}$$

$$\begin{aligned} \Pi = {} & \frac{1}{2} k_1 (y_1 - z_1)^2 + \frac{1}{2} k_2 (y_2 - y_1)^2 + \frac{1}{2} k_3 (y_{00})^2 + \frac{1}{2} k_4 (y_{00} + y_2 - z_2)^2 \\ & + \frac{1}{2} k_5 (y_{00} + y_2 - y_1)^2 + \frac{1}{2} k_6 (y_2 - z_2)^2, \end{aligned} \tag{5.62}$$

from which the equations of motion of the system can be obtained (see Fig. 5.5) in the system of coordinates y_1, y_2, y_{00}. The corresponding values of the coefficients of the unified system of equations are given in Table 5.5.

Table 5.6 Values of the coefficients of the system of equations in coordinates y_1, y_2, y_{00}

a_{11}	a_{12}	a_{13}
$m_1p^2+k_1+k_2+k_5$	$-k_2+k_5$	$-k_5$
a_{21}	a_{22}	a_{23}
$-k_2+k_5$	$(m_2+m_3)p^2+k_2+k_4+k_5+k_6$	$k_4+k_5+m_3p^2$
a_{31}	a_{32}	a_{33}
$-k_5$	$m_3p^2+k_4+k_5$	$m_3p^2+k_3+k_4+k_5$

The generalized forces of the system with coordinates y_1, y_2, y_{00} have the form

$$b_2 = k_6z_2 + k_4z_2. \tag{5.63}$$

In comparison with the traditional system of coordinates y_1, y_2, y_3 from Table 5.6, the partial system in the matrix a_{22} has a sum of masses $m_2 + m_3$. In turn, the forms of constraints a_{23} and a_{32} also change, acquiring an inertial-elastic character. At frequency $\omega_2^2 = \frac{k_4+k_5}{m_3}$ in this system, the decoupling of motions occurs between the partial systems a_{22} and a_{23}.

To consider the case of the joint of three bodies, we introduce the system of generalized coordinates y_1, y_2, y_{000}, where

$$y_{000} = y_3 - y_1. \tag{5.64}$$

In this case, the expressions for the kinetic and potential energies (5.50) and (5.51) are transformed to the form

$$\mathrm{T} = \frac{1}{2}m_1\dot{y}^2 1 + \frac{1}{2}m_2\ddot{y}^2 2 + \frac{1}{2}m_3(\dot{y}_1 + \dot{y}_{000})^2; \tag{5.65}$$

$$\begin{aligned}\Pi = {} & \frac{1}{2}k_1(y_1 - z_1)^2 + \frac{1}{2}k_2(y_2 - y_1)^2 + \frac{1}{2}k_3(y_1 + y_{000} - y_2)^2 \\ & + \frac{1}{2}k_4(y_1 + y_{000} - z_2)^2 + \frac{1}{2}k_5(y_{000})^2 + \frac{1}{2}k_6(y_2 - z_2)^2\end{aligned} \tag{5.66}$$

.

Table 5.7 gives the values of the coefficients of the unified system of equations that can be obtained by analogy to the above methods.

Table 5.7 Values of the coefficients of the equations of motion in coordinates y_1, y_2, y_{000}

a_{11}	a_{12}	a_{13}
$(m_1+m_3)p^2+k_1+k_2+k_3+k_4$	$-k_2-k_3$	$m_3p^2+k_3+k_4$
a_{21}	a_{22}	a_{23}
$-k_2-k_3$	$m_2p^2+k_2+k_3+k_6$	k_3
a_{31}	a_{32}	a_{33}
$m_3p^2+k_3+k_4$	$-k_3$	$m_3p^2+k_3+k_4+k_5$

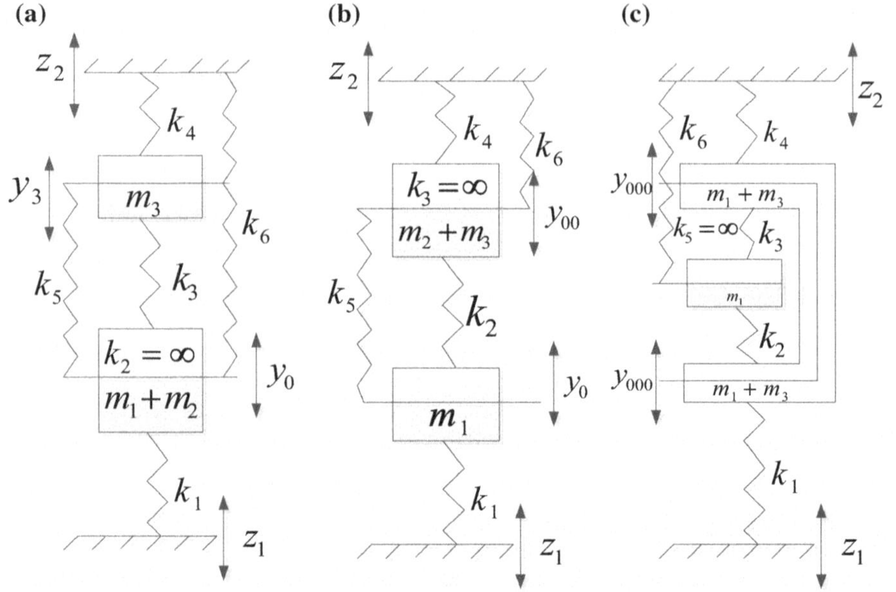

Fig. 5.17 Computational schemes for the VPS with joints. **a** $y_2 - y_1 = 0$ $(y_0 = 0)$; **b** 0 $y_2 - y_3 = 0$ $(y_{00} = 0)$; **c** $y_1 - y_3 = 0$ $(y_{000} = 0)$

The generalized forces for a system with coordinates y_1, y_2, y_{000} have the form

$$b_1 = k_1 z_1 + k_4 z_2; \quad b_2 = k_6 z_2; \quad b_3 = k_4 z_2. \tag{5.67}$$

The introduction of relative coordinates y_1, y_2, y_{000} allows us to obtain the corresponding particular types of computational schemes with respect to the initial system shown in Fig. 5.16.

Figure 5.17 presents the corresponding computational schemes. In this case, when "zeroing" y_0, y_{00}, y_{000}, in a certain way, the columns and rows of the coefficient matrix are "zeroed", which simplifies the construction.

5.3.3 *Structural Interpretations of Systems: Complex Modes*

Structural diagrams of dynamically equivalent ACSs are shown in Fig. 5.18. The initial data for constructing the corresponding structural diagrams can be taken from Tables 5.5, 5.6 and 5.7. The structural diagram in Fig. 5.18a reflects the properties of the original mechanical system (see Fig. 5.16), in which the dynamic state is described by three coordinates: y_1, y_2, y_3. If m_1 and m_2 are supposed to be jointed, then the structural diagram has the form, as shown in Fig. 5.18b. In turn, at the joint of m_2 and m_3, we have the structural diagram shown in Fig. 5.18c; at a joint of m_1

and m_3, we accordingly obtain the structural diagram shown in Fig. 5.18d. Note that the joints change the structure of the system; each joint eliminates one degree of freedom. The remaining dynamic constraints are determined by the coefficient matrix after excluding the corresponding row and column in Tables 5.5, 5.6 and 5.7. Considering the "zeroing" of the motion $\left(y_i = 0\ (i = \overline{1,3})\right)$ as a joint, we can simplify the computational schemes, shown in Fig. 5.18, to a system with one degree of freedom.

With the development of the proposed method of simplifying (or synthesizing) systems, it is of interest to consider the motion in the system of coordinates $y_1, y_0, y_{00}(y_0 = y_2 - y_1, y_{00} = y_2 - y_3)$. In this case, the expressions for the kinetic and potential energies (5.50) and (5.51) are transformed to the form

$$T = \frac{1}{2}m_1\dot{y}_1^2 + \frac{1}{2}m_2(\dot{y}_0 + \dot{y}_1)^2 + \frac{1}{2}m_3(\dot{y}_{00} + \dot{y}_0 + \dot{y}_1)^2; \tag{5.68}$$

$$\begin{aligned}\Pi = {} & \frac{1}{2}k_1(y_1 - z_1)^2 + \frac{1}{2}k_2(y_2 - y_1)^2 + \frac{1}{2}k_3(y_3 - y_2)^2 \\ & + \frac{1}{2}k_4(y_3 - z_2)^2 + \frac{1}{2}k_5(y_3 - y_1)^2 + \frac{1}{2}(y_2 - z_2)^2.\end{aligned} \tag{5.69}$$

Making a series of transformations analogous to the above, we obtain a system of differential equations of motion

$$\left.\begin{aligned} & (m_1 + m_2 + m_3)\ddot{y}_1 + y_1(k_1 + k_4 + k_6) + (m_2 + m_3)\ddot{y}_0 + (-k_4)y_0 \\ & \quad + \ddot{y}_{00}m_3 + y_{00}(k_6 + k_4) = k_1z_1 + k_4z_2 + k_6z_2; \\ & (m_2 + m_3)\ddot{y}_0 + y_0(k_2 + k_5) + m_3\ddot{y}_{00} + y_{00}(k_4 + k_5) + \ddot{y}_1(m_2 + m_3) + y_1(-k_4) = k_4z_2; \\ & m_3\ddot{y}_{00} + y_{00}(k_3 + k_4 + k_5 + k_6) + \ddot{y}_0(m_3) + y_0(k_4 + k_5) + m_3\ddot{y}_1 + (k_4 + k_6)y_1 = k_4z_2 + k_6z_2; \end{aligned}\right\} \tag{5.70}$$

The values of the coefficients of the equations from the system (5.70), reduced to a unified form, are presented in Table 5.8.

The generalized forces of the system with coordinates y_1, y_0, y_{00} have the form

$$b_1 = k_1z_1 + k_4z_2 + k_6z_2; \quad b_2 = k_4z_2; \quad b_3 = k_4z_2 + k_6z_2. \tag{5.71}$$

If we assume that $y_0 = 0$ and $y_{00} = 0$, the system takes the form, as shown in Fig. 5.19.

With two joints, the original system is transformed into a system with one degree of freedom, whose natural frequency is determined by the expression

$$\omega_{\text{nat}}^2 = \frac{k_1 + k_4 + k_6}{m_1 + m_2 + m_3}. \tag{5.72}$$

To assess the possibilities of using joints as a way of changing the structure and its subsequent simplification, let us consider the structural diagram of a dynamically

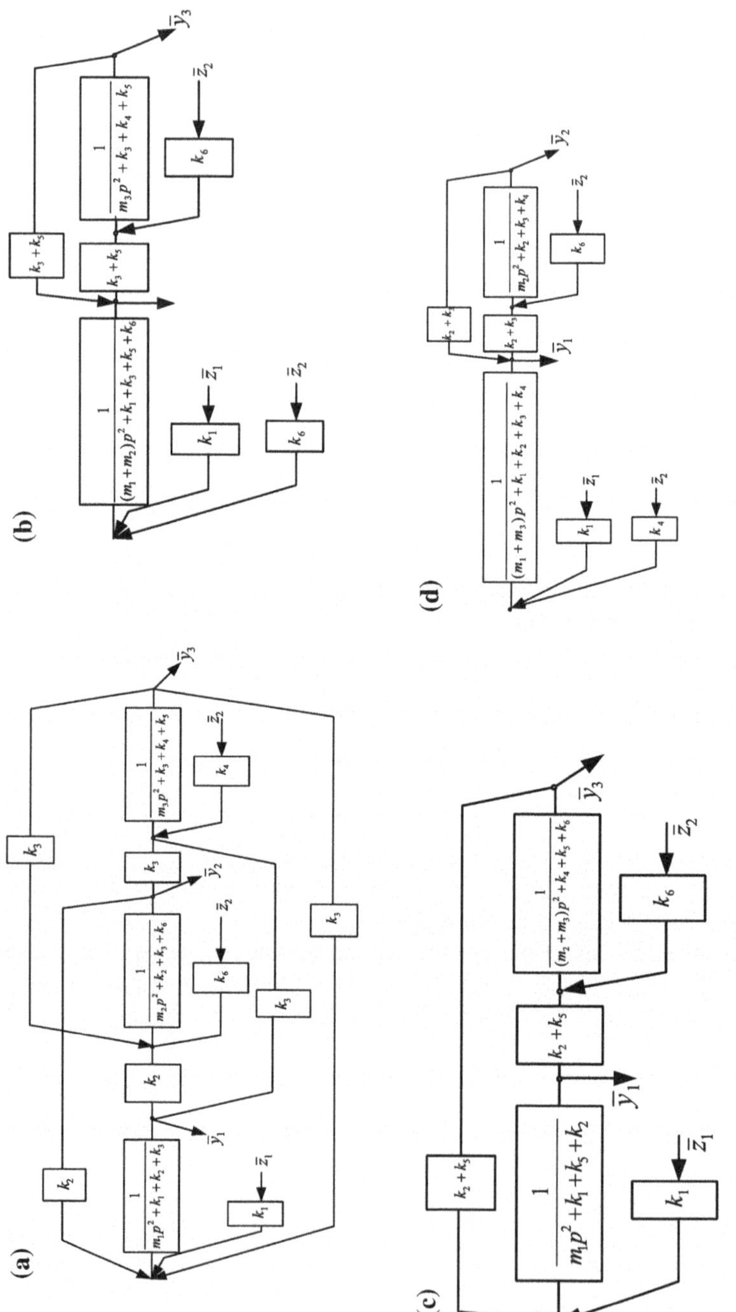

Fig. 5.18 Structural diagram of the VPS for different cases of joint for the system of coordinates: **a** y_1, y_2, y_3 (no joints); **b** y_0, y_1, y_3 $(y_0 = y_2 - y_1 = 0)$; **c** y_1, y_2, y_{00} $(y_{00} = y_2 - y_3 = 0)$; **d** y_1, y_2, y_{000} $(y_{000} = y_1 - y_3 = 0)$

equivalent automatic control system (Fig. 5.20) in the system of coordinates y_1, y_0, y_{00}, which corresponds to the mathematical model in the form of a system of equations (5.70). For comparison, Fig. 5.21 shows the structural diagrams of the equivalent ACS for vibration protection systems in coordinates y_1, y_0, y_3 (Fig. 5.21a) y_1, y_2, y_{00} (Fig. 5.21b) and y_1, y_2, y_{000} (Fig. 5.21c).

Note that the choice of a system of generalized coordinates changes not only the form of the partial systems, but also the cross-couplings. In the system of coordinates y_1, y_0, y_3 (Fig. 5.21a), between the motions along, an inertial-elastic coupling arises, which implies the possibility of the emergence of dynamic absorbing modes: this depends on what kind of selected system of external kinematic effects z_0, z_1 is going to be in the end. Between the coordinates y_0, y_3 there is an elastic constraint, determined by the elastic unit k_3. In the system of coordinates y_1, y_2, y_{00} (Fig. 5.21b), with the same external kinematic effects between the coordinates, there is an elastic constraint, and between y_2, y_{00} there is an inertial-elastic constraint that is "zeroed" at a certain frequency, which excludes the direct relationship of motions between partial systems. Between the coordinates y_1, y_{00}, y_{000} (Fig. 5.21c), there is a system of elastic cross-couplings, which excludes the emergence of modes of decoupling of oscillations between partial systems.

All this proves the possibility of forming a joint by means of an appropriate choice of the coordinate system. The subsequent procedures are carried out in a formalized manner and ensure that the corresponding model is obtained. The evidential basis of the approach is connected with the transition of a system with a greater number of degrees of freedom to a system with a smaller number of degrees, which does not affect the conditions for the solvability of the equations. Mathematical models of systems with joints can be obtained. And physically this is understandable if the parameters of the elements connecting certain points of the system (elastic elements and any others from the extended set of typical VPS) will take the limiting values (or very large in comparison with others).

5.4 Change of Dynamic Properties at Introduction of Joints Between the Units

Different forms of joints presuppose the presence of peculiarities in coupling methods, taking into account the complexity of their constructive design. The motion of all elements of the system occurs in the same plane. In the mechanical oscillatory system, two classes of units are distinguished: fixed and mobile. In connection with this, one of the most common joints is the joint in the form of a kinematic rotational pair. In the simplest versions, units connected by a hinge allow rotating and oscillating motions relative to each other. In this case, the joint of the total number of degrees of freedom produces an "exception" of one degree of freedom in the motions. An increase in the number of joints corresponding to the number of hinges causes a decrease in the total number of degrees of freedom (or the number of independent variables).

Table 5.8 The values of the coefficients of the equations of motion in the coordinates y_1, y_0, y_{00}

a_{11}	a_{12}	a_{13}
$(m_1+m_2+m_3)p^2+k_1+k_4+k_6$	$-(m_2+m_3)p^2-k_4$	$m_3p^2+k_6+k_4$
a_{21}	a_{22}	a_{23}
$(m_2+m_3)p^2-k_4$	$(m_2+m_3)p^2+k_2+k_5$	$m_3p^2+k_4+k_5+k_6$
a_{31}	a_{32}	a_{33}
$m_3p^2-k_4+k_6$	$m_3p^2+k_4+k_5+k_6$	$m_3p^2+k_3+k_4+k_5+k_6$

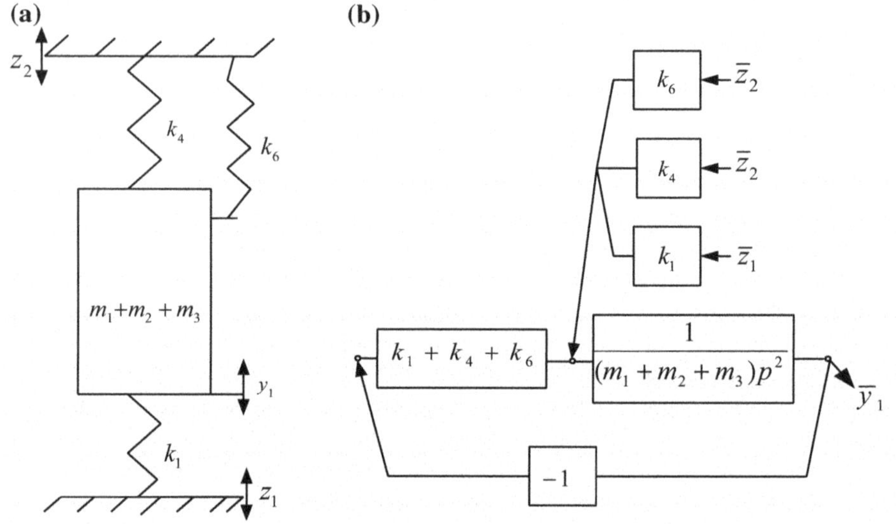

Fig. 5.19 Computational scheme of the initial system (see Fig. 5.16) for a case with joint of three bodies (**a**); structural diagram corresponding to a scheme with three joints (**b**)

In addition to connecting the mobile units to each other, often there are compounds of solids with fixed units or with a base (or conditionally fixed system). Such a joint in Fig. 5.22 is shown by a subshading.

Mechanical oscillatory systems can have joints of various types, which provides features of the structure of the system and the so-called "metric". The mathematical model of the system can be represented in the form of a system of ordinary inhomogeneous second-order differential equations with constant coefficients:

$$Ay = b, \tag{5.73}$$

where A is the coefficient matrix; y is the vector-column of variables; b is the vector-column of external influences.

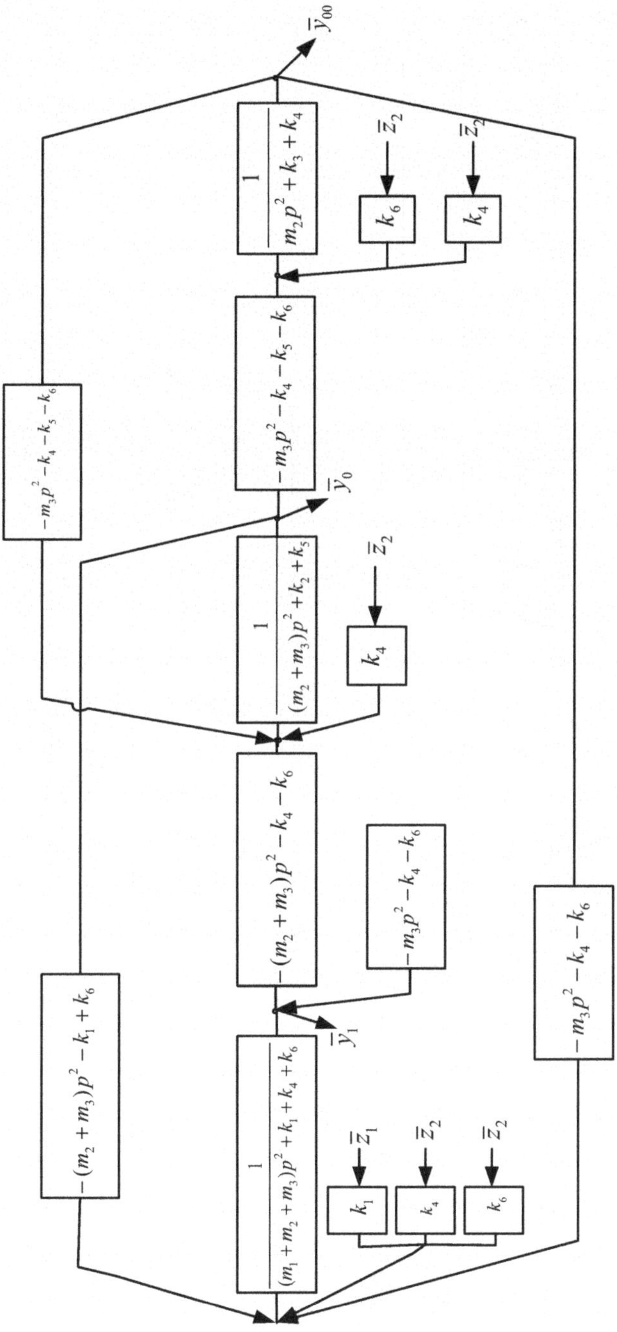

Fig. 5.20 Structural diagram of the equivalent ACS in the system of coordinates y_1, y_0, y_{00}

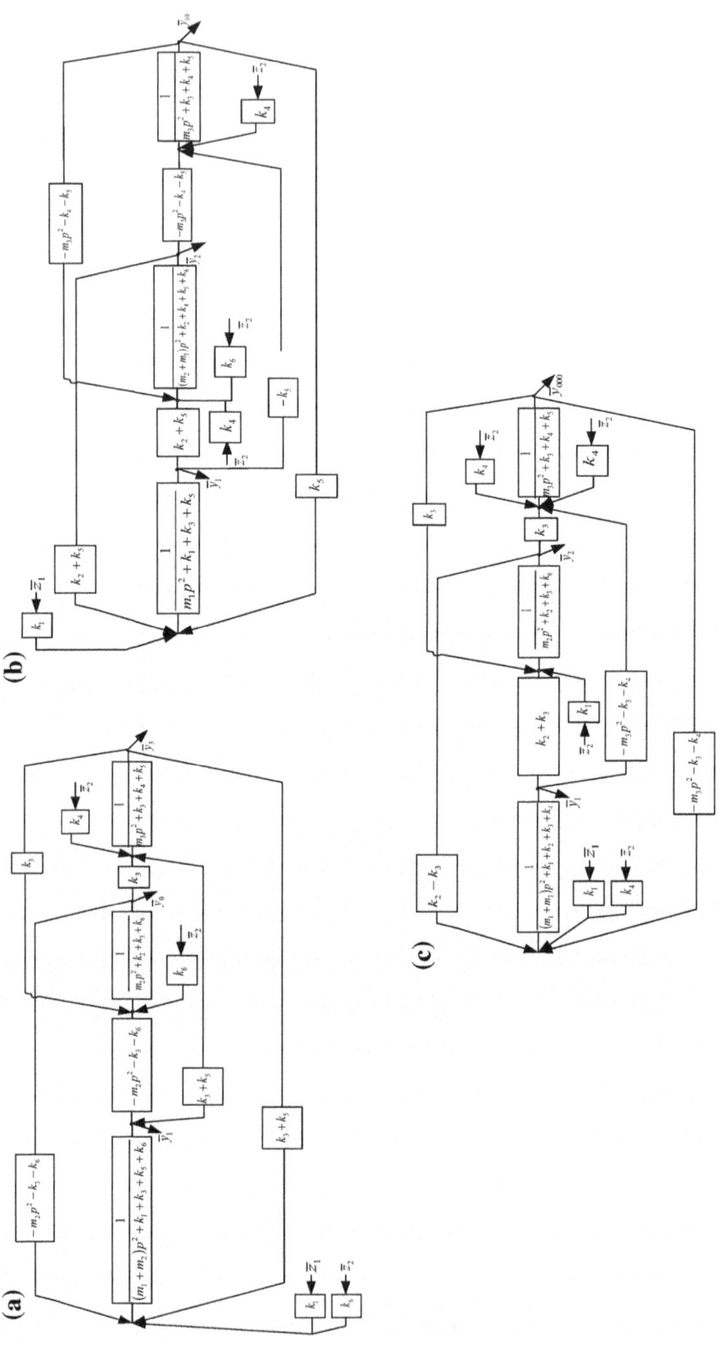

Fig. 5.21 Structural diagrams for the VPS (see Fig. 5.19) in different coordinate systems: **a** y_1, y_0, y_3 $(y_0 \neq 0)$; **b** y_1, y_2, y_{00} $(y_{00} \neq 0)$; **c** y_1, y_2, y_{000} $(y_{000} \neq 0)$

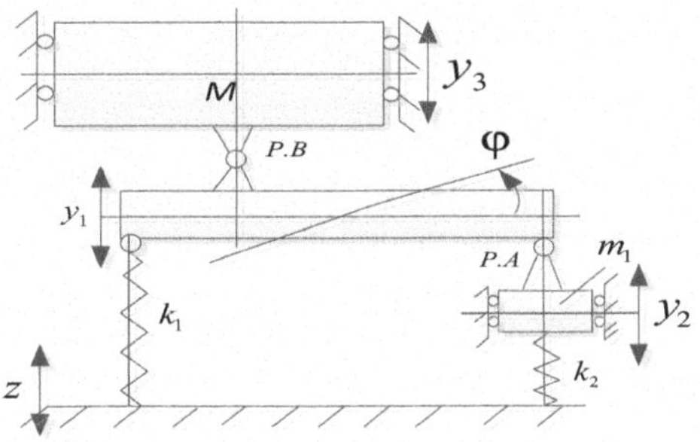

Fig. 5.22 The computational scheme of a mechanical system having joints: between mobile units, as well as between mobile and stationary units (at points *A* and *B* are joints in the form of rotational kinematic pairs)

In the general case, the matrix *A* is of order $n \times n$ and is symmetric:

$$A = \begin{vmatrix} a_{11} & a_{12} & \cdots & a_{1n} \\ a_{21} & a_{22} & \cdots & a_{2n} \\ \vdots & \vdots & \vdots & \vdots \\ a_{n1} & a_{n2} & \cdots & a_{nn} \end{vmatrix}. \tag{5.74}$$

When constructing mathematical models of systems with joints, various systems of generalized coordinates can be used, mainly those in which the coordinates reflect the relative motion. The joint can be implemented with respect to the element performing the "absolute" motion, i.e. in a fixed coordinate system. Naturally, in this case, the coordinate systems allow corresponding mutual transformations.

The introduction of an joint means the exclusion of certain columns and rows of the coefficient matrix, including the "exceptions" of the corresponding generalized forces [20].

The physical meaning of the operation is that the joint represented by the difference of the corresponding coordinates is excluded in the physical sense; Together with the variable, the matrix coefficients that determine the relationships between the retracted partial system and the rest of the systems are excluded. The right-hand member of the equation determined by the string is also excluded, since the point of application of forces "disappears" physically. The external action in this case is redistributed appropriately in the selection of systems of generalized coordinates, where it is necessary to observe the conditions for the equality of virtual works of generalized forces in various systems of generalized coordinates. In this case, we consider a number of specific examples of the use of procedures for constructing mathematical models, as well as examples of joints. A set of possible joints can provide even more complex forms of interactions, including those on the basis of kinematic pairs of classes IV and III. It is important to note that the possibility of taking into account the features of the joint in terms of constructing mathematical models can be extended to other systems with holonomic constraints [1].

5.4.1 *Joints in Systems of Combined Type*

A great influence on the development of research related to the search and development of dynamic vibration absorbers was the introduction of the practice of motion transformation devices. First of all, these devices were articulated linkages, later—non-locking screw and gear mechanisms, etc. At the same time, the emergence of motion transformation devices can also be viewed from the point of a general transformation of the structures of mechanical oscillatory systems based on the use of joint as an approach to obtaining simplified structures, and as a way of constructing non-traditional elements for the theory of vibrations.

I. Figure 5.23 presents the computational scheme of the VPS with two blocks, the presence of which is reflected by circuits I and II. At the heart of the blocks there is a solid body with a mass and moment of inertia. The structure of the VPS includes elastic elements. It is assumed that the displacement of the center of gravity of the block I does not have a significant effect on the dynamics of the system as a whole, and the resistance forces are small enough.

The computational scheme in the form of an oscillatory system with three degrees of freedom $(y - y_2)$ can be considered as a fragment of the VPS in which the block I (contour I, Fig. 5.23) and block II (contour II, Fig. 5.23), consisting of a solid body resting on elastic support. The contours I and II (Fig. 5.23) are in interaction through an elastic binding element k_{01}.

In turn, the solid body not only rests on elastic supports k_1 and k_2, but still has an elastic constraint k_0, the line of action of which passes through the center of gravity of the beam at p. O. We assume that $A_1O = l_1$, $B_2O = l_2$ and the attached mass m does not cause significant changes in the mass-inertial parameters of the system. Resistance forces are also assumed small. Developing the provisions, previously

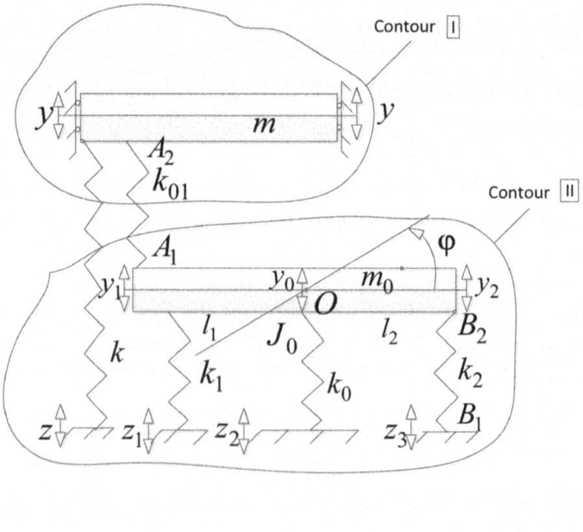

Fig. 5.23 The computational scheme of an oscillatory VPS having two interaction loops

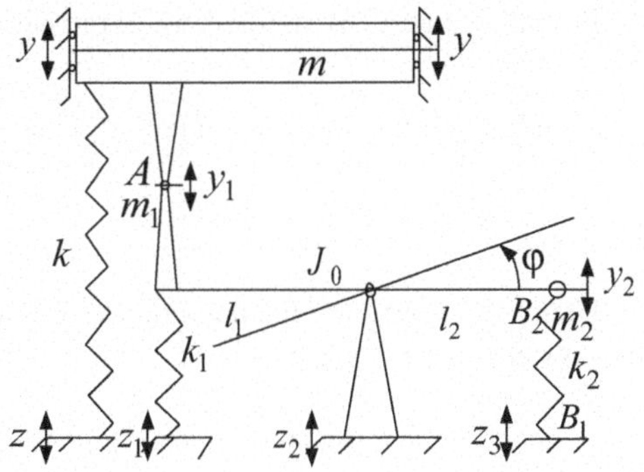

Fig. 5.24 Transformed computational scheme containing joints

stated by the authors, on the joints of solids as compounds of solids taking the form of a rotational hinge, we note that, assuming stiffness k_{01} and k_0 sufficiently large, we can transform the computational scheme to the form, as shown in Fig. 5.24.

Being the result of the simplification of the circuit in Fig. 5.23, the computational scheme (Fig. 5.24) was used, for example, in [23]. However, the question of the peculiarities of joints in this work was not considered. The computational scheme in Fig. 5.24 reflects the possibility of obtaining computational schemes with two joints.

II. Let us write the expressions for the kinetic and potential energy for the computational scheme of the VPS, shown in Fig. 5.24.

$$\mathrm{T} = \frac{1}{2}(m + m_1)\dot{y}^2 + \frac{1}{2}m_2\dot{y}_2; \tag{5.75}$$

$$\Pi = \frac{1}{2}k(y - z)^2 + \frac{1}{2}k_1(y - z_1)^2 + \frac{1}{2}k_2(y_2 - z_3)^2, \tag{5.76}$$

where $\dot{y}_2$ is the velocity of the element with mass m_2 in absolute motion, which is determined by expression

$$\dot{y}_2 = -\left(\frac{\dot{y}}{l_1} \cdot l_2\right) + z_2.$$

The minus sign reflects the motion change caused by the levers of the second kind. Thus, we find that (5.75) is transformed to the form

$$T = \frac{1}{2}(m + m_1)\dot{y}^2 + \frac{1}{2}m_2[-\dot{y}i + \dot{z}_2(1 + i)]^2, \tag{5.77}$$

where $i = \frac{l_2}{l_1}$ is the transfer ratio of the lever.

Potential energy is defined by expression

$$\begin{aligned} \Pi &= \frac{1}{2}k(y-z)^2 + \frac{1}{2}k_1(y_1 - z_1)^2 + \frac{1}{2}k_2(y_{2ab} - z_3)^2 \\ &= \frac{1}{2}k(y-z)^2 + \frac{1}{2}k_1(y_1 - z_1)^2 + \frac{1}{2}k_2[-yi + z_2(1+i) - z_3]^2. \end{aligned} \tag{5.78}$$

Using the Lagrange formalism, we obtain the differential equation of motion for the system in Fig. 5.24:

$$\begin{aligned} &\ddot{y}(m_1 + m_2 + m_2 i^2) + y(k + k_1 + k_2 i^2) \\ &= m_2(1+i)\ddot{z}_2 + z_2 k_2(1+i) + k_1 z_1 + kz - k_2 i z_3 \end{aligned} \tag{5.79}$$

For further calculations, we assume that $z_1 = z_2 = z_3; k_1 = 0$ and $k_2 = 0$, then the transfer function of the system takes the form

$$W(p) = \frac{\bar{y}}{\bar{z}(p)} = \frac{m_2(1+i)ip^2 + k}{(m_1 + m + m_2 i^2)p^2 + k}, \tag{5.80}$$

where $p = j\omega$ is the Laplace variable $\left(j = \sqrt{-1}\right)$. As an example, Fig. 5.25 shows a family of amplitude-frequency characteristics constructed on the basis of (5.80) with a parameter i ranging from 0 to 3 in increments of 0.5. The following system parameters are accepted as the initial ones: $m = 100$ kg; $m_1 = m_2 = 20$ kg; $k = 10,000$ N/m.

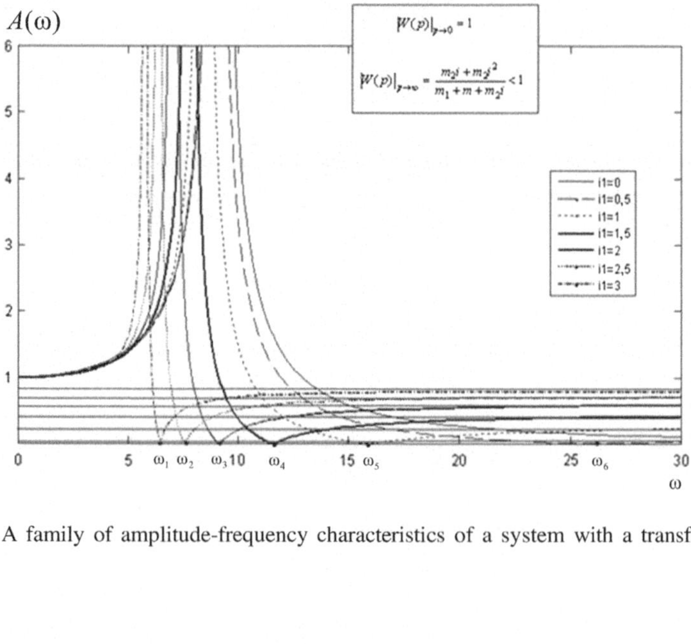

Fig. 5.25 A family of amplitude-frequency characteristics of a system with a transfer function (5.80)

Table 5.9 The values of the frequencies of natural oscillations in comparison with the frequencies ω_1–ω_6 for the system in Fig. 5.25

i	0.5	1	1.5	2	2.5	3
The value of the dynamical absorbing frequencies	ω_1	ω_2	ω_3	ω_3	ω_5	ω_6
	25.81	15.81	11.54	9.12	7.55	6.45
The value of the frequencies of natural oscillations	ω_{nat1}	ω_{nat2}	ω_{nat3}	ω_{nat4}	ω_{nat5}	ω_{nat6}
	8.94	8.45	7.78	7.07	6.38	5.77

In Fig. 5.25, ω_1–ω_6 denote the frequencies of dynamic absorbing. The corresponding values of the frequencies of the natural oscillations in comparison with the frequencies are given in Table 5.9.

The peculiarity of the amplitude-frequency characteristics is that, with increasing of i, the dynamic absorbing frequencies shift to the left; the $\omega_{nat} - \omega_{dyn}$ frequency difference decreases. The frequency of natural oscillations also decreases, but much more slowly. At high frequencies, the amplitude transfer coefficient after the dynamic absorbing mode tends to the limiting value $|W(p)|_{p\to\infty} = \frac{m_2 i + m_2 i^2}{m_1 + m + m_2 i} < 1$. The more i is, the more the value of $|W(p)|_{p\to\infty}$ becomes when $i \to \infty$—$|W(p)|_{p\to 0} = 0$.

III. We make a number of calculations that contribute to the development of a method for obtaining mathematical models of systems with joints, based on the simplifications of some more general systems using features that arise when constraining joints. Let us return to the full scheme shown in Fig. 5.23. Then the expressions (5.75) and (5.76) for the kinetic and potential energy can be transformed to the form

$$\mathrm{T} = \frac{1}{2} m_0 \dot{y}_0^2 + \frac{1}{2} J_0 \dot{\varphi} + \frac{1}{2} m \dot{y}^2, \tag{5.81}$$

$$\begin{aligned} \Pi &= \frac{1}{2} k(y - z)^2 + \frac{1}{2} k_1 (y_1 - z_1)^2 + \frac{1}{2} k_0 (y_0 - z_2)^2 \\ &+ \frac{1}{2} k_2 (y_2 - z_3)^2 + \frac{1}{2} k_{01} (y - y_1)^2. \end{aligned} \tag{5.82}$$

The following designations are accepted in the scheme (see Fig. 5.23): J_0 is the moment of inertia; m_0 is the mass of the intermediate body (can be converted into m_1 and m_2 connected by a lever); m is the mass of the object; k, k_1, k_0, k_2 are elastic elements of the intermediate body resting on the base; z–z_3 are kinematic perturbations. The coordinates of the points A_1 and A_2 (see Fig. 5.23) are defined as follows: $y_{A_1} = y_1$ and $y_{A_2} = y$.

To obtain joints between elements, it is necessary to satisfy the conditions $y_1 - y = 0$, $y_{10} = 0$, where it is assumed that $y_{10} = y_0 - z_2$.

Consider the motion of the system in coordinates y_0, φ. We write down a number of notation and relations: $y_0 = ay_1 + by_2$; $\varphi = c(y_2 - y_1)$:

$$a = \frac{l_2}{l_1 + l_2}; \quad b = \frac{l_1}{l_1 + l_2}; \quad c = \frac{1}{l_1 + l_2}.$$

When the object of protection moves in the system of coordinates y_0, y_2 and y_1, and the expression for the kinetic energy can be reduced to the form

$$T = \frac{1}{2} m_0 (a\dot{y}_1 + b\dot{y}_2)^2 + \frac{1}{2} J_0 c(\dot{y}_2 - \dot{y}_1)^2 + \frac{1}{2} m\dot{y}^2. \tag{5.83}$$

The expression for the potential energy in this case is determined by

$$\begin{aligned} \Pi = {} & \frac{1}{2} k(y - z)^2 + \frac{1}{2} k_1 (y_1 - z_1)^2 + \frac{1}{2} k_0 (ay_1 + by_2 - z_2)^2 \\ & + \frac{1}{2} k_2 (y_2 - z_3)^2 + \frac{1}{2} k_{01} (y - z_1)^2. \end{aligned} \tag{5.84}$$

The system of differential equations of motion of the system can be represented in the following form:

$$\ddot{y}_1 \left(m_0 a^2 + J_0 c^2\right) + y_1 \left(k_1 + k_0 a^2\right) + \ddot{y}_2 \left(m_0 ab - J_0 c^2\right) + y_2 k_0 ab = k_1 z_1 + k_0 a z_2, \tag{5.85}$$

$$\ddot{y}_1 \left(m_0 ab - J_0 c^2\right) + y_1 (-k_0 ab) + \ddot{y}_2 (m_0 b^2 + J_0 c^2) + y_2 (k_0 b_2 + k_2) = k_0 b z_2 + k_2 z_3, \tag{5.86}$$

$$\ddot{y}_1(0) + y_1(0) + \ddot{y}_2(0) + y_2(0) + \ddot{y}m + y(k + k_{01}) = kz + k_{01} z_1. \tag{5.87}$$

Table 5.10 shows the values of the coefficients of the equations in the coordinates y_1, y_2 and y.

Table 5.10 The values of the coefficients of the equations (5.85)–(5.87) in the coordinates y_1, y_2 and y

a_{11}	a_{12}	a_{13}
$(m_0 a^2 + J_0 c^2)p^2 + k_1 + k_0 a^2$	$(m_0 ab - J_0 c^2)p^2 + k_0 ab$	0
a_{21}	a_{22}	a_{23}
$(m_0 ab - J_0 c^2)p^2 + k_0 ab$	$(m_0 b^2 + J_0 c^2)p^2 + k_0 b^2 + k_2$	0
a_{31}	a_{32}	a_{33}
0	0	$mp^2 + k + k_{01}$

Generalized forces with respect to coordinates y_0, y_2 and y_1 have the form

$$Q_{y_1} = k_1 z_1 + k_0 a z_2; \quad Q_{y_2} = k_0 b z_2 + k_2 z_3; \quad Q_y = kz + k_{01} z_1. \tag{5.88}$$

Equations (5.85)–(5.87) describe the motion in a system of coordinates reflecting the vertical displacements of the mass-inertia elements of the VPS. Such a mathematical model can be called basic. The peculiarity of the matrix lies in the fact that $a_{13} = a_{31} = 0$, $a_{23} = a_{32} = 0$; this depends on the features of the dynamic interactions determined by the structure of the VPS (see Fig. 5.23) and the choice of the system of generalized coordinates.

IV. We proceed to the system of coordinates $x = y - y_1$, y_0 and φ. Let us write down the series of relations $y = x + y_1 = x + y_0 - l_1\varphi$ and obtain an expression for the kinetic energy

$$T = \frac{1}{2} m_0 (\dot{y}_0)^2 + \frac{1}{2} J_0 \dot{\varphi}^2 + \frac{1}{2} m (\dot{x} + \dot{y}_0 - l_1 \dot{\varphi})^2. \tag{5.89}$$

We write the expression for the potential energy in the expanded form:

$$\begin{aligned} \Pi = {} & \frac{1}{2} k (x + y_0 - l_1 \varphi - z)^2 + \frac{1}{2} k_1 (y_0 - l_1 \varphi - z_1)^2 \\ & + \frac{1}{2} k_0 (y_0 - z_2)^2 + \frac{1}{2} k_2 (y_0 + l_2 \varphi - z_3)^2 + \frac{1}{2} k_{01} (x)^2. \end{aligned} \tag{5.90}$$

The equations of motion of the system in coordinates φ, y_0, x are as follows:

$$\begin{aligned} & \ddot{\varphi}(J_0 + m l_1^2) + \varphi(k l_1^2 + k_1 l_1^2 + k_2 l_2^2) + \ddot{y}_0(-m l_1) + y_0(-k l_1 + k_1 l_1 + k_2 l_2) \\ & \quad + \ddot{x}(-m l_1) + x(-k l_1) = -k l_1 z + k_1 l_1 z_1 + k_2 l_2 z_3; \end{aligned} \tag{5.91}$$

$$\begin{aligned} & \ddot{\varphi}(-m l_1) + \varphi(-k l_1 + k_1 l_1 + k_2 l_2) + \ddot{y}_0(m + m_0) \\ & \quad + y_0(k + k_1 + k_0 + k_2) + \ddot{x}(m) + x(k) = kz - k_1 z_1 + k_0 z_2 - k_2 z_3; \end{aligned} \tag{5.92}$$

$$\ddot{\varphi}(-m l_1) + \varphi(-k_1 l_1) + \ddot{y}_0(m) + y_0(k) + \ddot{x}(m) + x(k + k_{01}) = kz. \tag{5.93}$$

The values of the coefficients of equations (5.91)–(5.93) in the coordinates φ, y_0 and x, respectively, are presented in Table 5.11.

Generalized forces with respect to the coordinates φ, y_0 and x have the form

$$Q_\phi = -k l_1 z + k_1 l_1 z_1 + k_2 l_2 z_3; \quad Q_{y_0} = kz + k_1 z_1 + k_0 z_2 - k_2 z_3; \quad Q_x = kz. \tag{5.94}$$

The mathematical model in the system of coordinates φ, y_0 and x differs from the previous model (coordinate y_1, y_2 and y) in that there are no zero cells

Table 5.11 The values of the coefficients of the equations (5.91)–(5.93) in the coordinates φ, y_0 and x

a_{11}	a_{12}	a_{13}
$(ml_1^2+J_0)p^2+kl_1^2+k_1l_1^2+k_2l_2^2$	$(-ml_1)p^2-kl_1+k_1l_1+k_2l_2$	$(-ml_1)p^2-kl_1$
a_{21}	a_{22}	a_{23}
$(-ml_1)p^2-kl_1+k_1l_1+k_2l_2$	$(m_0+m)p^2+k+k_1+k_0+k_2$	$(m)p^2+k$
a_{31}	a_{32}	a_{33}
$(-ml_1)p^2-k_1l_1$	$(m)p^2+k$	$(m)p^2+k+k_{01}$

(see Table 5.11) (that is, the constraints exist between all partial systems). As for the coordinate $x=y-y_1$, it can be "nullified" by the assumption that $k_{01}\to\infty$, the joint of solids of m and m_1 is formed. Physically, this means that the mass m is attached to the element of the VPS at p. A with mass-and-inertia parameters m_0, J_0. The motion of the system will be described in this case by the coordinates y_0 and φ. The necessary data for determining the transfer functions can be obtained by excluding in the matrix (see Table 5.3) a column and a string containing x (the variable is eliminated and the order of the matrix is reduced by one).

V. For further calculations, we introduce a system of coordinates y_0, y_1 and x. Then the expressions for the kinetic and potential energy in the coordinates y_0, y_1 and x are transformed to the form

$$T=\frac{1}{2}m_0(\dot{y}_0)^2+\frac{1}{2}J_0a_0^2(\dot{y}_0-\dot{y}_1)^2+\frac{1}{2}m(\dot{y}_1+\dot{x})^2. \tag{5.95}$$

Let us change the coordinate system with respect to the expression for (5.95)

$$T=\frac{1}{2}m_0(\dot{y}_0)^2+\frac{1}{2}J_0(\dot{\varphi})^2+\frac{1}{2}m(\dot{y})^2, \tag{5.96}$$

where $y=y_1+x$; $x=y-y_1$, then $\varphi=c(y_2-y_1)$; $y_0=ay_1+by_2$; $y_2=\frac{y_0-ay_1}{b}$.

We write down that $\varphi=cy_2-cy_1=\frac{c\cdot(y_0-ay_1)}{b}-cy_1=\frac{cy_0-cay_1-bcy_1}{b}=\frac{c}{b}(y_0-y_1)$, and find that $\varphi=a_0(y_0-y_1)$, where $a_0=\frac{c}{b}$; $y_0=ay_1+by_2$; $y_2=\frac{y_0-ay_1}{b}=\frac{1}{2}y_0-\frac{a}{b}y_1=a_{10}y_0-a_{01}y_1$.

The expression for the potential energy in the expanded form in coordinates y_0, y_1 and x and takes the form

$$\begin{aligned}\Pi=&\frac{1}{2}k(y_1+x-z)^2+\frac{1}{2}k_1(y_1-z_1)^2+\frac{1}{2}k_0(y_0-z_2)^2\\&+\frac{1}{2}k_2(a_{10}y_0-a_{01}y_1-z_3)^2+\frac{1}{2}k_{01}(x)^2,\end{aligned} \tag{5.97}$$

where $a_{10}=\frac{1}{b}$ and $a_{01}=\frac{a}{b}$.

Table 5.12 The values of the coefficients of the equations (5.98)–(5.100) in the coordinates y_0, y_1 and x

a_{11}	a_{12}	a_{13}
$(m + J_0 a_0^2)p^2 + k + k_1 + k_2 l a_{01}^2$	$(-J_0 a_0^2)p^2 - k_2 a_{01} a_{10}$	$(-m)p^2 + k$
a_{21}	a_{22}	a_{23}
$(-J_0 a_0^2)p^2 - k_2 a_{01} a_{10}$	$(m + J_0 a_0^2)p^2 + k + k_2 a_{10}^2$	0
a_{31}	a_{32}	a_{33}
$(-m)p^2 + k$	0	$(m)p^2 + k + k_{01}$

We make a number of auxiliary calculations and find the equations of motion of the system in coordinates y_0, y_1 and x:

$$\begin{aligned}&\ddot{y}_1\left(Ja_0^2 + m\right) + y_1\left(k + k_1 + k_2 a_{01}^2\right) + \ddot{y}_0\left(-Ja_0^2\right) + y_0(-k_2 a_{01} a_{10}) + \ddot{x}(-m) + x(k)\\ &= kz - k_2 a_{01} z_3 + k_1 z_1;\end{aligned} \tag{5.98}$$

$$\begin{aligned}&\ddot{y}_1\left(-Ja_0^2\right) + y_1\left(-k_2 a_{01} a_{10}\right) + \ddot{y}_0\left(m_0 + J_0 a_0^2\right) + y_0\left(k + k_2 a_{10}^2\right) + \ddot{x}(0) + x(0)\\ &= k_0 z_2 + k_2 a_{01} z_3;\end{aligned} \tag{5.99}$$

$$\ddot{y}_1(-m) + y_1(k) + \ddot{y}_0(0) + y_0(0) + \ddot{x}(m) + x(k + k_{01}) = -kz. \tag{5.100}$$

Table 5.12 shows the values of the coefficients of equations (5.98)–(5.100) in the coordinates y_0, y_1 and x.

Generalized forces with respect to coordinates y_0, y_1 and x have the form

$$Q_{y_1} = kz - k_2 a_{01} z_3 + k_1 z_1;\quad Q_{y_0} = k_0 z_2 + k_2 a_{01} z_3;\quad Q_x = -kz. \tag{5.101}$$

In the system of coordinates y_0, y_1 and x it is also possible to introduce a joint along the coordinate x. To obtain the transfer function of a system that has two degrees of freedom y_0 and y_1, you need to exclude the corresponding columns and string. We note that if $x = 0$, then $y_1 = y$ and the motion of the system with one hinge at p. A will be described by the coordinates y_0 and y_1. In this case, the lever has an elastic support for its center of rotation. Of greatest interest is the case with two joints.

VI. Consider the system of coordinates y_{10}, y and x. Introducing a series of relations $(y_{10} = y_0 - z_2)$, we write the expression for the kinetic energy of the system:

$$T=\frac{1}{2}m_0(\dot{y}_{10}+\dot{z}_2)^2+\frac{1}{2}J_0a_0^2(\dot{y}_{10}+\dot{z}_2-\dot{y}+\dot{x})^2+\frac{1}{2}m(\dot{y})^2, \tag{5.102}$$

where $a_0=\frac{c}{b}$, and also write the expression for the potential energy of the system

$$\Pi=\Pi_1+\Pi_2+\Pi_3+\Pi_4+\Pi_5, \tag{5.103}$$

where

$$\begin{aligned}
\Pi_1 &=\frac{1}{2}k(y-z)^2=\frac{1}{2}k(y^2-2yz+z^2);\\
\Pi_2 &=\frac{1}{2}k_1(y-x-z_1)^2=\frac{1}{2}k_1(y^2-2yx+x^2-2z_1y+2z_1x+z_1^2);\\
\Pi_3 &=\frac{1}{2}k_0(y_{10})^2;\\
\Pi_4 &=\frac{1}{2}k_2(a_{10}y_{10}-a_{01}y+a_{01}x+z_0)^2\\
&=\frac{1}{2}k_2(a_{10}^2y_{10}^2-2a_{10}a_{01}y_{10}y+a_{01}^2y^2+2a_{01}^2x+2a_{01}xz_0\\
&\quad+z_0^2+2a_{01}a_{10}xy_{10}-2a_{01}^2xy+2a_{10}z_0y_{10}-2a_{01}z_0y);\\
\Pi_5 &=\frac{1}{2}k_{01}x^2.
\end{aligned}$$

where $z_0=a_{10}z_2-z_3$.

The equations of motion of the system in the coordinates of y_{10}, y and x and take the form:

$$\begin{aligned}
&\ddot{y}_{10}(m_0+J_0a^2)+y_{10}(k_0+k_2a_{10}^2)+\ddot{y}(-Ja_0^2)+y(-k_2a_{10}a_{01})\\
&\quad+\ddot{x}(J_0a_0^2)+x(k_2a_{10}a_{01})=-k_2a_{10}z_0-m_0\ddot{z}_2-J_0a^2\ddot{z}_2;
\end{aligned} \tag{5.104}$$

$$\begin{aligned}
&\ddot{y}_{10}(-J_0a_0^2)+y_{10}(-k_2a_{10}a_{01})+\ddot{y}(m+Ja_0^2)+y(k+k_1+k_2a_{01}^2)\\
&\quad+\ddot{x}(-J_0a_0^2)+x(-k_1-k_2a_{01}^2)=J_0a_0^2\ddot{z}_2+kz+k_1z_1+k_2a_{01}z_0;
\end{aligned} \tag{5.105}$$

$$\begin{aligned}
&\ddot{y}_{10}(J_0a_0^2)+y_{10}(k_2a_{10}a_{01})+\ddot{y}(-Ja_0^2)+y(-k_1-k_2a_{01}^2)\\
&\quad+\ddot{x}(-J_0a_0^2)+x(k_1+k_2a_{01}^2+k_{01})=-J_0a_0^2\ddot{z}_2-k_1z_1-k_2a_{01}z_0.
\end{aligned} \tag{5.106}$$

Table 5.13 gives the corresponding coefficients of equations (5.104)–(5.106). Generalized forces with respect to coordinates y_{10}, y and x have the form

$$\begin{aligned}
Q_{y_{10}} &=-(m_0+J_0a_0^2)\ddot{z}_2-k_2a_{10}z_0;\\
Q_y &=(J_0a_0^2)\ddot{z}_2+kz+k_1z_1+k_2a_{01}z_0;\\
Q_x &=-(J_0a_0^2)\ddot{z}_2-k_1z_1-k_2a_{01}z_0.
\end{aligned} \tag{5.107}$$

Table 5.13 The values of the coefficients of the equations (5.104)–(5.106) in the coordinates y_{10}, y and x

a_{11}	a_{12}	a_{13}
$(m_0 + J_0 a_0^2)p^2 + k_0 + k_2 a_{10}^2$	$(-J_0 a_0^2)p^2 - k_2 a_{10} a_{01}$	$(J_0 a^2)p^2 + k_2 a_{10} a_{01}$
a_{21}	a_{22}	a_{23}
$(-J_0 a_0^2)p^2 - k_2 a_{10} a_{01}$	$(m + J_0 a_0^2)p^2 + k + k_2 a_{01}^2 + k_1$	$(-J_0 a_0^2)p^2 - k_1 - k_2 a_{01}^2$
a_{31}	a_{32}	a_{33}
$(J_0 a_0^2)p^2 + k_2 a_{10} a_{01}$	$(-J_0 a_0^2)p^2 - k_1 - k_2 a_{01}^2$	$(J_0 a_0^2)p^2 + k_1 + k_2 a_{01}^2 + k_{01}$

For $z_0 = a_{10} z_2 - z_3$ we obtain the following expressions for the generalized forces:

$$\begin{aligned} Q_{y_{10}} &= -\left(m_0 + J_0 a_0^2\right)\ddot{z}_2 - k_2 a_{10} z_0 \\ &= -(m_0 + J_0 a_0^2)p^2 - k_2 a_{10}^2 z_2 - k_2 a_{10} z_3; \end{aligned} \tag{5.108}$$

$$\begin{aligned} Q_y &= -\left(J_0 a_0^2\right)\ddot{z}_2 + kz + k_1 z_1 + k_2 a_{01} z_0 \\ &= J_0 a_0^2 p^2 + kz + k_1 z_1 + k_2 a_{10} a_{01} z_2 - k_2 a_{01} z_3; \end{aligned} \tag{5.109}$$

$$\begin{aligned} Q_x &= -\left(J_0 a_0^2\right)\ddot{z}_2 - k_1 z_1 - k_2 a_{10} z_0 \\ &= -J_0 a_0^2 p^2 - k_1 z_1 - k_2 a_{10} a_{01} z_2 + k_2 a_{10} z_3. \end{aligned} \tag{5.110}$$

In this coordinate system, it is possible to access two joints: by coordinate y_{10} and by coordinate x. Using the matrix (see Table 5.13) and excluding the corresponding rows and columns, we obtain the equation of motion for a system with two joints:

$$(m_0 + J_0 a_0^2)p^2 + k + k_2 a_{01}^2 + k_1 = J_0 a_0^2 \ddot{z}_2 + kz + k_1 z_1 + k_2 a_{01} z_0. \tag{5.111}$$

For simplicity, we assume that $k_2 = 0$, $k_1 = 0$, $z_1 = 0$, $z_3 = 0$; then $z = z_2$. Then Eq. (5.111) is transformed to the form

$$\bar{y}(m_0 + J_0 a_0^2)p^2 + k = (J_0 a_0^2 p^2 + k)\bar{z}, \tag{5.112}$$

from which the transfer function of the system takes the form

$$W(p) = \frac{\bar{y}}{\bar{z}_2} = \frac{J_0 a_0^2 + k}{m + J_0 a_0^2 + k}. \tag{5.113}$$

Comparison (5.113) and (5.80) shows that the structure of transfer functions is general and the proposed approach allows us to construct the necessary mathematical models.

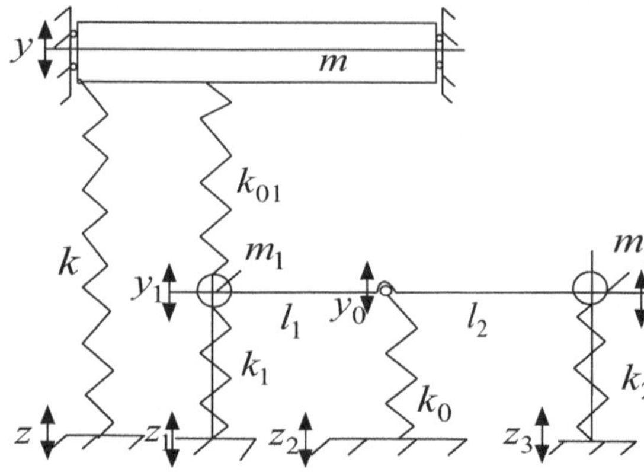

Fig. 5.26 The computational scheme of the VPS, shown in Fig. 5.23, but with spaced-apart masses m_1 and m_2

In order to obtain a complete coincidence of the results, it is required to present the computational scheme in more detail. To derive (5.80), we used the scheme shown in Fig. 5.24. The peculiarity of this computational scheme is taking into account the mass-and-inertia properties of the lever linkages. Let us consider, in connection with this, the computational scheme of the system in Fig. 5.24, which reflects the mass-inertial properties of the system. The computational scheme (Fig. 5.26) shows the masses m_1 and m_2; taking into account the features of their motion is an essential factor for the coincidence of expressions (5.80) and (5.113).

We choose a system of coordinates for further calculations y, y_{10} and x, assuming in this case that $y_{10} = y_0 + z_2$, $x = y - y_1$, where $y_0 = ay_1 + by_2$. We write the expressions for the kinetic and potential energy of the system

$$T = \frac{1}{2}m\dot{y}^2 + \frac{1}{2}m_1\dot{y}_1^2 + \frac{1}{2}m_2\dot{y}_2^2, \tag{5.114}$$

$$\begin{aligned} \Pi = {} & \frac{1}{2}k(y-z)^2 + \frac{1}{2}k_1(y_1 - z_1)^2 + \frac{1}{2}k_2(y_2 - z_3)^2 \\ & + \frac{1}{2}k_{01}(y - y_1)^2 + \frac{1}{2}k_0(y_0 - z_2)^2. \end{aligned} \tag{5.115}$$

Let us make a number of transformations:

$$y_2 = y_0 - ay_1 = a_0y_0 - iy_1, \quad i = \frac{a}{b} = \frac{l_2}{l_1};$$

$$a_0 = \frac{1}{b};\ y_2 = a_0y_{10} + a_0z_2 - iy + ix \cdot y_0 = y_{10} + x,$$

then we get

$$T = \frac{1}{2}m\dot{y}^2 + \frac{1}{2}m_1(\dot{y} - \dot{x})^2 + \frac{1}{2}m_2(a_0\dot{y}_{10} + a_0\dot{z}_2 - i\dot{y} + i\dot{x})^2, \qquad (5.116)$$

$$\begin{aligned} \Pi &= \frac{1}{2}k(y - z)^2 + \frac{1}{2}k_1(y - x - z_1)^2 \\ &+ \frac{1}{2}k_2(a_0y_{10} - iy + ix + z_0)^2 + \frac{1}{2}k_{01}(x)^2 + \frac{1}{2}k_0(y_{10})^2, \end{aligned} \qquad (5.117)$$

where $z_0 = a_0z_2 - z_3$. Having made a number of auxiliary calculations analogous to those given above for the derivation of Lagrange equations of the second kind, we obtain the equations of motion

$$\begin{aligned} &\ddot{y}(m_1 + m_2i^2 + m) + y(k + k_1 + k_2i^2) + \ddot{x}(-m_1 - m_2i^2) + x(-k_1 - k_2i^2) \\ &\quad + \ddot{y}_{10}(-m_2ia_0) + y_{10}(-k_2a_0i) = m_2a_0iz_2 + kz + k_1z_1 + k_2z_0i; \end{aligned} \qquad (5.118)$$

$$\begin{aligned} &\ddot{y}(-m_1 - m_2i^2) + y(-k_1 - k_2i^2) + \ddot{x}(m_1 + m_2i^2) + x(k_1 + k_{01} + k_2i^2) \\ &\quad + \ddot{y}_{10}(m_2ia_0) + y_{10}(k_2a_0i) = -m_2a_0i\ddot{z}_2 - k_1z_1 - k_2z_0i; \end{aligned} \qquad (5.119)$$

$$\begin{aligned} &\ddot{y}(-m_2ia_0) + y(-k_2i^2a_0) + \ddot{x}(m_2ia_0) + x(k_2ia_0) \\ &\quad + \ddot{y}_{10}(m_2a_0^2) + y_{10}(k_2a_0^2 + k_0) = -m_2a_0^2\ddot{z}_2 - k_2a_0z_0. \end{aligned} \qquad (5.120)$$

Table 5.14 presents the coefficients of equations (5.118)–(5.120).
In this case the generalized forces have the form

$$\begin{aligned} Q_y &= m_2a_0i\ddot{z}_2 + kz + k_1z_1 + k_2z_0i; \\ Q_x &= -a_0im_2\ddot{z}_2 - k_1z_1 - k_2iz_0; \\ Q_{y_{10}} &= -m_2a_0^2\ddot{z}_2 - k_2a_0z_0. \end{aligned} \qquad (5.121)$$

Excluding columns and rows with respect to the coordinates x and y_{10} from the matrix, we obtain the equation of motion for a system with coordinate y

Table 5.14 The values of the coefficients of the equations (5.118)–(5.120) in the coordinates y, x and y_{10}

a_{11}	a_{12}	a_{13}
$(m_1 + m + m_2i^2)p^2 + k + k_1 + k_2i^2$	$(-m_1 - m_2i^2)p^2 - k_1 - k_2i^2$	$-m_2ia_0p^2 - k_2a_9i$
a_{21}	a_{22}	a_{23}
$(-m_1 - m_2i^2)p^2 - k_1 - k_2i^2$	$(m_1 + m_2i^2)p^2 + k_1 + k_{01} + k_2i^2$	$m_2ia_0p^2 + k_2a_0i$
a_{31}	a_{32}	a_{33}
$-m_2ia_0p^2 - k_2a_0i$	$m_2ia_0p^2 + k_2a_0i$	$m_2a_0^2p^2 + k_2a_0^2 + k_0$

$$\ddot{y}\left(m_1+m+m_2 i^2\right)+y\left(k+k_1+k_2 i^2\right)=m_2 a_0 i \ddot{z}_2+kz+k_1 z_1+k_2 z_0 i. \qquad (5.122)$$

To construct the transfer function "displacement of y by input z_2," we assume that $z=z_2,\ z_1=0,\ z_3=z_2,\ k_1=0,\ k_2=0$.

In this case $k_{01}\to\infty,\ k_0\to\infty$, that said,

$$W=\frac{\bar{y}}{\bar{z}_2}=\frac{m_2 a_0 i p^2+k}{(m+m_1+m_2 i^2)p^2+k}. \qquad (5.123)$$

Note that $\frac{1}{b}=\frac{l_1+l_2}{l_1}=1+i$; in this case

$$W=\frac{\bar{y}}{\bar{z}_2}=\frac{m_2 i(i+1)p^2+k}{(m+m_1+m_2 i^2)p^2+k}. \qquad (5.124)$$

Expressions (5.80) and (5.124) completely coincide, which is exactly what we had to prove.

Figure 5.27 shows the family of amplitude-frequency characteristics corresponding to (5.124) for $m=100$ kg; $m_1=m_2=10$ kg; $k=10{,}000$ N/m and the transfer ratios i = 0–3 (the step of changing i is 1). An increase of i in the physical plane means that with a change in i, the values of the reduced mass change in a system, which introduces corresponding changes in the parameters of the system.

The frequency of dynamic absorbing of oscillations is defined by the expression

$$\omega_{\text{dyn}}^2=\frac{k}{m_2 i(i+1)}, \qquad (5.125)$$

and the frequency of natural oscillations is defined by

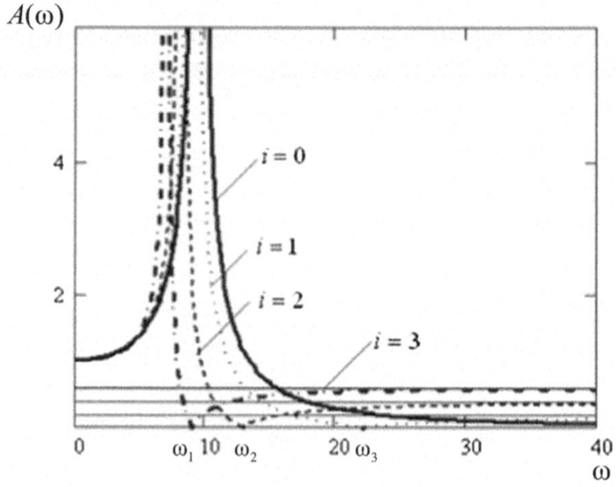

Fig. 5.27 The family of amplitude-frequency characteristics of a system with a transfer function (5.124)

Table 5.15 The values of the frequencies of natural oscillations in comparison with the frequencies ω_1–ω_3 for the system in Fig. 5.27

i	1	2	3
The value of the dynamic absorbing frequencies	ω_1	ω_2	ω_3
	22.36	12.9	9.1
The value of natural frequencies	ω_{nat1}	ω_{nat2}	ω_{nat3}
	9.53	8.16	7.07

$$\omega_{\text{nat}}^2 = \frac{k}{m + m_1 + m_2 i^2}. \tag{4.126}$$

“Locking” at high frequencies is determined from expression

$$R = \frac{m_2 i(i+1)}{m + m_1 + m_2 i^2}. \tag{5.127}$$

The frequencies of dynamic absorbing are denoted by $\omega_1 - \omega_3$. The corresponding values of the frequencies of the natural oscillations in comparison with the frequencies are given in Table 5.15.

Thus, choosing a system of generalized coordinates in an appropriate way, one can construct a mathematical model of a mechanical system with joints. In this case, the system with joints obtains a smaller number of degrees of freedom than the original system. The joint is possible between two bodies when they are connected to a kinematic pair of rotational type (class V). However, it is possible to connect the solid body with another body with the loss of the possibility of relative motion.

5.5 Dynamic Absorbing in Vibration Protection Systems with Joints

I. Consider a dynamic vibration absorber (Fig. 5.28) as part of a vibration protection system that protects the object from vibrations on the side of the base with joint at p. *A*.

In Fig. 5.28 the following designations are accepted: $P(t)$ is the external force disturbance; $z(t)$ is the external kinematic perturbation; m_1 is the mass of the protection object; m_2 and m_3 are the masses of the customizable elements; k_1, k_2, k_3 are the stiffness coefficients of the elastic elements; φ is the angle of rotation of the lever relative to the object of protection; l_1, l_2 are the lengths of the lever arms; y_1, y_2, y_3 are the coordinates of the mass-and-inertia elements in absolute motion.

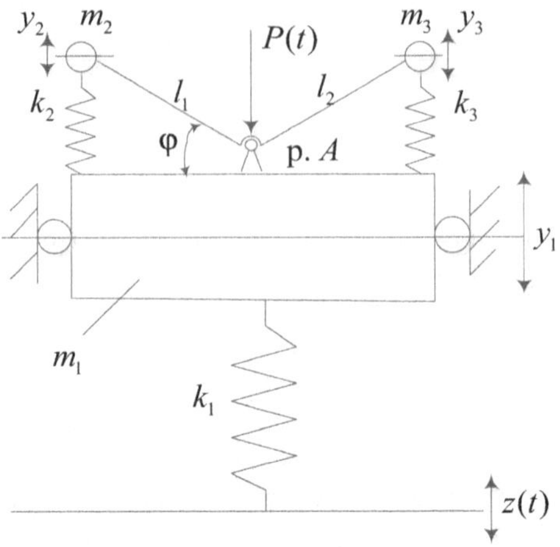

Fig. 5.28 Computational scheme of the dynamic absorber of oscillations of lever type

It is assumed that the vibrational motions in the system relative to the equilibrium position are sufficiently small, which makes it possible to use simplified linear representations; it is also assumed that the frictional forces are small. The purpose of the study is to assess the possibilities to create dynamic absorbing modes in the system, which are determined by tuning parameters. These could be the lengths of the arms of the lever and the masses of the elements and. Constructive options for developing systems of changing these parameters seem quite feasible, as well as schemes for collecting and processing information on the dynamic state of the system. We write the expressions for the kinetic and potential energy:

$$T = \frac{1}{2}m_1\dot{y}_1^2 + \frac{1}{2}m_2\dot{y}_2^2 + \frac{1}{2}m_3\dot{y}_3^2, \tag{5.128}$$

$$\Pi = \frac{1}{2}k_1(y_1 - z)^2 + \frac{1}{2}k_2(y_2 - y_1)^2 + \frac{1}{2}k_3(y_3 - y_1)^2. \tag{5.129}$$

We introduce a series of relations between the coordinates

$$y_2 = y_1 + \varphi l_1; \quad y_3 = y_1 - \varphi l_2, \tag{5.130}$$

which takes into account the characteristics of the lever of the second kind with respect to changing the input signal in both magnitude and direction. We assume that the elements m_2 and m_3 have vertical motion, and the bending of the lever is not taken into account (although this is not so and the configuration of the arrangement l_1 and l_2 is also of importance). Taking into account (5.130), the expressions (5.128) and (5.129) can be written in the following form

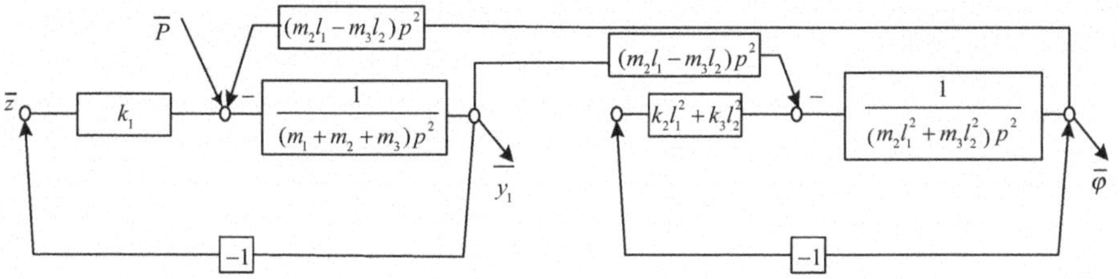

Fig. 5.29 The structural diagram corresponding to the system in Fig. 5.28

$$\mathrm{T} = \frac{1}{2}m_1\dot{y}_1^2 + \frac{1}{2}m_2(\dot{y}_1 + \dot{\varphi}l_1)^2 + \frac{1}{2}m_3(\dot{y}_1 - \dot{\varphi}l_2)^2; \tag{5.131}$$

$$\Pi = \frac{1}{2}k_1(y_1 - z)^2 + \frac{1}{2}k_2(-\varphi l_1)^2 + \frac{1}{2}k_3(\varphi l_2)^2. \tag{5.132}$$

Using the generalized Lagrange equation of the second kind, we obtain the equations of motion of the system

$$\begin{aligned} &\ddot{y}_1(m_1 + m_2 + m_3) + \ddot{\varphi}(m_2l_1 - m_3l_2) + k_1y_1 = k_1z + P, \\ &\ddot{\varphi}(m_2l_1^2 + m_3l_2^2) + \ddot{y}_1(m_2l_1 - m_3l_2) + \varphi(k_2l_1^2 + k_3l_2^2) = 0. \end{aligned} \tag{5.133}$$

A structural diagram of the equivalent automatic control system (ACS) is shown in Fig. 5.29. From her analysis it follows that there is an elastic constraint between partial systems, which, under the conditions of symmetry, can be "zeroed" and make the motions of partial systems independent.

Let us find the transfer function of the system for a kinematic perturbation

$$W_1(p) = \frac{\overline{y_1}}{\overline{z}} = \frac{\left[\left(m_2l_1^2 + m_3l_2^2\right)p^2 + k_2l_1^2 + k_3l_2^2\right] \cdot k_1}{\left[(m_1 + m_2 + m_3)p^2 + k_1\right]\left[\left(m_2l_1^2 + m_3l_2^2\right)p^2 + k_2l_1^2 + k_3l_2^2\right] - (m_2l_1 - m_3l_2)^2p^4}. \tag{5.134}$$

From expression (5.134) it is possible to find the frequency of dynamic absorbing under a kinematic perturbation

$$\omega_{dyn}^2 = \frac{k_2l_1^2 + k_3l_2^2}{m_2l_1^2 + m_3l_2^2} = \frac{k_2 + k_3i^2}{m_2 + m_3i^2} \tag{5.135}$$

where $i = l_2/l_1$ is the transfer ratio of the lever of the second kind (the sign is taken into account in the formulation of the expression for the potential energy). The frequency of the natural oscillations of the system can be found from the characteristic equation

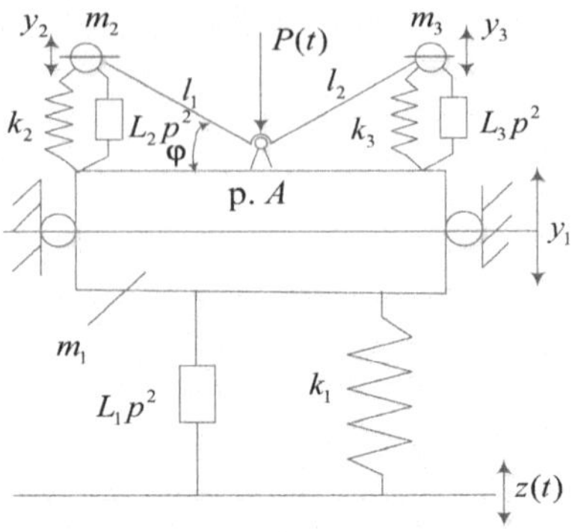

Fig. 5.30 Computational scheme of the system with additional constraints

$$\left[(m_1+m_2+m_3)p^2+k_1\right]\left[(m_2l_1^2+m_3l_2^2)p^2+k_2l_1^2+k_3l_2^2\right]-(m_2l_1-m_3l_2)^2p^4=0. \tag{5.136}$$

II. In order to expand the possibilities for changing the dynamic state, additional constraints can be introduced into the system in the form of elementary units of double differentiation, as shown in Fig. 5.30 [20].

In this case, the expression for the kinetic energy takes the form

$$\mathrm{T}=\frac{1}{2}m_1\dot{y}_1^2+\frac{1}{2}L_1(\dot{y}_1-\dot{z})^2+\frac{1}{2}m_2\dot{y}_2^2+\frac{1}{2}L_2(\dot{y}_2-\dot{y}_1)^2+\frac{1}{2}m_3\dot{y}_3^2+\frac{1}{2}L_3(\dot{y}_3-\dot{y}_1)^2, \tag{5.137}$$

and the potential energy can be determined from expression (5.129).

Taking into account relations (5.130), we write the expression for the kinetic energy of the system

$$\begin{aligned}T=&\frac{1}{2}m_1\dot{y}_1^2+\frac{1}{2}L_1(\dot{y}_1-\dot{z})^2+\frac{1}{2}m_2(\dot{y}_1+\dot{\varphi}l_1)^2\\&+\frac{1}{2}L_2(-\dot{\varphi}l_1)^2+\frac{1}{2}m_3(\dot{y}_1-\dot{\varphi}l_2)^2+\frac{1}{2}L_3(\dot{\varphi}l_2)^2.\end{aligned} \tag{5.138}$$

The system of differential equations of motion takes the form

$$\begin{aligned}\ddot{y}_1(m_1+L_1+m_2+m_3)+\ddot{\varphi}(m_2l_1-m_3l_2)+k_1y_1&=\ddot{z}L_1+k_1z;\\ \ddot{\varphi}(m_2l_1^2+L_2l_1^2+m_3l_2^2+L_1l_2^2)+\ddot{y}_1(m_2l_1-m_3l_2)+\varphi(k_2l_1^2+k_3l_2^2)&=0.\end{aligned} \tag{5.139}$$

The structural diagram of an equivalent automatic control system is shown in Fig. 5.31.

From the structural diagram it follows that the introduction of motion transformation devices L_1, L_2, L_3 change the properties of the system: L_1 affects the

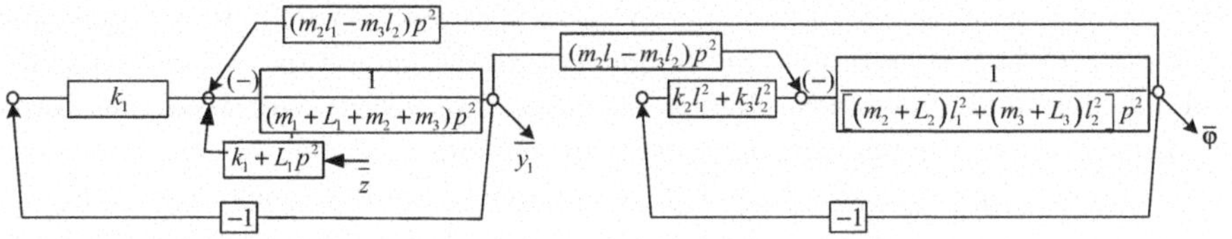

Fig. 5.31 The structural diagram of the equivalent ACS, corresponding to Fig. 5.30

nature of the external impact, and the system acquires an additional mode of dynamic absorbing and "locking" at high frequencies; L_2 and L_3 reduce the frequencies of the natural oscillations of the partial systems.

The transfer function of the system under the kinematic perturbation of the system has the form:

$$W(p)_1 = \frac{\overline{y_1}}{\overline{z}} = \frac{(k_1 + L_1p^2)\{[(m_2 + L_2)l_1^2 + (m_3 + L_3)l_2^2]p^2 + k_2l_1^2 + k_3l_2^2\}}{[[(m_1 + L_1) + m_2 + m_3]p^2 + k_1]\{[(m_2 + L_2)l_1^2 + (m_3 + L_3)l_2^2]p^2 + k_2l_1^2 + k_3l_2^2\} - (m_2l_1 - m_3l_2)^2p^4}. \tag{5.140}$$

For the investigation, we transform (5.106) and obtain

$$W(p)_1 = \frac{\overline{y_1}}{\overline{z}} = \frac{(k_1 + L_1p^2)\{[m_2 + L_2 + (m_3 + L_3)i^2]p^2 + k_2 + k_3i^2\}}{[[(m_1 + L_1) + m_2 + m_3]p^2 + k_1]\{[m_2 + L_2 + (m_3 + L_3)i^2]p^2 + k_2 + k_3i^2\} - (m_2 - m_3i^2)^2p^4}, \tag{5.141}$$

where $i = l_2/l_1$ is the ratio of the arms of the lever of the second kind.

We introduce some notation. Let $r_2 = m_2 + L_2 + (m_3 + L_3)i^2$; $r_1 = m_1 + m_2 + m_3 + L_1$; $r_3 = m_2 - m_3i^2$, then $A_1 = (k_1 + L_1p^2)(r_2p^2 + k_2 + k_3i^2)$ is the numerator (5.140).

Let us investigate the characteristic equation of the transfer function (5.141):

$$\begin{aligned} A_2 &= (r_1p^2 + k_1)(r_2p^2 + k_2 + k_3i^2) - r_3^2p^4 \\ &= (r_1r_2 - r_3^2)p^4 + [r_2k_1 + r_1(k_2 + k_3i^2)]p^2 + k_1(k_2 + k_3i^2) = 0, \end{aligned}$$

from which we find the frequencies of the natural oscillations:

$$\omega_{1,2}^2 = \frac{r_2k_1 + r_1(k_2 + k_3i^2)}{2(r_1r_2 - r_3^2)} \pm \sqrt{\frac{[r_2k_1 + r_1(k_2 + k_3i^2)]^2 - 4(r_1r_2 - r_3^2)[k_1(k_2 + k_3i^2)}{4(r_1r_2 - r_3^2)^2}}. \tag{5.142}$$

If $r_1r_2 - r_3^2 = [m_2 + L_2 + (m_3 + L_3)i^2](m_1 + m_2 + m_3 + L_1) - (m_2 - m_3i)^2$, then the difference will have the form $\Delta = -m_2^2 + 2m_2m_3i - m_3^2i^2 + m_2^2 + m_3^2i^2 + R$, where is the positive remainder, i.e. $\Delta > 0$ is always satisfied. If the radicand expression (5.142) is zero, then the frequencies of the natural oscillations coincide and the amplitude-frequency characteristics of the system will have the form characteristic for systems with one degree of freedom.

It follows from expression (5.141) that a regime is possible when $\omega^2_{\text{dyn}1} = \omega^2_{\text{dyn}2}$, what is being carried out at $\frac{k_1}{L_1} = \frac{k_2 + k_3 i^2}{m_2 + L_2 + (m_3 + L_3)i^2}$, then the condition of coincidence of the frequencies of dynamic extinction has the form

$$k_1 = \frac{L_1(k_2 + k_3 i^2)}{m_2 + L_2 + (m_3 + L_3)i^2} \tag{5.143}$$

or

$$L_1 = \frac{k_1[m_2 + L_2 + (m_3 + L_3)i^2]}{k_2 + k_3 i^2}. \tag{5.144}$$

When $i \to \infty$ is done, we obtain the limiting relations. In this case, between the values of the parameters, relations

$$k_1 = \frac{k_3 L_1}{m_3 + L_3}; \tag{5.145}$$

$$L_1 = \frac{k_1(m_3 + L_3)}{k_3}. \tag{5.146}$$

In turn, when $i \to 0$, we get

$$k_1 = \frac{k_2 L_1}{m_2 + L_2}; \tag{5.147}$$

$$L_1 = \frac{k_1(m_2 + L_2)}{k_2}. \tag{5.148}$$

If conditions (5.143) and (5.144) are satisfied, then the system with two degrees of freedom will have the form of AFC, as shown in Fig. 5.32 will introduce itself as a system with one degree of freedom. The values (5.144) corresponding to the graphs in Fig. 5.32 for m_1 = 100 kg; $m_2 = m_3$ = 10 kg; $i = 2, 4, 6$; L_2 = 10 kg, L_3 = 10 kg, k_2 = 600 N/m; k_3 = 700 N/m; k_1 = 1000 N/m.

Table 5.16 shows the corresponding values of the frequencies of natural oscillations and dynamic absorbing.

To estimate the dynamic properties of a system whose transfer function is represented by expression (5.141), with the parameters of the model problem m_1 = 100 kg; $m_2 = m_3$ = 10 kg; $i = 2, 4, 6$; L_2 = 10 kg, L_3 = 10 kg; k_2 = 600 N/m; k_3 = 700 N/m; k_1 = 1000 N/m.

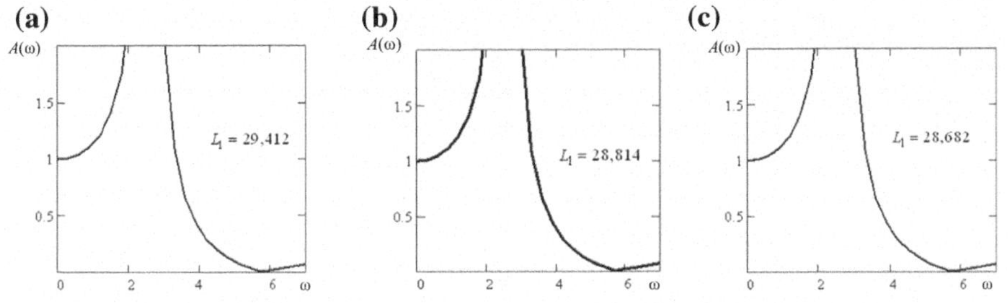

Fig. 5.32 Amplitude-frequency characteristics under condition of equality of expressions (5.143) and (5.144) at corresponding values of L_1; i is chosen as a variable parameter. **a** $i = 2$; **b** $i = 4$; **c** $i = 6$

Table 5.16 The values of the frequencies of natural oscillations and dynamic absorbing (see Fig. 5.32) with the corresponding values of L_1

Value of L_1	Frequency of natural oscillations	Dynamic absorbing frequencies
	ω_{nat}	ω_{dyn}
$L_1 = 29,412;\ i = 2$	2585 5855	5831 5831
$L_1 = 28,682;\ i = 6$	2586 5989	5907 5907
$L_1 = 28,814;\ i = 4$	2587 5957	5891 5891

In Fig. 5.33 a diagram of the behavior of the frequencies of dynamic absorbing and natural oscillations is constructed. In the general case, considering the same order of the frequency equations of the numerator and denominator (5.141), it can be assumed, depending on the values of the parameters (in particular, L_1) that the relationships between the frequencies, as well as the shape of the amplitude-frequency characteristics of the system will vary significantly. In Fig. 5.34 it is shown that with a change of L_1, characteristics with two modes of dynamic absorbing and two resonances are possible. However, cases of coincidence of the frequencies of dynamic absorbing with one another are possible in the system, as well as coincidence with the frequencies of natural oscillations. Software package MathCAD 11 was used for the calculations. In the high-frequency region, the system is "locked". The value of the transfer coefficient of the amplitude of the oscillations with increasing frequency is determined from expression (5.141), provided that $p \to \infty$. If we denote this value by a (infinity), then the values of this parameter will depend on L_1. In Fig. 5.34 gives the corresponding information on the example of several variants. A specific feature of the amplitude-frequency characteristics, presented in Fig. 5.34 in various versions, is the fact that a system with two degrees of freedom practically acts like a system with one degree of freedom, which is ensured by certain values of the reduced mass of the inertial

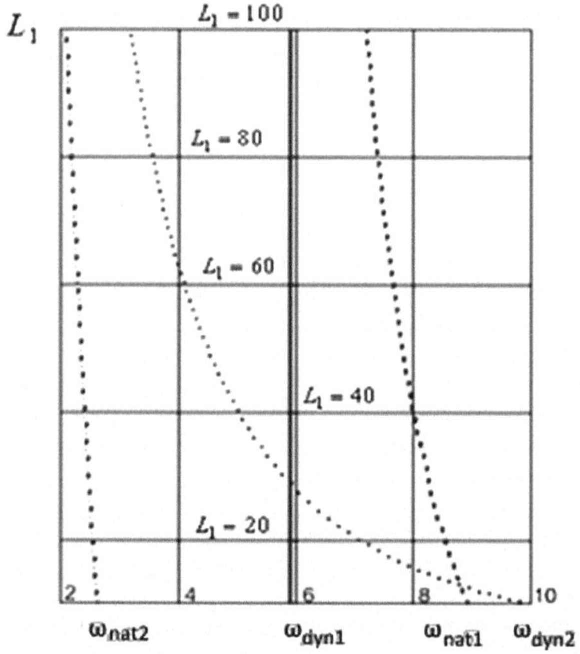

Fig. 5.33 Diagram of the behavior of the frequencies of dynamic absorbing and natural oscillations with a change in the parameter L_1

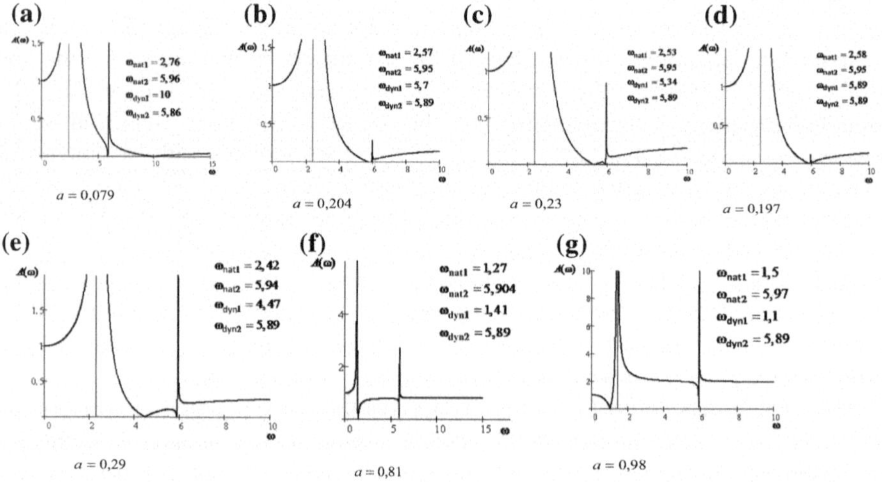

Fig. 5.34 The AFCs for expression (5.140) for different values of L_1: **a** $L_1 = 10$; **b** $L_1 = 13.5$; **c** $L_1 = 25$; **d** $L_1 = 28.8$—critical value ($\omega_{\mathrm{dyn}1} = \omega_{\mathrm{dyn}2}$); **e** $L_1 = 50$; **f** $L_1 = 200$; **g** $L_1 \to \infty$

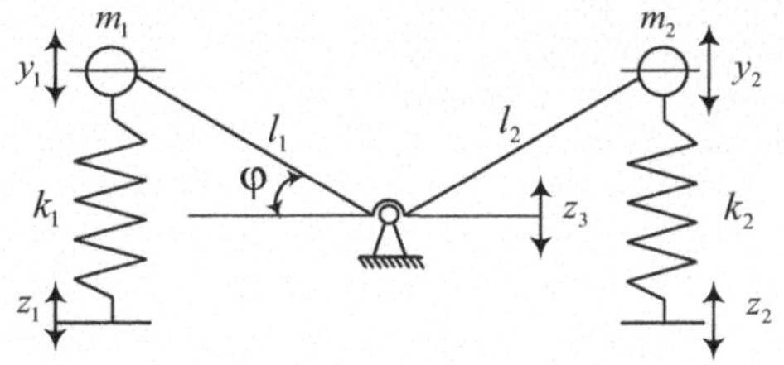

Fig. 5.35 The computational scheme of a system with spaced-apart masses at a combined kinematic perturbation

characteristic L_1 found from expression (5.144). The proximity of resonant frequencies creates a zone of unstable motions of an elevated level beyond which the system has the form of an AFC with one degree of freedom. As a variable parameter, in particular, the transfer ratio of the arms of the lever linkage i was chosen. In the model example $L_1 = 29,412$ corresponds to $i = 2$, $L_1 = 28,814$ corresponds to $i = 4$, $L_1 = 28,682$—$i = 6$ (see Fig. 5.32). For AFC, two regions are characteristic, on which the transfer coefficient of the amplitude of the oscillations is equal to 1 in the frequency range 0–$\omega_{1\text{nat}}$ $1/c$. In the frequency range $\omega_{1\text{nat}}$–∞, the transfer coefficient of the oscillation amplitude is less than unity, which determines the possible direction of using the system in vibration protection problems.

III. The nature of the external impact on the object of protection is important, because it helps to change the system of dynamic constraints. Consider a system consisting of two mass-and-inertia elements m_1 and m_2, spaced apart by a L-shaped arm with arms l_1 and l_2, as shown in Fig. 5.35. Such a computational scheme can be attributed to one of the two mass-and-inertia element s m_1 and m_2, spaced apart by a L-shaped arm with arms l_1 and l_2 (see Fig. 5.35), as well as to one of the variants of the above-discussed dynamic absorber, provided that it can either be attached to the object of protection, or to be removed from it.

The constructive use of such an attachment can be based on the use of a magnetic support. In this case, the expression for the kinetic and potential energy of the system (see Fig. 5.35) can be written in the following form:

$$\begin{aligned} T &= \frac{1}{2}m_1(\dot{y}_1 - \dot{z}_3)^2 + \frac{1}{2}m_2(\dot{y}_2 - \dot{z}_3)^2; \\ \Pi &= \frac{1}{2}k_1(y_1 - z_1)^2 + \frac{1}{2}k_2(y_2 - z_2)^2. \end{aligned} \tag{5.149}$$

Using the relation $y_2 = -iy$, where $i = l_2/l_1$ represents the ratio of the lever arms at small angles of φ and without taking into account the inclination of the rods, we write the differential equation of motion of the system:

$$\ddot{y}_1\left(m_1 + i^2 m_2\right) + y_1\left(k_1 + i^2 k_2\right) = \ddot{z}_3(m_1 - m_2 i) + k_1 z_1 - k_2 i z_2. \tag{5.150}$$

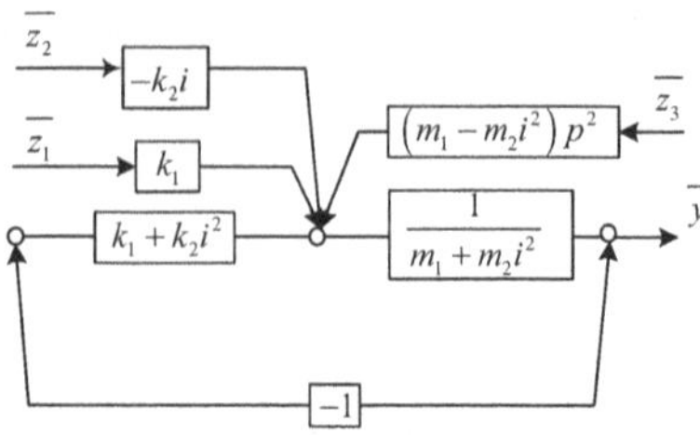

Fig. 5.36 The structural diagram of a mechanical system with spaced-apart masses

The structural diagram of the system is shown in Fig. 5.36, from which the frequency of natural oscillations can be found:

$$\omega_{nat}^2 = \frac{k_1 + k_2 i^2}{m_1 + m_2 i^2}. \tag{5.151}$$

Having considered the structural diagram in Fig. 5.36, it can be noted that in this case, external influences form a system in which, by virtue of design features, they can act in antiphase and create a zero effect at any frequency for $k_1 - k_2 i = 0$ at $z_1 = z_2$. In addition, for $m_1 - m_2 i^2 = 0$, a "blocking" condition is created for an external action z_3. As for the modes of dynamic absorbing and natural frequencies, it is necessary to take into account the relationship between the parameters of the external kinematic perturbation. So, if $z_3 = z_1 = z_2 = z$, then the frequency of dynamic absorbing should be determined by the formula

$$\omega_{dyn}^2 = \frac{k_1 - k_2 i}{m_1 - m_3 i} \tag{5.152}$$

Assuming that the dynamic absorbing modes are associated with the estimation of the numerator of the transfer function obtained from the structural diagram shown in Fig. 5.36, let us present the possible variants in Table 5.17.

Analysis of the data given in Table 5.17 allows us to conclude that dynamic absorbing modes are encountered quite often, but their emergence depends on the design features of the vibration protection system and the characteristics of the system of external influences.

IV. If additional constraints $L_1 p^2$ and $L_2 p^2$ are introduced into the vibration protection system (see Fig. 5.35), that is, elementary units with transfer functions of differentiation of the second kind, then the differential equation of motion takes the form

$$\begin{aligned} &\bar{y}_1 \left(m_1 + i^2 m_2 + L_1 + L_2 i^2\right) p^2 + k_1 + i^2 k_2 \\ &= z_3 (m_1 - m_2 i) p^2 + \left(k_1 + L_1 p^2\right) z_1 - \left(L_2 p^2 + k_2\right) z_2. \end{aligned} \tag{5.153}$$

Table 5.17 Transfer functions for various types of external perturbation

S/p No.	Combination of the parameter of external disturbance	Type of transfer function	Note
1	$z_3 = z_2 = z_1 = z$	$W(p) = \frac{\bar{y}_1}{\bar{z}} = \frac{(m_1 - m_2 i)p^2 + k_1 - k_2 i}{(m_1 + m_2 i^2)p^2 + k_1 + k_2 i^2}$	$\omega_{dyn}^2 = \frac{k_1 - k_2 i}{m_1 - m_2 i}$
2	$z_3 = 0, z_1 = z_2 = z$	$W(p) = \frac{\bar{y}_1}{\bar{z}} = \frac{k_1 - k_2 i}{(m_1 + m_2 i^2)p^2 + k_1 + k_2 i^2}$	Special mode: $k_1 = k_2 i$
3	$z_1 = 0, z_2 = 0, z_3 \neq 0$	$W(p) = \frac{\bar{y}_1}{\bar{z}} = \frac{(m_1 - m_2 i)p^2}{(m_1 + m_2 i^2)p^2 + k_1 + k_2 i^2}$	Special mode: $m_1 = m_2 i$
4	$z_1 = 0, z_3 = z_3 \neq 0$	$W(p) = \frac{\bar{y}_1}{\bar{z}} = \frac{(m_1 - m_2 i)p^2 - k_2 i}{(m_1 + m_2 i^2)p^2 + k_1 + k_2 i^2}$	$\omega_{dyn}^2 = \frac{k_2 i}{m_1 - m_2 i}$
5	$z_2 = 0, z_1 = z_3 \neq 0$	$W(p) = \frac{\bar{y}_1}{\bar{z}} = \frac{(m_1 - m_2 i)p^2 + k_1}{(m_1 + m_2 i^2)p^2 + k_1 + k_2 i^2}$	$\omega_{dyn}^2 = \frac{k_2}{m_1 - m_2 i}$
6	$z_3 = 0, z_1 = 0, z_2 \neq 0$	$W(p) = \frac{\bar{y}_1}{\bar{z}} = \frac{-k_2 i}{(m_1 + m_2 i^2)p^2 + k_1 + k_2 i^2}$	There is no dynamic absorbing mode
7	$z_3 = 0, z_2 = 0, z_1 \neq 0$	$W(p) = \frac{\bar{y}_1}{\bar{z}} = \frac{k_1}{(m_1 + m_2 i^2)p^2 + k_1 + k_2 i^2}$	There is no dynamic absorbing mode

A possible range of situations in which the properties of the dynamic absorbing modes are reflected in one way or another can be estimated using Eq. (5.153). For example, when $z_3 = z_1 = z_2 = z$, we obtain

$$\omega_{dyn}^2 = \frac{k_1 - k_2 i}{(m_1 - m_2 i) + (L_1 - iL_2)}. \tag{5.154}$$

The difference between expression (5.153) and (5.154) is that in the first case the dynamic absorbing mode is determined by the parameters L_1 and L_2 of motion transformation devices, which expands the possibilities of the corresponding adjustment of the vibration protection systems.

Summarizing some results, we note that it is possible to construct mathematical models of mechanical oscillatory systems with joints based on the choice of generalized coordinates that determine the relative motion of selected points (joint elements), which makes it possible to obtain matrices of the coefficients of the differential equations of motion of systems with joints.

The formation of joints can follow two directions. The first one is related to the construction of a matrix of coefficients of the equations of motion with subsequent formal methods of excluding from the matrix of rows and columns of those coordinates of relative motion that are equated to zero. The second technique is characterized by the presence of a joint, which can be formally introduced and formed on the basis of assigning limiting values to the corresponding parameters of the connecting elements. The latter can be carried out using the transfer functions. The joints play an important role in the kinematics of mechanical oscillatory systems, participating in the processes of redistribution of vibrational energy flows in interactions, which makes them useful in constructing various kinds of dynamic vibration absorbers. Studies show that the introduction of joints based on the proposed method (more accurately, the formation of mathematical models of joints) is most effective in systems of a combined type in which the reciprocating motions interact with reciprocating rotational ones. The latter is particularly interesting in that, when supporting a rotational pair on a vibrating base, it is possible to use inertial forces of moving space to reduce the oscillations of the object.

References

1. Lurie, A.I.: Analytical Mechanics, 516 pp. Nauka Publication, Moscow (1986)
2. Artobolevsky, I.I.: Theory of Mechanisms and Machines, 638 pp. Nauka Publication, Moscow (1988)
3. Khomenko, A.P., Eliseev, S.V., Ermoshenko, Y.V.: System Analysis and Mathematical Modeling in Mechatronics of Vibration Protection Systems, 288 p. IrGUPS Publication, Irkutsk (2012)
4. Belokobylsky, S.V., Eliseev, S.V., Sitov, I.S.: Dynamics of Mechanical Systems. Lever Linkages and Inertial-Elastic Constraints, 319 pp. Polytekhnika Publication, St. Petersburg (2013)
5. Eliseev, S.V., Artyunin, A.I., Kinash, N.Z., Kaimov, E.V.: Lever linkages and mechanisms in vibration protection systems of technical objects. In: Theoretical and Applied Aspects of Modern Science: Coll. of Scientific Papers, vol. 3, pp. 51–59. Belgorod (2014)
6. Klimov, A.V.: Dynamics of a lever relaxation suspension with intermittent damping. Ph.D. (Engineering) Thesis, 186 pp. Orlov. State Tech. University Publication, Oryol (2001)
7. Khomenko, A.P., Eliseev, S.V.: Vibration protection systems with articulated links. A method for constructing mathematical models. News Transsib. **2**(10), 36–45 (2012)
8. Eliseev, S.V., Volkov, L.N., Kukharenko, V.P.: Dynamics of Mechanical Systems with Additional Constraints, 214 pp. Nauka Publication, Novosibirsk (1990)
9. Loitsyansky, L.G., Lurie, A.I.: The course of theoretical mechanics: in 2 vols. In: Dynamics, vol. 2, 640 pp. Nauka Publication, Moscow (1983)
10. Eliseev, S.V., Gordeeva, A.A., Fomina, I.V.: Dynamic properties of oscillatory systems with limiting values of parameters. Bull. ESSTU **4**(35), 45–53 (2011)
11. Eliseev, S.V., Kashuba, V.B., Savchenko, A.A.: Lever oscillation damper in a mechanical system with the object of protection against vibrations in the form of a solid on elastic supports (Electronic resource). In: Science and Education: Electron. Scientific and Technical ed., vol. 10. Access mode: http://technomag.edu.ru/doc/465518.html. Access date 25 Apr 2014 (2012)
12. Weits, V.L., Kochura, A.E., Martynenko, A.M.: Dynamic Calculations of Drive Machines, 352 pp. Mashinostroenie Publication, Leningrad (1971)
13. Den-Hartog, J.P.: Mechanical Oscillations, 530 pp. Publishing House of Foreign Literature, Moscow (1962)
14. Parshuta, E.A.: Evaluation of the dynamic properties of vibration protection systems in the study and design of engineering objects. Author's Abstract of Ph.D. (Engineering) Thesis, 17 pp. BrSU Publication, Bratsk (2014)
15. Alabuzhev, P.M., Gritchin, A.A., Kim, L.I., et al.: Vibration systems with quasi-zero stiffness. In: Ragulskis, K.M. (ed.), 96 pp. Mashinostroenie Publication, Leningrad (1986)
16. Eliseev A.V., Selvinsky V.V., Eliseev S.V. The Dynamics of Vibrational Interactions of Elements of Technological Systems, Taking into Account Unilateral Constraints, 332 pp. Nauka Publication, Novosibirsk (2015)

17. Eliseev, S.V., Reznik, Y.N., Khomenko, A.P.: Mechatronic Approaches in the Dynamics of Mechanical Oscillatory Systems, 384 pp. Nauka Publication, Novosibirsk (2011)
18. Eliseev, S.V., Reznik, Y.N., Khomenko, A.P., et al.: Dynamic Synthesis in Generalized Problems of Vibration Protection and Vibration Isolation of Technical Objects, 523 pp. ISU Publication, Irkutsk (2008)
19. Eliseev, S.V., Khomenko, A.P.: Dynamic Damping. Feedback Concept and Structural Methods of Mathematical Modeling, 357 pp. Nauka Publication, Novosibirsk (2014)
20. Fomina, I.V.: Development of methods for constructing mathematical models with joints of units. Author's Abstract of Ph.D. (Engineering) Thesis, 18 pp. IrGUPS Publication, Irkutsk (2011)
21. Druzhinsky, I.A.: Mechanical Chains, 240 pp. Mashinostroenie Publication, Moscow (1977)
22. Timoshenko, S.P., Gere, J.: Mechanics of Materials, 672 pp. Lan' Publication, St. Petersburg (2002)
23. Khomenko, A.P., Eliseev, S.V.: Joints in vibration protection systems as a process of reducing the number of degrees of freedom of movement. Mod. Technol. Syst. Anal. Model. **4**(28), 8–14 (2010)
24. Bolshakov, R.S.: Determination of dynamic interactions between elements of vibration protection systems based on the structural transformation method. Ph.D. (Engineering) Thesis, 167 pp. SWSU Publication, Kursk (2014)

Chapter 6
Reduced Characteristics in Assessing Properties of Mechanical Oscillatory Systems: Generalized Approaches in the Construction of Mathematical Models

Controlling the dynamic state of technical objects in theoretical and practical aspect has been reflected in a wide range of problems of dynamics of modern machines [1, 2]. At the same time, much attention is paid to vibrations in various forms of manifestation that are characteristic both for transitional operating modes and for stable ones [3, 4]. Problems of dynamics of machine assemblies belong to rather developed direction of the modern theory of mechanisms and machines with the established theoretical basis [1, 5, 6].

Features of the generalized approach while solving various problems of dynamics of machines are the methodological positions based on a possibility of constructing mathematical models of machines and mechanisms to the basic model in the form of a reduction unit with the corresponding formation of the reduced parameters characterizing force factors and mass-and-inertia properties of the system. This approach in formation of the equations of motion of machine assemblies has great opportunities in the research and assessment of parameters of original systems in the periodic motions of machine assemblies, taking into account their nonlinear properties and constructive and engineering features.

Problems of vibration protection of machines, equipment, devices and instrumentation as one of the relevant directions of modern dynamics of machines, are methodologically solved on similar bases, within which ideas of basic systems are widely used [7, 8]. Distinguishing an object of protection against vibration influences predetermines the possibilities of general concepts about dynamic constraints in the mechanical oscillatory systems considered in the form of computational schemes for assessment, control and necessary measurements of a dynamic state of technical systems in the specific modes [9–12].

Further, a generalized approach develops in constructing mathematical models for the machines and mechanisms working in the mode of small oscillations with relation to the established dynamic state. This assumes combining opportunities of using models of a unit of reduction of machine assemblies with definition of mass-and-inertia parameters of the original system within one turn of the driving

© Springer Nature Switzerland AG 2020
S. V. Eliseev and A. V. Eliseev, *Theory of Oscillations*, Studies in Systems, Decision and Control 252, https://doi.org/10.1007/978-3-030-31295-4_6

member and the models of elastic interactions concerning the chosen mechanism configuration.

6.1 The Reduced Parameters of Dynamic Interactions

The feature of the generalized approach in dynamics of machines is that the dynamic system which the technological machine is reduced to the main or basic model in the form of a reduction unit. Such unit has reduced mass-and-inertia parameters; all forces acting on the car are reduced to this unit that finally allows us to obtain the equation of motion of the machine assembly [6, 13].

The similar situation develops while solving the problems of vibration protection. Features of such problem are that dynamic interactions in the mechanical oscillatory system containing an object of protection, elastic and dissipative and mass-and-inertia elements are considered. In the simplified form, the task can be reduced to consideration of basic model (by analogy with the dynamics of the machine assembly). The basic model represents a mechanical system from the object of protection. It is the unit having mass-and-inertia properties. Elastic elements are considered as linear, with the lumped parameters or with distributed ones which can be reduced to the equivalent lumped ones. The rack or supporting surfaces are included in the system. The basic model of vibration protection system in the form of mechanical oscillatory system with one degree of freedom can be also considered as some mechanism having certain features. In particular, the object of protection is capable to have a reduced mass, and an elastic element of the basic model can have the reduced stiffness reflecting elastic properties of all the system. The basic model within the linear theory can also reflect the detailed ideas of the vibration protection system in respect of accounting of forces of resistance and constructive and engineering features of vibration protection and vibration insulation that is implemented through introduction of additional constraints to the original system [14].

At the same time in the determination of the reduced parameters of dynamic systems of different function there are formulated ideas of equivalent ratios and conditions of reduction of parameters.

6.1.1 Reduction of Forces and Masses

For the vibration protection systems with developed structure incorporating multiunit formations or contours, various methods of mathematical modeling among which, owing to specifics of a dynamic state of the object of protection by one or two coordinates, the method of reduction of forces and masses to the object of protection as to a reduction unit can be used.

In many cases, reduced masses (or the inertia moments) arise due to variability of the transfer ratios if the mass of units do not change. The reduced mass is a specially introduced concept which is not related to the ideas of physical change of mass of solid bodies.

The reduced mass is formed proceeding from the equivalence of kinetic energy of reducible or reduced systems. In particular, this explains wide use of the second kind Lagrange equation in the form obtained for systems with a constant weight to create mathematical models of mechanical systems.

A vibration protection system (VPS) represents a construction with the distributed and lumped parameters. The object of protection and elements of the system have mechanical inertia which is distributed on the VPS units, as the mass of the system is also distributed on units. At the same time there can be elements at the system which are identified with concepts of the concentrated weight, which corresponds to constructive and engineering features of the VPS.

For simplification of researches and drawing up mathematical models, the distributed parameters are represented in the form of lumped ones. At the same time the masses of units are considered lumped in the centers of gravity. At the same time inertia in rotary motions of units is considered, which is connected with definition of the moments of inertia of units concerning the axes passing through the centers of gravity. In general, the choice of approaches for simplification of representation of units depends on research problems and has a heuristic character [5, 15].

Usually the VPS consists of several units loaded with various forces and pairs of forces. To research the motion of the VPS, it is possible to compose for each unit the equation of motion as for the free solid body with the known weight making the plane-parallel motion taking into account external forces and forces of reactions of kinematic pairs of rejected units. Thus, the system of the joint equations of motion can be created. While solving such number of the equations, there are a number of problems which are shown when accounting features of the reactions of constraint imposed by kinematic pairs [16].

Instead of drawing up and solving a system of the equations whose number corresponds to number of mobile units of the VPS while researching systems with one degree of freedom, one can apply the approaches based on use of the reduced forces and masses. Instead of a multiunit VPS in this case, the simplified model is considered in which the object of protection is chosen as a reduction unit.

If the reduction unit is included into a rectilinear pair with a rack (or a motionless basis), which is characteristic of the VPS, then it is possible to take any point of a unit for a point of reduction (Fig. 6.1a). However, other forms of the choice of the unit of reduction (Fig. 6.1c), depending on features of the object of protection, are also possible.

Having chosen one of the schemes shown in Fig. 6.1, it is possible to find the reduced force Q_{red}, the moment of the reduced force, the reduced mass m_{red} and the reduced moment of inertia J_{red}. Reduction of the force and pair of forces is carried out on the basis of the principle of possible motions which consists that the sum of elementary works of all external forces applied to the VPS on all possible motions

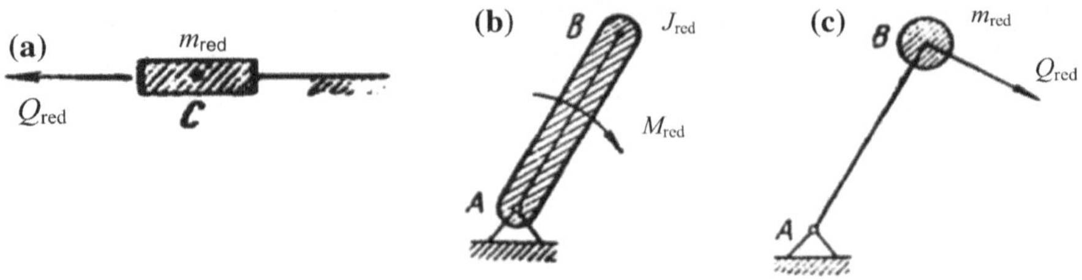

Fig. 6.1 Schemes of reduction of forces and masses: **a** is the rectilinear moving unit with a reduction point; **b** is the rotating unit with the reduced moment of pair of forces and the reduced inertia moment; **c** is the same with a reduction point

of points of application of these forces is equal to elementary work of the reduced force or the reduced moment at the corresponding possible displacement. In vibration protection system with one degree of freedom, each its point can make only one motion which is its possible displacement.

Possibilities of the analytical tools of the theory of circuits or the theory of automatic control [9, 10, 17], allow using not elementary work of forces and the moments of pairs of forces, but their powers, which, in particular, makes it possible to use a concept of impedances [18, 19]. As in systems with one degree of freedom possible motions coincide with the valid motions determining the valid speeds of separate points of the VPS, the force of any force *of* Q_i can be counted on a formula:

$$N_i = Q_i \cdot v_i \cdot \cos\alpha_i, \tag{6.1}$$

where N_i is the force power Q_i; v_i is the speed of a point of application of force; α_i is an angle between the direction of the force and speed.

If the value M_j of the moment of pair of forces applied to *j* unit and angular speed ω_j the same unit are known, then power can be determined as follows:

$$N_j = M_j \cdot \omega_j. \tag{6.2}$$

As the sum of powers of the reduced forces and the moments of pairs of forces is equal to the power of the reduced force and power of the reduced moment of pair of forces,

$$Q_{red} = \sum_{i=1}^{i=m} Q_i \cdot \frac{v_i \cdot \cos\alpha_i}{v_B} + \sum_{j=1}^{j=n} M_j \cdot \frac{\omega_j}{v_B}, \tag{6.3}$$

$$M_{red} = \sum_{i=1}^{i=m} Q_i \cdot \frac{v_i \cdot \cos\alpha_i}{\omega} + \sum_{j=1}^{j=n} M_j \cdot \frac{\omega_j}{\omega}. \tag{6.4}$$

where *m* is the number of all reduced forces; *n* is the number of all specified pairs of forces applied to VPS units; v_B is the speed of the point of reduction located on the rotating unit or speed of any point of translationally moving unit; ω is the angular speed of a unit of reduction.

The reduced forces Q_{red} and the moments of forces M_{red} depend not only on the reduced forces, but also on the speed relation, and the relation of powers of separate points of the VPS with one degree of freedom or can be constants, or depend only on position of the mechanism. At the same time the relation of speeds should not depend on the speed of the object of protection. Such approach which features are described in systems with one degree of freedom [5, 6] allows us to reduce forces without knowing the valid law of the motion of units, and then to use these reduced characteristics of the law of the motion.

6.1.2 The Reduced Mass of Units and Mechanisms

Reduction of masses is carried out on the basis of equality of kinetic energy, i.e. the reduced system has to have the same kinetic energy, as the set system. The reduced mass or the reduced inertia moment is defined, kinetic energy of all units of the VPS is counted, and these values are equated. Kinetic energy of a unit of reduction contains either the required reduced inertia moment, or the required reduced mass which are defined from the obtained equality.

In a flat VPS each unit can either move translationally, or rotate, or make the plane-parallel motion.

At the plane-parallel motion of a unit its kinetic energy is defined as follows:

$$\mathrm{T} = \frac{J_S\omega^2}{2} + \frac{mv_S^2}{2}, \tag{6.5}$$

where first term of the right part represents the kinetic energy in a rotary motion of the unit, and the second term is the kinetic energy in translational motion together with the center of gravity of the same unit. In equality (6.5) J_S is the moment of inertia of the unit concerning the axis passing through its center of gravity; ω is the angular speed of the unit; m is the mass of the unit; v_S is the speed of the center of gravity.

If the unit rotates around a motionless axis, then the forward speed v_S of the center of gravity is equal to zero and for calculation of kinetic energy of the unit, one first member of the right part (6.5) is enough. In this case the inertia moment J_0 of the unit is calculated with relation to the motionless axis of rotation. If the unit moves translationally, then its angular speed ω is equal to zero and for calculation of its kinetic energy enough one second member of the right part (6.5).

Kinetic energy of all VPS is equal to the sum of kinetic energy of all its units and is defined as

$$T = \sum_{i=1}^{i=n} \left| \frac{J_{S_i}\omega_i^2}{2} + \frac{m_i v_{S_i}^2}{2} \right|, \tag{6.6}$$

where i = 1, 2,…, n (n is the number of mobile units of the VPS).

Expressions of kinetic energy for a point and for a unit of reduction can be written as

$$T = \frac{m_{\Pi \text{p}} v_B^2}{2}, \; T = \frac{J_{\Pi \text{p}} \omega^2}{2},$$

where m_{red} is the reduced mass of the VPS; v_B is the reduction point speed; J_{red} is the reduced inertia moment concerning the axis that coincides with an axis of rotation of a unit; ω is the angular speed of a unit of reduction.

Equating at first kinetic energy of the point of reduction, and then the unit of reduction of kinetic energy of the VPS, we will obtain that

$$m_{\Pi \text{p}} = \sum_{i=1}^{i=n} \left[J_{S_i} \cdot \left(\frac{\omega_i}{v_B} \right)^2 + m_i \cdot \left(\frac{v_{S_i}}{\omega} \right)^2 \right]; \tag{6.7}$$

$$J_{\Pi \text{p}} = \sum_{i=1}^{i=n} \left[J_{S_i} \cdot \left(\frac{\omega_i}{\omega} \right)^2 + m_i \cdot \left(\frac{v_{S_i}}{\omega} \right)^2 \right]. \tag{6.8}$$

Data of the expression are identical by structure, therefore, it is enough to analyze one of them, for example, equality (6.8).

The reduced moment of inertia depends on a square of the relation of speeds; this value is the variable depending on the provision of the VPS elements. In that specific case, when the transfer ratio between the VPS elements does not change (additional constraints in the form of toothed gearings with round wheels, friction and belt drives, etc.), the reduced moment of inertia remains constant. The reduced inertia moment is always positive. As the relations of speeds of separate points of the VPS depend only on the mutual provision of elements, the reduced moment of inertia does not depend on the speed of the motion of the object of protection. The reduced inertia moment is generally variable because of the variable transfer ratio, the mass of units are most often constant. As the reduced moment of inertia is calculated from the equality of kinetic energies, in dynamic calculations with the reduced moment of inertia one can use the Lagrange's equation of the second kind and the equation of kinetic energy in differential and integrated forms. For example, for the planetary mechanism (Fig. 6.2) the first moment of inertia reduced to a unit is defined by expression:

$$T = T_1 + 3T_2 + T_H = \frac{J_1\omega_1^2}{2} + 3 \cdot \left[\frac{J_2\omega_2^2}{2} + \frac{m_2 \cdot (R_1 + R_2)^2 \cdot \omega_H^2}{2} \right] + \frac{J_H\omega_H^2}{2}, \tag{6.9}$$

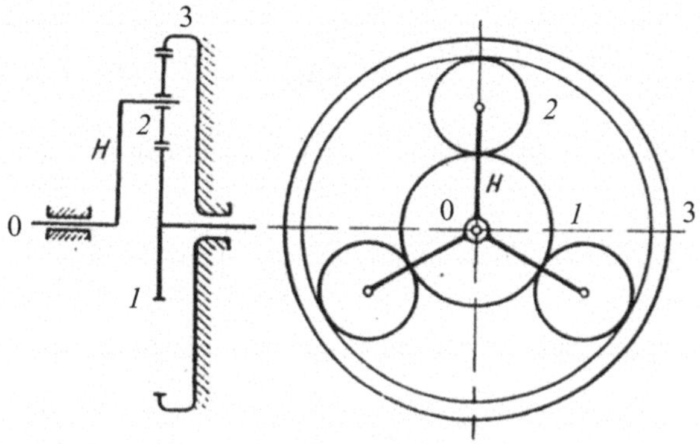

Fig. 6.2 Reduction of masses in the planetary mechanism

where R and R_1, R_2 are radii of initial circles of wheels 1 and 2; ω_H are the angular speed of the guide H (Fig. 6.2).

As the transfer ratios of the considered mechanism are constants, the reduced moment of inertia *of* J_{red} turns out constant and can be easily calculated. The reviewed example is characteristic by interactions of several rotary motions in which ratios between radii of units are implemented, which represents a form of manifestation of lever linkages in interactions of unit. Works [7, 8, 14] review the detailed examples of interaction of units of the VPS making angular oscillations. Of certain interest are the interactions of elements in mechanisms with different types of the motion of units, for example, for the crank-and-rod mechanism (Fig. 6.3).

With specified values of l_{AB}, l_{BC} and l_{BS_2}, moments of inertia J_{1A} and J_{S_2} of units 1 and 2; the masses m_2 and m_3 of units 2 and 3 (Fig. 6.3), the kinetic energy of all the mechanism will be equal to the sum of kinetic energy of all its units:

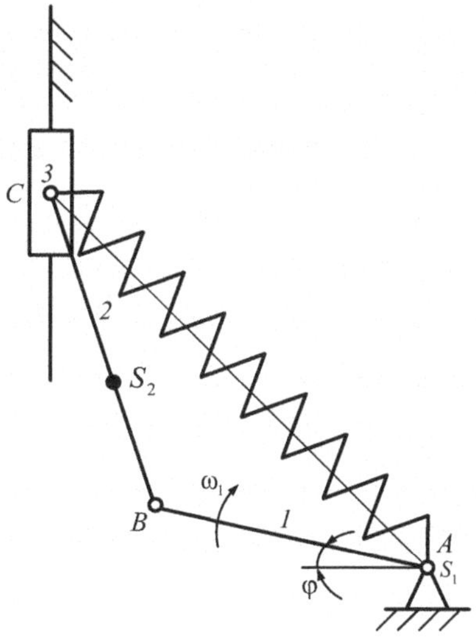

Fig. 6.3 Reduction of masses in the crank-and-rod mechanism

$$J_{\Pi\,\mathrm{p}} = J_{1A} + J_{S_2} \cdot \frac{l_{AB}^2}{l_{BC}^2}\left(\frac{v_{CB}}{v_B}\right)^2 + m_2 \cdot l_{AB}^2 \cdot \left(\frac{v_{S_2}}{v_B}\right)^2 + m_3 \cdot l_{AB}^2 \cdot \left(\frac{v_C}{v_B}\right)^2,$$

from where the reduced inertia moment can be found:

$$J_{\Pi\,\mathrm{p}} = J_{1A} + J_{S_2} \cdot \left(\frac{\omega_2}{\omega_1}\right)^2 + m_2 \cdot \left(\frac{v_{S_2}}{\omega_1}\right)^2 + m_3 \cdot \left(\frac{v_C}{\omega_1}\right)^2. \qquad (6.10)$$

Expression (6.10) can be presented in the form

$$J_{\Pi\,\mathrm{p}} = J_{1A} + J_{S_2} \cdot \frac{l_{AB}^2}{l_{BC}^2}\left(\frac{v_{CB}}{v_B}\right)^2 + m_2 \cdot l_{AB}^2 \cdot \left(\frac{v_{S_2}}{v_B}\right)^2 + m_3 \cdot l_{AB}^2 \cdot \left(\frac{v_C}{v_B}\right)^2. \qquad (6.11)$$

The last expression demonstrates that the reduced moment of inertia of masses of units depends on an arrangement of the VPS elements, as (6.11) includes the relations of linear speeds of separate points of the mechanism. In the considered mechanism the reduced moment of inertia is the periodic function of the position of the mechanism, i.e. the function of its generalized coordinate which is the angle φ of the crank turn to the datum line. While considering the mechanism in Fig. 6.3 as the basic vibration protection system, in which the unit 3 with mass m_3 is the object of protection, the methodical basis of creation of ratios between mass-and-inertia parameters of separate units remains invariable, though features of the motion of units, uncharacteristic for mechanisms, are formed in the system in general. In particular, the mechanical system in Fig. 6.3 in the presence of additional elastic constraints k, for example between points A and C, has an opportunity to form small oscillating motions in relation to some position of static balance. The object of protection m_C $(m)_3$ makes small oscillations concerning balance position at external disturbances which can represent oscillations of the supporting surface (p. A), or it can be force applied directly to the object of protection. In this case the crank does not implement its opportunities in making the final rotation, but defines a structure or configuration of the mechanical system, creating possibilities of variable changes of transfer properties of all the system in solving specific problems of vibration protection [20, 21].

The commonness of techniques in the construction of mathematical models for machine assemblies and vibration protection systems creates certain advantages when solving problems of dynamic synthesis of vibration protection systems.

The research problem, proceeding from the developed starting positions, is to substantiate an opportunity and to develop approaches to the construction of the generalized mathematical models of the vibration protection systems, the structure of which comprises, except traditional elements, additional constraints in the form of mechanical chains, including the mechanisms that make small oscillating motions in relation to established dynamic state.

6.1.3 Construction of the Generalized Mathematical Model

The vibration protection system can be considered in the form of a mechanism with one degree of freedom which rests on the vibrating basis (Fig. 6.4).

In view of vibrations of the basis, i.e. under kinematic influences, the object of protection will execute small oscillations, and the configuration of the mechanism gains elastic and mass-and-inertia properties. It is supposed that the angles of installation of units can be considered conditionally constant owing to the smallness of deviations. We believe that the supporting surface generally has forward two-component and rotary oscillations in one plane.

While constructing the mathematical model, the motions of the vibration protection system are considered around the positions of the stable dynamic equilibrium.

To compose the differential equation of motion of the VPS, Lagrange's equation of the second kind is used:

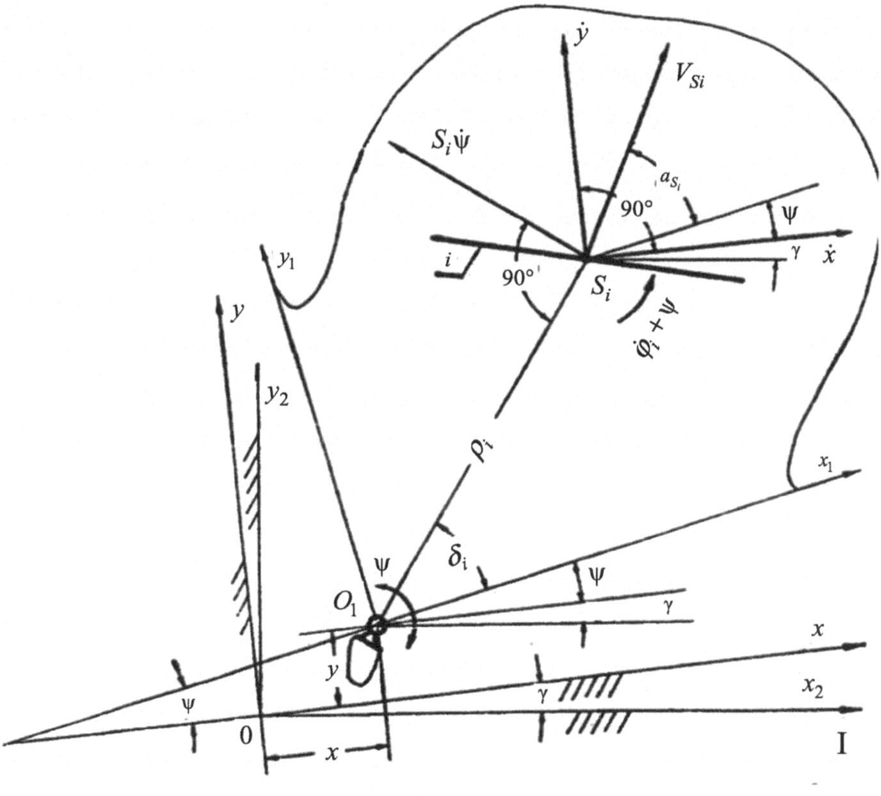

Fig. 6.4 The ith VPS unit supported by the basis vibrating along the coordinates x, y and ψ

$$\frac{d}{dt}\left(\frac{\partial T}{\partial \dot{\varphi}}\right) - \frac{\partial T}{\partial \dot{\varphi}} + \frac{\partial \Pi}{\partial \dot{\varphi}} = Q,$$

where φ is the generalized coordinate (an angle of rotation of a unit of reduction of the VPS in the relative motion).

According to Fig. 6.4, the system of coordinates $x_1O_1y_1$ is rigidly connected with the supporting surface *I* and can have angular vibration around p. O_1 under the law ψ = ψ(t). Translational motion, in turn, has vibrations in two perpendicular directions: $x = x$ (t) and $y = y(t)$. These motions are parallel to axes *Ox* and *Oy* of motionless system of coordinates. Figure 6.4 also shows the second motionless system of coordinates x_2Oy_2, whose axis *Ox* is horizontal, and Oy_2 is vertical. The system of coordinates *xOy* is turned counterclockwise by the angle γ concerning xOy_{22}. The systems of coordinates $x_1O_1y_1$ and *xOy* are chosen so that in the absence of angular motions of the basis at ψ = 0 of their axes coincide.

The center of gravity S_i of the considered *i*th unit of the VPS has the absolute speed determined by expression:

$$\overline{V} = \overline{V}_{S_i} + \rho_i \frac{d\psi}{dt} + \frac{dx}{dt} + \frac{dy}{dt}, \tag{6.12}$$

where $\overline{V}_{S_i} = \frac{dS_i}{d\varphi} \cdot \frac{d\varphi}{dt}$ is the relative speed (in relation to the basis) of the center of gravity; S_i is the motion of the center of gravity in the relative motion; $\rho_i \cdot \frac{d\psi}{dt}$ is the translational speed from angular motions of the basis; $\rho_i = O_1S_i$, $\frac{dx}{dt}$ and $\frac{dy}{dt}$ are translational speeds from progress of the supporting surface.

Let us note that *i*th unit has also the absolute speed of the angular motion:

$$\frac{d\varphi_1}{dt} + \frac{d\psi}{dt} = \frac{d\varphi_i}{d\varphi} \cdot \frac{d\varphi}{dt} + \frac{d\psi}{dt}\varphi. \tag{6.13}$$

Here φ_i is the angle of rotation of the *i*th unit in the relative motion. The angle of relative speed V_{S_i} with the axis x_1 is designated through α_{S_i}, and angle $\delta_i = \angle S_iO_1x_2$. Values ρ_i, $\frac{dS_i}{d\varphi}$, α_{S_i} and δ_i depend on an angle of rotation φ of a unit of reduction of the VPS in the relative motion and do not depend on time *t*. As for *x*, *y* and ψ, they are the set functions of time.

$$\mathrm{T} = \frac{1}{2} J_{\Pi \mathrm{p}}(\varphi) \cdot \left(\frac{d\varphi}{dt}\right)^2 + \frac{1}{2}\left(\frac{d\psi}{dt}\right)^2 \sum_{i=1}^{j} (J_i + m_i\rho_i^2)$$

$$+ \frac{d\psi}{dt} \cdot \frac{d\varphi}{dt} \sum_{i=1}^{j} \left[J_i \frac{d\varphi_i}{d\varphi} + m_i\rho_i \frac{dS_i}{d\varphi} \sin(\alpha_{S_i} - \delta_i) \right]$$

$$+\sum_{i=1}^{j} m_i \begin{bmatrix} -\frac{d\psi}{dt}\cdot\frac{dx}{dt}\cdot\rho_i\cdot\sin(\delta_i+\psi)+\frac{d\psi}{dt}\cdot\frac{dy}{dt}\rho_i\cos(\delta_i+\psi) \\ +\frac{dx}{dt}\cdot\frac{d\varphi}{dt}\cdot\frac{dS_i}{d\varphi}\cdot\cos(\alpha_{S_i}+\psi) \\ +\frac{dy}{dt}\cdot\frac{d\varphi}{dt}\cdot\frac{dS_i}{d\varphi}\cdot\sin(\alpha_{S_i}+\psi)+\frac{1}{2}\cdot\left(\frac{dx}{dt}\right)^2+\frac{1}{2}\cdot\left(\frac{dy}{dt}\right)^2 \end{bmatrix}, \tag{6.14}$$

where $i = 1, 2, 3, \ldots$ are numbers of units of the mechanism, j is the number of units; mobile in the relative motion; J_{red} (φ) is the reduced moment of inertia of the VPS determined by expression

$$J_{\Pi\text{p}}(\varphi)=\sum_{i=1}^{j}\left[J_i\left(\frac{d\varphi_i}{d\varphi}\right)^2+m_i\left(\frac{dS_i}{d\varphi}\right)^2\right]. \tag{6.15}$$

Here J_i is the moment of inertia *of* ith unit of the mechanism; j is the number of units concerning the center of gravity; m_i is the mass of the ith unit. According to Fig. 6.4, a potential energy of the VPS will be defined

$$\Pi=g\cdot\left[\sum_{i=1}^{j} m_i\rho_i\sin(\gamma+\delta_i+\psi)+\Pi(t)\right]+\Pi_0(\varphi)+C, \tag{6.16}$$

where g is the acceleration of gravity; $\Pi(t)$ is the a part of potential force that depends on time t and does not depend on the angle of rotation φ; $\Pi_0(\varphi)$ is the potential energy of elastic elements whose deformation depends on φ; C is some constant determining a static shift.

If the plane of operation of the mechanism makes angle α_0 with a vertical, then in the Eq. (6.16) it is necessary to accept $g\cdot\cos\alpha_0$ instead of g.

From expression (6.11), using (6.14) and (6.16), we make a number of transformations and find the differential equation of motion of the mechanism in a form of:

$$\begin{aligned} &J_{\Pi\text{p}}(\varphi)\frac{d^2\varphi}{dt^2}+\frac{1}{2}\frac{dJ_{np}(\varphi)}{d\varphi}\cdot\left(\frac{d\varphi}{dt}\right)^2 \\ &+\frac{d^2\psi}{dt^2}\sum_{i=1}^{j}\left[J_i\frac{d\varphi_i}{d\varphi}+m_i\rho_i\frac{dS_i}{d\varphi}\sin(\alpha_{S_i}-\delta_i)\right] \\ &-\left(\frac{d\psi}{dt}\right)^2\sum_{i=1}^{j} m_i\rho_i\frac{d\rho_i}{d\varphi}-\frac{d\psi}{dt}\cdot\frac{dx}{dt} \\ &\times\sum_{i=1}^{j} m_i\cdot\left[\frac{dS_i}{d\varphi}\sin(\alpha_{S_i}+\psi)-\frac{d\rho_i}{d\varphi}\sin(\psi+\delta_i)-\rho_i\frac{d\delta_i}{d\varphi}\cos(\psi+\delta_i)\right] \\ &+\frac{d\psi}{dt}\cdot\frac{dy}{dt}\times\sum_{i=1}^{j} m_i\cdot\left[\frac{dS_i}{d\varphi}\sin(\alpha_{S_i}+\psi)-\frac{d\rho_i}{d\varphi}\cos(\psi+\delta_i)+\rho_i\frac{d\delta_i}{d\varphi}\sin(\psi+\delta_i)\right] \\ &+\frac{d^2x}{dt^2}\times\sum_{i=1}^{j} m_i\cdot\frac{dS_i}{d\varphi}\cdot\cos(\psi+\alpha_{S_i}) \\ &+\frac{d^2y}{dt^2}\times\sum_{i=1}^{j} m_i\cdot\frac{dS_i}{d\varphi}\cdot\sin(\psi+\alpha_{S_i}) \\ &+g\cdot\sum_{i=1}^{j} m_i\cdot\left[\frac{d\rho_i}{d\varphi}\sin(\gamma+\delta_i+\psi)+\rho_i\frac{d\delta_i}{d\varphi}\cos(\gamma+\delta_i+\psi)\right]+\frac{d\Pi_0}{d\varphi}=Q. \end{aligned} \tag{6.17}$$

In practice, there are simpler cases that those described by the Eq. (6.17). In particular, the basis or a supporting surface of the VPS can have only forward or torsional vibrations. The equation of motion at forward vibrations from the Eq. (6.17) at $\psi = 0$:

$$J_{\Pi p}(\varphi)\frac{d^2\varphi}{dt^2} + \frac{1}{2}\frac{dJ_{np}(\varphi)}{d\varphi}\cdot\left(\frac{d\varphi}{dt}\right)^2 + \frac{d^2x}{dt^2}\cdot F(\varphi) + \frac{d^2y}{dt^2}\cdot f(\varphi) + P(\varphi) + \frac{d\Pi_0}{d\varphi} = Q, \tag{6.18}$$

where

$$\left.\begin{aligned} F(\varphi) = \sum_{i=1}^{j} m_i \tfrac{dS_i}{d\varphi}\cos\alpha_{S_i};\; f(\varphi) = \sum_{i=1}^{j} m_i \tfrac{dS_i}{d\varphi}\sin\alpha_{S_i}, \\ P(\varphi) = g\cdot\sum_{i=1}^{j} m_i\cdot\left[\tfrac{d\rho_i}{d\varphi}\sin(\gamma+\delta_i) + \rho_i\tfrac{d\delta_i}{d\varphi}\cdot\cos(\gamma+\delta_i)\right]. \end{aligned}\right\} \tag{6.19}$$

$\Pi_0(\varphi)$ is the potential energy of elastic elements (springs) which deformation is defined by the small angular motions along the coordinate φ, causing changes of location of the system.

In the last equation, when considering specific VPSs, $P(\varphi)$ takes a simpler form, as well as $\Pi_0(\varphi)$.

From (6.17) at $\frac{d\psi}{dt} = \frac{d\gamma}{dt} = 0$ we have:

$$\begin{aligned} J_{\text{red}}(\varphi)\frac{d^2\varphi}{dt^2} + \frac{1}{2}\frac{dJ_{\text{red}}(\varphi)}{d\varphi}\cdot\left(\frac{d\varphi}{dt}\right)^2 + \frac{d^2\psi}{dt^2} \\ \times H(\varphi) - \left(\frac{d\psi}{dt}\right)^2 h(\varphi) + K(\varphi,\psi) + \frac{d\Pi_0}{d\varphi} = Q, \end{aligned} \tag{6.20}$$

where

$$\left.\begin{aligned} H(\varphi) = \sum_{i=1}^{j}\left[J_i\tfrac{d\varphi_i}{d\varphi} + m_i\rho_i\tfrac{dS_i}{d\varphi}\sin(\alpha_{S_i}-\delta_i)\right]; \\ h(\varphi) = \sum_{i=1}^{j} m_i\rho_i\tfrac{d\rho_i}{d\varphi}, \\ K(\varphi,\psi) = g\cdot\sum_{i=1}^{j} m_i\cdot\left[\tfrac{d\rho_i}{d\varphi}\sin(\gamma+\delta_i+\psi) + \rho_i\tfrac{d\delta_i}{d\varphi}\cdot\cos(\gamma+\delta_i+\psi)\right]. \end{aligned}\right\} \tag{6.21}$$

$\Pi_0(\varphi)$ is defined by a configuration of the system of elastic constraints.

Let us note that, according to the Eq. (6.17), influence of forward and torsional vibrations on the relative motion is connected. Forward vibrations of the basis x (t) and $y(t)$ influence the motion independently of each other.

Forward vibrations of the basis do not exert action on potential energy of the elements with gravity. However, forward and torsional oscillations of supporting surfaces form the translational motions operating on deformation of elastic elements and on their potential energy.

Features of such interactions are considered in the works of Eliseev and others [22, 23]. While solving many problems of vibration protection, a vertical motion of the basis is selected; at the same time it is supposed that the orthogonal motions have no significant effect on the motions of the system in general, if the problems of vibration protection are not among special.

In the solution of specific objectives of vibration protection, in many cases constructing equivalent mathematical models, in the dynamic sense, requires taking consideration of features of the arrangement of kinematic pairs in mechanical chains and reductions of characteristic points to a new basis. Some opportunities of transfers of characteristic points of units with formation of the compensating adjustments of parameters of mass-and-inertia characteristics are analyzed.

Possible simplifications of the original system of location of masses. In the Eqs. (6.15)–(6.21) a methodical approach is based on obtaining exact inertial parameters of units of mechanical systems that finally results in certain difficulties arising because ρ_i, α_{S_i} and some other values included in (6.15)–(6.21) are in a lengthy dependence on the law of the motion of the basis and the angle φ of the turn of the VPS reduction unit. While simplifying calculations, there is a need to place the mass of units of the VPS elements not in their centers of gravity, but in such points where it is possible to define, in simpler way, the angle α_{S_i} between their relative and translational speeds. To maintain a dynamic equivalence by drawing up the equations of the motion by the simplified way, fictitious additional massless moments of inertia for the corresponding units can be introduced. In particular, this approach is offered by Ragulskis [24]. In this case at impossibility of dynamic equilibration as it is treated in [5, 6], the mass of units are placed in convenient points and proceeding from a condition of maintaining the invariance of the moments of inertia of separate units the additional moments of inertia are introduced.

Then the first two Eqs. (6.15) and (6.19) after such location of masses take a form

$$\left.\begin{aligned} F(\varphi) = \sum_{i=1}^{j}\sum_{p_i=1}^{q_i} m_{ip_i}\frac{dS_{ip_i}}{d\varphi}\cos\alpha_{ip_i}; f(\varphi) = \sum_{i=1}^{j}\sum_{p_i=1}^{q_i} m_{ip_i}\frac{dS_{ip_i}}{d\varphi}\sin\alpha_{ip_i}; \\ J_{\Pi\mathrm{p}}(\varphi) = \sum_{i=1}^{j}\sum_{p_i=1}^{q_i}\left[m_{ip_i}\left(\frac{dS_{ip_i}}{d\varphi}\right)^2 + \Delta J_i\left(\frac{d\varphi_i}{d\varphi}\right)^2\right], \end{aligned}\right\} \tag{6.22}$$

where $p_i = 1, 2, \ldots, q_i$ are numbers of the replacing points of mass of the ith unit; q_i is the number of these points for the ith unit; m_{ip_i} is the mass of the ith unit placed at point S_{ip_i}; ΔJ_i is the fictitious additional moment of inertia of the ith unit concerning its center of gravity; α_{ip_i} is the angle between vectors of relative and translational speeds of the point S_{ip_i}.

Features of determination of additional masses. As an example, we will consider location of mass of a unit of the VPS (Fig. 6.5) at two set points A and B lying on a straight line with the center of gravity S.

The equations of static location can be presented in the form:

$$\left.\begin{aligned} m_A + m_B &= \gamma \cdot \int_0^{L_1+L+L_2} y dx, \\ m_A l + m_B(l - L) &= 0, \end{aligned}\right\} \tag{6.23}$$

where m_A and m_B are mass in points of A and B respectively; l is the distance from A point to the center of gravity S; γ is the mass of unit of the unit volume; $y = y(x)$ is the cross-sectional area of the unit (it is assumed that the unit is from uniform material).

The introduced fictitious additional moment of inertia can be defined by expression

$$\Delta J = \gamma \int_0^{L_1+l} (L_1 + l - x)^2 y dx + \gamma \int_{L_1+l}^{L_1+L+L_2} (L_1 + l - x)^2 y dx - \left(1 - \frac{m_B}{m}\right) \cdot m_B L^2, \tag{6.24}$$

where

$$\left.\begin{aligned} l = \tfrac{m_B}{m} L, \; m = \gamma_0 \int_0^{L_1+L+L_2} y dx, \\ m_B = \tfrac{\gamma_0}{L} \cdot \left[\int_0^{L_1+L+L_2} (x - L_1) y dx - \int_0^{L_1} (L_1 - x) y dx\right] \end{aligned}\right\}. \tag{6.25}$$

Depending on features of the unit, ΔJ can take on not only positive values, but also negative ones. Being set by the law of change $y(x)$, it is possible to obtain expressions for approximation. A relative error of the moment of inertia, if to be

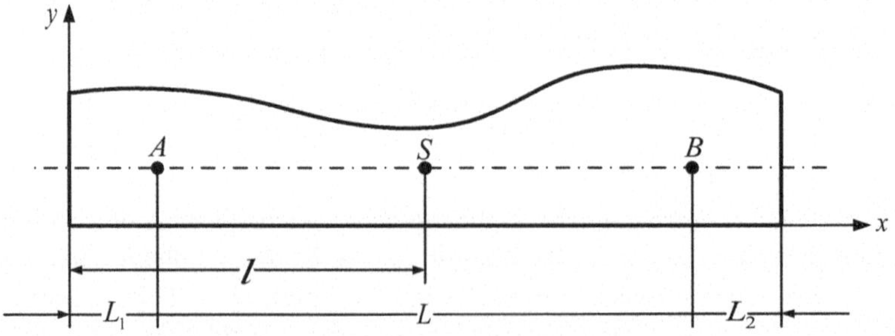

Fig. 6.5 Distribution of the mechanism unit mass along the length

limited only to static placement of mass of the unit and to assume that $y = a + bx$ at $L_1 = L_2 = 0$, will be equal:

$$\frac{\Delta J}{J_S} = -\frac{3 \cdot (2+\delta)^2}{\delta^2 + 6\delta + 6}. \tag{6.26}$$

6.1.4 The Vibration Protection System with a Two-Rail Assur Group

In this case (Fig. 6.6) the vibration protection system, in a sense, has properties of the crank-and-rod mechanism at the object of protection with mass m_C.

Let us compose a simplified differential equation of motion of the system when the basis makes translational rectilinear oscillations under the law $x = x(t)$. Let us place the mass of units of the VPS according to (6.23)–(6.25) at points A, B, C (m_A, m_B, m_C). By 1 and 2 we will designate the fictitious additional moments of inertia for units ΔJ_1, ΔJ_2, respectively. From Fig. 6.6 it follows that

$$\left.\begin{aligned} \alpha_B &= 90^\circ + \varphi + \sigma - \gamma, \alpha_C = \sigma - \gamma; \\ v_C &= \left(1 + \frac{\lambda \cdot \cos\varphi}{\sqrt{1-\lambda^2 \sin^2\varphi}}\right) \cdot R \cdot \dot{\varphi} \cdot \sin\varphi, \end{aligned}\right\} \tag{6.27}$$

where $R = AB$, $L = BC$.

According to (6.18), the differential equation of motion will take a form

$$J_{\text{red}}(\varphi)\frac{d^2\varphi}{dt^2} + \frac{1}{2}\frac{dJ_{\text{red}}(\varphi)}{d\varphi} \cdot \left(\frac{d\varphi}{dt}\right)^2 + \frac{d^2x}{dt^2} \cdot F(\varphi) + P(\varphi) + \Pi_0(\varphi) = Q, \tag{6.28}$$

where

$$\left.\begin{aligned} J_{\text{red}}(\varphi) &= \Delta J_1 + m_B R^2 + \Delta J_2 \cdot \frac{\lambda^2 \cos^2\varphi}{1-\lambda^2 \sin^2\varphi} + m_C R^2 \cdot \left(1 + \frac{\lambda \cdot \cos\varphi}{\sqrt{1-\lambda^2 \sin^2\varphi}}\right)^2 \cdot \sin^2\varphi; \\ F(\varphi) &= -R \cdot \left[m_B \cdot \sin(\varphi + \sigma - \gamma) + m_C \cdot \left(1 + \frac{\lambda \cdot \cos\varphi}{\sqrt{1-\lambda^2 \sin^2\varphi}}\right)^2 \sin\varphi \cdot \cos(\sigma - \gamma)\right]; \\ P(\varphi) &= gR \cdot \left[m_B \cdot \cos(\varphi + \sigma) - m_C \cdot \left(1 + \frac{\lambda \cdot \cos\varphi}{\sqrt{1-\lambda^2 \sin^2\varphi}}\right) \cdot \sin\varphi\right]; \\ \lambda &= \frac{R}{L}, \end{aligned}\right\} \tag{6.29}$$

$\Pi_0(\varphi)$ is defined by the stiffness of spring k (Fig. 6.6) on change of distance AC.

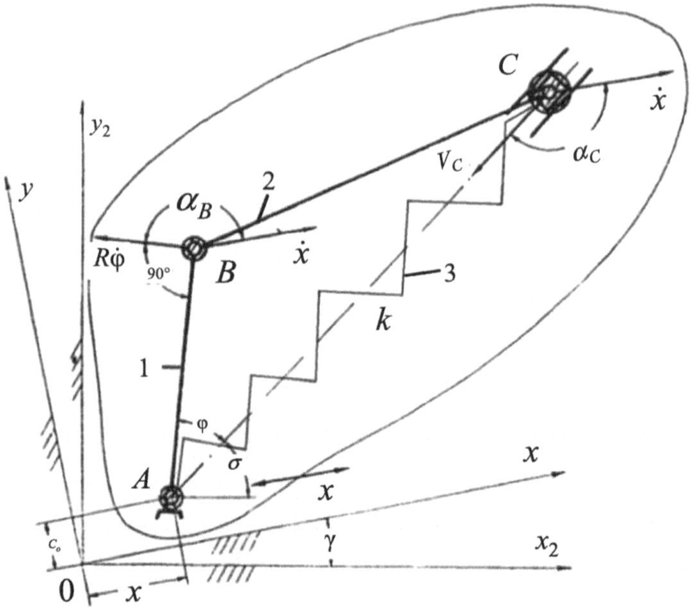

Fig. 6.6 The crank-and-rod mechanism which is on the translationally vibrating basis. 1, 2 are mechanism levers; 3 are elastic element stiffness k

Transformation of the Eq. (6.28) from a coordinate to the coordinate of the motion in the direction of AC can be executed from obvious geometrical ratios in the scheme (see Fig. 6.6).

Thus, the solution of problems of dynamics of machine assemblies on the basis of the choice of a unit of reduction, with the reduced masses, moments of inertia and external influences subsequently determined, can be viewed as the system of dynamic analogies in relation to problems of vibration protection. In this regard, choosing the driving member in the machine assembly is similar to choosing and determining parameters of the object of protection with the corresponding reduced characteristics of external influences. The suggested scheme of constructing the generalized model of a mechanical chain with a focus on accounting for kinematic disturbances from the basis or supporting surfaces expands ideas about the features of dynamic interaction processes between the elements of mechanical oscillatory systems. It is shown that the reduced parameters can be used for equivalent transformations of original systems when there is a need of the convenient choice for location of points of reduction or connection of elements. The possibilities of using the negative reduced masses, or the inertia moments in the reduction schemes do not contradict the principles of dynamics, since they reflect the directions of inertial forces in the interactions of the system elements. At the same time the reduced parameters define the frequency opportunities of dynamic systems and the modes of dynamic states at the forced oscillations.

Kinematic disturbances of mechanical systems possess a wide range of dynamic disturbances to the original system owing to an initial possibility to form translational and relative motions in comparison with force disturbances. In this regard, in the developed structures of mechanical oscillatory systems in the presence of several types of the motions caused by vibrations of supporting surfaces, manifestations of new forms of self-organization of the motion of elements are possible, including modes of single, narrow- and broadband dynamic absorbing of oscillations.

6.2 Quasielements in Mechanical Oscillatory Systems: The Features of Systems When Excluding Variables of Dynamic States

Dynamic absorbing of oscillations is one of the directions of the theory and practice of vibration protection of machines, the equipment and devices [18, 25, 26]. The detailed idea of properties of dynamic absorbers of oscillations are connected with tasks in which the object of protection in the form of a solid body with one degree of freedom is reduced to the mode providing considerable reduction of oscillations of the object at some fixed frequency of external harmonic disturbance. In such cases the dynamic absorber is considered as the attached structure (in the simplest case, as mass-and-inertia and elastic elements) adding one more degree of freedom of motion to the system.

The dynamic absorber can have more complex design and possess several degrees of freedom, whereas the object of protection represents a solid body with one degree of freedom. Similar questions are considered in Eliseev's works, etc. [9, 25, 27].

Questions of protection of objects whose condition is described by several variables are developed to a lesser extent. In particular, dynamics of transport devices can be related to such problems. In the simplest versions of these devices the object of protection represents a solid body on elastic supports. Its motion can be described in several systems of coordinates.

The computational scheme and structural diagram of a mechanical oscillatory system with three degrees of freedom are provided in Fig. 6.7. The schemes have a fairly general view and reflects possibilities of force influences Q_1–Q_3 along the coordinates y_1–y_3, and also kinematic disturbances $z_1(t)$ and $z_2(t)$ from the supporting surfaces of I and II (Fig. 6.7a). The structural diagram consists of three partial systems. The motion of the system is described in the generalized coordinates connected with motionless basis (Fig. 6.7b). Such a system can be used for modeling of effects of dynamic absorbing of oscillations in the form of an object with one or two degrees of freedom.

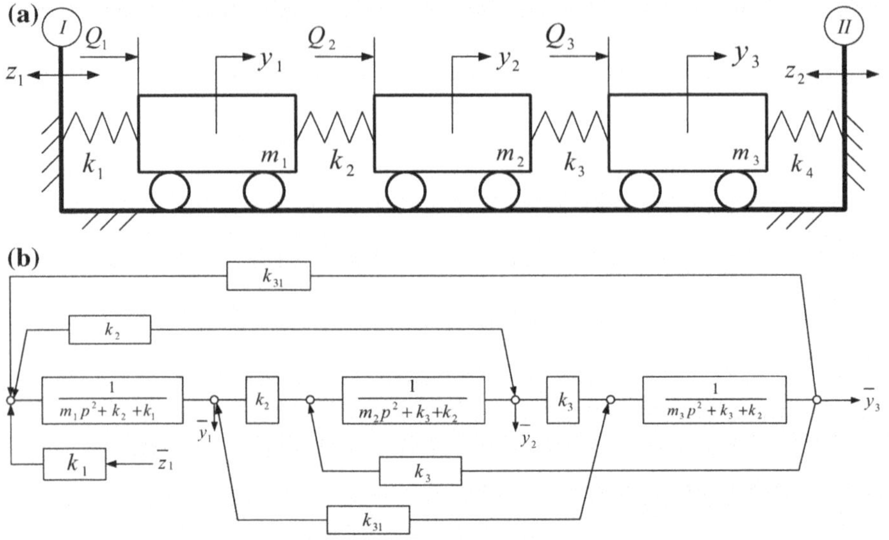

Fig. 6.7 The computational scheme and structural diagram in the form of the mechanical oscillatory system with three degrees of freedom

In Fig. 6.7a mass-and-inertia elements are designated as m_1–m_3; k_0–k_4 reflect elastic properties of the system; y_1–y_3 are connected with a motionless system of coordinates. The system makes small motions; forces of resistance are considered as small. External influences can be considered as required as kinematic (z_1 and z_2) or the force harmonic disturbances attached to the corresponding masses.

The possibilities of change of dynamic properties of vibration protection systems on the basis of use of additional constraints and a variation of structural constructions are considered.

6.2.1 *Features of the Mechanical System*

If the object of protection—the system—has two degrees of freedom, then the structural diagram in Fig. 6.7, can be reduced to a form, as shown in Fig. 6.8. At the same time, coordinate y_3 as a separate factor of observation is excluded.

Believing that an object has a complex structure, we will find transfer functions:

$$W_1(p) = \frac{\bar{y}_2}{\bar{z}_1} = \frac{k_1k_2(m_3p^2 + k_3 + k_4)}{A_0}, \tag{6.30}$$

where

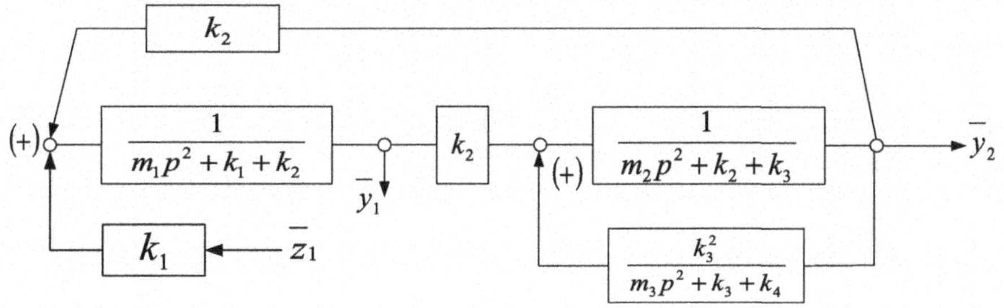

Fig. 6.8 The structural diagram of the vibration protection system with two degrees of freedom (y_1, s_2)

$$A_0 = \left(m_1p^2 + k_1 + k_2\right) \cdot \left[\left(m_2p^2 + k_2 + k_3\right) \cdot \left(m_3p^2 + k_3 + k_4\right) - k_3^2\right] - k_2^2 \cdot \left(m_2p^2 + k_2 + k_3\right) \quad (6.31)$$

is the characteristic equation of the system.

Let us transform the structural diagram in Fig. 6.8 to a form, as shown in Fig. 6.9.

In turn, the structural diagram in Fig. 6.9 can be transformed to a form, as shown in Fig. 6.10.

Let us note that in Fig. 6.10 the second partial system (coordinate y_2), unlike traditional elements, contains the generalized spring in structure of the scheme (or a quasispring) with stiffness

$$k_{\Pi p}(p) = \frac{k_3\left(m_3p^2 + k_4\right)}{m_3p^2 + k_3 + k_4}. \quad (6.32)$$

Quasisprings have stiffness which depends on the frequency of external influence. With frequency

$$\omega_1^2 = \frac{k_4}{m_3} \quad (6.33)$$

such a spring has zero stiffness and the partial system along the coordinate y_2 will work with participation of the elastic element k_2. At frequency

$$\omega_2^2 = \frac{k_3 + k_4}{m_3} \quad (6.34)$$

the stiffness of the spring becomes infinitely big and influence of this partial system on the motion along the coordinate y_1 decreases. Thus, while excluding the coordinate y_3, properties of the system will differ from properties of a usual system with two degrees of freedom.

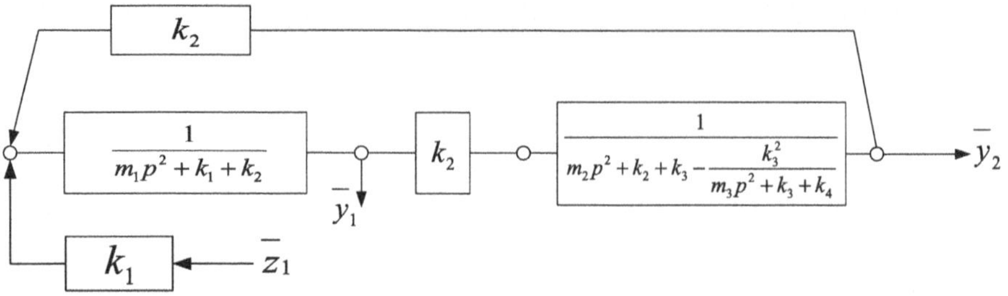

Fig. 6.9 The structural diagram of the system with three degrees of freedom while excluding the coordinate y_3

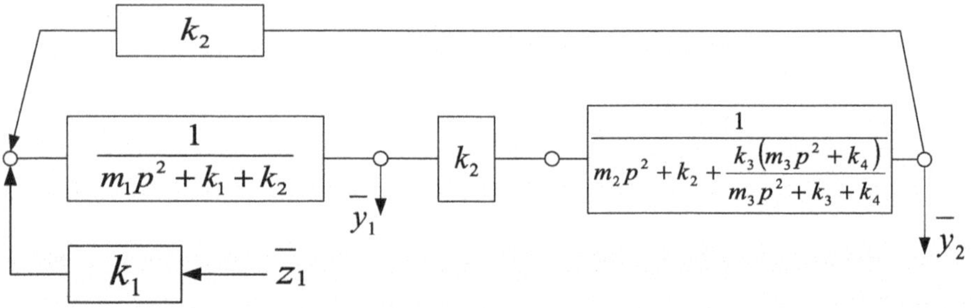

Fig. 6.10 The structural diagram of the system (according to Fig. 6.9, with an additional spring, possessing dynamic stiffness)

Let us note also that at $r = 0$ ($r = j\omega$ is a complex variable [8]) the dynamic stiffness will correspond to the stiffness of the usual spring defined by (6.30)

$$k'_{\text{red}}(p) = \frac{k_3k_4}{k_3 + k_4}. \tag{6.35}$$

At $r \to \infty$ we will obtain respectively:

$$\underset{p\to\infty}{k''_{\text{red}}}(p) = k_3. \tag{6.36}$$

As the stiffness of the spring tends to take on value k_3, the system turns into interaction with a more rigid spring. Features of systems in formation of dynamic properties while excluding one of coordinates with the subsequent reduction of total number of degrees of freedom and emergence of additional constraints in structure of the system are considered.

6.2.2 Methods of Creation of Mathematical Models

The system in Fig. 6.10 has transfer function

$$W(p) = \frac{\bar{y}_2}{\bar{z}_1} = k_1k_2/(m_1p^2 + k_1 + k_2) \cdot \left[m_2p^2 + k_2 + \frac{k_3(m_3p^2 + k_4)}{m_3p^2 + k_3 + k_4}\right] - k_2^2, \tag{6.37}$$

which can be brought to a form

$$W(p) = \frac{\bar{y}_2}{\bar{z}_1} = \frac{k_1 \cdot (m_3p^2 + k_3 + k_4) \cdot k_2}{(m_1p^2 + k_1 + k_2) \cdot [(m_2p^2 + k_2)(m_3p^2 + k_3 + k_4) + k_3(m_3p^2 + k_4)] - k_2^2(m_3p^2 + k_3 + k_4)}. \tag{6.38}$$

From (6.38) it follows that at the kinematic disturbance $z_1 = z_{10}$ sinωt along the coordinate y_2 the mode of dynamic absorbing is possible

$$\omega^2_{\text{dyn}} = \frac{k_3 + k_4}{m_3}. \tag{6.39}$$

In turn, from the frequency characteristic equation, i.e. the denominator (6.38), is follows that in system along the coordinate y_2 the existence of three modes of resonance is possible. Thus, the exclusion of the coordinate does not mean that dynamic properties of the system will change. The distinction consists in reduction of number of observed coordinates. At the same time a generalized spring (or a quasispring) with the stiffness determined by expression (6.30) is included in the elements of a mechanical oscillatory system. At such approach it is necessary to take into account the circumstance that the form of the partial system itself changes, i.e. the partial system can represent more complex formation. In essence, this partial system is a system with two degrees of freedom. So, from the structural diagram in Fig. 6.10 one can find the transfer function of a partial system along the coordinate y_2

$$W''(p) = \frac{(m_2p^2 + k_2) \cdot (m_3p^2 + k_3 + k_4) + k_3 \cdot (m_3p^2 + k_4)}{m_3p^2 + k_3 + k_4}. \tag{6.40}$$

Properties of the partial system will change appropriately.

As well as in simple forms, in the partial system of expanded type there is a resonance mode with the corresponding influence on the general properties of system. At the same time the partial system gets two frequencies at which expression "is zeroed" (6.40). In this case, connection with the motion along the coordinate y_1 breaks up, which corresponds to specific modes of dynamic interactions between partial systems when the motions between coordinates y_1 and y_2 are not connected.

The developed approach allows us to formulate a statement of problems of vibration protection and vibration insulation when the object of protection can represent a system with two degrees of freedom; at the same time a part of coordinates is excluded, and the external forces acting on the excluded coordinates "are reduced" to the observed motions.

The features of properties: possibilities of creating symmetric structures.

We believe that the mathematical model of the system with three degrees of freedom is characterized by the system of the equations in the field of Laplace's transformations:

$$a_{11}\bar{y}_1 + a_{12}\bar{y}_2 + a_{13}\bar{y}_3 = \overline{Q}_1; \tag{6.41}$$

$$a_{21}\bar{y}_1 + a_{22}\bar{y}_2 + a_{23}\bar{y}_3 = \overline{Q}_2; \tag{6.42}$$

$$a_{31}\bar{y}_1 + a_{32}\bar{y}_2 + a_{33}\bar{y}_3 = \overline{Q}_3. \tag{6.43}$$

Consecutive exclusion of coordinates $\bar{y}_1$, $\bar{y}_2$, $\bar{y}_3$ from the Eqs. (6.41)–(6.43) can be carried out only in a certain sequence, otherwise structural diagrams obtained on the basis of transformations (6.41)–(6.43) will not have symmetry in cross-couplings between partial systems. Let us note that, in view of a computational scheme of the initial mechanical system of the same form, as in Fig. 6.7a, coefficients of the system of the equations (6.41)–(6.43) a_{11}, a_{22}, a_{33} reproduce properties of the corresponding partial systems:

$$a_{11} = m_1p^2 + k_1 + k_2;\ a_{22} = m_2p^2 + k_2 + k_3;\ a_{33} = m_3p^2 + k_3 + k_4.$$

It was reflected in creation of the structural diagram provided in Fig. 6.7b. To obtain more exact results in the subsequent calculations, we will believe that elastic constraint k_{13} is considered between the coordinates y_1 and y_3. In this case coordinates of the system of the equations (6.41)–(6.43) will take a form

$$\begin{aligned} &a_{11} = m_1p^2 + k_1 + k_2 + k_{13};\ a_{22} = m_2p^2 + k_2 + k_3;\\ &a_{33} = m_3p^2 + k_3 + k_4 + k_{13};\ a_{12} = a_{21} = k_2;\ a_{23} = a_{32} = k_3;\\ &a_{13} = a_{31} = k_{13}. \end{aligned}$$

The type of transfer functions of partial systems along the coordinates y_1 and y_3 will change appropriately.

Assuming that from (6.41):

$$\bar{y}_1 = \frac{1}{a_{11}}\left(\overline{Q}_1 - a_{12}\bar{y}_2 - a_{13}\bar{y}_3\right) \tag{6.44}$$

and substituting (6.44) into (6.42) and (6.43), we will obtain a system of two equations of motion in the coordinates $\bar{y}_2$ and $\bar{y}_3$:

$$\bar{y}_2\left(a_{22}-\frac{a_{12}^2}{a_{11}}\right)+\bar{y}_3\left(a_{23}-\frac{a_{13}\cdot a_{21}}{a_{11}}\right)=\bar{Q}_2-\bar{Q}_1\frac{a_{12}}{a_{11}}; \quad (6.45)$$

$$\bar{y}_2\left(a_{32}-\frac{a_{31}a_{12}}{a_{11}}\right)+\bar{y}_3\left(a_{33}-\frac{a_{13}^2}{a_{11}}\right)=\bar{Q}_3-\bar{Q}_1\frac{a_{31}}{a_{11}}. \quad (6.46)$$

On the basis of the system of the equations (6.45) and (6.46) the structural diagram (Fig. 6.11) can be constructed.

To prove that the exclusion of $\bar{y}_1$ is necessary to be introduced only with using the transformations made respectively not on the basis of (6.41) but of Eqs. (6.42) and (6.43), we will respectively consider in Fig. 6.12 structural diagrams of the system with the excluded coordinates $\bar{y}_1$.

From structural diagrams in Figs. 6.11 and 6.12 it is seen that the transformations connected with the exclusion of y_1 and based on use of the Eqs. (6.42) and (6.43) yield negative result; the systems lose properties of symmetry of constraints. Thus, the assumption on rules of the choice of the equations for an exclusion of coordinates above made is proved by transformations and the corresponding structural constructions in Fig. 6.12.

Coordinates y_2 and y_3 can be excluded in a similar way, which is presented by the systems of the equations to describe the motion along the coordinates y_1 and y_3 (at this, the coordinate y_2 is excluded):

$$\bar{y}_1\left(a_{11}-\frac{a_{12}a_{21}}{a_{22}}\right)+\bar{y}_3\left(\frac{a_{13}-a_{12}a_{23}}{a_{22}}\right)=-\bar{Q}_1-\bar{Q}_2\frac{a_{12}}{a_{22}}; \quad (6.47)$$

$$\bar{y}_1\left(a_{31}-\frac{a_{32}a_{21}}{a_{22}}\right)+\bar{y}_3\left(a_{33}-\frac{a_{12}a_{23}}{a_{22}}\right)=\bar{Q}_3-\bar{Q}_2\frac{a_{32}}{a_{22}}. \quad (6.48)$$

Respectively, the structural diagram has a form, as shown in Fig. 6.13.

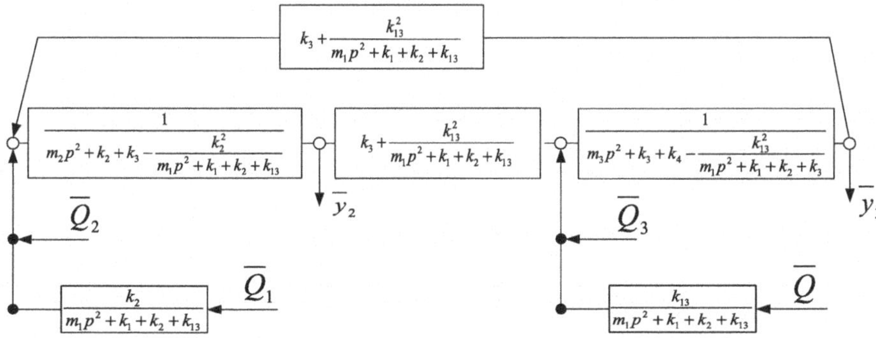

Fig. 6.11 The structural diagram of the system at $a_{13}=a_{31}\neq 0$ and the exclusion of $\bar{y}_2$

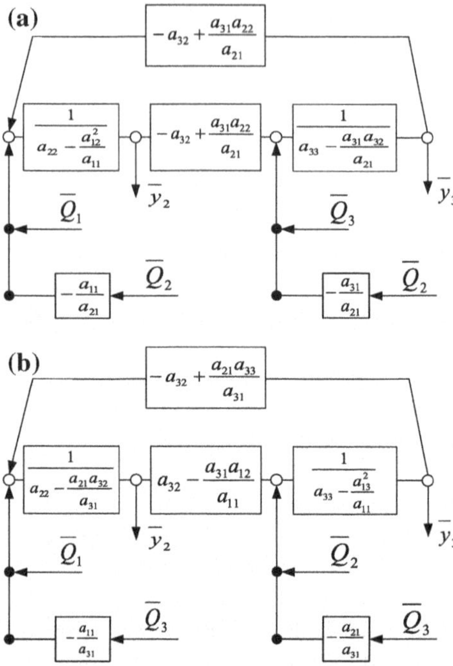

Fig. 6.12 Structural diagrams in coordinates y_2 and y_3: **a** is y_1 from the Eq. (6.42); **b** is y_1 from the Eq. (6.43)

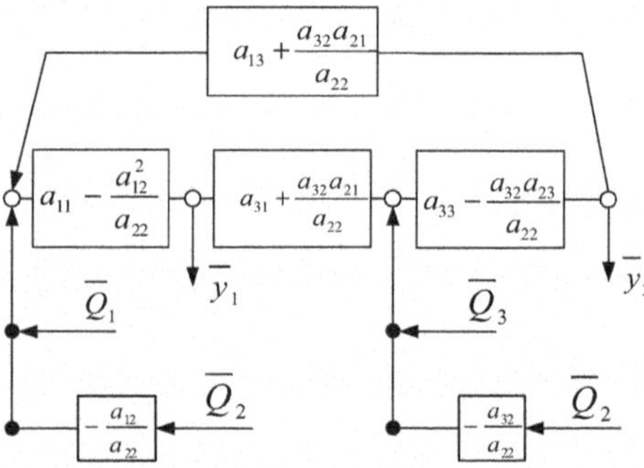

Fig. 6.13 The structural diagram of the system in coordinates y_1 and y_3 (with a coordinate exclusion y_2)

Let us note that while excluding the coordinate y_1 the expression (6.44) was applied. Respectively, for the coordinate y_2 exclusion the Eq. (6.42) is used, from where it follows that

$$\bar{y}_2 = \frac{\bar{Q}_2 - a_{21} \cdot \bar{y}_1 - a_{31} \cdot \bar{y}_3}{a_{22}}. \tag{6.49}$$

Comparing (6.44) and (6.49) allows us to define that the exclusion of the intermediate coordinate is carried out with the application of the corresponding equations giving the opportunity to transfer the operator a_{11} and also a_{22} to a denominator with ratio (6.44) and (6.49).

The same can be obtained while excluding the coordinate y_3. From (6.43) it follows that

$$\bar{y}_3 = \bar{Q}_3 - \frac{a_{31} \cdot \bar{y}_1 - a_{32} \cdot \bar{y}_2}{a_{33}}. \tag{6.50}$$

Substituting (6.50) into (6.41) and (6.42) with the corresponding transformations allows us to find the required system of the equations of motion.

Excluding the coordinate y_3, we will obtain a mathematical model:

$$\bar{y}_1\left(a_{11} - \frac{a_{13}a_{31}}{a_{33}}\right) + \bar{y}_2\left(a_{21} - \frac{a_{13}a_{32}}{a_{33}}\right) = \bar{Q}_1 - \bar{Q}_3\frac{a_{13}}{a_{33}}, \tag{6.51}$$

$$y_1\left(a_{31} - \frac{a_{32}a_{21}}{a_{22}}\right) + y_3\left(a_{33} - \frac{a_{12}a_{23}}{a_{22}}\right) = \bar{Q}_3 - \bar{Q}_2\frac{a_{32}}{a_{22}}. \tag{6.52}$$

On the basis of (6.51) and (6.52) we can respectively construct a structural diagram (Fig. 6.14).

Thus, the exclusion of variables in systems with several degrees of freedom leads to the fact that the "simplified" system (by number of degrees of freedom), keeps all properties of the oscillatory systems displayed by structural diagrams [28]. It should be noted that "simplification" leads to introduction of new concepts which are connected with ideas of a possibility of use of units of new type. Such units are constructed as some blocks, structures or compacts of the known standard elements in the form of mass-and-inertia units, springs, dampers and also motion transformation devices on the basis of rules of structural transformations and the theory of automatic control. In the considered cases such formations are a part of structural diagrams in Figs. 6.10, 6.11, 6.12, 6.13 and 6.14, and the analysis shows that complex or compound units of similar look, except springs, can also include other standard elements into their structure, finally forming a compact or a quasispring. The last names seem to be quite appropriate, as the considered formations are the

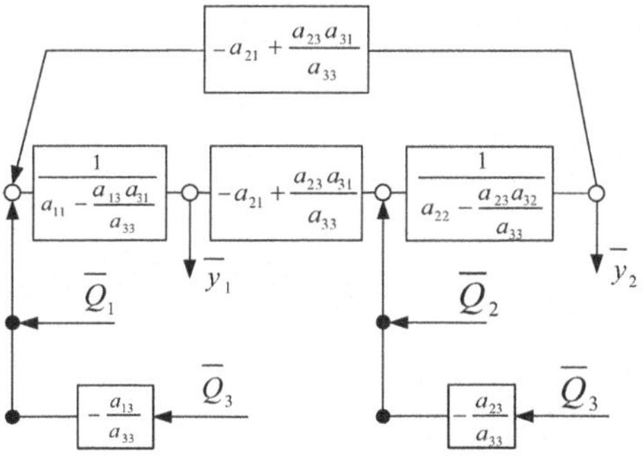

Fig. 6.14 The structural diagram in the coordinates y_1, y_2 (when excluding the coordinate y_3)

reduced springs with the dynamic stiffness depending on the frequency of the excited oscillations. The concept of a "quasispring" is connected, in particular, with the fact that the quasispring, if to mean its opportunity to be a part of blocks, behaves as a usual standard element (a spring, a damper, etc.). Some details of these approaches are stated in the work of Eliseev and others [29].

The transformations given above reflect some general properties of mechanical oscillatory systems based on ideas that partial systems are defined as necessary stages of creating more complex structures, whose algorithm of development relies on the Dalamber principles. Properties of symmetry of couplings between partial systems are also connected with it, in particular. Statement of such tasks can be found, for example, in [30].

6.2.3 Options of Display of Quasisprings in Computational Schemes Mechanical Oscillatory Systems

Believing that Fig. 6.7a represents a basic computational scheme of mechanical oscillatory system with three degrees of freedom, it is possible to show that the exclusion of coordinates is connected with introduction of new structural elements.

The computational scheme developed in Fig. 6.15a contains an additional element—the spring k_{13} connecting mass-and-inertia elements m_1 and m_3 which can be transformed as the spring k_{13} works in parallel connection (Fig. 6.15c) with a chain of k_3, m_2, k_2. At the same time such parallel connection can be transformed or curtailed into a quasispring with the dynamic stiffness determined by expression:

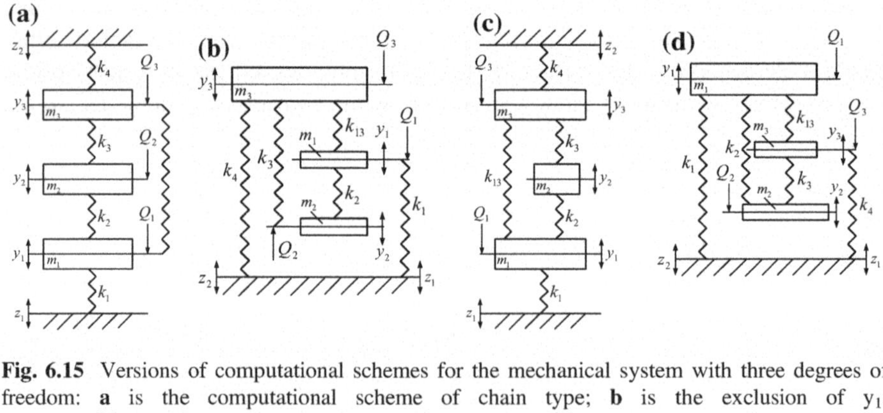

Fig. 6.15 Versions of computational schemes for the mechanical system with three degrees of freedom: **a** is the computational scheme of chain type; **b** is the exclusion of y_1: $k_3 + \frac{k_2 k_{13}}{m_1 p^2 + k_1 + k_2 + k_{13}}$; **c** is the exclusion of y_2: $k_{13} + \frac{k_3 k_2}{m_1 p^2 + k_2 + k_3}$; **d** is the exclusion of y_3: $k_2 + \frac{k_{13} k_3}{m_3 p^2 + k_3 + k_4 + k_{13}}$

$$k_{\mathrm{red}_i} = k_{13} + \frac{k_2 k_3}{m_1 p^2 + k_2 + k_{13}}. \quad (6.53)$$

A version of the computational scheme according to Fig. 6.15a corresponds to a scheme in Fig. 6.15c. The computational scheme in Fig. 6.15c in this case will take a form, as shown in Fig. 6.16b. Transformations of the system lead to the fact that external forces change in the computational scheme (they are reduced to coordinates). Besides, quasisprings emerge in the scheme. These are new parallel additional springs (in relation to k_4) between the supporting surface II and mass m_3 (Fig. 6.16b).

The transfer function of the quasispring has a form of

$$W_3(p) = \frac{m_2 p^2 \cdot k_3}{m_2 p^2 + k_2 + k_3}. \quad (6.54)$$

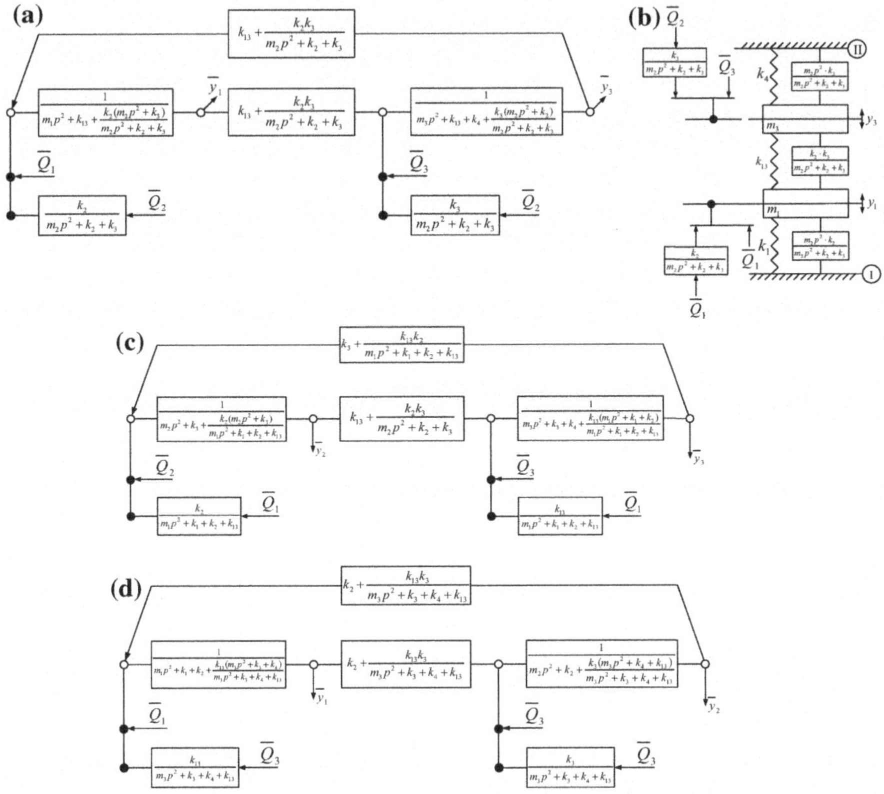

Fig. 6.16 Structural diagrams of the system with the computational scheme according to Fig. 6.7a in the coordinates y_1 and y_3 (**a**) and with the computational scheme with quasisprings (**b**), structural diagrams of systems along the coordinates y_2, y_3 (**c**) and to coordinates y_1 and y_2 (**d**)

Between a mass-and-inertia element m_1 and a supporting surface *I* the quasispring (Fig. 6.16b) with transfer function also appears

$$W_1(p) = \frac{m_2p^2 \cdot k_2}{m_2p^2 + k_2 + k_3}. \tag{6.55}$$

Let us note also that changes happen in transfer function of interpartial constraint (Fig. 6.16b) which takes a form

$$W(p) = \frac{k_3 \cdot k_2}{m_2p^2 + k_2 + k_3}. \tag{6.56}$$

Between the structural diagram (Fig. 6.16a) and its computational scheme (Fig. 6.16b) schemes there can be noted one-to-one correspondence. Parameters of the structural diagram according to Fig. 6.16a can be obtained by the direct transformations based on rules of the theory of mechanical chains or the structural theory of vibration protection systems. Figure 6.16c shows structural diagrams of systems in coordinates y_2, y_3 (Fig. 6.16c) and y_1, y_2 (Fig. 6.16d).

From the block and computational diagrams provided in Fig. 6.16 it is visible that quasisprings introduce parallel connections with the standard elements k_1–k_4. In this case, necessary transformations are possible. In particular, it is possible to show compliance of the structural diagram in Fig. 6.16a to the structural diagram in Fig. 6.8. Both the diagrams reflect properties of systems with two degrees of freedom. In Fig. 6.8, the existence of k_2 in transfer functions of partial systems and also of the element k_2 in a direct chain is characteristic of the structural diagram. Similar can be observed in the structural diagram provided in Fig. 6.16a and also in the computational scheme in Fig. 6.16, where a unit with parameters $k_{13} + \frac{k_2k_3}{m_2p^2 + k_2 + k_3}$ is distinguished. As for partial systems, the corresponding elements are easily distinguished with simple transformations.

6.2.4 *Influence of Additional Elasticities*

While solving a number of the tasks connected with assessment of influence of different types of connection of elastic elements, there arise problems of accounting of features and forms of interactions. Figure 6.17 shows the computational scheme with three degrees of freedom which is similar to the computational scheme in Fig. 6.7a but differs in a set of elastic elements.

In Table 6.1, coefficients of the system of the equations of motion are specified in the coordinates y_1, y_2, y_3. It follows from Table 6.1 that the introduced springs k_{01}, k_{04} do not influence directly the interpartial constraints determined by stiffnesses of elastic elements k_2, k_3, k_{13}. The noted features are important when carrying out transformations of structural diagrams, including the exclusion of variables.

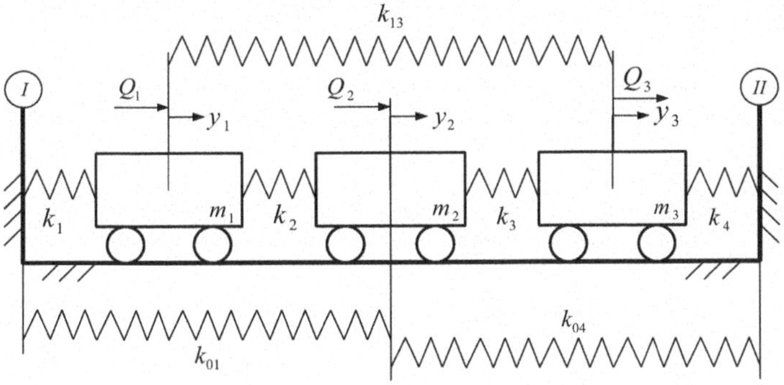

Fig. 6.17 The computational scheme of the system with three degrees of freedom with additional elastic elements

Table 6.1 Coefficients of the equations of motion in the coordinates y_1, y_2, y_3

a_{11}	a_{12}	a_{13}
$m_1p^2 + k_1 + k_2 + k_{13}$	$-k_2$	$-k_{13}$
a_{21}	a_{22}	a_{23}
$-k_2$	$m_2p^2 + k_2 + k_3 + k_{01}$	$-k_3$
a_{31}	a_{32}	a_{33}
$-k_{13}$	$-k_3$	$m_3p^2 + k_3 + k_4 + k_{04}$

Note Q_1, Q_2, Q_3 are the generalized forces along the coordinates y_1, y_2, y_3

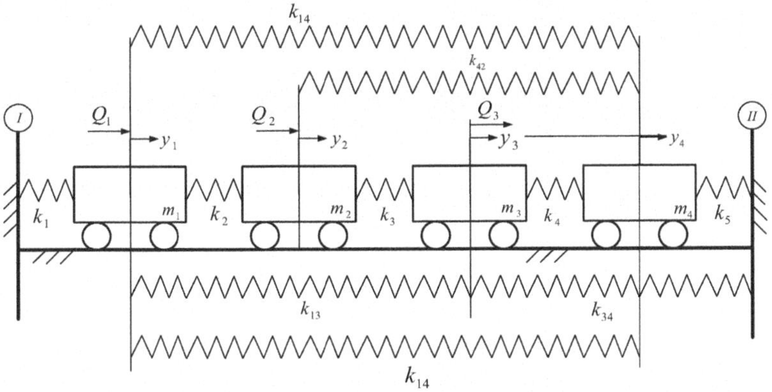

Fig. 6.18 The computational scheme of the system with four degrees of freedom

Let us review a more complex example in the form of a computational scheme of the system with four degrees of freedom having broader set of elastic constraints (Fig. 6.18).

Table 6.2 Coefficients of the equations of motion in the coordinates y_1–y_4

a_{11}	a_{12}	a_{13}	a_{14}
$m_1p^2 + k_1 + k_2 + k_{13} + k_{14}$	$-k_2$	$-k_{13}$	$-k_{14}$
a_{21}	a_{22}	a_{23}	a_{24}
$-k_2$	$m_2p^2 + k_2 + k_3 + k_{42}$	$-k_3$	$-k_{42}$
a_{31}	a_{32}	a_{33}	a_{34}
$-k_{13}$	$-k_3$	$m_3p^2 + k_3 + k_4 + k_{13} + k_{35}$	$-k_4$
a_{41}	a_{42}	a_{43}	a_{44}
$-k_{14}$	$-k_{42}$	$-k_4$	$m_4p^2 + k_4 + k_5 + k_{42} + k_{41}$

In Table 6.2 coefficients of the equation of motion (linear system) are specified in the coordinates y_1–y_4.

From Table 6.2 it follows that, as well as in the previous case, introduction of additional constraints leads to the filling of the matrix and it does not take a form of banded structure, i.e. the mechanical system is not chain, and there are non-planar constraints in it, which makes transformations more bulky. As for additional constraints in general, they are subdivided into two forms: constraints between the coordinates (k_{13}, k_{14}, k_{42}) and between coordinates and supporting surfaces (k_{35}). Let us note that constraints of the first kind are included into structure of the partial system (a_{11}, a_{22}, a_{33}, a_{44}) and also take positions in the matrix which define interpartial constraints.

Constraints of the second kind are, for example, k_{35}, and in general there can also be any other constraints with supporting surfaces, but such constraints are included only into partial blocks (in our case, a_{33}).

The exclusion of coordinates in mechanical oscillatory systems with several degrees of freedom is a formal technique of simplification of systems which leads to reduction of number of variables, but do not change the mathematical model in general. At the same time the offered approaches can be useful in several directions. The first of them is connected with opportunities of representation of partial systems in the form of more complex structures having a form of systems with two degrees of freedom. It allows us to consider problems of vibration protection and vibration insulation of machines and the equipment in a more complex formulation providing control of a dynamic state of the system on two independent variables. At the same time, possibilities of using techniques, methods and means of the structural theory of vibration protection systems remain.

The second direction for further applications is connected with the fact that technologies of an exclusion of coordinates (or variables) introduce into consideration the blocks or compacts consisting of the standard elements interconnected by certain rules; it is possible to call the obtained "formation" a quasispring. Such a component behaves as an ordinary spring if to take into consideration rules of its interaction and compatibility when transforming structural diagrams. Actually, the quasispring has the reduced (or dynamic) stiffness depending on the frequency. At the same time quasisprings have the parameters uniting them with ordinary springs. These parameters are shown under conditions: $r \to 0$ or $r \to \infty$.

And at last, the convenience of the offered approaches is based that accounting of dissipative factors and also in addition introduced mass-and-inertia devices or units, is connected with simple methods of addition of the introduced elements in parallel to the available elastic units that is detailed adequately in works on the structural theory of vibration protection systems.

In general, the offered approach can be considered as a methodological basis of a new method of transformation of mechanical oscillatory systems and development of evidential base in the representations that a set of standard units can be expanded at the expense of quasielements which are characterized by a large diversity but have simple principles of construction and the rules of structural transformations.

6.3 Creation of Compacts of Elastic Elements. Interactions and Forms of Connections

Elastic devices in the theory of oscillations and its applications are one of the basic elements providing oscillatory character of a dynamic state of the system. Constructive and engineering forms of implementation of elastic elements are widely presented in many works (for example, [7, 31, 32]). Features of elastic systems are fairly diverse and are represented in the works that are related to taking into account non-linearities and structural damping [33–35]. At the same time many problems of specification of ideas of complex properties of springs still attract attention. In particular, it is defined by the fact that real elements of mechanical systems differ from ideal ones, and physical and mechanical properties of springs in many cases represent manifestation of some integration of properties. Such a perception of elastic elements is especially characteristic in the theory of suspensions of transportation vehicles, problems of creating vibration equipment and also in the problems of protection of machines and equipment from vibration influences [36–38].

With all the attention to the use of elastic elements in solving problems of design, calculation and operation of technical objects, the assessment of the properties of elastic systems turned out to be less studied, that is, the elastic properties of structures formed on the basis of compounds of elastic elements, as such, and in interactions with other elements, in particular. Interactions taking into account the constraints introduced by lever mechanisms [39] are of particular interest. The possibilities of the generalized approach to the assessment of elastic properties of systems which contain heterogeneous elements and constraints are considered.

6.3.1 *The Description of System Properties*

Let us consider a number of standard computational schemes of technical objects, believing that the simplest forms of physical models are mechanical systems with one and two degrees of freedom (Fig. 6.19). As external influence in systems harmonic force Q is analyzed. To describe dynamic properties, the operator method is used that assumes Laplace's transformations in relation to the initial mathematical models in the form of ordinary differential equations with constant coefficients and the subsequent construction of structural diagrams and the corresponding transfer functions.

Consecutive connection of elastic elements with the stiffnesses k_1 and k_2 (Fig. 6.19c) can be considered as the simplification of the system with two degrees of freedom.

Such a simplification (Fig. 6.20) is provided with "zeroing" of the mass-and-inertia element ($m_1 = 0$).

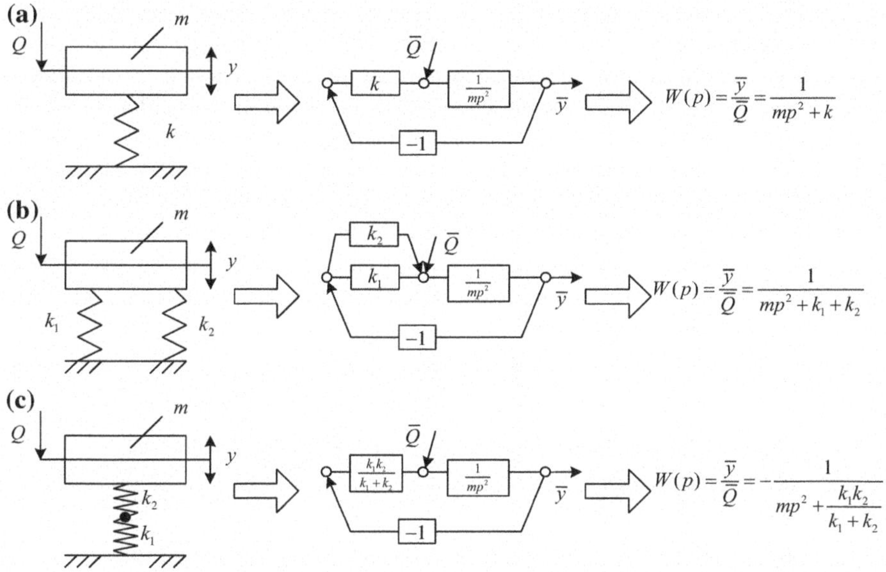

Fig. 6.19 Structural diagrams of representation of elastic elements in systems with one degree of freedom: **a** is the single spring; **b** is the parallel connection of springs; **c** is the consecutive connection of springs

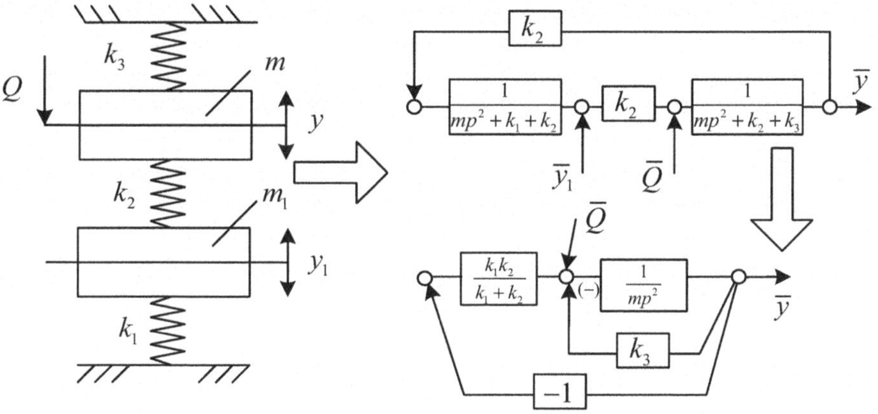

$$W(p) = \frac{\bar{y}}{\bar{Q}} = -\frac{m_1 p^2 + k_1 + k_2}{(m_1 p^2 + k_1 + k_2)(mp^2 + k_2 + k_3) - k_2^2} \rightarrow m_1 = 0$$

$$m_1 = 0 \rightarrow W(p) = \frac{\bar{y}}{\bar{Q}} = -\frac{k_1 + k_2}{(k_1 + k_2)(mp^2 + k_2 + k_3) - k_2^2} = -\frac{1}{mp^2 + k_3 + \dfrac{k_1 k_2}{k_1 + k_2}}$$

Fig. 6.20 The scheme of obtaining consecutive connection of elastic elements k_1 and k_2 by the "zeroing" of the intermediate mass m_1

When considering the connection a larger number of elements, a method similar to that considered above is used. It relies on the definition of transfer function of the system from which, in turn, the required general elasticity is found by simplification. In essence, properties of some complex of elements, in particular, of elastic elements in combination with other standard elements of mechanical systems are defined. Thus, when considering the elastic properties of some complex (or a compact) in the mechanical system, it is possible to distinguish an object, i.e. a weight to which a force is applied, and the rest of the system is considered as a part of the general system, complementary in relation to the object. After determining the corresponding transfer function by simplifications, the necessary static and dynamic characteristics are found.

6.3.2 *Possible Forms of Connection of Elastic Elements into Structures*

Let us note that the transfer function considered above in the form of the relation of transform images [input (force Q) and output (displacement y)] represents dynamic compliance [8]. To obtain dynamic stiffness, it is necessary to invert the transfer function. If the system consists only of ideal springs, then properties of the elastic complex will not depend on the complex variable p. However, if the structure of the elastic complex (or a compact) contains dissipative elements and motion transformation devices called the differentiating units of the first and second order, the dynamic stiffness will depend on the complex variable p. That is, the dynamic stiffness of the system, in the generalized sense, is supposed to be depending on the disturbing force frequency. The transfer function, whose examples of definition are given in Fig. 6.20, gives the chance to find stiffness of the system under static loading. For this purpose, it is necessary to invert the transfer function and to take $p = 0$; the stiffness of the system can thus be considered as the reciprocal of the modulus of the transfer function in its definition done above.

Consideration of elasticity of the system is usually connected with force which is applied to the mass-and-inertia element of the system. In this case the displacement from action of force is defined in the same point. If the force is applied to an elastic element, then situations may be different (see Fig. 6.20).

The elasticity of the system, as is seen from Fig. 6.21, depends on options of a relative positioning of points of application of forces and points of observation of the shift of coordinates, since situations are possible to determine elasticity of the system when a force is applied (for example, Fig. 6.21a) to the mass m_1, and the elasticity of the system is estimated on the y_2 coordinate shift. The structural diagram of the system in Fig. 6.21a is provided in Fig. 6.22.

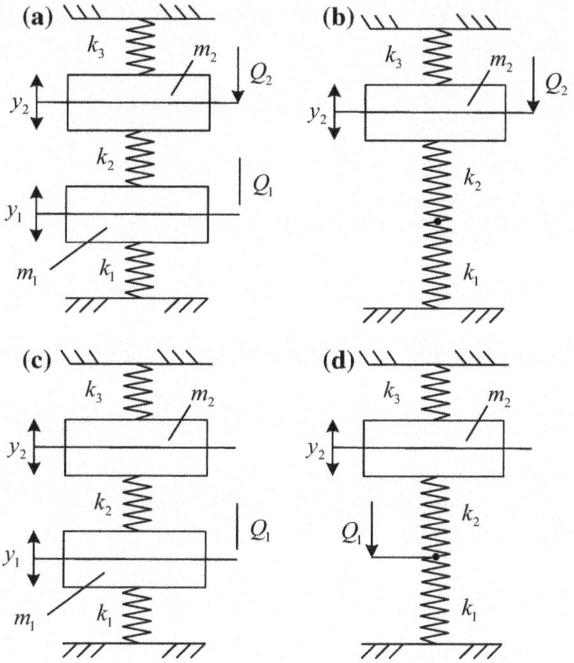

Fig. 6.21 Schemes for consideration of various options of application of forces

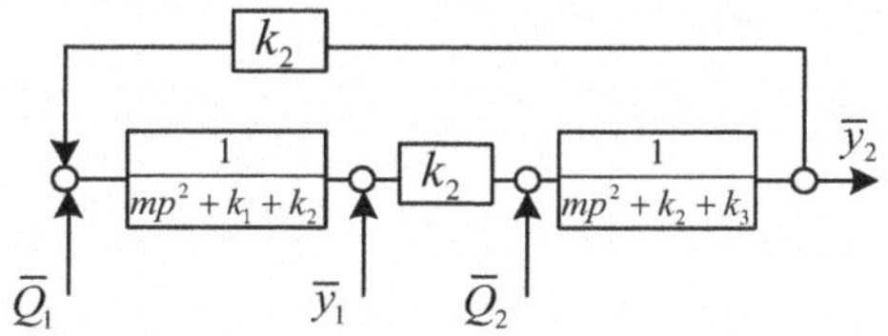

Fig. 6.22 The structural diagram the system corresponding to Fig. 6.21a

From the structural diagram of the system one can define transfer functions $W_1(p) = \frac{\bar{y}_1}{\bar{Q}_1}; W_2(p) = \frac{\bar{y}_1}{\bar{Q}_2}; W_3(p) = \frac{\bar{y}_2}{\bar{Q}_1}; W_4(p) = \frac{\bar{y}_2}{\bar{Q}_2}$, which allows us to make necessary estimates. Let us accept that $Q_1 = 0$, $Q_2 \neq 0$, then

$$W_1(p) = \frac{\bar{y}_2}{\bar{Q}_2} = \frac{m_1p^2 + k_1 + k_2}{A}, \tag{6.57}$$

where

$$A = (m_1p^2 + k_1 + k_2)(m_2p^2 + k_2 + k_3) - k_2^2 \tag{6.58}$$

is the characteristic equation.

From (6.57) it is possible to find static stiffness of the system which will make up (Fig. 6.21b)

$$k' = \frac{k_1 k_2}{k_1 + k_2} + k_3. \tag{6.59}$$

For the case $Q_2 = 0$, $Q_1 \neq 0$

$$W(p) = \frac{\bar{y}_1}{\bar{Q}_1} = \frac{m_2 p^2 + k_2 + k_3}{A}, \tag{6.60}$$

from where we will find that at $m_2 = 0$ (Fig. 6.21c)

$$k'' = k_1 + \frac{k_2 k_3}{k_2 + k_3}. \tag{6.61}$$

If to accept that at $m_1 = 0$ (Fig. 6.21d), then

$$\begin{aligned} W(p) = \frac{\bar{y}_2}{\bar{Q}_1} &= \frac{k_2}{(m_2 p^2 + k_2 + k_3)(k_1 + k_2) - k_2^2} \\ &= \frac{k_2}{(m_2 p^2 + k_2 + k_3)k_1 + m_2 p^2 + k_2 k_3}, \end{aligned} \tag{6.62}$$

from where

$$k''' = \frac{k_2}{k_2 k_1 + k_3 k_1 + k_2 k_3}, \quad k''' = k_3 + k_1 + \frac{k_3 k_1}{k_2}. \tag{6.63}$$

If in Fig. 6.21d to accept that $k_3 = 0$, then $k''' = k_1$.
In turn, in Fig. 6.21, *a* we will accept that $m_2 = 0$, $k_3 = 0$, $Q_2 \neq 0$, $m_1 \neq 0$.

$$W(p) = \frac{\bar{y}_1}{\bar{Q}_2} = \frac{k_2}{(k_2 + k_3)(m_1 p^2 + k_1 + k_2) - k_2^2}, \tag{6.64}$$

Whence it follows that

$$k^{IV} = k_1 + k_3 + k_1 k_3, \tag{6.65}$$

from where at $k_3 = 0$ we will find $k^{\mathrm{IV}} = k_1$.

This case demonstrates what while applying the force Q_2 immediately to the spring k_2 corresponds to the direct application of Q_2 to the mass m_1. The given outcome is in accordance to results [17].

6.3.3 The System with Three Degrees of Freedom

Let us consider a more complex structure consisting of three mass-and-inertia elements forming a mechanical chain (Fig. 6.23).

In the analysis of the chain mechanical system (Fig. 6.23) a constraint between m_1, m_2, m_3 is provided with elastic elements $k_2 \neq 0$, $k_3 \neq 0$. If $k_1 = 0$ and $k_4 = 0$, then the system is open and has a cyclic coordinate. Formation of a certain motion requires the connection of one of the elements m_1 and m_3 with a motionless basis. Then the following situations are possible $k_1 = 0$, $k_4 \neq 0$, $k_1 \neq 0$, $k_4 = 0$, $k_1 \neq 0$, $k_4 \neq 0$. In the latter case $k_1 \neq 0$, $k_4 \neq 0$; the chain in Fig. 6.23 can be considered as closed. Figure 6.24 shows some configurations of closed mechanical chains.

To construct mathematical models in Fig. 6.22 we will introduce an additional elastic constraint k_{13} and will write down expressions for kinetic and potential energy:

$$\mathrm{T} = \frac{1}{2}m_1\dot{y}_1^2 + \frac{1}{2}m_2\dot{y}_2^2 + \frac{1}{2}m_3\dot{y}_3^2, \tag{6.66}$$

$$\Pi = \frac{1}{2}k_1y_1^2 + \frac{1}{2}k_2(y_2 - y_1)^2 + \frac{1}{2}k_3(y_3 - y_2)^2 + \frac{1}{2}k_4y_3^2 + \frac{1}{2}k_{13}(y_3 - y_1)^2. \tag{6.67}$$

The equations of motion of the system given in Fig. 6.24 are of the form of

$$m_1\ddot{y}_1 + k_1y_1 + k_2y_1 + k_{13}y_1 - k_2y_2 - k_{13}y_3 = Q_1; \tag{6.68}$$

$$m_2\ddot{y}_2 + k_2y_2 + k_2y_3 + k_2y_1 - k_3y_2 = Q_2; \tag{6.69}$$

$$m_3\ddot{y}_3 + k_3y_3 + k_{13}y_3 + k_4y_3 - k_{13}y_1 - k_{13}y_1 - k_3y_2 = Q_3. \tag{6.70}$$

Coefficients of the equations (6.68)–(6.70) are specified in Table 6.3.

A mathematical model for the system in Fig. 6.24 can be constructed a little differently. In this case the kinetic energy corresponds to the expression (6.66), and the expressions for the potential energy, respectively, are of the form of

$$\Pi = \frac{1}{2}k_1y_1^2 + \frac{1}{2}k_{13}(y_3 - y_1)^2 + \frac{1}{2}k_4y_3^2 + \frac{1}{2}k_2(y_2 - y_1)^2 + \frac{1}{2}k_3(y_3 - y_2)^2; \tag{6.71}$$

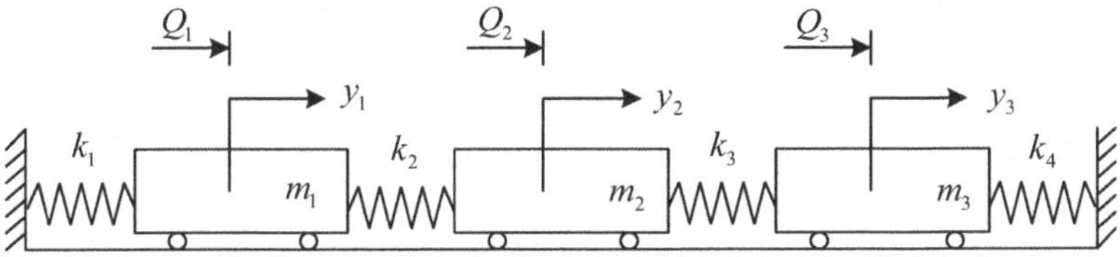

Fig. 6.23 The mechanical chain with three mass-and-inertia elements

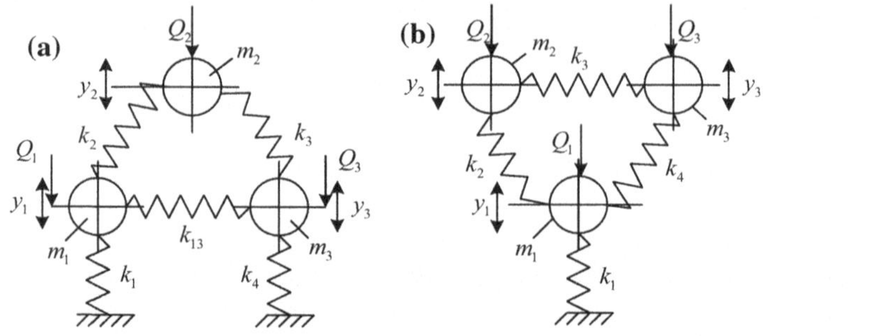

Fig. 6.24 Mechanical systems with three degrees of freedom in the presence of closed contours: **a** while introducing the elastic constraint k_{13}; **b** is the introduction of three elastic constraints k_1, k_2, k_4 on one element m_1

Table 6.3 Coefficients of the system of equations in the coordinates y_1, y_2, y_3

a_{11}	a_{12}	a_{13}	Variants
$m_1p^2 + k_1 + k_2 + k_{13}$	$-k_2$	k_{13}	Figure 6.24 $k_{13} \neq 0$
$m_1p^2 + k_1 + k_2$	$-k_2$	0	$k_{13} = 0$
a_{21}	a_{22}	a_{23}	Variant
$-k_2$	$m_2p^2 + k_2 + k_3$	$-k_3$	Figure 6.24 $k_{13} \neq 0$
$-k_2$	$m_2p^2 + k_2 + k_3$	$-k_3$	$k_{13} = 0$
a_{31}	a_{32}	a_{33}	Variant
$-k_{13}$	$-k_3$	$m_3p^2 + k_3 + k_4 + k_{13}$	Figure 6.24 $k_{13} \neq 0$
0	$-k_3$	$m_1p^2 + k_3 + k_4$	$k_{13} = 0$

Note Table 6.3 can be used for cases of an open chain ($k_4 = 0$) and a free motion ($k_4 = 0$, $k_4 = 0$)

$$\Pi = \frac{1}{2}k_1y_1^2 + \frac{1}{2}k_2(y_2 - y_1)^2 + \frac{1}{2}k_3(y_3 - y_2)^2 + \frac{1}{2}k_4(y_3 - y_2)^2. \tag{6.72}$$

Table 6.4 gives coefficients of the equations of motion for the system in Fig. 6.24a.

In turn, the system in Fig. 6.24b has the system of the equations whose coefficients are presented in Table 6.5.

It follows from the analysis of Tables 6.3, 6.4 and 6.5 that structural diagrams of all systems in Figs. 6.22 and 6.24 will be different, as well as the system transfer functions used to assess elastic properties under static and dynamic loadings. Cases of joint loading on several coordinates are not considered.

The corresponding structural diagrams are provided in Fig. 6.25 where changes of partial systems and cross-couplings between them are shown.

Table 6.4 Coefficients of the equations of motion for the system in Fig. 6.24a

a_{11}	a_{12}	a_{13}
$m_1p^2 + k_1 + k_2 + k_{13}$	$-k_2$	$-k_{13}$
a_{21}	a_{22}	a_{23}
$-k_2$	$m_2p^2 + k_2 + k_3$	$-k_3$
a_{31}	a_{32}	a_{33}
$-k_{13}$	$-k_3$	$m_3p^2 + k_3 + k_4 + k_{13}$

Table 6.5 Coefficients of the equations of motion for the system in Fig. 6.24b

a_{11}	a_{12}	a_{13}
$m_1p^2 + k_1 + k_2$	$-k_2$	0
a_{21}	a_{22}	a_{23}
$-k_2$	$m_2p^2 + k_2 + k_3 + k_4$	$-k_3$ to k_4
a_{31}	a_{32}	a_{33}
0	$-k_3$ to k_4	$m_3p^2 + k_3 + k_4$

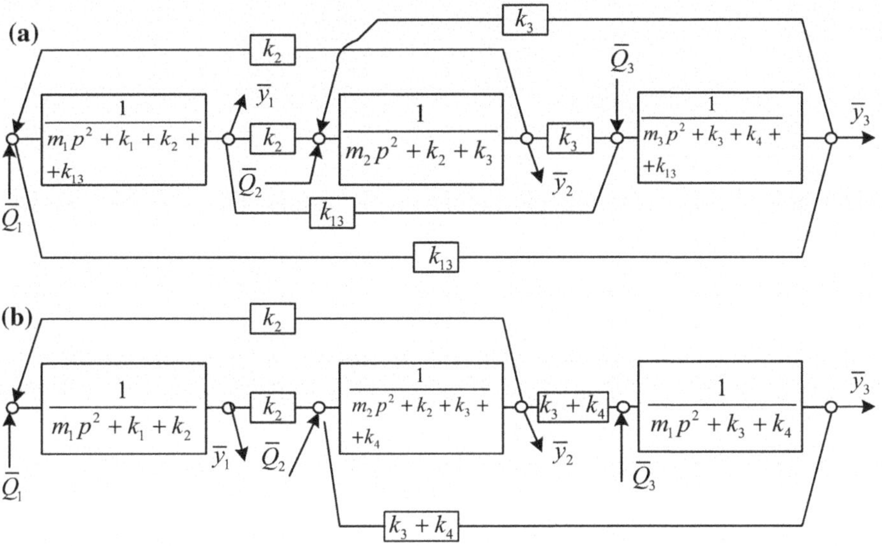

Fig. 6.25 Structural diagrams of systems: a the diagram corresponds to Fig. 6.22; **b** the diagram corresponds to Fig. 6.24

To find transfer functions on three coordinates y_1–y_3 from input disturbances Q_1–Q_3, it is possible to use Kramer [17] formulas, as structural transformations are quite bulky.

$$\bar{y}_1 = \frac{Q_1(a_{22}a_{33} - a_{23}^2) + Q_2(a_{13}a_{32} - a_{12}a_{33}) + Q_3(a_{12}a_{23} - a_{13}a_{22})}{A}, \quad (6.73)$$

$$\bar{y}_2 = \frac{Q_1(a_{23}a_{31} - a_{21}a_{33}) + Q_2(a_{11}a_{33} - a_{13}^2) + Q_3(a_{13}a_{21} - a_{11}a_{23})}{A}, \quad (6.74)$$

$$\bar{y}_3 = \frac{Q_1(a_{21}a_{32} - a_{22}a_{31}) + Q_2(a_{12}a_{31} - a_{11}a_{32}) + Q_3(a_{11}a_{22} - a_{12}^2)}{A}, \quad (6.75)$$

where

$$A = a_{11}a_{22}a_{33} - a_{11}a_{23}a_{32} + a_{13}a_{21}a_{32} - a_{33}a_{12}^2 + a_{12}a_{23}a_{31} - a_{22}a_{13}a_{31}$$

is the characteristic equation.

Let us consider a number of the special cases reflecting characteristics of elastic structures. Let us assume that $k_4 = 0$, $Q_2 = 0$, $Q_3 = 0$, $Q_1 \neq 0$, $k_{13} = 0$, then transfer function of the system (Fig. 6.24a) will take a form

$$W_1(p) = \frac{\bar{y}_1}{\bar{Q}_1} = \frac{(k_2 + k_3)(k_3 + k_4 + k_{13}) - k_3^2}{(m_1p^2 + k_1 + k_2 + k_{13})(k_2 + k_3)(k_3 + k_4 + k_{13}) - (m_1p^2 + k_1 + k_2 + k_{13})}$$
$$\cdots \overline{\times k_3^2 + (-k_{13})(-k_2)(-k_3) - (k_3 + k_4 + k_{13})k_2^2 + (-k_{13})(-k_2)(-k_3) - (k_2 + k_3)(-k_{13})^2}. \quad (6.76)$$

Let us make a number of transformations for expression (6.76) and we will obtain the reduced stiffness:

$$k_{\text{red}} = k_1. \quad (6.77)$$

It follows from the analysis that the reduced stiffness of the system is defined by the value of stiffness of the spring k_1. If $k_4 \neq 0$, then the reduced stiffness is defined by expression

$$k'_{\text{red}} = k_1 + \frac{k_2k_3k_4 + k_{13}k_4(k_2 + k_3)}{(k_2 + k_3)(k_{13} + k_4) + k_2k_3}. \quad (6.78)$$

Values of the reduced stiffnesses when accounting the loading of elements m_1, m_2, m_3 are respectively presented in Table 6.6 by forces Q_1, Q_2, Q_3 (options are $k_4 = 0$, $k_4 \neq 0$).

Thus, the reduced stiffness at a point of application of force is formed by elastic constraints which are from two parts of the distinguished mass-and-inertia element. At the same time, the general meaning of the reduced stiffness depends on the features of structure of the chain, in particular, such properties as openness and isolation, which can be compared with abilities of certain reflections if processes are wave-like. Though in the theory of elastic oscillations of bodies, for example, with

Table 6.6 Values of reduced stiffness coefficients

$k'_{red}(Q_1 \neq 0, Q_2 = 0, Q_3 = 0)$	$k''_{red}(Q_1 = 0, Q_2 \neq 0, Q_3 = 0)$	k'''_{red} *for* $y_3(Q_1 = 0, Q_2 = 0, Q_3 \neq 0)$	Variants
$k_1 + \frac{k_2k_3k_4 + k_{13}k_4(k_2+k_3)}{(k_2+k_3)(k_{13}+k_4)+k_2k_3}.$	$\frac{k_1(k_2k_3+(k_2+k_3)(k_{13}+k_4))+k_2k_3k_4+k_{13}k_4(k_2+k_3)}{(k_1+k_2)(k_3+k_4+k_{13})+k_{13}(k_3+k_4)}$	$\frac{k_1(k_2k_3+(k_2+k_3)(k_{13}+k_4))+k_2k_3k_4+k_{13}k_4(k_2+k_3)}{k_2k_3+(k_1+k_{13})(k_2+k_3)}$	Figure 6.24a $k_4 \neq 0$
k_1	$\frac{k_1(k_2k_3+(k_2+k_3)k_{13})}{(k_1+k_2)(k_3+k_{13})+k_{13}k_3}$	$\frac{k_1(k_2k_3+(k_2+k_3)k_{13})}{k_2k_3+(k_1+k_{13})(k_2+k_3)}$	$k_4 = 0$
	$k_{13} = 0$		Variants
$k_1 + \frac{k_2k_3k_4 + k_4(k_2+k_3)}{(k_2+k_3)k_4+k_2k_3}$	$\frac{k_1(k_2k_3+(k_2+k_3)k_4)+k_2k_3k_4}{(k_1+k_2)(k_3+k_4)}$	$\frac{k_1(k_2k_3+(k_2+k_3)k_4)+k_2k_3k_4}{k_2k_3+k_1(k_2+k_3)}$	Figure 6.24a $k_4 \neq 0$
k_1	$\frac{k_1k_2k_3}{(k_1+k_2)k_3}$	$\frac{k_1k_2k_3}{k_2k_3+k_1(k_2+k_3)}$	$k_4 = 0$
	$k_3 = 0, k_{13} = 0$		Variants
$k_1 + \frac{k_4k_2}{k_2+k_4}$	$\frac{k_1k_2k_4}{(k_1+k_2)k_4}$	$\frac{k_1k_2k_4}{k_1k_2}$	Figure 6.24a $k_4 \neq 0$
k_1	0	0	$k_4 = 0$
	$k_3 = 0$		Variants
$k_1 + \frac{k_{13}k_4k_2}{k_2(k_{13}+k_4)}$	$\frac{k_1k_2(k_{13}+k_4)+k_{13}k_4k_2}{(k_1+k_2)(k_4+k_{13})+k_{13}k_4}$	$\frac{k_1k_2(k_{13}+k_4)+k_{13}k_4k_2}{(k_1+k_{13})k_2}$	Figure 6.24a $k_4 \neq 0$
k_1	$\frac{k_1k_2k_{13}}{(k_1+k_2)k_{13}}$	$\frac{k_1k_2k_{13}}{(k_1+k_{13})k_2}$	$k_4 = 0$
	$k_2 = 0$		Variants
$k_1 + \frac{k_{13}k_4k_3)}{k_3(k_{13}+k_4)}$	$\frac{k_1k_3(k_{13}+k_4)+k_{13}k_4k_3}{k_1(k_3+k_4+k_{13})+k_{13}(k_3+k_4)}$	$\frac{k_1k_3(k_{13}+k_4)+k_{13}k_4k_3)}{(k_1+k_{13})k_3}$	Figure 6.24a $k_4 \neq 0$
k_1	$\frac{k_1k_3k_{13}}{k_1(k_3+k_{13})+k_{13}k_3}$	$\frac{k_1k_3k_{13}}{(k_1+k_{13})k_3}$	$k_4 = 0$
	$k_1 = 0$		Variants
$\frac{k_2k_3k_4+k_{13}k_4(k_2+k_3)}{(k_2+k_3)(k_{13}+k_4)+k_2k_3}$	$\frac{k_2k_3k_4+k_{13}k_4(k_2+k_3)}{k_2(k_3+k_4+k_{13})+k_{13}(k_3+k_4)}$	$\frac{k_2k_3k_4+k_{13}k_4(k_2+k_3)}{k_2k_3+k_{13}(k_2+k_3)}$	Figure 6.24a $k_4 \neq 0$

(continued)

Table 6.6 (continued)

$k'_{red}(Q_1 \neq 0, Q_2 = 0, Q_3 = 0)$	$k''_{red}(Q_1 = 0, Q_2 \neq 0, Q_3 = 0)$	k'''_{red} *for* $y_3(Q_1 = 0, Q_2 = 0, Q_3 \neq 0)$	Variants
k_1	0	0	$k_4 = 0$
	$k_1 = 0, k_2 = 0$		Variants
$\frac{k_{13}k_4k_3}{k_3(k_{13}+k_4)}$	$\frac{k_{13}k_4k_3}{k_{13}(k_3+k_4)}$	$\frac{k_{13}k_4k_3}{k_{13}k_3}$	Figure 6.24a $k_4 \neq 0$
0	0	0	$k_4 = 0$
	$k_1 = 0, k_{13} = 0$		Variants
$\frac{k_2k_3k_4}{(k_2+k_3)k_4+k_2k_3}$.	$\frac{k_2k_3k_4}{k_2(k_3+k_4)}$	$\frac{k_2k_3k_4}{k_2k_3}$	Figure 6.24a $k_4 \neq 0$
0	0	0	$k_4 = 0$

Note Cases when $Q_1 \neq 0,\ Q_2 = Q_3 = 0; Q_2 \neq 0,\ Q_1 = Q_3 = 0;\ Q_3 \neq 0,\ Q_1 = Q_2 = 0$ are considered a form of beams, longitudinal oscillations can be considered as wave processes with reflection from the free ends [40].

6.3.4 Features of Models of Chain Systems

Believing that in the scheme (see Fig. 6.24) the elasticities k_1–k_4 are not equal to zero but k_{13}= 0, we will write down consecutive values of the reduced stiffnesses, passing from m_1 to m_3 with the same value of the external force Q, then along the coordinate y_1

$$k_{\text{red}} = k_1 + \frac{k_2 k_3 k_4}{k_2 k_3 + k_3 k_4 + k_2 k_4}; \tag{6.79}$$

along the coordinate y_2

$$k_{\text{red}} = \frac{k_1 k_2}{k_1 + k_2} + \frac{k_3 k_4}{k_3 + k_4}; \tag{6.80}$$

along the coordinate y_3

$$k_{\text{red}} = k_4 + \frac{k_1 k_2 k_3}{k_1 k_2 + k_2 k_3 + k_1 k_3}. \tag{6.81}$$

At $k_4 = 0$ we have respectively:
along the coordinate y_1

$$k = k_{\text{red}1}; \tag{6.82}$$

along the coordinate y_2

$$k_{\text{red}} = \frac{k_1 k_2}{k_1 + k_2}, \tag{6.83}$$

along the coordinate y_3

$$k_{\text{red}} = \frac{k_1 k_2 k_3}{k_1 k_2 + k_2 k_3 + k_1 k_3}. \tag{6.84}$$

If to assume that $k_{13} \neq 0$, then under loading conditions by force Q we will obtain:

along the coordinate y_1

$$k = k_{\text{red}1}, \tag{6.85}$$

along the coordinate y_2

$$k_{\text{red}} = \frac{k_{13}(k_1k_3 + k_1k_2) + k_1k_2k_3}{k_{13}(k_1 + k_2 + k_3) + k_2k_3 + k_1k_3}, \tag{6.86}$$

along the coordinate y_3

$$k_{\text{red}} = \frac{k_{13}(k_1k_2 + k_1k_3) + k_1k_2k_3}{(k_1 + k_{13})(k_2 + k_3) + k_2k_3}. \tag{6.87}$$

Introduction of intercoordinate constraints changes properties of the elastic structure (or the elastic compact). Let us note that when moving a point of application of force with increase in number of elements, the reduced stiffness will decrease. Distinctions of the reduced stiffness for the opened and closed chains are important for creation of mathematical models of distribution of waves.

The peculiarity of the above-stated reasons is that the motion of the system, even in the presence of several degrees of freedom, does not assume spatial forms of interaction and transformation of motions. The algorithm of constructing reduced stiffnesses in static situations is based on application of rules of consecutive and parallel connection of springs, which is characteristic of the theory of circuits in general and of mechanical chains in particular. Structures made of elastic elements (or compacts) can be considered as quasisprings, as the complication of their forms does not change the rule of their connection, creation and transformation. The similar situation arises also when considering problems connected with assessment of dynamic stiffness. However, direct distribution of the above-stated approach has to be conducted taking into account features of systems (linearity, existence of solid bodies and their joints, planarity) [41].

6.3.5 Lever Linkages in Systems with Elastic Elements

Introduction and consideration of lever linkages in mechanical oscillatory systems resulted in the expansion of ideas of properties of systems in static and dynamic states of interaction of the constituent elements [42–44]. At the same time, lever mechanisms in works on the theory of circuits and operator methods of the analysis and synthesis were not considered at the level of standard elements, since ideas of dual elements of chains having two points of a joint did not extend to levers that use three points to ensure interaction with other elements. Works of recent years showed that lever mechanisms fit seamlessly into processes of formation of structures of elastic elements (or compacts).

Let us consider a number of examples of interaction of elastic elements and levers when forming the reduced stiffness of the system (Fig. 6.26).

To create mathematical models of systems, we will use the scheme in Fig. 6.19c, since schemes in Fig. 6.19a, b it would be possible to consider as simplifications of the scheme in Fig. 6.19c. Expressions for kinetic and potential energy have the form of

$$\mathrm{T} = \frac{1}{2} m\dot{y}^2, \tag{6.88}$$

$$\Pi = \frac{1}{2} k(yi)^2 + \frac{1}{2} k_1 y^2 + \frac{1}{2} k_2 y^2, \tag{6.89}$$

where $i = \frac{l_2}{l_1}$.

Let us write down the equation of motion along the coordinate y

$$m_1 \ddot{y} + y(k_1 + k + ki^2) = Q, \tag{6.90}$$

The structural diagram of the system is provided in Fig. 6.27.

According to the structural diagram in Fig. 6.19c, the transfer function will be defined by expression

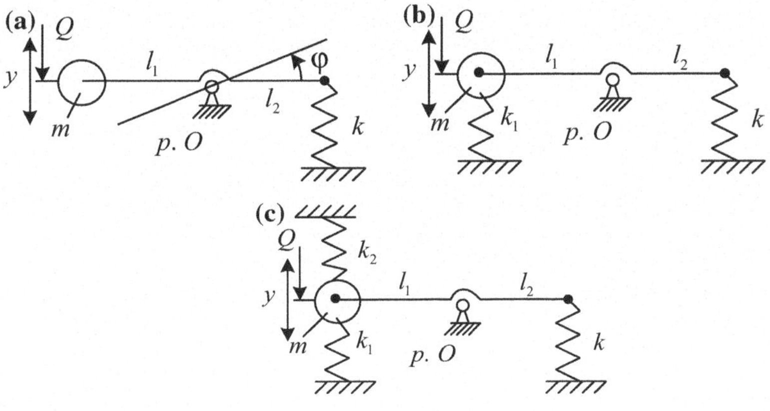

Fig. 6.26 Computational schemes of systems with elastic—lever linkages

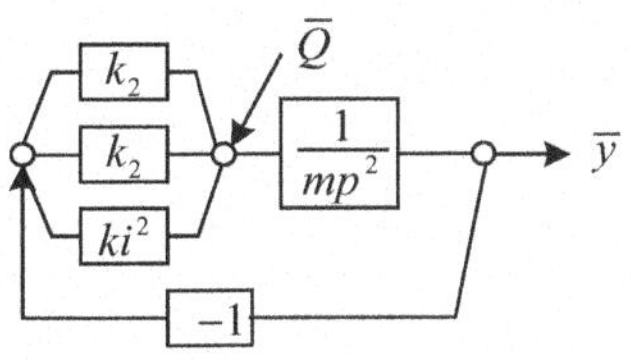

Fig. 6.27 The structural diagram of the mechanical oscillatory system

$$W(p) = \frac{\bar{y}}{\bar{Q}} = \frac{1}{mp^2 + k_1 + k_2 + ki^2}. \tag{6.91}$$

Let us make inversion of (6.91) and we will find that the scheme in Fig. 6.26c can be transformed, as shown in Fig. 6.28.

The reduced stiffness of the system has the form of

$$k = ki2_{\text{red}} + k_1 + k_2. \tag{6.92}$$

From expression (6.92) one can obtain reduced stiffnesses—for the scheme in Fig. 6.26a $k_{red} = ki^2$; for the scheme in Fig. 6.26b $k_{red} = k_1 + ki^2$. Thus, the lever mechanism of the second kind (Fig. 6.26) in the elastic system of the reduced spring (or the elastic compact) is considered through the transfer ratio of the lever $i = l_2/l_1$. In this case, the property of the lever to change the direction of motion is not taken into account, since the coordinate y is included squared into the expression (6.89) for potential energy. Generally for the lever of the second kind the transfer ratio i is taken with the negative sign.

When using lever mechanisms of the first kind, as shown in Fig. 6.29, it is possible to obtain similar results.

According to Fig. 6.21,

$$\mathrm{T} = \frac{1}{2} m\dot{y}^2; \tag{6.93}$$

$$\Pi = \frac{1}{2} k_1 y^2 + \frac{1}{2} k_2 y^2 + \frac{1}{2} k (iy)^2, \tag{6.94}$$

where $i = \frac{l_2}{l_1} = \frac{OB}{OA}$.

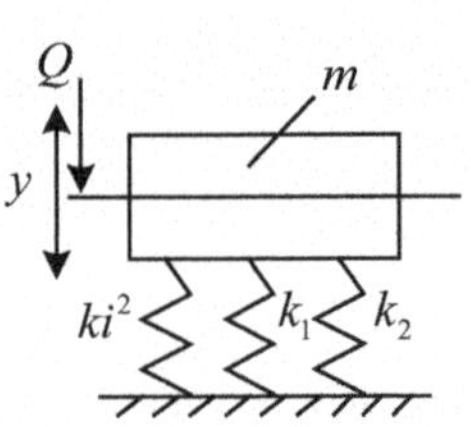

Fig. 6.28 The reduced scheme with lever linkages

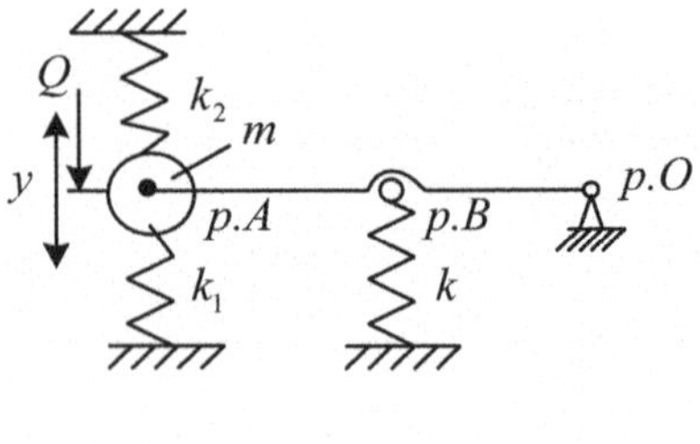

Fig. 6.29 The computational scheme of the system with the lever of the first kind ($OA = l_1$, $OB = l_2$)

The reduced stiffness of the system will have the same value ($k_{red}\ ki^2 + k_1 + k_2$) as in case of the lever of the second kind. Let us note that the transfer ratio of the lever of the first kind has the positive sign, since it does not change the direction of motion. The mentioned distinctions are noted in dynamic interactions, as the type of a lever linkage can also define a type of a feedback in a structural diagram (the feedback can be positive or negative, which influences parameters of transfer function).

The mechanical oscillatory systems consisting of the mass-and-inertia elements connected by elastic and any other elements from the expanded set of the standard elements having the general property (an input in the form of a displacement, and an output in the form of an effort), can be reduced to the simplified option with use of a concept of a quasi-elastic compact, or a quasispring. When considering systems with several degrees of freedom, it is possible to build an ensemble from several systems with one degree of freedom. At the same time, the number of such systems will correspond to the number of degrees of freedom of the original system. Parameters of an elastic compact, or a quasispring, are defined from the corresponding transfer function of the system during the "zeroing" of intermediate masses and adoption of zero value of the complex variable in the transfer function expression. This situation reflects static properties of the system with several degrees of freedom and opens opportunities to building models of the coupled oscillations using assumptions of various nature of connectivity, including stochastic one. If the complex variable is not equal to zero, then the dynamic stiffness of standard elements is taken into account with the corresponding similar approach to the determination of parameters of a quasispring, or a quasi-elastic compact.

6.4 About a Ratio of Parameters During the Transition in Mechanical Oscillatory Systems from Star Connections to Triangle Connections

When transforming computational schemes and structural diagrams of mechanical oscillatory systems, in particular, when considering problems of dynamic absorbing with two elastically interconnected masses, non-planar chains become the subject of study. It is possible to obtain compact expressions for transfer functions of feedback chains with the help of the equivalent transformations. Separate issues on this problem are covered in [17, 45, 46].

6.4.1 Some General Provisions

Let us consider the computational scheme which shows mechanical oscillatory "star" systems (Fig. 6.30a) and "triangle" systems (Fig. 6.30b). Oscillations of the bases act as perturbing factors.

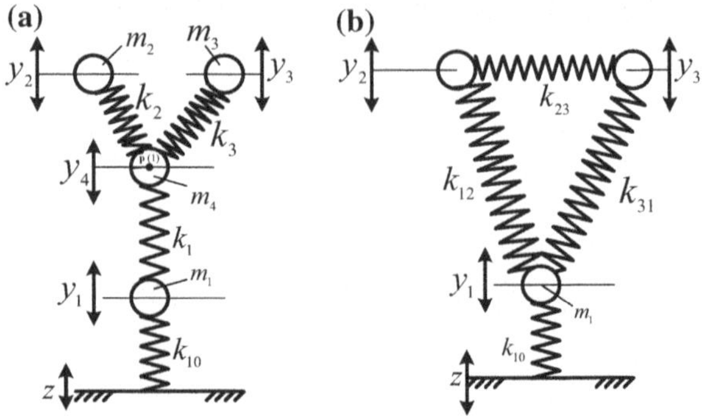

Fig. 6.30 Schematic computational diagrams of the "star" (**a**) and "triangle" (**b**) connections: z—kinematic disturbance; k_1, k_2, k_3—coefficients of stiffness of springs in the star connection; the same in the "triangle" connection; k_{10}—coefficient of stiffness of the elastic element connecting mass m_1 with the basis; m_1, m_2, m_3—mass-and-inertia elements

Schemes presented in Fig. 6.30a would reflect only the vertical motions in the absence of resistance forces.

Computational schemes in Fig. 6.30 are of fundamental nature, which is reflected in images of inclinations of springs; actually, all the oscillations are executed vertically, and trajectories of motion of the elements m_1, m_2, m_3 are located on one vertical line. In essence, we propose a variant of the proof of the theorem on a possibility of constructing equivalent ratios on the basis of special methods of drawing up mathematical models and using some assumptions.

1. To simplify procedures of defining the reduced stiffnesses in the scheme (Fig. 6.30a), moving intermediate mass m_4 can be introduced with the coordinate y_4. Further it is assumed that $m_4 = 0$, which allows us to return to a class of systems with three degrees of freedom.
2. The pole of the "star" scheme is an element with mass m_4, which then is represented by point (1) at $m_4 = 0$.
3. Since transformations are of equivalent nature (ratios are found between k_{12}, k_{23}, k_{31} and k_1, k_2, k_3), with both types of connections kinetic and potential energy, when considering motion in the status of a three-mass system, have to remain invariable.

6.4.2 *Properties of the "Star" Connection of Elements*

Let us consider the "star" connection (see Fig. 6.30a), assuming that kinetic and potential energy of the system are of the form of

$$\mathrm{T} = \frac{1}{2}m_1\dot{y}_1^2 + \frac{1}{2}m_2\dot{y}_2^2 + \frac{1}{2}m_3\dot{y}_3^2 + \frac{1}{2}m_4\dot{y}_4^2, \tag{6.95}$$

$$\Pi = \frac{1}{2}k_{10}(y_1 - z)^2 + \frac{1}{2}k_1(y_4 - y_1)^2 + \frac{1}{2}k_2(y_2 - y_4)^2 + \frac{1}{2}k_3(y_3 - y_4)^2. \tag{6.96}$$

Table 6.7 Coefficients of the equations (6.97)–(6.99) in the coordinates y_1, y_2, y_3 ("star")

a_{11}	a_{12}	a_{13}
$m_1p^2+k_{10}+\frac{k_1(k_2+k_3)}{a}$	$\frac{-k_2k_1}{a}$	$\frac{-k_1k_3}{a}$
a_{21}	a_{22}	a_{23}
$\frac{-k_2k_1}{a}$	$m_1p^2+\frac{k_2(k_1+k_3)}{a}$	$\frac{-k_2k_3}{a}$
a_{31}	a_{32}	a_{33}
$\frac{-k_1k_3}{a}$	$\frac{-k_2k_3}{a}$	$m_3p^2+\frac{k_3(k_1+k_2)}{a}$
Q_1	Q_2	Q_3
k_1z	0	0

Note Q_1, Q_2, Q_3-generalized forces in the coordinate system y_1, y_2, y_3

We assume that

$$\frac{\partial T}{\partial \dot{y}_1}=m_1\dot{y}_1;\frac{\partial T}{\partial \dot{y}_2}=m_2\dot{y}_2;\frac{\partial T}{\partial \dot{y}_3}=m_3\dot{y}_3;\frac{\partial T}{\partial \dot{y}_4}=m_4\dot{y}_4;\frac{\partial \Pi}{\partial y_1}=k_{10}y_1-k_{10}z+k_1y_1-k_1y_4;$$
$$\frac{\partial \Pi}{\partial y_2}=k_2y_2-k_2y_4;\frac{\partial \Pi}{\partial y_3}=k_3y_3-k_3y_4;\frac{\partial \Pi}{\partial y_4}=k_1y_4-k_1y_1+k_2y_4-k_2y_2+k_3y_4-k_3y_3.$$

Let us find the system of the differential equations of motion:

$$m_1\ddot{y}_1+y_1(k_{10}+k_1)-k_1y_4=k_{10}z; \tag{6.97}$$

$$m_2\ddot{y}_2+k_2y_2-k_2y_4=0; \tag{6.98}$$

$$m_3\ddot{y}_3+k_3y_3-k_3y_4=0; \tag{6.99}$$

$$-k_1y_1-k_2y_2-k_3y_4+m_4\ddot{y}_4+y_4(k_1+k_2+k_3)=0. \tag{6.100}$$

Let us accept that $m_4=0$ in the Eq. (6.100), then we will obtain

$$y_4=\frac{k_1y_1+k_2y_2+k_3y_3}{k_1+k_2+k_3}. \tag{6.101}$$

Let us substitute (6.101) into the Eqs. (6.97)–(6.99); coefficients of the equations of the system in the coordinates y_1, y_2, y_3 are provided in Table 6.7.

6.4.3 *The Mathematical Model of the "Triangle" Connection*

For the triangle connection (Fig. 6.30b) we can carry out similar calculations, believing that kinetic energy is of the form of (6.95) at $m_4=0$, and potential can be presented in the form

$$\Pi = \frac{1}{2}k_{10}(y_1 - z)^2 + \frac{1}{2}k_{12}(y_2 - y_1)^2 + \frac{1}{2}k_{23}(y_2 - y_3)^2 + \frac{1}{2}k_{13}(y_3 - y_1)^2. \tag{6.102}$$

Let us make a number of intermediate calculations and write down the equations of motion for the scheme in Fig. 6.30 as triangle connections:

$$m_1\ddot{y}_1 + y_1(k_{10} + k_{12} + k_{13}) - y_2(k_{12}) - y_3k_{13} = k_{10}z; \tag{6.103}$$

$$m_2\ddot{y}_2 + y_2(k_{12} + k_{23}) - y_1k_{12} - k_{23}y_3 = 0; \tag{6.104}$$

$$m_2\ddot{y}_2 + y_2(k_{12} + k_{23}) - k_{12}y_1 - k_{23}y_3 = 0. \tag{6.105}$$

Coefficients of the equations (6.103)–(6.105) for the system of "triangle" connections are specified in Table 6.8.

In the basis of a transformation method, as shown in [47], there is a procedure of preliminary increase in number of degrees of freedom of the "star" connection system involving additional mass m_4. After accepting that $m_4 = 0$ and aligning the dimensions of matrixes, it is possible to compare two systems of the Eqs. (6.97)–(6.99) and (6.103)–(6.105). Comparison is legitimate when taking into consideration the invariance of the system of coordinates and kinetic energy.

As for values of potential energy, it is possible to assume that at equivalent transformations they have to remain equal. Let us write down expression for potential energy for the "triangle" connection.

$$\Pi = \frac{1}{2}k_{10}(y_1 - z)^2 + \frac{1}{2}k_{12}(y_2 - y_1)^2 + \frac{1}{2}k_{23}(y_3 - y_2)^2 + \frac{1}{2}k_{13}(y_3 - y_1)^2. \tag{6.106}$$

In this case, coefficients for variables (6.106) y_1, y_2, y_3 can be found:

- at y_1^2 we obtain: $k_{12} + k_{13} + k_{10}$; (6.107)
- at y_2^2, respectively, $k_{12} + k_{23}$; (6.108)

Table 6.8 Coefficients of the equations (6.103)–(6.105) in a "triangle" connection

a_{11}	a_{12}	a_{13}
$m_1p^2 + k_{10} + k_{12} + k_{13}$	$-k_{12}$	$-k_{13}$
a_{21}	a_{22}	a_{23}
$-k_{12}$	$m_1p^2 + k_{12} + k_{23}$	$-k_{23}$
a_{31}	a_{32}	a_{33}
$-k_{13}$	$-k_{23}$	$m_3p^2 + k_{23} + k_{13}$
Q_1	Q_2	Q_3
$k_{10}z$	0	0

- at y_3^2 we find $k_{13} + k_{23}$; (6.109)

- at y_1y_2 we have $- k_{12}$; (6.110)

- at $y_2y_3 - k_{23}$; (6.111)

- at $y_1y_3 - k_{13}$. (6.112)

Let us write down similar ratios for the "star" connection, believing that $y_4 = (k_1y_1 + k_2y_2 + k_3y_3)/a$, where $a = k_1 + k_2 + k_3$. Thus, similar to the aforesaid, we will obtain:

- at $y_1^2 - k_{10} + \frac{k_1(k_2 + k_3)^2}{a^2} + \frac{k_2k_1^2}{a^2} + \frac{k_3k_1^2}{a^2}$; (6.113)

- at $y_2^2 - \frac{k_1k_2^2}{a^2} + \frac{k_2(k_1 + k_3)^2}{a^2} + \frac{k_3k_2^2}{a^2}$; (6.114)

- at $y_3^2 - \frac{k_3(k_1 + k_2)^2}{a^2} + \frac{k_2k_3^2}{a^2} + \frac{k_1k_3^2}{a^2}$; (6.115)

- at $y_1y_2 - \frac{k_1k_2}{a}$; (6.116)

- at $y_2y_3 - \frac{k_3k_2}{a}$; (6.117)

- at $y_1y_3 - \frac{k_1k_3}{a}$. (6.118)

Comparing (6.107)–(6.111) and (6.113)–(6.118), we will obtain the system of ratios:

$$k_{12} + k_{13} = k_1k_2 + \frac{k_3(k_1k_3 + k_2k_3 + k_3^2)}{a^2}; \tag{6.119}$$

$$k_{12} + k_{23} = \frac{k_1k_3}{a} + \frac{k_3(k_1k_2 + k_1k_3 + k_2^2)}{a^2}; \tag{6.120}$$

$$k_{13} + k_{23} = k_1k_3 + \frac{k_3(k_2k_3 + k_2k_1 + k_2^2)}{a^2}; \tag{6.121}$$

$$k_{12} = \frac{k_1k_2}{a}; \tag{6.122}$$

$$k_{23} = \frac{k_2 k_3}{a}; \tag{6.123}$$

$$k_{13} = \frac{k_1 k_3}{a}. \tag{6.124}$$

From the expressions given above, knowing parameters of "star" k_1, k_2 and k_3, it is possible to determine parameters of "triangle" k_{12}, k_{13}, k_{23}.

Using the above-stated approach, it is also possible to solve a problem of the construction of an n-sided polygon from a n—beam "star".

6.4.4 *Transition from the Triangle Connection to the Star Connection*

For further calculations we use a concept of compliance, believing that $s = 1/k$, i.e. the compliance is an inverse value in relation to the coefficient of stiffness of a spring. Let us transform (6.122), accepting

$$k_{12} = \frac{1}{s_{12}}, \quad k_1 = \frac{1}{s_1}, \quad k_2 = \frac{1}{s_2}, \quad k_3 = \frac{1}{s_3},$$

then

$$s_{12}^{-1} = \frac{s_1^{-1} \cdot s_2^{-1}}{s_1^{-1} + s_2^{-1} + s_3^{-1}} = \frac{(s_1 s_2)^{-1}}{\frac{s_2 s_3 + s_1 s_3 + s_1 s_2}{s_1 s_2 s_3}} = \frac{s_1 s_2 s_3}{s_1 s_2 (s_2 s_3 + s_1 s_3 + s_1 s_2)}. \tag{6.125}$$

From (6.125) it follows, in particular, that

$$s_{12} = \frac{(s_1 s_2 + s_2 s_3 + s_1 s_3) s_1 s_2}{s_1 s_2 s_3}. \tag{6.126}$$

Let us accept that

$$s_1 s_2 + s_1 s_3 + s_2 s_3 = a_1, \tag{6.127}$$

then

$$s_3 = \frac{a_1}{s_{12}}; \; s_2 = \frac{a_1}{s_{13}}, \quad s_1 = \frac{a_1}{s_{23}}. \tag{6.128}$$

Substituting (6.128) into a_1, we will obtain

$$a_1 = \frac{a_1^2}{s_{12}s_{13}} + \frac{a_1^2}{s_{12}s_{23}} + \frac{a_1^2}{s_{13}s_{23}}, \text{or} \quad a_1 = \frac{s_{12}s_{13}s_{23}}{s_{12}+s_{13}+s_{23}}. \tag{6.129}$$

Using (6.129), it is possible to find that

$$s_1 = \frac{s_{12}s_{13}}{s_{12}+s_{23}+s_{13}}; \tag{6.130}$$

$$s_2 = \frac{s_{12}s_{23}}{s_{12}+s_{23}+s_{13}}; \tag{6.131}$$

$$s_3 = \frac{s_{23}s_{31}}{s_{12}+s_{23}+s_{13}}. \tag{6.132}$$

After a number of transformations we will finally obtain

$$k_1 = k_{12} + k_{13} + \frac{k_{12}k_{13}}{k_{23}}; \tag{6.133}$$

$$k_2 = k_{12} + k_{23} + \frac{k_{12}k_{23}}{k_{13}}; \tag{6.134}$$

$$k_3 = k_{23} + k_{13} + \frac{k_{23}k_{13}}{k_{12}}. \tag{6.135}$$

Thus, knowing value of k_{12}, k_{23}, k_{13}, it is possible, using certain techniques of transformations, to find the corresponding expressions to determine the parameters of the "star" connection: k_1, k_2, k_3.

Let us note that the transformation of the "star" to "triangle" was carried out by introduction to the computational scheme (Fig. 6.30a) of the additional mass m_4 and y_4, the corresponding coordinate of the point (1). Subsequently, it was assumed that expression $m_4 = 0$, and y_4 finally took a form of (6.101). This technique was used while deriving the equations of motion for mechanical oscillatory systems with joints [47, 48].

Connection of elements of the "star–triangle" type (Fig. 6.31) is of interest to further researches of the dynamic properties of mechanical systems.

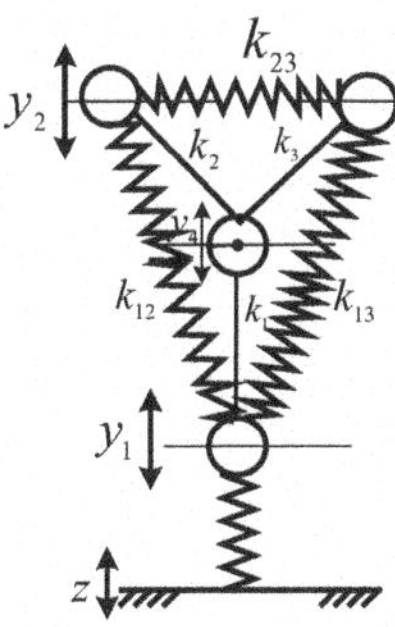

Fig. 6.31 The schematic computational diagram for the "star–triangle" mechanical system

Kinetic energy of the system is defined by the expression (6.95); potential energy will register in a form of

$$\Pi = \frac{1}{2}k_{10}(y_1 - z)^2 + \frac{1}{2}k_{12}(y_2 - y_1)^2 + \frac{1}{2}k_1(y_4 - y_1)^2 + \frac{1}{2}k_{23}(y_3 - y_2)^2 + \frac{1}{2}k_2(y_2 - y_4)^2 + \frac{1}{2}k_3(y_3 - y_4)^2 + \frac{1}{2}k_{13}(y_3 - y_1)^2. \quad (6.136)$$

Making a number of calculations, we will write down the system of the differential equations of motion in the coordinates y_1–y_4 as follows:

$$m_1\ddot{y}_1 + y_1(k_{10} + k_{12} + k_1 + k_{13}) + y_2(-k_{12}) + y_3(-k_{13}) + y_4(-k_1) = k_1 z_1; \quad (6.137)$$

$$m_2\ddot{y}_2 + y_2(k_{12} + k_{23} + k_2) + y_1(-k_{12}) + y_3(-k_{23}) + y_4(-k_2) = 0; \quad (6.138)$$

$$m_3\ddot{y}_3 + y_3(k_{23} + k_3 + k_{13}) + y_1(-k_{13}) + y_2(-k_{13}) + y_2(-k_{23}) + y_4(-k_3) = 0; \quad (6.139)$$

$$m_4\ddot{y}_4 + y_4(k_1 + k_2 + k_3) + y_1(-k_1) + y_2(-k_2) + y_3(-k_3) = 0. \quad (6.140)$$

Coefficients of the equations (6.137)–(6.140) are specified in Table 6.9.

Assuming that $m_4 = 0$, we will perform a number of transformations. Results are presented in Table 6.10, in which it is accepted that $m_4 = 0$, and the coordinate y_4 is excluded. The latter reduces a number of degrees of freedom of the system in Fig. 6.31 up to 3.

Table 6.10 (positions a_{12}, a_{13}, a_{23}) clearly shows constraints between the parameters of systems for both types of connections.

Table 6.9 Coefficients of the equations (6.137)–(6.140) (the scheme in Fig. 6.31)

a_{11}	a_{12}	a_{13}	a_{14}
$m_1p^2 + k_{10} + k_1 + k_{12} + k_{13}$	$-k_{12}$	$-k_{13}$	$-k_1$
a_{21}	a_{22}	a_{23}	a_{24}
$-k_{12}$	$m_2p^2 + k_2 + k_{12} + k_{23}$	$-k_{23}$	$-k_2$
a_{31}	a_{32}	a_{33}	a_{34}
$-k_{13}$	$-k_{23}$	$m_3p^2 + k_3 + k_{23} + k_{13}$	$-k_3$
a_{41}	a_{42}	a_{43}	a_{44}
$-k_1$	$-k_2$	$-k_3$	$m_4p^2 + k_1 + k_2 + k_3$
Q_1	Q_2	Q_3	Q_4
$k_1 z$	0	0	0

Note Q_1–Q_4 are the generalized forces of the system

Table 6.10 Coefficients in the system of the equations for the computational scheme in Fig. 6.31

a_{11}	a_{12}	a_{13}
$m_1p^2+k_{10}+k_1+k_{12}+k_{13}-\frac{k_1^2}{a}$	$-k_{12}-\frac{k_1k_2}{a}$	$-k_{13}-\frac{k_1k_3}{a}$
a_{21}	a_{22}	a_{23}
$-k_{12}-\frac{k_1k_2}{a}$	$m_2p^2+k_{12}+k_{23}+k_2-\frac{k_2^2}{a}$	$-k_{23}-k_2k_3$
a_{31}	a_{32}	a_{33}
$-k_{13}-\frac{k_3k_1}{a}$	$-k_{23}-\frac{k_2k_3}{a}$	$m_3p^2+k_{23}+k_{13}+k_3-\frac{k_3^2}{a}$
Q_1	Q_2	Q_3
k_1z	0	0

6.4.5 *Properties of Connections*

In the problems of vibration protection and vibration insulation, the "star" connection has the form shown in Fig. 6.30a, which assumes connection of an element of mass m_1 through a spring with stiffness coefficient k_{10}. However, it is also possible that an element mass m_4 will also be connected with the basis through a spring with stiffness k_4: in this case, the law of motion $z_4(t)$ will work.

Then the system of the equations (6.137)–(6.140) will change. In this case the Eqs. (6.137)–(6.140) remain, and the Eq. (6.140) can be written as

$$m_4\ddot{y}_4+y_4(k_1+k_2+k_3+k_4)-y_1k_1-k_2y_2-k_3y_3=k_4z_4. \tag{6.141}$$

If to assume that $m_4=0$, then

$$y_4=\frac{y_1k_1+y_2k_2+y_3k_3+k_4z_4}{k_1+k_2+k_3+k_4}. \tag{6.142}$$

Coefficients of the equations of "star" system, taking into account external influence z_4, are specified in Table 6.11. Let us note that force disturbance in the form of the force factor k_4z, applied at point (1), results in essential changes of the right parts of the system of the equations (6.137)–(6.140).

There is a difference in coefficients $a_{ij}(i=\overline{1,3},\ j=\overline{1,3})$ in Tables 6.7 and 6.1 due to the emergence of the additional member k_4. As for the generalized forces Q_1,Q_2,Q_3, they also change: with disturbance z_4, external influence is transferred to all elements. If to accept that $k_4=0$, Tables 6.7 and 6.11 become identical.

Comparing Tables 6.7 and 6.8, we will note that taking into account the external disturbance when k_4z_4 is applied to the point (1) also allows us to obtain a number of ratios:

$$k_{12}=\frac{k_1k_2}{a_2}; \tag{6.143}$$

Table 6.11 Coefficients of the equations for the "star" connection with external disturbance z_4

a_{11}	a_{12}	a_{13}
$m_1p^2 + k_{10} + \frac{k_1(k_2+k_3+k_4)}{a_2}$	$\frac{-k_1k_2}{a_2}$	$\frac{-k_1k_3}{a_2}$
a_{21}	a_{22}	a_{23}
$\frac{-k_3k_1}{a_2}$	$m_2p^2 + \frac{k_2(k_1+k_3+k_4)}{a_2}$	$\frac{-k_2k_3}{a_2}$
a_{31}	a_{32}	a_{33}
$\frac{-k_3k_1}{a_2}$	$\frac{-k_3k_2}{a_2}$	$m_3p^2 + \frac{k_3(k_1+k_2+k_4)}{a_2}$
Q_1	Q_2	Q_3
$k_1z_1 + \frac{k_1k_4z_4}{a_2}$	$\frac{k_4k_2z_4}{a_2}$	$\frac{k_4k_3z_4}{a_2}$

$$k_{23} = \frac{k_2k_3}{a_2}; \tag{6.144}$$

$$k_{13} = \frac{k_1k_3}{a_2}; \tag{6.145}$$

$$k_{12} + k_{13} = \frac{k_1(k_2 + k_3 + k_4)}{a_2}; \tag{6.146}$$

$$k_{23} + k_{12} = \frac{k_2(k_1 + k_3 + k_4)}{a_2}; \tag{6.147}$$

$$k_{23} + k_{13} = \frac{k_3(k_1 + k_2 + k_4)}{a_2}. \tag{6.148}$$

Using expression (6.143)–(6.145) can be formulated k_{12}, k_{13}, k_{23} through stiffness parameters k_1, k_2, k_3, k_4, i.e. taking into account the external factor, introducing the spring k_4 into the structure of the mechanical oscillatory system, does not prevent the formation of a "triangle" connection from a three-beam "star".

If $k_4 = 0$, then $a = k_1 + k_2 + k_3$, and the problem is solved with obtaining expressions k_{12}, k_{13}, k_{23} through parameters of "star" k_1, k_2, k_3, such parameters can be determined through k_{12}, k_{23}, k_{13}. In case when $k_4 \neq 0$, a situation arises when you can find parameters of a "triangle", but you cannot unambiguously express k_{12}, k_{13}, k_{23} through k_1, k_2, k_3 and k_4. This case is discussed, for example, in [17, 49].

Problems of transfer of disturbance to the system through a point of connection of elastic elements, assuming that the point of connection of springs has no mass, can be related to insufficiently studied issues, though these situations occur quite often in practice. Let us also note that a "star" system configuration, after the conjunction of some additional unit at p. 1 (Fig. 6.30a) can be considered as the change of the initial structure. Other situation arises when at p. 1 (Fig. 6.30a) an external force is applied. Then we can assume that the additional elastic element does not seem to join, but the coordinate of point 1 cannot be excluded due to application of force. Anyway, it is necessary to create some condition of constraint or definition of k_4 through $k_1,\ k_2,\ k_3$.

6.4.6 Possibilities of Transfer of Forces

Let us consider a question of application of force to a point of connection of springs with stiffnesses k_1 and k_2, as shown in Fig. 6.32.

If at $k_3 = 0$ force is at p. 1, in a static state shift at p. 1 it will be equal to

$$y_1 = \frac{Q}{k_1}. \tag{6.149}$$

Shift of the upper mass (p. 2) will be same as at p. 1. Let us assume that force Q it can be moved to p. 2, and designate force through Q_1, believing that during the transfer the shift of the p. 1 will remain the same. Then during the action of force Q_1 the shift of the p. 2 has to be equal to y_1, found from (6.149). However, the shift will already be defined (Fig. 6.32b) taking into consideration the consecutive connection of k_1 and k_2. Then $\frac{Q}{k_1} = \frac{Q_1(k_1+k_2)}{k_1 k_2}$, from where $Q_1 = \frac{k_2}{k_1+k_2}Q$. The similar result can be obtained, transforming the structural diagrams, dynamically equivalent to the computational schemes of mechanical systems in Fig. 6.32a, b. Thus, if force Q is applied not to the mass, but to the p. (1), i.e. to a point of consecutive connection of springs k_1 and k_2, then force Q can be transferred to the mass m_2 (p. 2). However, in doing this, the obtained force Q_1 will no longer be equal to Q: there will be a change of force. When transforming, we must comply to the following condition. Transfer of force from p. 1 to p. 2 is followed by change of force, but when it is applied at the point (2), the shift at the point (1) should not change.

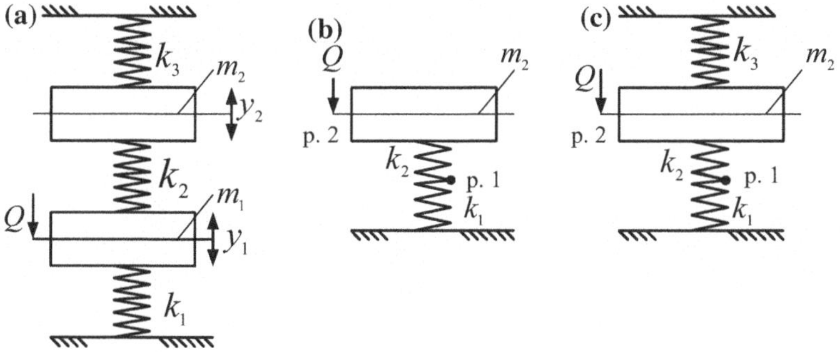

Fig. 6.32 Computational schemes of the system with two degrees of freedom and their transformations: **a** is the scheme of a general view ($m_1 \neq 0;\ m_2 \neq 0;\ k_1 \neq 0;\ k_2 \neq 0;\ k_3 \neq 0$); **b** is the introduction of a consecutive connection of springs ($m_1 = 0;\ k_3 = 0$); **c** is the accounting of the additional constraint ($m_4 = 0;\ k_3 \neq 0$)

Let us consider a case corresponding to Fig. 6.32c (here $k_3 \neq 0$) and we will find that the condition

$$\frac{Q}{k_1} = \frac{Q_1}{k_1 k_2/(k_1 + k_2) + k_3}. \tag{6.150}$$

is satisfied.

Thus, generally (Fig. 6.32), while transferring Q from p. 1 to p. 2, we have $Q_1 = \frac{Q[(k_1 + k_2) + k_1 k_3 + k_2 k_3]}{k_1 (k_1 + k_2 + k_3)} = \frac{k_1 (k_2 + k_3)}{k_1 + k_2 + k_3}$.

6.4.7 Features of Connections. Special Cases

If to consider mechanical oscillatory systems in general, bearing in mind the possibilities of their equivalent transformation, then the approach based on taking into account the identity of systems of differential equations seems to be the simplest. As the observed conditions, at least two can be singled out: equality in compounds of two types of kinetic energies and equality of potential energies.

Different types of external influences result in various mathematical models. If force factors are applied directly to the system elements with masses, then the situation remains the same when transforming. In such cases the structure of the connection, for example, for a "triangle", generally has the form, as shown in Fig. 6.33.

Figure 6.33 presents the general case when all external disturbances are either applied, or connected with mass-and-inertia elements by simple techniques. In particular, when using the structural approaches relying on general concepts of forces, kinematic and force disturbances can be reduced to each other [8].

In "star" connections external influences are considered in the same way as for "triangle" connections. However, the systems have differences in that the "star" has a special pole point (p. 1 in Fig. 6.30a). If an external influence is applied to this point, then options are possible. Figure 6.34 shows the schematic diagram of external impacts of the "star" connection.

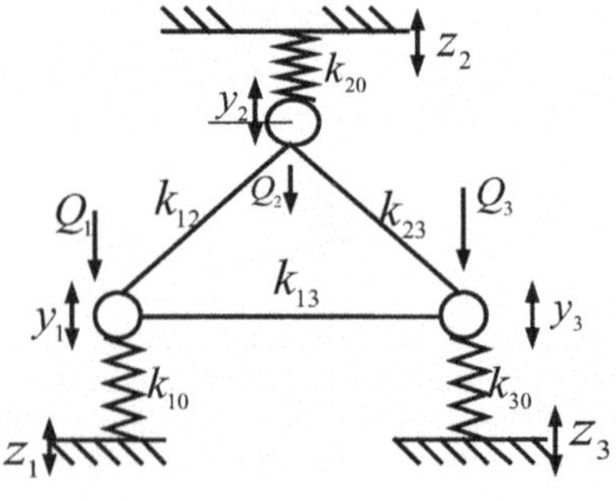

Fig. 6.33 The computational scheme of the mechanical "triangle" system

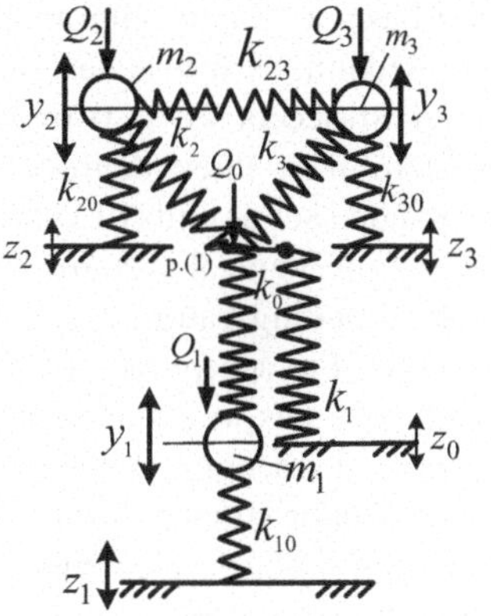

Fig. 6.34 Schematic diagram of external impacts on the star system

The application of loading through the elastic element k_0 at the kinematic influence z_0, or the application of the force Q_0 to the p. 1, is characteristic in Fig. 6.34. In the cases considered earlier, the mass m_4 was considered at the p. 1, and for the elastic element it was accepted that $k_0 = k_4$. Taking account of the elastic constraint of the pole (p. 1) complicates a problem of equivalent transformation to a "triangle", however, such unambiguous transition is possible. Given that, a characteristic circumstance is that if there are perturbations in the pole, the corresponding external perturbations Q_1, Q_2 and Q_3 will appear in the equivalent "triangle" on all coordinates. In these circumstances parameters $k_{12},\ k_{13},\ k_{23}$ can be found through $k_1,\ k_2,\ k_3$ and k_4, however, the reverse transition becomes ambiguous. Some opportunities are accompanied by simplifications in the system on the basis of transfer of forces as it is considered in Sect. 6.4.6, which can lead to some conditions of constraint between $k_1,\ k_2,\ k_3$ and k_4 in the form of the algebraic equations.

The equivalence of transformations when their unambiguity is observed, may result in more compact structural diagrams that expand possibilities of the physical interpretation of processes of dynamic oscillations and formation of certain forms of self-organization of motion of the system. Let us note one of features in use of various forms of connections. For example, the three-beam "star" can be transformed into four-beam, if to believe that k_0 is supported not by the basis but by an element with an infinitely large mass.

In a class of the systems allowing equivalent transformations, the "triangle" connection is the simplest. More complex structures, for example, "n-squares", can be reduced to "triangles" if constraints between separate elements take on extreme values $\left(k_{i,j} = \infty\right)$ [48]. In turn, when $k_{23} \to \infty$, the "triangle" in Fig. 6.30b would turn into a system with two degrees of freedom. In this case k_{12} and k_{13} have to be in the certain relations to each other that are characteristic of parallel connection of springs. If $k_{23} = 0$, the system turns into one of its versions with three degrees of freedom. In the presence of the elastic constraint k_{10} (Fig. 6.30b) and considering the basis as a unit with an extremely large value of mass, we can obtain a “star” connection, as it was designated above.

If to accept $k_{12} = 0$ or $k_{13} = 0$, the “triangle” turns into a chain mechanical system with three degrees of freedom. Using standard elementary units of an expanded set of units of vibration protection systems allows us to obtain a new class of the mechanical systems having properties of equivalent transformations and the corresponding ratios between parameters. A multibeam “star” can also become simpler if to accept radial stiffnesses as zero. In each case we will pass from a “star” with n-beams to a star with n-1-beams, etc., until we come to a system with one degree of freedom. When performing conditions of zeroing masses, there will also be a similar simplification. However, in these cases an essential role is played by additional elastic elements $k_{10}, k_{20}, \ldots k_{n0}$, which can also yield other options of creating structures.

When considering structures of an n-gon type in the presence of one constraint of some node through elastic element $k_{i0(i=\overline{1,n})}$ with the basis, the n-gon can be turned into a triangle with parameters k_{12}, k_{23}, k_{13} the generalized springs [49].

6.5 Nonplanarity in Structural Analogues of Mechanical Systems with Intercoordinate Constraints

Dynamic interactions in mechanical systems that have closed contours in the structure have their specific features. A number of issues are considered in [17, 50], but the problem requires taking into account interactions in systems of elements that differ from simple elements in parameters and structure. Such elements of mechanical systems can be units in the form of a solid body making the flat (not elementary) motion. Dynamics of such an element is defined by two coordinates, which presents a more complex scheme of interactions. The theory of mechanical chains can be also distributed to systems with solid bodies. At the same time, with all possible coincidences there are also some fundamental differences. One can assume that characteristic features are manifested in lever linkages, more precisely, in their physical forms of implementation. These issues were partially addressed in [7, 8, 51].

6.5.1 Features of Computational Schemes

Consider characteristic properties of dynamic constraints in the systems of various types. Computational schemes of chain (Fig. 6.35a) and mixed (Fig. 6.35b) mechanical systems are provided. To describe the motion, the coordinates y_1, y_2, y_3 and y_1, y_2, φ in a motionless reference system are used; p. 0 is the center of gravity of a solid body with the mass M and the moment of inertia J; m_1, m_2, m_3

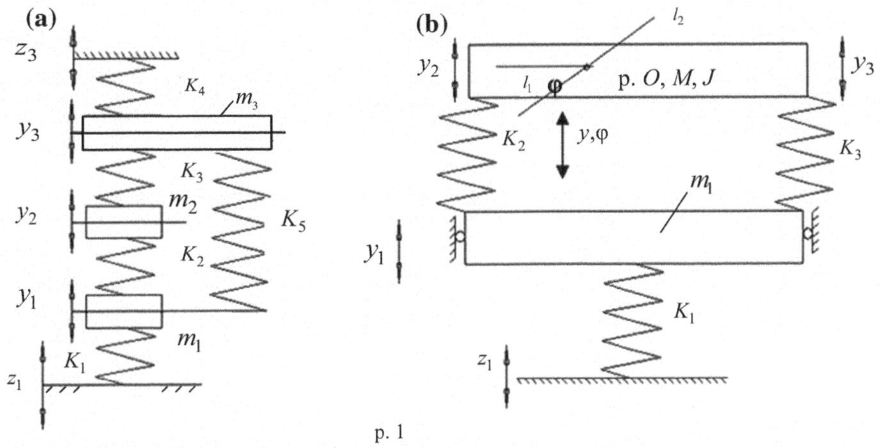

Fig. 6.35 Computational schemes of mechanical systems of the chain and mixed types

are masses, and K_1, K_2, K_3, K_4, K_5 are coefficients of stiffness of separate elements.

In Fig. 6.35a an intercoordinate constraint between m_1 and m_3 is designated through. For a solid body (Fig. 6.35b), the intercoordinate constraint (between y_2 and y_3) is provided with its lever properties in the flat motion.

For the computational scheme (Fig. 6.35) expressions for kinetic and potential energy have the form of

$$\mathrm{T} = \frac{1}{2} m_1 \dot{y}_1^2 + \frac{1}{2} m_2 \dot{y}_2^2 + \frac{1}{2} m_3 \dot{y}_3^2; \tag{6.151}$$

$$\begin{aligned} \Pi = {} & \frac{1}{2} K_1 (y_1 - z_1)^2 + \frac{1}{2} K_2 (y_2 - y_1)^2 + \frac{1}{2} K_3 (y_3 - y_2)^2 \\ & + \frac{1}{2} K_4 (y_3 - z_3)^2 + \frac{1}{2} K_5 (y_3 - y_1)^2. \end{aligned} \tag{6.152}$$

The equations for the mechanical system (Fig. 6.35a) can be expressed as

$$m_1 \ddot{y}_1 + y_1 (K_1 + K_2 + K_3) - K_2 y_2 - K_5 y_3 = K_1 z_1, \tag{6.153}$$

$$m_2 \ddot{y}_2 + y_2 (K_2 + K_3) - K_2 y_1 - K_3 y_3 = 0, \tag{6.154}$$

$$m_3 \ddot{y}_3 + y_3 (K_3 + K_4 + K_5) - K_5 y_1 = K_4 z_3. \tag{6.155}$$

For the mechanical system (Fig. 6.35b) expressions for kinetic and potential energy take a form

$$T = \frac{1}{2} m_1 \dot{y}_1^2 + \frac{1}{2} M \dot{y}^2 + \frac{1}{2} J \dot{\varphi}^2; \tag{6.156}$$

$$\Pi = \frac{1}{2} K_1 (y_1 - z_1)^2 + \frac{1}{2} K_2 (y_2 - y_1) + \frac{1}{2} K_3 (y_3 - y_1)^2. \tag{6.157}$$

The system of the equations of motion (Fig. 6.35b) can be written as:

$$m_1 \ddot{y}_1 + y_1 (K_1 + K_2 + K_3) - K_2 y_2 - K_3 y_3 = K_1 z_1; \tag{6.158}$$

$$(Ma^2 + Jc^2) \ddot{y}_2 + y_2 K_2 + (Mab - Jc^2) \ddot{y}_3 - K_2 y_1 = 0; \tag{6.159}$$

$$(Mb^2 + Jc^2) \ddot{y}_3 + y_3 K_3 + (Mab - Jc^2) \ddot{y}_2 - K_3 y_1 = 0, \tag{6.160}$$

where $a = \frac{l_2}{l_1 + l_2}; b = \frac{l_1}{l_1 + l_2}; c = \frac{1}{l_1 + l_2}$. In Fig. 6.36a, structural diagrams of the system whose motion is described by the Eqs. (6.153)–(6.155)—Fig. 6.36a and also by the Eqs. (6.158)–(6.160)—Fig. 6.36b are provided.

In structural diagrams a number of ratios is introduced:

$$y = ay_2 + by_3; \; \varphi = (y_3 - y_2)c; \; y_2 = y - l_1 \varphi; \; y_3 = y + l_2 \varphi. \tag{6.161}$$

Coefficients of the equations (6.153)–(6.155) and (6.158)–(6.160) are presented in Table 6.12 for comparison.

Comparison shows that the mixed type system (Fig. 6.35b) differs from the chain system (Fig. 6.36a) in that intercoordinate constraints between y_2 and y_3 are different. If in the chain type system the constraint has an elastic nature, then for the mixed type system (Fig. 6.36b) it has an inertial nature. Taking into account suggestions on the expansion of the system of standard units of mechanical oscillatory systems [8], it is possible to believe that in the mixed type systems the generalized springs are used in the form of standard elements with transfer function of a second order differentiation unit. Given that, let us note that in such situation the absorber can be considered as a first order differentiating unit.

The scheme in Fig. 6.35, *a* can be transformed to a form, as shown in Fig. 6.37a, b. If to believe that in Fig. 6.37a $K_4 = 0$, then the elastic element K_5 in comparison with the scheme in Fig. 6.37b would perform the same function, as well as the elastic element. Let us accept that the block I in Fig. 6.37b would correspond to an element with the mass $m_{\text{I}} = Ma^2 + Jc^2$ in the Eq. (6.159), and $m_{\text{II}} = Mb^2 + Jc^2$ can be respectively related to the Eq. (6.160); in turn, $m_{\text{III}} = Mab - Jc^2$. Let us note that an element m_{III} carries out in the system of the mixed type (Fig. 6.37b) the same function, as K_3 in the scheme in Fig. 6.37a. However, these constraints have various physical nature. In the scheme in Fig. 6.37b this constraint is inertial, and in Fig. 6.37a it is elastic. Owing to the identity of functions (force interaction), comparison gives the grounds to believe that the inertial constraint can be referred to a class of generalized springs [51].

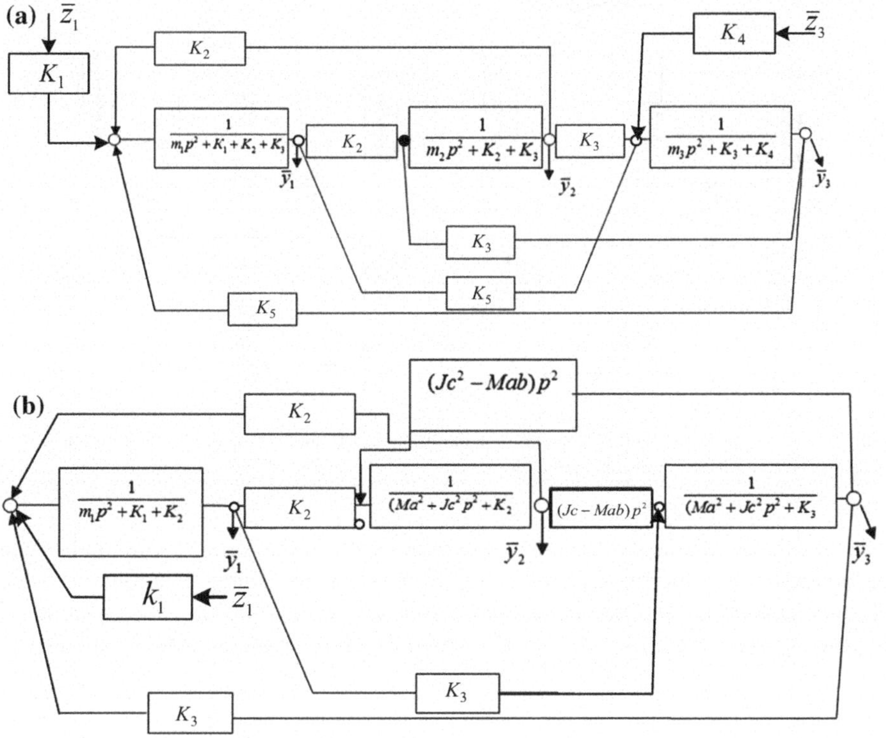

Fig. 6.36 Structural diagrams for the chain system (**a**) and for the system of the mixed type (**b**)

Table 6.12 Coefficients of the equations (6.153)–(6.155) and (6.158)–(6.160)

a_{11}	a_{12}	a_{13}
$m_1p^2+k_1+k_2+k_3$	$-k_2$	$-k_5$
$*m_1p^2+k_1+k_2+k_3$	$-k_2$	$-k_3$
a_{21}	a_{22}	a_{23}
$-k_2$	$m_1p^2+k_2+k_3$	$-k_3$
$*\ -k_2$	$(Ma^2+Jc^2)p^2+k_2$	$(Mab+Ic^2)p^2$
a_{31}	a_{32}	a_{33}
$-k_5$	$-k_3$	$m_3p^2+k_3+k_4$
$*\ -k_3$	$(Mab+Jc^2)p^2$	$(Mb^2+Jc^2)p^2+k_3$
Q_1	Q_2	Q_3
k_1z_1	0	k_4z_3
$*\ k_1z_1$	0	0

Note $*$ designate the parameters of the system of the equations (6.158)–(6.160)

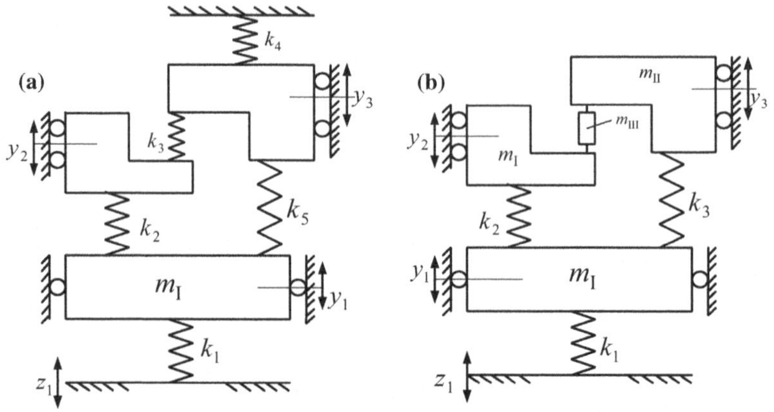

Fig. 6.37 The computational scheme of systems of chain (**a**) and mixed (**b**) types, transformed to demonstrate the closed contour on m_I elements

Thus, the mechanical systems incorporating units in the form of solid bodies can be considered as the singular elements performing the same functions as usual elastic elements i.e. provide connection. However, it not absolutely so, since these inertial elements introduce certain features through their inherent lever linkages.

6.5.2 Forms of Dynamic Interactions Between Partial Systems

Let us consider features of dynamic interactions in the mechanical system presented in Fig. 6.35b. Using (6.158)–(6.160), we will write down that

$$\bar{y}_2 = \frac{K_2\bar{y}_1 + \bar{y}_3(Jc^2 - Mab)p^2}{(Ma^2 + Jc^2)p^2 + K_2} = a_1\bar{y}_1 + b_1y_3, \tag{6.162}$$

where $p = j\omega$. Then

$$a_1 = \frac{K_2}{(Ma^2 + Jc^2)p^2 + K_2}; \tag{6.163}$$

$$b_1 = \frac{(Jc^2 - Mab)p^2}{(Ma^2 + Jc^2)p^2 + K_2}. \tag{6.164}$$

The system of the equations (6.158)–(6.160), taking into account (6.162), can be reduced to a form

$$\bar{y}_1(m_1p^2 + K_1 + K_2 + K_3 - K_2a_1) - \bar{y}_3(K_2b_1 + K_3) = K_1z_1, \tag{6.165}$$

$$\bar{y}_3[(Mb^2 + Jc^2)p^2 + K_3 - (Jc^2 - Mab)p^2b_1] - \bar{y}_1[(-Jc^2 + Mab)a_1 + K_3) = 0. \tag{6.166}$$

The structural diagram provided in Fig. 6.38 corresponds to the Eqs. (6.165) and (6.166).

Cross-couplings in the system have the transfer function

$$W_1(p) = K_2b_1 + K_3 = \frac{K_2(Jc^2 - Mab)p^2}{(Ma^2 + Jc^2)p^2 + K_2} + K_3. \tag{6.167}$$

Let us similarly obtain that

$$(Jc^2 - Mab)p^2a_1 + K_3 = \frac{K_2(Jc^2 - Mab)p^2}{(Ma^2 + Jc^2)p^2 + K_2} + K_3 = K_2b_1 + K_3.$$

Let us separately consider blocks of partial subsystems in the structural diagram (Fig. 6.38):

$$\frac{1}{m_1p^2 + K_1 + K_2 + K_3 - \frac{K_2^2}{(Ma^2 + Jc^2)p^2 + K_2}} = \frac{1}{m_1p^2 + K_1 + \frac{(K_2 + K_3)(Ma^2 + Jc^2)p^2 - K_2K_3}{(Ma^2 + Jc^2)p^2 + K_2}}. \tag{6.168}$$

Let us introduce into a denominator (6.168) the expression

$$-\frac{K_2(Jc^2 - Mab)p^2}{(Ma^2 + Jc^2)p^2 + K_2} + \frac{K_2(Jc^2 - Mab)p^2}{(Ma^2 + Jc^2) + K_2},$$

then

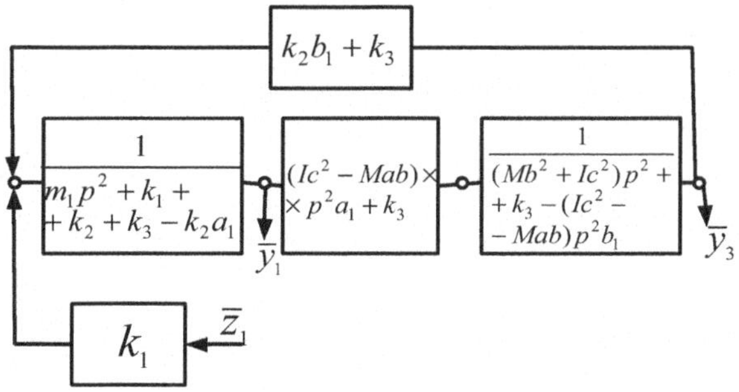

Fig. 6.38 The structural diagram corresponding to the system of the equations (6.165) and (6.166)

$$\frac{1}{m_1p^2 + K_1 + \frac{K_2(Ma^2+Jc^2)p^2 + K_3(Ma^2+Jc^2)p^2 + K_2K_3 - K_2(Jc^2-Mab)p^2}{(Ma^2+Jc^2)p^2+K_2}} \\ \cdots \frac{1}{\frac{+K_2(Jc^2-Mab)p^2}{(Ma^2+Jc^2)p^2+K_2}} = \frac{1}{m_1p^2 + K_1 + K_3 + \frac{K_2(Ma^2+Jc^2)p^2 + K_2Map^2}{(Ma^2+Jc^2)p^2+K_2}}. \tag{6.169}$$

The denominator (6.169) determines structure of the partial system by the coordinate y_1. Let us consider a partial system along the coordinate y_3:

$$\frac{1}{(Mb^2+Jc^2)p^2 + K_3 - \frac{(Jc^2-Mab)^2p^4 + K_2(Jc^2-Mab)p^2 - K_2(Jc^2-Mab)p^2}{(Ma^2+Jc^2)p^2+K_2}} \\ = \frac{1}{(Mb^2+Jc^2)p^2 + K_3 + \frac{K_2(Jc^2-Mab)p^2 - (Mab+Jc^2)p^2[(Jc^2-Mab)p^2) + K_2]}{(Ma^2+Jc^2)p^2+K_2}}. \tag{6.170}$$

The structural diagram for system (Fig. 6.38) can be transformed to a form, as shown in Fig. 6.39.

The obtained structural model of the original system (Fig. 6.35b) in the coordinates $y_1 - y_3$ has the form of mathematical model of the system with two degrees of freedom y_1 and y_3. External influence is carried out through the mass-and-inertia element m_1 (in this case it is a kinematic disturbance z_1), at the same time the coordinate y_2 is excluded as a result of transformations, as it was shown above. The exclusion of y_2 leads to emergence of additional constraints, whose physical sense is connected with implementation of functions of elastic elements with the parameters depending on frequency. As well as earlier, such additional elements can be called generalized springs. However, these elements differ by functional purpose. For comparison, let us consider a structural model from simple dual elements [7] reflecting dynamic properties of the mechanical system with two degrees of freedom and consisting of two mass-and-inertia elements M_1 and M_2, as shown in Fig. 6.40.

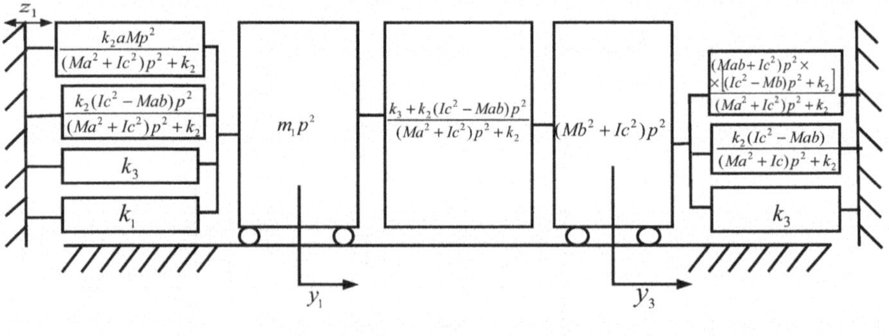

Fig. 6.39 The structural diagram for the system in Fig. 6.38 using ideas of the generalized dual elements

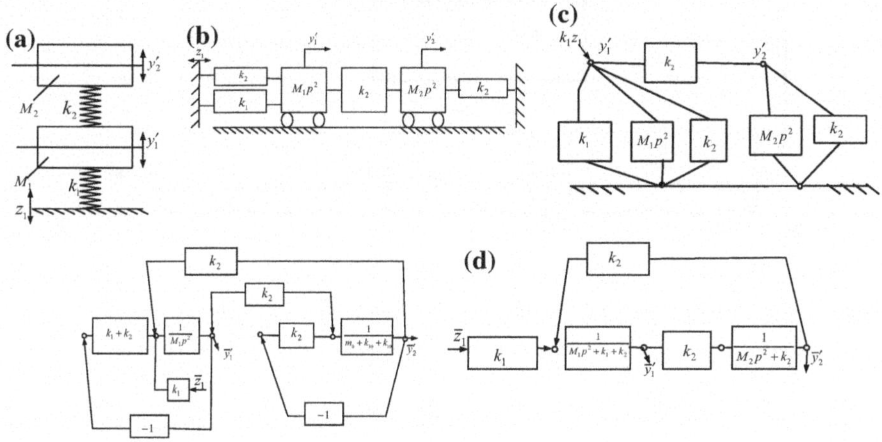

Fig. 6.40 Options of structural representations of the system with two degrees of freedom

When comparing mathematical models in Figs. 6.39 and 6.40, details of their coincidence are visible. It concerns elements K_1 and K_2 in Fig. 6.40, and elements with transfer functions K_1, K_3

$$W^*(p) = \frac{K_2(Jc^2 - Mab)p^2}{(Ma^2 + Jc^2)p^2 + K_2}. \tag{6.171}$$

As for an element with transfer function $W^*(p)$, at $p = 0 - W^*(p) = 0$ at $p \to 0 - W^*(p) \to \frac{K_2(Jc^2 - Mab)}{Ma^2 + Jc^2}$, what reflects lever interactions between the motions along the coordinates y_1 and y_3. If $Mab = Jc^2$, that action of lever linkages collapses. Noted circumstance is important meaning that existence of a solid body as more complex unit (in comparison with a material point) changes ideas of space of the motion of the system as the motion along the coordinates y_1 and y_3 are geometrically carried. Constraint between the motions has inertial character and is reflected through forces formed by the lever. Therefore in transfer function $W^*(p)$ are present stiffness of a spring K_2 and dimensionless coefficient $\frac{Jc^2 - Mab}{Ma^2 + Jc^2}$.

6.5.3 *Systems of Coordinates and Their Influence on Forms of Constraints*

If to be limited to considering constraints only between the coordinates y_2 and y_3, then a constraint between partial systems is displayed, as shown in Fig. 6.41, on the example of a ordinary system with two degrees of freedom and the coordinates y_1' and y_2'.

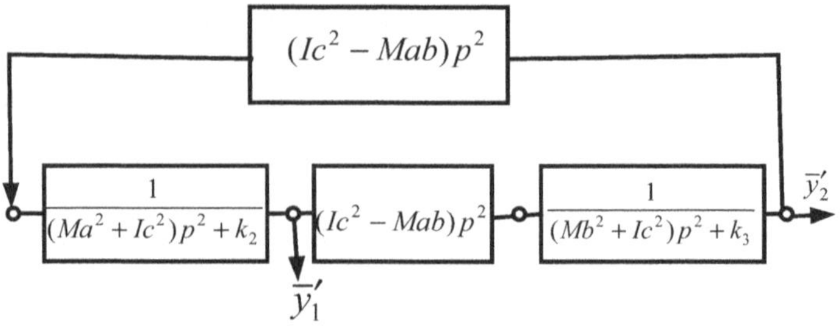

Fig. 6.41 The structural diagram of a block of the system (Fig. 6.35b) in the coordinates y_2, y_3

y_2, y_3.

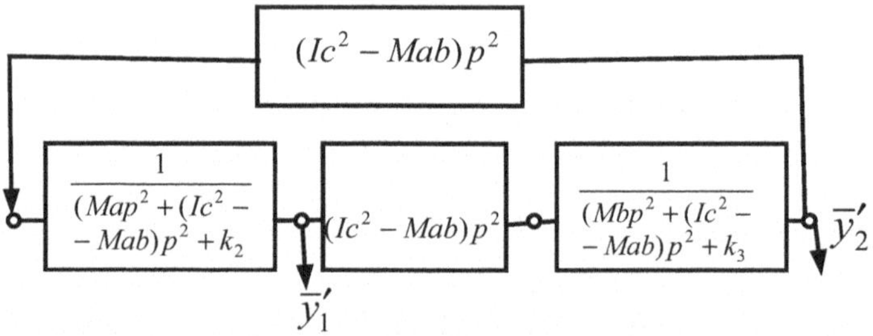

Fig. 6.42 The reduced structural diagram in relation to the inertial force of interaction between mass-and-inertia elements

Let us note that $(Jc^2 - Mab)p^2$ represents a force which works only between mass-and-inertia elements $(Ma^2 - Jc^2)$ and $(Mb^2 + Jc^2)$, which corresponds to interactions in an equal-arm lever. A lever like this provides the spacing-apart of forces in space, however, the force of inertial constraint depends on frequency. The computational scheme in Fig. 6.41 can be provided to a form, as shown in Fig. 6.42.

Let us consider structure of inertial force $(Jc^2 - Mab)p^2$. As $a = \frac{l_2}{l_1 + l_2}$; $b = \frac{l_1}{l_1 + l_2}$; $c = \frac{1}{l_1 + l_2}$, $(Jc^2 - Mab)p^2 = p^2\left(J\frac{1}{(l_1 + l_2)^2} - \frac{Ml_1 l_2}{(l_1 + l_2)^2}\right)$.

If we consider parameter $l_1 + l_2$ in relation to a rod (or to a lever), then parameter $\frac{J}{(l_1 + l_2)}$, can be physically correlated to the rotary motion relative to the center of gravity (J) and we can finally pass to the consideration of inertial properties of a material point with a mass $\frac{J}{(l_1 + l_2)}$, spaced apart to the end of the lever $l_1 + l_2$. In this case the rotary motion forms the dynamic interaction arising during the rotary motion of the rod with two material points on the ends (the part of a dynamic reaction arises, equal to Jc^2p^2). A translational component of the motion of a solid body is specifically provided with the inertia force applied to the center of gravity.

A part of the inertial force falling along the coordinate y_1', in particular, is defined from the condition $Ml_2 = M_1(l_1 + l_2)$, from where $M_1 = M\frac{l_2}{l_1 + l_2}$. Here M_1 means the reduced mass to the coordinate y_2, in particular, $M_1 = Ma$ (Fig. 6.42). To define reaction to the mass $M_2 = M\frac{l_1}{l_1 + l_2} = Mb$ reduced to the coordinate y_3, it is necessary to transfer from components of force of the inertia determined by the mass M_1, in parallel to the center of gravity and to define the arising moment from pair of forces. If to relate this moment of pair of forces to the length $l_1 + l_2$, a reaction to the inertia force along the coordinate y_2' (Fig. 6.42) it will be equal to $\frac{Ml_1l_2}{(l_1 + l_2)^2}$ or Mab, i.e. the lever existing in a solid body, which is associated with ideas of spatial or geometrical forms of the solid body and provides connection in force interactions of elements. Dynamic interaction along the coordinate y_1' is provided similarly (see Fig. 6.42). The provided example on the system with two degrees of freedom gives an indication of how dynamic interactions according to laws of mechanics are formed.

Those elements that are presented in partial systems have more irregular shapes of interaction (Fig. 6.39). There are no special difficulties with elements K_1, K_3 and $W^*(p)$ and their existence has analogies to usual systems (Figs. 6.41 and 6.42), but other elements having transfer functions

$$W_1^*(p) = \frac{K_2aMp^2}{(Ma^2 + Jc^2)p^2 + K_2}; \tag{6.172}$$

$$W_2^*(p) = -\frac{p^2(Ma^2 + Jc^2)[(Jc^2 - Mab)p^2 + K_2]}{(Ma^2 + Jc^2)p^2 + K_2}, \tag{6.173}$$

are the generalized springs which mutually interact with mass-and-inertia elements of the partial system in the same way as the spring K_1. Physically, the generalized spring with the stiffness $|W_1^*(p)|$ reflects the dynamic influence of the reduced mass Ma, belonging to the coordinate y_2, on the element with mass m_1. The latter is connected with a property of mass-and-inertia elements to form not only interactions among each other, but also with the basis. As for the element $W_2^*(p)$ in the structure, the partial system by y_3 is also a generalized spring which forms a reaction from the mass-and-inertia element along the coordinate y_2. Its specific feature is that at $p = 0$ it is a spring whose stiffness depends on the frequency, has zero value. In this regard (6.172) and (6.173) coincide, however, at $p \to \infty$ their properties will be different. With the frequency

$$\omega^2 = \frac{K_2}{Jc^2 - Mab} \tag{6.174}$$

the generalized spring (6.174) will have zero stiffness. If to expand (6.174) into components, then one can get:

$$W_2^{*\prime}(p) = -\frac{p^4(-Mab + Jc^2)(Jc^2 - Mab)}{(Ma^2 + Jc^2)p^2 + K_2} = \frac{p^4(Jc^2 - Mab)^2}{(Ma^2 + Jc^2)p^2 + K_2}; \quad (6.175)$$

$$W_2^{*\prime\prime}(p) = -\frac{K_2(-Mab + Jc^2)p^2}{(Ma^2 + Jc^2)p^2 + K_2}. \quad (6.176)$$

As for an element whose stiffness is defined (6.176), its emergence is caused by the same reasons as (6.172). The generalized spring (6.172) forms the same influence from a mass-and-inertia element along the coordinate y_2 on an element along the coordinate y_3, as well as on an element m_1 along the coordinate y_1.

The spring (6.175) reflects influence on a mass-and-inertia element along the coordinate y_3, taking into account inertial constraints. As the coordinate y_2 is excluded when transforming from direct contact with the element m_1, then influence on the partial system along the coordinate y_3 has an asymmetrical nature. Let us note that this asymmetry is characteristic of non-planar systems.

Let us consider the original system (Fig. 6.35b), but instead of coordinates y_2 and y_3 we use y and φ. Let us introduce a number of ratios:

$$\begin{gathered} y_2 = y - l_1\varphi;\ y_3 = y + l_2\varphi;\ \varphi = \frac{y_3 - y_2}{l_1 + l_2} = c\varphi;\ c = \frac{1}{l_1 + l_2}; \\ y = a_2y_2 + b_2y_3;\ a_2 = \frac{l_2}{l_1 + l_2}; b_2 = \frac{l_1}{l_1 + l_2}. \end{gathered} \quad (6.177)$$

An expression for the kinetic and potential energy in this case takes a form

$$\mathrm{T} = \frac{1}{2}m_1\dot{y}_1^2 + \frac{1}{2}M\dot{y}^2 + \frac{1}{2}J\dot{\varphi}^2; \quad (6.178)$$

$$\Pi = \frac{1}{2}K_1(y_1 - z_1)^2 + \frac{1}{2}K_2(y - l_1\varphi - y_1)^2 + \frac{1}{2}K_3(y + l_2\varphi - y_1)^2. \quad (6.179)$$

Let us carry out a number of auxiliary transformations:

$$m_1\ddot{y}_1 + y_1(K_1 + K_2 + K_3) - y(K_2 + K_3) + \varphi(K_2l_1 - K_3l_2) = K_1z_1; \quad (6.180)$$

$$M\ddot{y} + y(K_2 + K_3) - y_1(K_2 + K_3) + \varphi(K_3l_2 - K_2l_1) = 0; \quad (6.181)$$

$$J\ddot{\varphi} + \varphi(K_2l_1^2 + K_2l_2^2) + y_1(K_2l_1 - K_3l_2) + y(K_3l_2 - K_2l_1) = 0. \quad (6.182)$$

Believing that $\bar{\varphi} = a_3y_1 + b_3y_3$, where

$$a_3 = \frac{K_3l_2 - K_2l_1}{Jp^2 + K_2l_1^2 + K_3l_2^2}; \quad (6.183)$$

$$b_3 = \frac{K_2 l_1 - K_3 l_2}{Jp^2 + K_2 l_1^2 + K_3 l_2^2}. \tag{6.184}$$

Let us transform (6.180)–(6.182) to a form

$$\begin{gathered}\bar{y}_1\left[m_1p^2 + K_1 + K_2 + K_3 - \frac{(K_3l_2 - K_2l_1)^2}{Jp^2 + K_2l_1^2 + K_3l_2^2}\right] \\ -y\left[K_2 + K_3 - \frac{(K_2l_1 - K_3l_2)^2}{Jp^2 + K_2l_1^2 + K_3l_2^2}\right] = K_1 z,\end{gathered} \tag{6.185}$$

$$y\left(Mp^2 + K_2 + K_3 - \frac{(K_3l_2 - K_2l_1)^2}{Jp^2 + K_2l_1^2 + K_3l_2^2}\right) - y_1\left[K_2 + K_3 - \frac{(K_3l_2 - K_2l_1)^2}{Jp^2 + K_2l_1^2 + K_3l_2^2}\right] = 0. \tag{6.186}$$

The structural diagram of the system according to (6.185) and (6.186) will take a form, as shown in Fig. 6.43.

With the chosen coordinates, the structural model with dual elements takes a form (Fig. 6.44) allowing us to estimate stiffnesses of the introduced constraints.

At the choice of coordinates $\bar{y}_1$ and $\bar{y}$ the scheme has a simpler form. Interaction between partial systems is carried out through elastic elements K_2 and K_3, what defines dynamic properties of progress. As for the generalized spring with transfer function

$$W^{**}(p) = -\frac{(K_3l_2 - K_2l_1)^2}{Jp^2 + K_2l_1^2 + K_3l_2^2}, \tag{6.187}$$

this element reflects the dynamic interaction of elements of the system during the rotary motion of a solid body. Let us transform using (6.187) and we will write

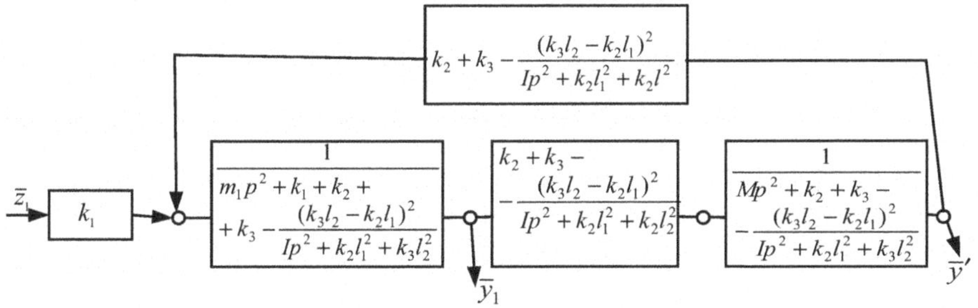

Fig. 6.43 The structural diagram of the system in the coordinates $\bar{y}_1$, $\bar{y}$

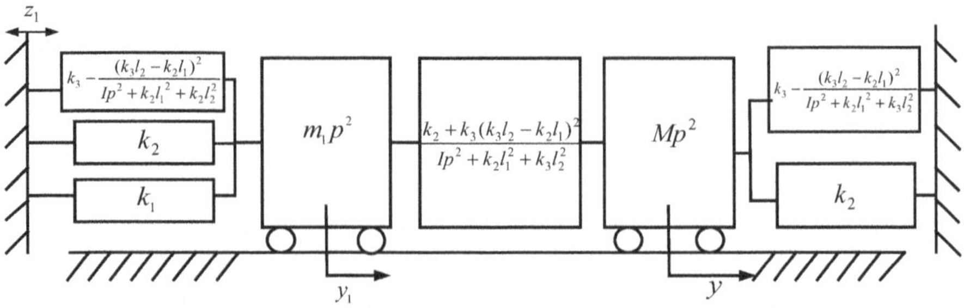

Fig. 6.44 The structural model of the system in the coordinates $\bar{y}_1$ and $\bar{y}$ with use of dual elements

$$\frac{(K_2 + K_3)(Jp^2 + K_2 l_1^2 + K_3 l_2^2) - (K_3 l_2 - K_2 l_1)}{Jp^2 + K_2 l_1^2 + K_3 l_2^2},$$

that results in the expression

$$\begin{aligned} W^{**}(p) &= \frac{Jp^2(K_2 + K_3) + K_3 K_2 (l_1 + l_2)^2}{Jp^2 + K_2 l_1^2 + K_3 l_2^2} = \frac{Jp^2(K_2 + K_3)}{Jp^2 + K_2 l_1^2 + K_3 l_2^2} + \frac{K_3 K_2 (l_1 + l_2)^2}{Jp^2 + K_2 l_1^2 + K_3 l_2^2} \\ &= \frac{Jp^2 \frac{1}{l_1^2}(K_2 + K_3)}{Jp^2 \frac{1}{l_1^2} + K_2 + K_3 i} + \frac{K_3 K_2 (l_1 + l_2)(l_1 + l_2)}{\frac{J}{l_1^2} p^2 + K_2 + K_3 i^2} = \frac{m_1^* p^2 (K_2 + K_3)}{m_1^* p^2 + K_2 + K_3 i^2} \\ &\quad + \frac{K_3 K_2 (1 + i)(1 + i)}{m_1^* p^2 + K_2 + K_3 i^2}, \end{aligned} \tag{6.188}$$

where

$$m^* = \frac{J}{(l_1 + l_2)^2}. \tag{6.189}$$

From (6.188), the existence of a lever with arms l_1 and l_2 and the point of support coinciding with the center of gravity is quite obvious in the formation of a dynamic constraint. At the same time $i = \frac{l_2}{l_1}$ serves as the transfer ratio of the lever. The question is how it would be possible to decide on the status of the lever in relation to a set of standard elementary units of mechanical oscillatory and, in particular, vibration protection systems. After transformations, the structural diagram in Fig. 6.44 can be presented in the final form, as shown in Fig. 6.45.

The generalized spring with transfer function

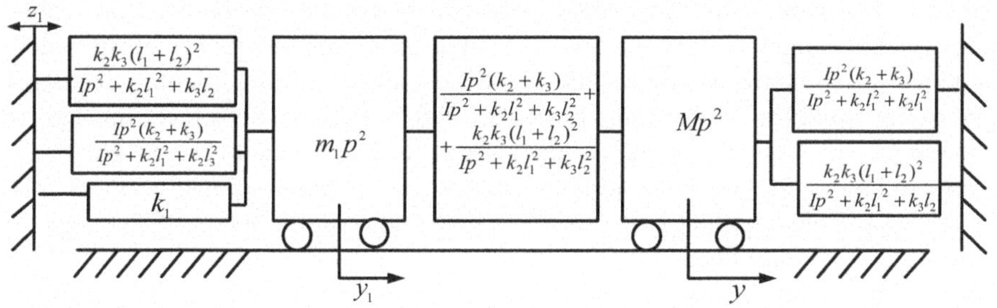

Fig. 6.45 The simplified structural diagram corresponding to the scheme in Fig. 6.44

Fig. 6.46 The auxiliary scheme for calculation of dynamic interactions between elements with mass m_1 and M

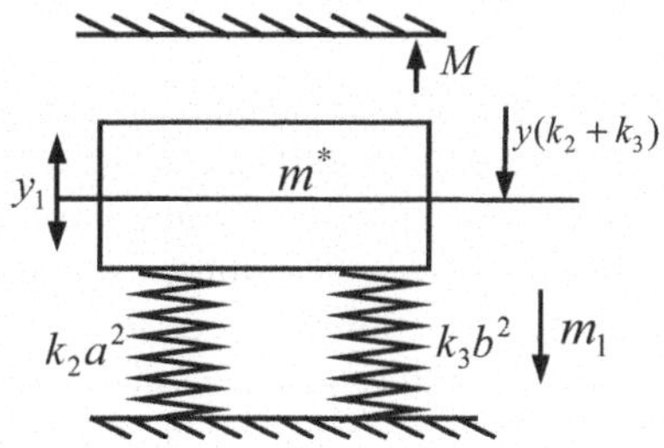

$$W'_{\mathrm{II}}(p) = \frac{Jp^2(K_2 + K_3)}{Jp^2 + K_2 l_1^2 + K_3 l_2^2} \tag{6.190}$$

is physically a dynamic reaction of an element with the mass m^* defined from (6.190). In this case the computational scheme with one degree of freedom (Fig. 6.46) is used, by means of which a dynamic reaction to the basis is defined. In turn, the basis is an element with mass m_1 and also an element with mass M, but when determining the dynamic reaction, the features of the scheme of the external disturbance are considered.

The generalized spring with transfer function

$$W'_{\mathrm{II}}(p) = \frac{K_2 K_3}{m^* p^2 + K_2 b_1^2 + K_3 c^2} \tag{6.191}$$

provides interaction between elements with weight m^* and m_1 and M according to the similar scheme in Fig. 6.46, but with external influence of a kinematic form, unlike a case with the generalized spring (6.189) where the dynamic reaction was defined. The details, mentioned in above transformations, are considered in [17].

Thus, the solid body included into the mechanical oscillatory system as a singular element, differs from the dual standard elements used in the theory of mechanical systems. However, some methodological basis can be proposed that is connected with the exclusion of one of the variables, which simplifies the system structure and makes the procedures of assessing dynamic properties quite transparent. Let us note that the solid body as a unit introduces non-planar constraints in systems, which demands certain efforts for the "outcome", in particular, excluding one of the variables. At the same time, the solid body introduces lever linkages in dynamic interactions. Parameters of the lever depend on a ratio of the weight M and the inertia moment J of the solid body and also of the gravity position center. The lever support predetermines its properties depending on the nature of the support location on the solid body.

6.6 Possibilities of Equivalent Representations of Systems with Angular Oscillations of Solid Bodies

Questions of equivalent transformations of mechanical oscillatory systems were considered quite in detail in the researches [52–54] connected with development of modern dynamics of machines and theories of mechanisms and machines. In recent years there were a number of publications [55, 56] in which much attention was paid to formation of lever linkages and their corresponding display by means of structural mathematical models in structures of mechanical systems.

In this respect, the research of the peculiar features of dynamics of systems with lever mechanisms can be viewed as the relevant areas of modern dynamics of machines. Evaluation of the opportunities of structural transformations of mechanical systems in which partial formations make oscillating and reciprocating motions still meets certain difficulties and requires the development of methodological basis.

Possibilities of transformation of rotary forms of motion in their relationship with structures of the mechanisms having partial blocks of forward type are studied.

6.6.1 Description of Interactions of Elements of Computational Schemes

Let us consider a mechanical oscillatory system consisting of two solid bodies, each of which has a motionless point of rotation (points O, O_{12}), in the presence of the constraints implemented by elastic elements (Fig. 6.47).

Let us write down expressions for a kinetic and potential energy:

$$\mathrm{T} = \frac{1}{2} J_1 \cdot (\dot{\varphi}_1)^2 + \frac{1}{2} J_2 \cdot (\dot{\varphi}_2)^2; \tag{6.192}$$

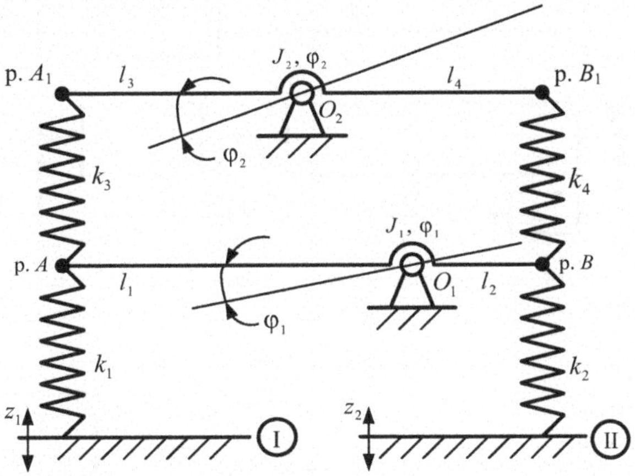

Fig. 6.47 The computational scheme of the system with partial blocks of rotary type (levers of the second kind)

$$\Pi = \frac{1}{2}k_1 \cdot (\varphi_1 l_1)^2 + \frac{1}{2}k_2 \cdot (-\varphi_1 l_2)^2 + \frac{1}{2}k_3 \cdot (\varphi_2 l_3 - \varphi_1 l_1)^2 + \frac{1}{2}k_4 \cdot (-\varphi_2 l_4 - \varphi_1 l_2)^2. \tag{6.193}$$

After the transformations, usual for Lagrange's formalism, we will write down the equations of motion:

$$J_1\ddot{\varphi}_1 + \varphi_1 \cdot \left(k_1 l_1^2 + k_2 l_2^2 + k_3 l_1^2 + k_4 l_2^2\right) - \varphi_2 \cdot (k_3 l_1 l_3 - k_4 l_2 l_4) = k_1 l_1 z_1; \tag{6.194}$$

$$J_2\ddot{\varphi}_2 + \varphi_2 \cdot \left(k_3 l_3^2 + k_4 l_4^2\right) - \varphi_1 \cdot (k_3 l_1 l_3 - k_4 l_2 l_4) = k_2 l_2 z_2. \tag{6.195}$$

The structural diagram of the system is of the form, as in Fig. 6.48.

In the case under consideration, constraints between partial systems, in physical sense, correspond to elastic interactions between solid bodies J_1 and J_2.

If to look for analogies for the system in Fig. 6.47 with a system with two degrees of freedom (for example, chain type), as shown in Fig. 6.49, then it would be expedient to pay attention to structure of partial blocks (Fig. 6.49b) which is reflected in the denominators of transfer functions of the k_2 parameter blocks.

This parameter reflects the nature of constraints for the system of a certain type. In this case (Fig. 6.49a) for coordinates y_1 and y_2 a constraint between partial systems is called elastic. Existence of k_2, repeating in structures in Fig. 6.49, predetermines rules of transformation of the structural diagrams obtained on the basis of Laplace's transformations that was used while deriving the Eqs. (6.194) and (6.195) and building the schemes (Figs. 6.48 and 6.49b). For the subsequent researches we will execute a number of transformations of the equations (6.194) and (6.195).

Let us change expressions for partial systems in the scheme provided in Fig. 6.48.

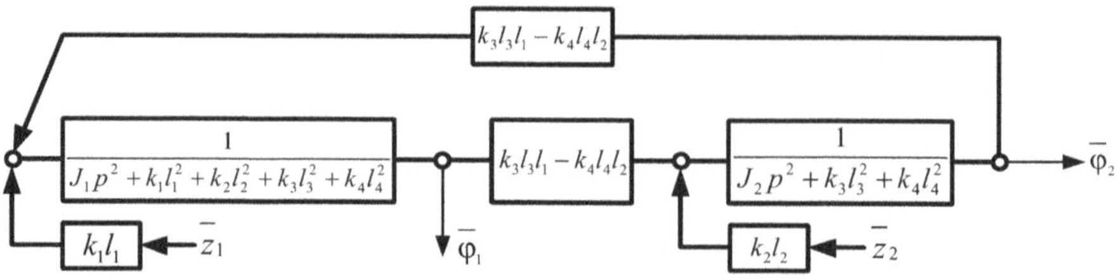

Fig. 6.48 The structural diagram of the system in the coordinates φ_1, φ_2 (rotary motions)

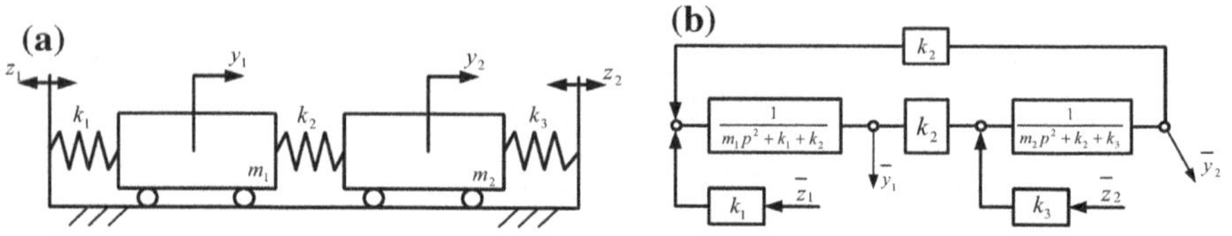

Fig. 6.49 The computational scheme (**a**) and structural diagram (**b**) for the system of chain type with forward partial blocks

$$J_1p^2 + l_1^2(k_1 + k_3) + l_2^2(k_2 + k_4) + (k_3l_1l_3 - k_4l_2l_4) - (k_3l_1l_3 - k_4l_4)$$
$$= J_1p^2 + k_1l_1^2 + k_2l_2^2 + k_3l_1 \cdot (l_1 - l_3) + k_4l_2 \cdot (l_2 + l_4) + (k_3l_1l_3 - k_4l_2l_4). \tag{6.195}$$

In turn, for the second partial system:

$$J_2p^2 + k_3l_3^2 + k_4l_4^2 + (k_3l_1l_3 - k_4l_2l_4) - (k_3l_1l_3 - k_4l_4)$$
$$= J_2p^2 + k_3l_3 \cdot (l_3 - l_1) + k_4l_4 \cdot (l_2 + l_4) + (k_3l_1l_3 - k_4l_2l_4). \tag{6.196}$$

Let us note that the structural diagram in Fig. 6.48 will be transformed to a form, as shown in Fig. 6.50.

Such a structural diagram displays interactions between the angles of rotation of the solid bodies $\bar{\varphi}_1$ and $\bar{\varphi}_2$, as if $\bar{\varphi}_1$ and $\bar{\varphi}_2$ were progress coordinates. In this case, the property of analogy between rotating and translational motion is visually shown. External influence is taken into account in the usual way and in the structural diagram, reduced to the solid body $\overline{M}_{\text{red}_1} = k_1l_1\bar{z}_1$, it corresponds to kinematic disturbance from the basis ($z_1 \neq 0$, $z_2 = 0$). Using the input influence $\overline{M}_{\text{red}_1}$, it is possible to construct the corresponding transfer functions. The computational scheme in Fig. 6.51 corresponds to the structural diagram in Fig. 6.49.

From Fig. 6.51 it is possible to conclude that the system in Fig. 6.47 is the same chain system as it is shown, in particular, in Fig. 6.49.

Let us note also that the computational scheme in Fig. 6.51 serves as an analog of the computational scheme in Fig. 6.49a. Using the scheme in Fig. 6.51, it is

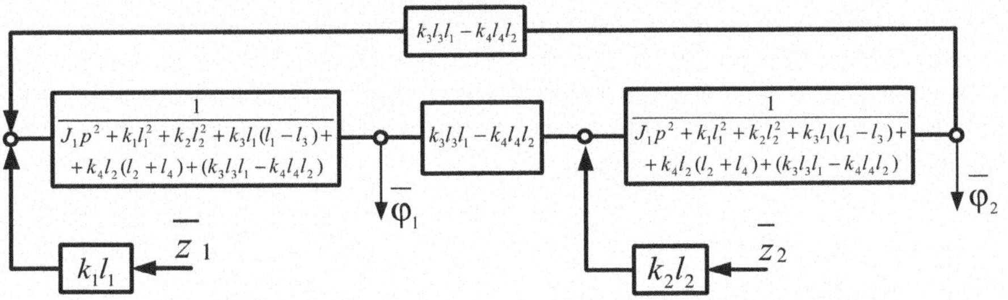

Fig. 6.50 The structural diagram of the system with two solid bodies, having rotation points according to the scheme of the lever of the second kind

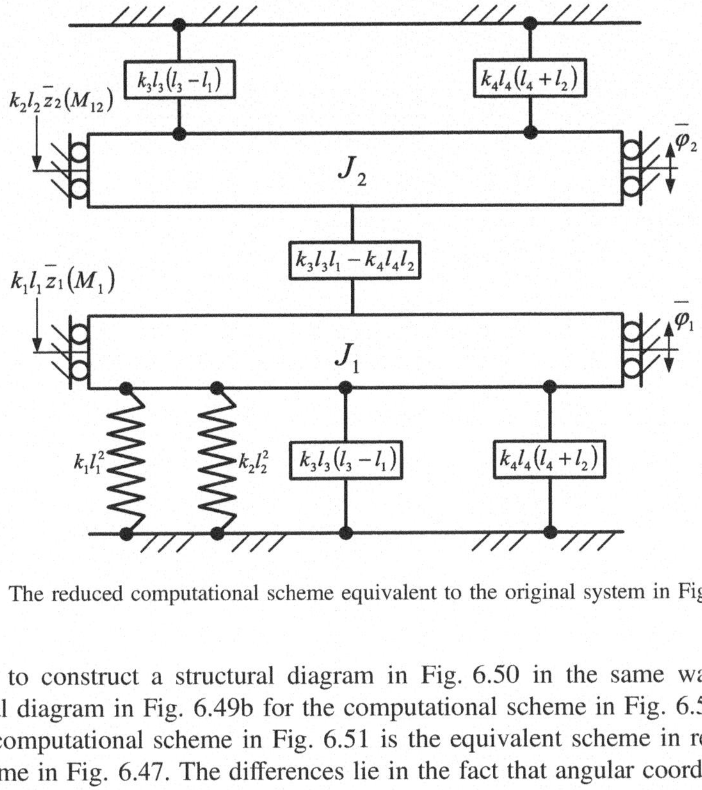

Fig. 6.51 The reduced computational scheme equivalent to the original system in Fig. 6.47

possible to construct a structural diagram in Fig. 6.50 in the same way as the structural diagram in Fig. 6.49b for the computational scheme in Fig. 6.50a.

The computational scheme in Fig. 6.51 is the equivalent scheme in relation to the scheme in Fig. 6.47. The differences lie in the fact that angular coordinates φ_1 and φ_2, in this case, reflect (in the conditional plan) progress, i.e. the angular motion acts as an analog of the translational one. With that, J_1 and J_2 in this scheme are analogs of masses in translational motion. The moment of force, created by the motion of the basis, acts as an external disturbance ($z_1 \neq 0$) (it is assumed that $z_2 = 0$ that does not reduce the generality of consideration of dynamic states).

The elastic elements in the original system designated by the coefficients of stiffness k_1, k_2, k_3, k_4 upon transition to the system analog are transformed to elastic elements analogs: $k_1 l_1^2$, $k_2 l_2^2$, which, in physical sense, define the elastic properties of interaction in rotary motion. The main difference of the approach in comparison with initial representations (see Fig. 6.47) is that the system implementing interaction of partial systems of rotary type is transformed to a scheme of the interrelations that are characteristic of progress.

6.6.2 *Features of Interaction of Systems with Lever Linkages*

Using the structural diagram in Fig. 6.50, we will find the transfer function of the system with the output signal $\bar{\varphi}_2$ and the input signal—$\overline{M}_1$ that is the moment of the external forces created by the motion of the basis z_1 ($z_2 = 0$). Thus:

$$W_1 = \frac{\bar{\varphi}_1}{\overline{M}_1} = \frac{J_2 p^2 + k_3 l_3 \cdot (l_3 - l_1) + k_4 l_4 \cdot (l_2 + l_4) + (k_3 l_1 l_3 - k_4 l_2 l_4)}{[A] \cdot [B] - (k_3 l_1 l_3 - k_4 l_2 l_4)}, \tag{6.198}$$

where

$$A = J_1 p^2 + k_1 l_1^2 + k_2 l_2^2 + k_3 l_1 \cdot (l_3 - l_1) + k_4 l_2 \cdot (l_2 + l_4) + (k_3 l_1 l_3 - k_4 l_2 l_4), \tag{6.199}$$

$$B = J_2 p^2 + k_3 l_3 \cdot (l_3 - l_1) + k_4 l_4 \cdot (l_2 + l_4) + (k_3 l_1 l_3 - k_4 l_2 l_4). \tag{6.200}$$

After transformations we will obtain that

$$W_1 = \frac{\bar{\varphi}_1}{\overline{M}_1} = \frac{J_2 p^2 + k_3 l_3^2 + k_4 l_4^2}{\left[J_1 p^2 + k_1 l_1^2 + k_2 l_2^2 + k_3 l_1^2 + k_4 l_2^2\right]\left[J_2 p^2 + k_3 l_3^2 + k_4 l_4^2\right] - (k_3 l_1 l_3 - k_4 l_2 l_4)^2}. \tag{6.201}$$

If to use the computational scheme in Fig. 6.51 which is the scheme analog in relation to rotary motion, then it is possible to obtain the same result. The coefficients of the equations of motion (6.193) and (6.194) transformed for obtaining necessary ratios between the analog schemes are specified in Table 6.13.

Using the computational scheme in Fig. 6.51 and the corresponding parameters of coefficients of the equations (Table 6.13), it is possible to obtain a similar (6.196) expression to define the transfer function $W_1(p)$.

In this case, it is of interest to consider the mechanical oscillatory system consisting of two partial systems that make rotary motions along the coordinates φ_1 and φ_2.

Table 6.13 The coefficients of equations of motion along the coordinates φ_1, φ_2 in the analog scheme of translational motion

a_{11}	a_{12}
$A = J_1p^2 + k_1l_1^2 + k_2l_2^2 + k_3l_1 \cdot (l_3 - l_1) + k_4l_2 \cdot (l_2 + l_4) + (k_3l_1l_3 - k_4l_2l_4)$	$-(k_3l_1l_2 - k_4l_2l_4)$
a_{21}	a_{22}
$-(k_3l_1l_2 - k_4l_2l_4)$	$J_2p^2 + k_3l_3 \cdot (l_3 - l_1) + k_4l_4 \cdot (l_2 + l_4) + (k_3l_1l_3 - k_4l_2l_4)$
Q_1	Q_2
$M_1 = k_1l_1z_1$	$k_2l_2z_2$

Note Q_1, Q_2 are the generalized forces corresponding to the generalized coordinates φ_1, φ_2

With the corresponding transformations, block and computational diagrams of this original system become analog schemes in relation to systems which have partial blocks making only translational motions. In this case the torsional stiffnesses of elastic elements correspond to the stiffnesses of usual springs, and the moments of inertia of the solid bodies J_1 and J_2 are analogs of solid bodies in translational motion. If to introduce into consideration the transfer ratios $i_1 = \frac{l_2}{l_1}$ and $i_2 = \frac{l_4}{l_3}$, then the expression for the transfer function can be transformed to a form

$$W_1(p) = \frac{\bar{\varphi}_1}{\bar{M}_1} = \frac{l_3^2 \cdot \left(J_3/l_3^2 + k_3 + k_4i_2^2\right)}{l_1^2 \cdot l_3^2 \cdot \left[J_1/l_1^2 + k_1 + k_2i_1^2 + i_{13}^2 \cdot \left(k_3 + k_4i_2^2\right)\right]\left(J_2/l_3^2 + k_3 + k_4i_2^2\right) - \left(k_3 - k_4i_1i_2\right)^2}, \tag{6.202}$$

where $i_{13} = \frac{l_3}{l_1}$.

In turn,

$$W_1(p) = \frac{\bar{y}_1}{\bar{z}_1} = \frac{k_1 \cdot \left(J_3/l_3^2 + k_3 + k_4i_2^2\right)}{\left[J_1/l_1^2 + k_1 + k_2i_1^2 + i_{13}^2 \cdot \left(k_3 + k_4i_2^2\right)\right]\left(J_2/l_3^2 + k_3 + k_4i_2^2\right) - \left(k_3 - k_4i_1i_2\right)^2}. \tag{6.203}$$

It is possible to transform the expressions (6.202) and (6.203) using ratios $y_1 = \varphi_1l_1$, $y_2 = \varphi_2l_3$, $m_1 = \frac{J_1}{l_1^2}$, $m_2 = \frac{J_2}{l_3^2}$. At this approach mass-and-inertia parameters are reduced to the ends of levers with the corresponding lengths l_1 and l_3:

$$W_1(p) = \frac{\overline{y}_1}{\overline{z}_1}$$
$$= \frac{k_1 \cdot \left(m_2p^2 + k_3 + k_4 i_2^2\right)}{\left[m_1 p^2 + k_1 + k_2 i_1^2 + i_{13}^2 \cdot \left(k_3 + k_4 i_2^2\right)\right]\left(m_2 p^2 + k_3 + k_4 i_2^2\right) - \left(k_3 - k_4 i_1 i_2\right)^2}. \tag{6.204}$$

The transformed scheme, which shows that the masses m_1 and m_2 are, on the one side of rotation points (points O, O_2), is provided in Fig. 6.52; with that, points A and B are connected, respectively, with masses m_1 and m_3. In turn, points A_1 and B_1 are used to attach elastic elements, creating a spatial system of element interaction.

The equivalent scheme in Fig. 6.52, if to use it for the derivation of equations of motion, yields the same result as the reduced expressions (6.194) and (6.195).

If to believe that $m_2 = 0$, then (6.203) will be transformed to the form:

$$W_1(p) = \frac{\overline{y}_1}{\overline{z}_1} = \frac{k_1 \cdot \left(k_3 + k_4 i_2^2\right)}{\left(\left[m_1 p^2 + k_1 + k_2 i_1^2 + i_{13}^2 \cdot \left(k_3 + k_4 i_2^2\right)\right]\left(k_3 + k_4 i_2^2\right) - \left(k_3 - k_4 i_1 i_2\right)^2\right)}. \tag{6.205}$$

Expression (6.205) can be reduced to the form:

$$W_1(p) = \frac{\overline{y}_1}{\overline{z}_1} = \frac{k_1}{\left[m_1 p^2 + k_1 + k_2 i_1^2 + i_{13}^2 \cdot \left(k_3 + k_4 i_2^2\right)\right] - \frac{\left(k_3 - k_4 i_1 i_2\right)^2}{\left(k_3 + k_4 i_2^2\right)}}. \tag{6.206}$$

Using a denominator (6.206), it is possible to construct a computational scheme of one-mass system ($m_1 \neq 0$, $m_2 = 0$) which has the form, as shown in Fig. 6.53.

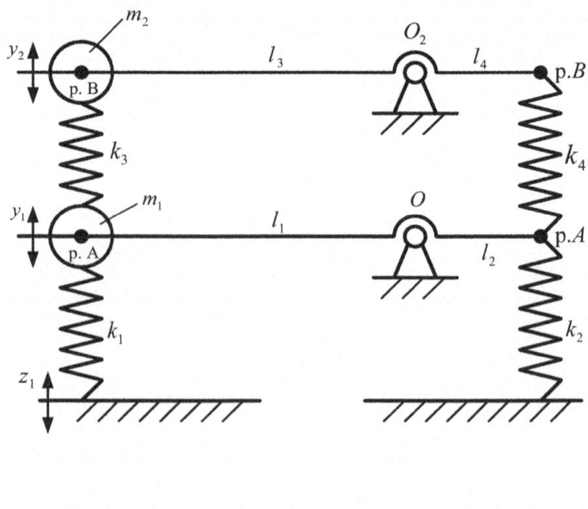

Fig. 6.52 The computational scheme of the equivalent system consisting of two material points m_1 and m_3 located at pp. A and B

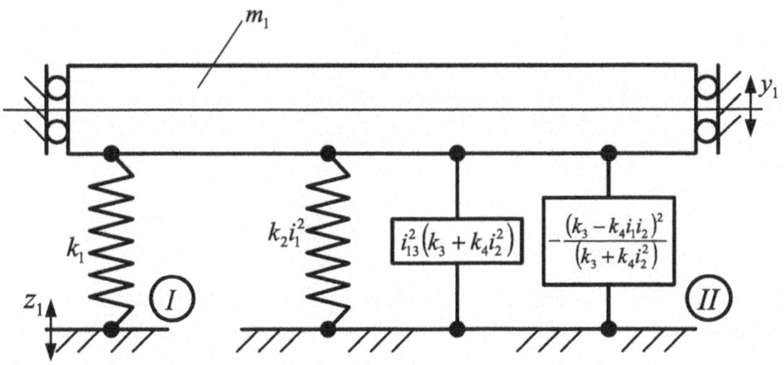

Fig. 6.53 The computational scheme with a quasispring

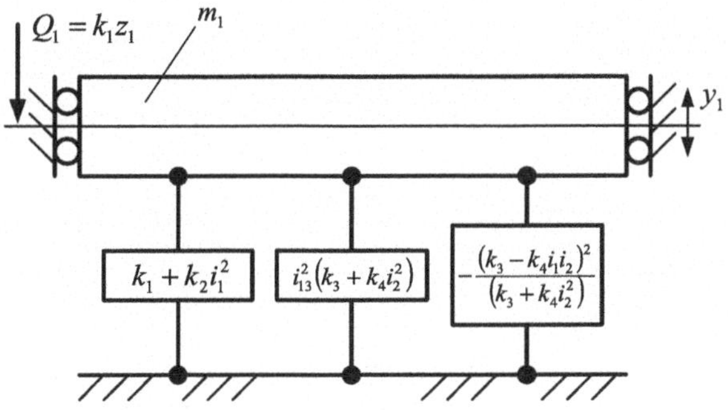

Fig. 6.54 The equivalent computational scheme with force disturbance and three quasisprings

Let us note that at $m_2 = 0$ the elastic system turns into the structure consisting of quasisprings: $(k_1 + k_2 i_1^2)$, $(k_3 + k_4 i_2^2) \cdot i_{13}^2$, $\frac{(k_3 - k_4 i_1 i_2)^2}{(k_3 + k_4 i_2^2)}$.

The quasisprings are in parallel connection. In this case the computational scheme in Fig. 6.53, in which there is a kinetic disturbance, can be reduced to the equivalent force (Fig. 6.54).

Each of the quasisprings (we mean expressions for the reduced stiffness) reflects features of the structure of lever linkages, i.e. the general reduced stiffness will be defined by expression

$$k_{\text{red}} = k_1 + k_2 i_1^2 + i_{13}^2 \cdot (k_3 + k_4 i_2^2) - \frac{(k_3 - k_4 i_1 i_2)^2}{k_3 + k_4 i_2^2}. \qquad (6.207)$$

At $i_1 = 1$, $i_2 = 1$, $i_{13} = 1$, $k_{\text{red}} = 4k$ (it is supposed that $k_1 = k_2 = k_3 = k_4 = k$). In the expression (6.207) one of elements of the quasispring has the negative sign. At the same time it should be noted that the general expression for the reduced stiffness cannot be equal. Therefore, the expression (6.207) can be enhanced concerning the elasticity coefficient k_4 that yields the equation at $k_{\text{red}} = 0$, whose research defines possibilities of ensuring static stability with the corresponding choice of parameters of the system (k_1–k_4, i_1, i_2, i_{13}).

At

$$k_3 = k_4 i_1 i_2 \tag{6.208}$$

the reduced stiffness takes a form

$$k_{\Pi p} = k_1 + k_2 i_1^2 + i_{13}^2 \cdot \left(k_3 + k_4 i_2^2\right). \tag{6.209}$$

From (6.208) it follows that it is possible to find missing parameters, setting the value of known

$$i_1 = \frac{k_3}{k_4 i_2}. \tag{6.210}$$

Then

$$k_{\Pi p} = k_1 + k_2 \frac{k_3^2}{k_4^2 \cdot i_2^2} + i_{13}^2 \cdot \left(k_3 + k_4 i_2^2\right) \cdot i_{13}^2.$$

The research of $k_{red} = 0$ is of the greatest interest.

$$k_{\Pi p} = \left[k_1 + k_2 i_1^2 + i_{13}^2 \cdot \left(k_3 + k_4 i_2^2\right)\right]\left(k_3 + k_4 i_2^2\right) - \left(k_3 - k_4 i_1 i_2\right)^2 = 0. \tag{6.211}$$

$$\begin{aligned} k_4^2 \cdot \left[i_2^4 - i_2^2 i_1^2 i_{13}^2\right] + k_4 \cdot \left[\left(k_1 + k_2 i_1^2\right) \cdot i_2^2 + i_{13}^2 k_3 i_2^2 + 2 k_3 i_1 i_2\right] \\ + \left(k_1 + k_2 i_1^2\right) \cdot k_3 + k_3^2 i_{13}^2 - k_3^2 = 0 \end{aligned}$$

or

$$\begin{aligned} k_4^2 i_2^2 \left(i_2^2 - i_1^2 i_{13}^2\right) + k_4 \left[i_2^2 \left(k_1 + k_2 i_1^2\right) + k_3 \left(i_{13}^2 i_2^2 + 2 i_1 i_2\right)\right] \\ + k_3^2 \left(i_{13}^2 - 1\right) + k_3 \left(k_1 + k_2 i_1^2\right) = 0. \end{aligned} \tag{6.212}$$

The Eq. (6.212) can be used to estimate static stability.

The expression $k_3 l_1 l_3 - k_4 l_4 l_2$ is, in essence, the torsional or rotary stiffness in interactions of rotations of the solid bodies J_1 and J_2 along the coordinates φ_1 and φ_2.

It can be noted that the corresponding torsional stiffnesses of the solid bodies J_1 and J_2, considered relative to quite motionless supporting surfaces, are quite complex emergence, which is defined by a spatial arrangement of the solid bodies and elastic elements making, in general, a complex of lever linkages.

6.6.3 The Choice of an Object of Protection in the Form of $\mathbf{J}_2$

Let us find the transfer function on $\overline{\varphi}_2$ from $\overline{M}_1$:

$$\underset{z_2=0}{W_2(p)} = \frac{\bar{\varphi}_2}{\overline{M}_1} = \frac{k_3 l_1 l_3 - k_2 l_2 l_4}{\left[J_1 p^2 + k_1 l_1^2 + + k_2 l_2^2 + k_3 l_3^2 + k_4 l_4^2\right]\left(J_2 p^2 + k_3 l_3^2 + k_4 l_4^2\right) - \left(k_3 l_1 l_3 - k_2 l_2 l_4\right)^2}. \tag{6.213}$$

As $y_2 = \varphi_2 \cdot l_3$:

$$\underset{J_1=0, z_2=0}{W_2'(p)} = \frac{\overline{y}_2}{\overline{z}_1} = \frac{k_1 l_1^2 l_3^2 \cdot (k_3 - k_2 i_1 i_2)}{l_1^2 l_3^2 \cdot \left[k_1 + k_2 i_2^2 + i_{13}^2 \cdot \left(k_3 + k_4 i_2^2\right)\right]\left[m_2 p^2 + k_3 + k_4 i_2^2\right] - \left(k_3 - k_4 i_1 i_2\right)^2}. \tag{6.214}$$

Let us transform (6.213):

$$W''(p) = \frac{\overline{y}_2}{\overline{z}_1} = \frac{\left(\frac{k_3 - k_2 i_1 i_2}{k_1 + k_2 i_2^2 + i_{13}^2 \cdot (k_3 + k_4 i^2)}\right)}{\left(m_2 p^2 + k_3 + k_4 i_2^2 - \frac{(k_3 - k_4 i_1 i_2)^2}{k_1 + k_2 i_1^2 + i_{13}^2 \cdot \left(k_3 + k_4 i_2^2\right)}\right)}. \tag{6.215}$$

The computational scheme of the system with the object J_2 at $J_1 = 0$ is provided in Fig. 6.55.

The specific feature of the computational scheme is that mass-and-inertia properties of all the system at $J_1 = 0$, $\mathrm{J}_2 \neq 0$ can be displayed by translational

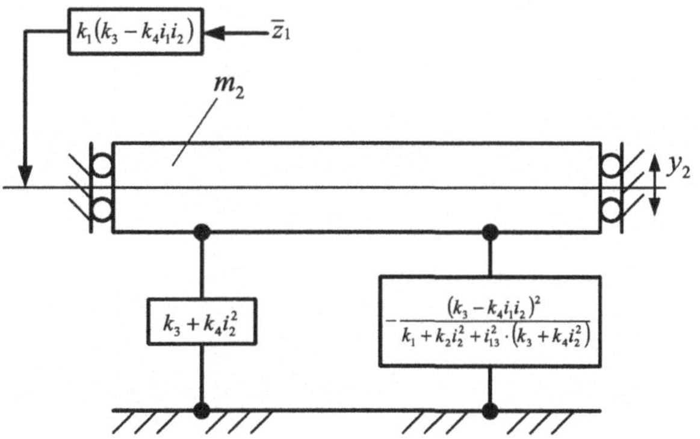

Fig. 6.55 The computational scheme of the system at $J_1 = 0$

motion of the element $m_2 = \frac{J_2}{l_3^2}$. In this case the kinematic disturbance is reduced to the mass m_2 (the material point attached to the mobile end of the lever l_3):

$$\overline{Q}_2 = k_1 \cdot (k_3 - k_4 i_1 i_2) \cdot \bar{z}_1.$$

In a simpler physical shape, the computational scheme is provided in Fig. 6.56.

In this scheme we can accept that the object of vibration protection makes oscillating and reciprocating motions around the point O_2. With that, the object is affected by the complex structure made of elastic elements.

Let us consider the possibilities of the system, using expression (6.212).

In the most general view at $J_2 \neq 0$ ($J_1 = 0$) the reduced stiffness will be defined by the expression

$$k_{\Pi p} = k_3 + k_4 i_2^2 - \frac{(k_3 - k_4 i_1 i_2)^2}{k_1 + k_2 i_1^2 + i_{13}^2 \cdot (k_3 + k_4 i_2^2)}. \tag{6.216}$$

After the transformations, the equation of determining the critical values of k_4 takes a form

$$\begin{aligned} & k_4^2 i_2^2 \cdot (i_{13}^2 i_2^2 - i_1^2) + k_4 \cdot \left[i_2^2 (k_1 + k_2 i_1^2) + k_3 i_{13}^2 i_2^2 + k_3 i_1 i_2\right] \\ & + k_3 (k_1 + k_2 i_1^2) + k_3 (k_1 + k_2 i_1^2) + k_3^2 i_{13}^2 - k_3^2 = 0, \end{aligned} \tag{6.217}$$

or:

$$\begin{aligned} & k_4^2 + \frac{k_4 \cdot \left[i_2^2 (k_1 + k_2 i_1^2) + k_3 i_{13}^2 i_2^2 + k_3 i_1 i_2\right]}{i_2^2 \cdot (i_{13}^2 i_2^2 - i_1^2)} \\ & + \frac{k_3 (k_1 + k_2 i_1^2) + k_3 (k_1 + k_2 i_1^2) + k_3^2 (i_{13}^2 - 1)}{i_2^2 \cdot (i_{13}^2 i_2^2 - i_1^2)} = 0. \end{aligned} \tag{6.218}$$

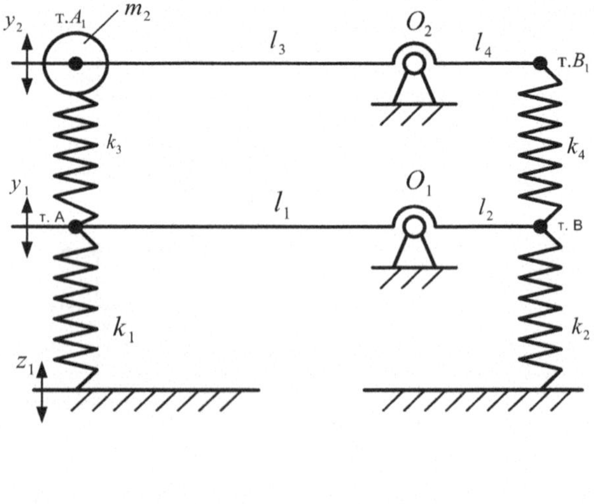

Fig. 6.56 The computational scheme of the system with one degree of freedom at $J_1 = 0$

A critical situation arises when the value of k_4 corresponds to zero.

The system can be unstable even in a static state if certain conditions are satisfied.

Thus, the rotary system with two degrees of freedom consisting of two solid bodies with rotation points, as shown in Fig. 6.47, i.e. according to the scheme of connections of levers of the second kind, can be reduced to the equivalent system in the form of a chain system with two degrees of translational motion. The systems are equivalent in the sense that they are reduced to computational schemes from which the transfer ratios can be obtained by the same rules. The method of coordinating the similarity of computational schemes is proposed, which is based on the observance of existence in the structural diagrams of partial systems of identical members, i.e. the ratios corresponding to the transfer function of a unit of inter-partial constraint. During the alignment of such ratios, the system of rotary type behaves the same as the system of forward type with the corresponding analogies.

With that, the dynamics of systems with partial formations of rotary type is based on more complex interactions.

It is shown that creation of mathematical models for problems of vibration protection and vibration insulation depends on the choice of location of the object of protection that can be chosen in the top (p. O_2) or bottom (p. O_1) cascade.

Ensuring the system stability depends on a ratio of parameters that requires checking by solving a second-order equation concerning the coefficients of stiffness of elastic elements of the top cascade. The corresponding analytical forms are offered. Let us note that properties of systems also depend on a ratio of types of lever linkages.

6.7 Dynamic Properties of Oscillatory Systems. Connectivity of Motions

Constraints between partial structures in mechanical oscillatory systems substantially define possibilities of dynamic states of the interacting elements that was represented in many known works relating to physics, molecular mechanics and the theory of oscillations [30, 40, 57]. It is noted that the choice of systems of the generalized coordinates can wield major influence on forms and content of interactions, which finally led to formation of the criteria of connectivity that gained considerable development in molecular mechanics.

To a lesser extent attention was paid to interactions of elements of the mechanical oscillatory systems reflecting dynamic properties of technical systems. In works of the recent years [3, 7, 27] some conceptual ideas of opportunities of expansion of standard set of elementary units of mechanical oscillatory systems has been developed which, in particular, found the application in the structural theory of vibration protection systems which include lever mechanisms and motion transformation devices. Implementation of an expanded set of the arising forms of interaction results in need of detailed attention to the forms of connectivity of partial systems. It is largely determined by ratios of different types of motions of the separate elements of the system, for example, rotating and translational motions of partial structures [58, 59]. With that, the greater focus was placed to criteria of connectivity, in determining which the symmetry of interactions and uniformity of structures of partial systems was assumed.

Along with this, there are various kinds of mechanical oscillatory systems in which partial systems have motions of different types, i.e. interactions occur between the solid bodies jointly performing translational and rotary motions.

Some questions of dynamic interactions in these systems are considered in works [60, 61], however, features of manifestation of lever linkages and their influence on properties of mechanical oscillatory systems demand detailed studying.

Approaches are being developed that make it possible to consider dynamic constraints, arising in mechanical oscillatory systems with different types of partial motions, in particular, in the presence of rotations, which results in the emergence of lever linkages and the corresponding dynamic features. Dynamic properties of systems are the cornerstone of statements of problems of the linear theory of vibration protection systems [3, 7, 12].

6.7.1 Systems with Solid Bodies as Binding Elements

A mechanical oscillatory system (Fig. 6.57) is considered in which mass-and-inertia elements can make rotating and translational motions. The system has two degrees of freedom of motion which can be described by several systems of the generalized coordinates. Let us note that when evaluating dynamic properties of systems, it is necessary to pay attention to uniformity of coordinates, which is related to the issues of coincidence of the dimension of constraint reactions between partial systems.

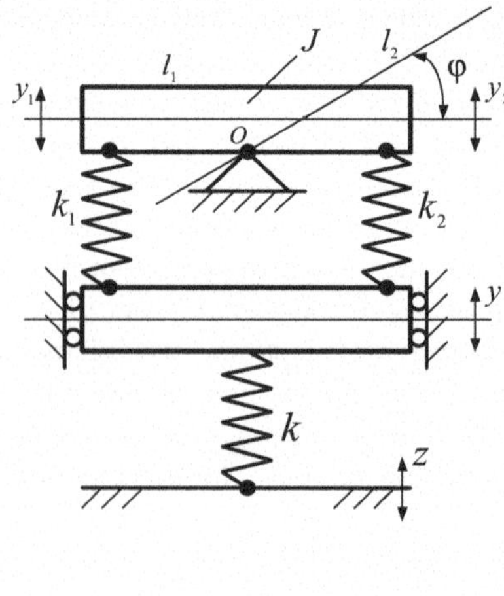

Fig. 6.57 The computational scheme of the system of the combined type (translational motion is along the coordinate y; rotary motion is relative to the motionless point O)

Let us carry out a number of auxiliary transformations, typical when using the Lagrange formalism. In the coordinates y and φ the system of the equations of motion for the computational scheme in Fig. 6.57 will take a form

$$m\ddot{y} + y \cdot (k + k_1 + k_2) + \varphi \cdot (k_2 l_2 - k_1 l_1) = kz; \tag{6.219}$$

$$J\ddot{\phi} + \phi \cdot \left(k_1 l_1^2 + k_2 l_2^2\right) + y \cdot (k_2 l_2 - k_1 l_1) = 0. \tag{6.220}$$

The structural diagram of the original system (taking into account Laplace's transformations) in the coordinates y, φ according to (6.218) and (6.220) can be presented, as shown in Fig. 6.58.

The structural diagram in Fig. 6.58 can be transformed, as shown in Fig. 6.59 and have several forms of display.

Transfer functions of the system can be found from structural diagrams. At the same time the same results can be obtained directly from the Eqs. (6.219) and (6.220) after Laplace's transformations:

$$\bar{\varphi} = \frac{(k_1 l_1 - k_2 l_2) \cdot \bar{y}}{Jp^2 + k_1 l_1^2 + k_2 l_2^2}. \tag{6.221}$$

We use (6.221) for the exclusion of $\bar{\varphi}$:

$$\bar{y} \cdot \left(mp^2 + k\right) + \frac{Jp^2 \cdot (k_1 + k_2) + k_1 k_2 \cdot (l_1 + l_2)^2}{Jp^2 + k_1 l_1^2 + k_2 l_2^2} = k\bar{z}. \tag{6.222}$$

If to believe that the partial system (along the coordinate y) is presented by the expression $mp^2 + k + k_1 + k_2$, then the Eq. (6.222) will take a form

$$\bar{y} \cdot \left(mp^2 + k + k_1 + k_2\right) - \frac{(k_2 l_2 - k_1 l_1)^2}{Jp^2 + k_1 l_1^2 + k_2 l_2^2} = kz. \tag{6.223}$$

Thus, the original system (see Fig. 6.57) can be reduced by an exclusion of a coordinate of rotary motion φ to system with one degree of freedom in which the

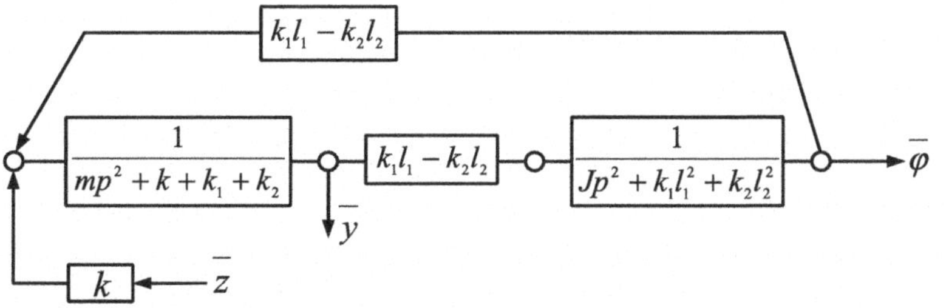

Fig. 6.58 The structural diagram of the original system (Fig. 6.57) in the coordinates y and φ

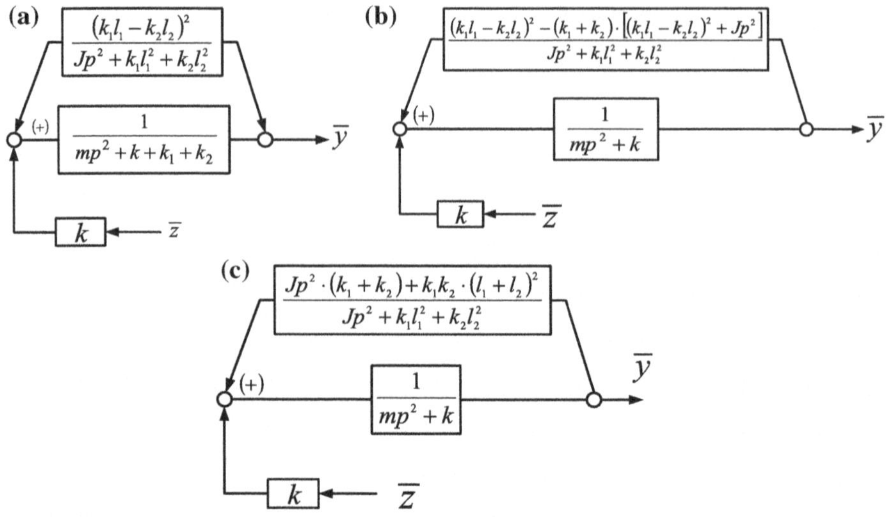

Fig. 6.59 Transformation of the original structural diagram: **a** is the coordinate φ exclusion; **b** is reduction to the partial system $mp^2 + k$ positive feedback; **c** is reduction of the system to a form with negative feedback

mass-and-inertia element m makes the forward rectilinear motion. In application to problems of vibration protection, the mass-and-inertia element m can be considered as the object of protection. In this case, in the structure of the vibration protection system, some motion transformation device is formed, having a form of the lever of the second kind. Such a lever has mass-and-inertia properties (has the moment of inertia J with respect to the point O (see Fig. 6.57).

In this case the original system (see Fig. 6.57) can be considered as a system with one degree of freedom. At the same time, as shown in Fig. 6.60 in the system there is an additional element which differs from the known standard elements (in this case, springs with stiffnesses k, k_1, k_2).

The computational scheme in Fig. 6.60 assumes the use of supporting surfaces I and II. The motion transformation device introduced into the scheme in Fig. 6.60 is characterized by the reduced dynamic stiffness which is written with use of the complex variable p ($p = j\omega$). At $p = 0$, i.e. in the absence of dynamic disturbance from the basis ($z = 0$), the dynamic stiffness of the device for transformation of the motion is transformed to the stiffness of the complex spring incorporating the lever of the second kind.

From the analysis of the computational scheme in Fig. 6.60a, it also follows that the original system (see Fig. 6.57) will be transformed to the dynamically equivalent system, but with another set of components. The coordinate φ is excluded, but the interactions introduced by a rotary motion along the coordinate φ remain and are reflected by the quasispring (or a compact) having the reduced stiffness:

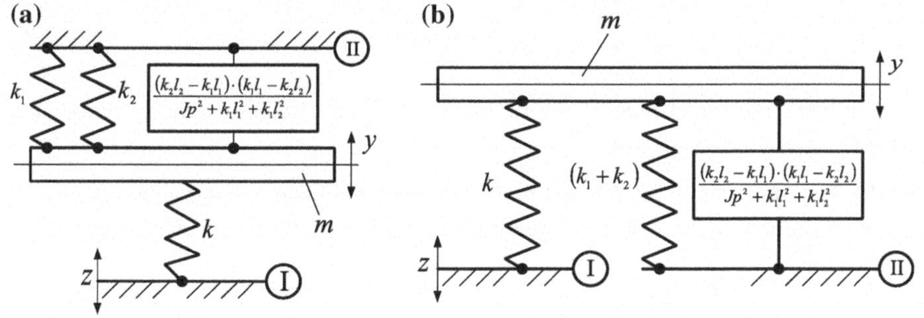

Fig. 6.60 The computational scheme of the combined system reduced to the system making translational motion along the coordinate y. **a** the supporting surfaces are spaced-apart vertically; **b** the supporting surfaces I and II are divided: disturbance z is along the supporting surface I

$$k_{\text{red}} = \frac{(k_2l_2 - k_1l_1)\cdot(k_1l_1 - k_2l_2)}{Jp^2 + k_1l_1^2 + k_1l_2^2}. \tag{6.224}$$

The quasispring has such a property that at certain ratios of parameters the reduced stiffness can become negative. In physical sense it means a change of the direction of the elastic force developed by a quasispring. The same effect can be gained with changing the frequency of external influence, since the denominator (6.224) is defined by the expression

$$A_0' = -J\omega^2 + k_1l_1^2 + k_2l_2^2, \tag{6.225}$$

which with increase of p ($p = j\omega$) also takes negative value.

Let us note that in structural transformations the quasispring behaves as an ordinary elastic element. A number of issues, related to the mentioned features of the properties, are considered in [29].

6.7.2 *Features of Transformation of Systems*

From the computational scheme in Fig. 6.60 a number of characteristics can be determined. If between the partial systems of the original system in Fig. 6.57 connectivity is determined by a unit with transfer function $W'(p) = k_1l_1 - k_2l_2$, then a constraint between partial systems in the coordinates y, φ becomes zero when performing the condition $k_1l_1 = k_2l_2$. In this case, with external disturbance z the system will execute a motion as an object with mass m, having one degree of freedom. The partial frequency of the system coincides with the frequency of natural oscillations of the reduced system:

$$\omega_1^2 = \frac{k + k_1 + k_2}{m}. \tag{6.226}$$

With that, the magnitude of the moment of inertia of the solid body J does not matter.

1. If $k_1 l_1 \neq k_2 l_2$, the case when $J = 0$ is of interest. Then the reduced stiffness of the system, formed with participation of a partial system of rotary type, will be defined:

$$k_{\text{red}} = k + k_1 + k_2 - \frac{(k_2 l_2 - k_1 l_1)^2}{k_1 l_1^2 + k_2 l_2^2}, \tag{6.227}$$

or

$$k_{\text{red}} = \frac{k \cdot (k_1 l_1^2 + k_2 l_2^2) + k_1 k_2 \cdot (l_1 + l_2)^2}{k_1 l_1^2 + k_2 l_2^2}. \tag{6.228}$$

With $J = 0$ in the system an emergence of the lever mechanism of the second kind stiffness is possible which not only forms the reduced stiffness for the object with mass m, but also creates quite a certain structure of constraints in space (in geometrical sense). The reduced scheme of the system (Fig. 6.60) in this case can be interpreted according to Fig. 6.61 where

$$k'_{\text{red}} = \frac{k_1 k_2 \cdot (l_1 + l_2)^2}{k_1 l_1^2 + k_2 l_2^2}. \tag{6.229}$$

We use a concept of the transfer ratio of the lever:

$$i = \frac{l_2}{l_1}.$$

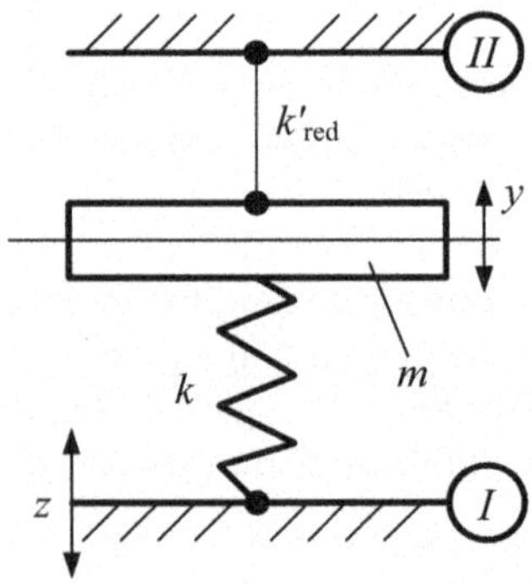

Fig. 6.61 The reduced original computational scheme at $J = 0$

The sign of the transfer ratio, i.e. a feature of the lever mechanism, is considered while deriving the equations of motion. Generally the transfer ratio of the lever of the second kind is negative:

$$k'_{\text{red}} = \frac{k_1 k_2 \cdot (1+i)^2}{k_1 + k_2 i^2}. \tag{6.230}$$

If $i = 0$, $k'_{\text{red}} = k_2$. At $i = \infty$ $k'_{\text{red}} = k_1$, which coincides with physical ideas of properties of the mechanical system with lever linkages.

2. The lever of the second kind with a motionless point of rotation (p. O) creates a spatial structure of arrangement of elements of the mechanical oscillatory system. If the lever has zero mass-and-inertia characteristics, then in addition to the main spring with stiffness k it creates a parallel elastic constraint, determined by the expression (6.230).

This constraint is an elastic connection of the object with mass m with the supporting surface II. In this case the frequency of natural oscillations will be determined as in the system with one degree of freedom:

$$\omega^2 = \frac{k_2 + k'_{\text{red}}}{m}. \tag{6.231}$$

$$\omega^2 = \frac{k_1 k_2 \cdot (1+i)^2 + k_2 \cdot (k_1 + k_2 i^2)}{(k_1 + k_2 i^2) \cdot m}. \tag{6.232}$$

Let us note that the transfer ratio i can act as adjusting parameter while solving various tasks connected with assessment and control of the dynamic state of mechanical oscillatory systems, in particular, vibration protection ones [3, 7].

3. If $J \neq 0$, i.e. the rotary unit has a sufficiently significant moment of inertia, the reduced stiffness of the quasispring, unlike expression (6.228), will depend on p. In this case the reduced stiffness of the quasispring can be called dynamic:

$$k''_{\text{red}} = \frac{(k + k_1 + k_2) \cdot \left(Jp^2 + k_1 l_1^2 + k_2 l_2^2\right) - (k_1 l_1 - k_2 l_2)^2}{Jp^2 + k_1 l_1^2 + k_2 l_2^2}, \tag{6.233}$$

$$k''_{\text{red}} = \frac{Jp^2 \cdot (k + k_1 + k_2) + k \cdot \left(k_1 l_1^2 + k_2 l_2^2 + k_1 + k_2\right) \cdot \left(k_1 l_1^2 + k_2 l_2^2 - (k_1 l_1 - k_2 l_2)^2\right)}{Jp^2 + k_1 l_1^2 + k_2 l_2^2}. \tag{6.234}$$

The expression (6.234) gives an idea of the complex nature of the constraints which are formed at the massive lever in structure of an object of vibration protection system in the case when the object makes a translational (rectilinear) motion.

Expression (6.234) can be also presented as

$$k_{\Pi p} = k + k_1 + k_2 - \frac{(k_2 l_2 - k_1 l_1)^2}{k_1 l_1^2 + k_2 l_2^2}. \tag{6.230}$$

4. Taking into account $J \neq 0$, the original system becomes a system with two degrees of freedom in which partial systems have various motions: one (with coordinate y) implements translational motion, another (with coordinate φ) implements the rotary one. Partial frequencies of the system can be determined from the characteristic frequency equation:

$$Jmp^4 + p^2 \cdot \left[J \cdot (k + k_1 + k_2) + m \cdot \left(k_1 l_1^2 + k_2 l_2^2\right)(k + k_1 + k_2) \cdot \left(k_1 l_1^2 + k_2 l_2^2\right)\right] - (k_1 l_1 - k_2 l_2)^2 = 0. \tag{6.236}$$

Frequencies of natural oscillations in this case can be determined in a form of

$$\omega_{1,2}^2 = \frac{J \cdot (k + k_1 + k_2) + m \cdot \left(k_1 l_1^2 + k_2 l_2^2\right)}{2Jm} \pm \sqrt{\frac{\left[J \cdot (k + k_1 + k_2) - m \cdot \left(k_1 l_1^2 + k_2 l_2^2\right)\right]^2 + 4Jm \cdot (k_1 l_1 - k_2 l_2)^2}{4(Jm)^2}}. \tag{6.237}$$

Let us write down that partial frequencies are determined:

$$n_1^2 = \frac{k + k_1 + k_2}{m}; \tag{6.238}$$

$$n_2^2 = \frac{k_1 l_1^2 + k_2 l_2^2}{J}. \tag{6.239}$$

In turn,

$$\omega_1^2 = \frac{n_1^2 + n_2^2}{2} - D; \tag{6.240}$$

$$\omega_2^2 = \frac{n_1^2 + n_2^2}{2} + D, \tag{6.241}$$

where

$$D = \frac{1}{2Jm}\sqrt{\left[J \cdot (k + k_1 + k_2) - m \cdot \left(k_1 l_1^2 + k_2 l_2^2\right)\right]^2 + 4Jm \cdot (k_1 l_1 - k_2 l_2)^2}. \tag{6.242}$$

Thus, the mechanical oscillatory system (Fig. 6.57) having two partial components, which condition is defined by coordinates y and φ, can be transformed and reduced to a simpler system characterized by translational motion coordinate y. With this "simplification" in the system structure, it is necessary to introduce an element of new type—it can be called a quasispring and have the corresponding reduced stiffness. Generally the reduced stiffness depends on the frequency of external influence and can be called dynamic.

In [1, 62–64] the quasispring of the mentioned form obtained the name of the generalized spring. Both names reflect the same physical essence, but their conceptual fields are used differently in various contexts. An important circumstance is that when $J = 0$ in the system with one degree of freedom it is possible to justify the emergence of lever linkages. In this case they are implemented by a lever mechanism of the second kind. Such constraints in mechanical oscillatory systems introduce new properties.

6.7.3 The Coordinate y Exclusion

We use the structural diagram in Fig. 6.58 and the system of the equations of motion:

$$mp^2\bar{y} + \bar{y} \cdot (k + k_1 + k_2) + \bar{\varphi} \cdot (k_2 l_2 - k_1 l_1) = k\bar{z}; \tag{6.243}$$

$$Jp^2\bar{\varphi} + \bar{\varphi} \cdot \left(k_1 l_1^2 + k_2 l_2^2\right) + \bar{y} \cdot (k_2 l_2 - k_1 l_1) = 0. \tag{6.244}$$

From (6.243) it follows that

$$\bar{y} = \frac{k\bar{z} + \bar{\varphi} \cdot (k_2 l_2 - k_1 l_1)}{mp^2 + k + k_1 + k_2}. \tag{6.245}$$

After substitution (6.245) into (6.244) we will obtain:

$$\bar{\varphi} \cdot \left(Jp^2 + k_1 l_1^2 + k_2 l_2^2\right) - \bar{\varphi} \cdot \frac{(k_1 l_1 - k_2 l_2)^2}{mp^2 + k + k_1 + k_2} = \frac{k \cdot (k_1 l_1 - k_2 l_2)}{mp^2 + k + k_1 + k_2}. \tag{6.246}$$

Figure 6.62b shows a structural diagram of the original system (Fig. 6.57) with the coordinate $\bar{y}$ exclusion: Fig. 6.62a presents the corresponding computational scheme in a symbolical form as a mechanical system with one degree of freedom determined by the coordinate $\bar{\varphi}$.

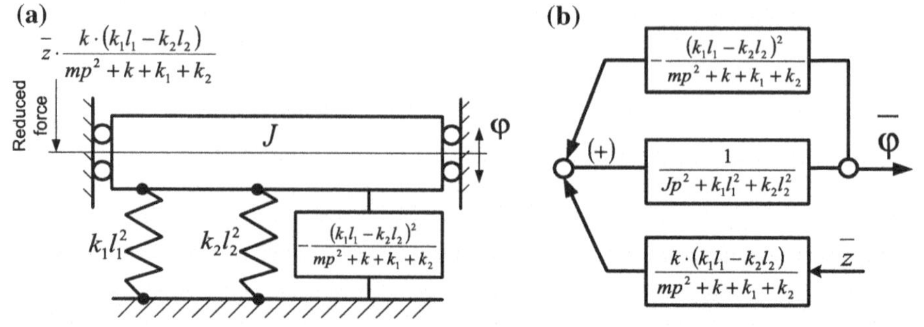

Fig. 6.62 The structural diagram (**a**) and computational scheme (**b**) with the coordinate *y* exclusion

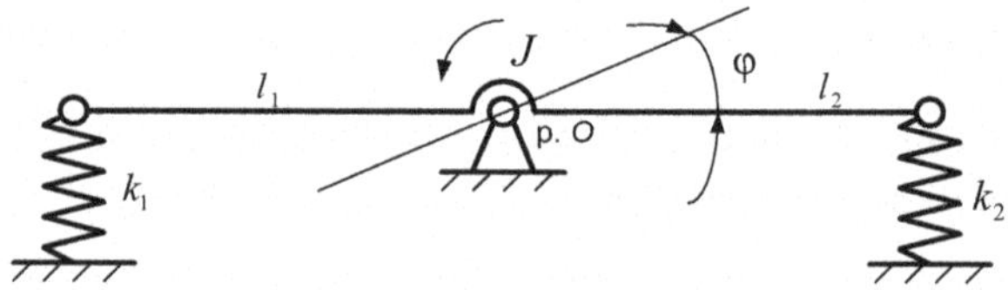

Fig. 6.63 The computational scheme of the partial system of rotary motion

If to consider the computational scheme of the partial system with the corresponding rotary motion, then it takes a form, as shown in Fig. 6.63.

Using the scheme in Fig. 6.63, we can find the partial frequency:

$$n_2^2 = \frac{k_1 l_1^2 + k_2 l_2^2}{J}. \tag{6.247}$$

These will be angular oscillations. Coming back to the computational scheme to Fig. 6.62a, we will find that the reduced stiffness of the quasispring is defined by the formula

$$k''_{\text{red}} = \frac{l_1^2 \cdot (k_1 - k_2 \cdot i)^2}{mp^2 + k + k_1 + k_2}. \tag{6.248}$$

The expression (6.243) corresponds to the torsional dynamic stiffness in a rotary motion with the coordinate φ.

If to reduce the original system (Fig. 6.57) to a system with one degree of freedom (that can be done with the coordinate *y* exclusion), then the expression (6.243) can be presented in the detailed form, believing that:

$$(k_2 l_2 - k_1 l_1)^2 = (k_2 l_2)^2 - 2k_1 k_2 l_1 l_2 + (k_1 l_1)^2. \tag{6.249}$$

To introduce a format of the computational scheme reflecting features of the motion of the solid body relative to the point O, we will return to the Eq. (6.246). Let us make a number of transformations with (6.246):

$$Jp^2\bar{\varphi} + \frac{A_1k_1l_1^2 + A_1k_2l_2^2 - k_1^2l_1^2 + 2k_1k_2l_1l_2 - k_2l_2^2}{A_1} = \frac{k(k_1l_1 - k_2l_2)}{A_1}\bar{z}. \quad (6.250)$$

Then (6.250) can be written as follows:

$$\begin{aligned} &Jp^2 + \frac{k_1l_1^2(mp^2 + k + k_2)}{A_1} + \frac{k_2l_2^2(mp^2 + k + k_1)}{A_1} \\ &+ \frac{k_1k_2l_1^2i}{A_1} + \frac{k_1k_2l_2^2}{iA_1} = \frac{k(k_1l_1 - k_2l_2)^2}{A_1}. \end{aligned} \quad (6.251)$$

In this case the computational scheme (Fig. 6.57) with the variable y exclusion will take a form, as in Fig. 6.64.

Let us note that in Fig. 6.64 the accepted designations of points A and B localize conditions of adjunction of standard elements and quasisprings to the object making the rotationally oscillatory motion along the coordinate φ. The reduced moment of forces, applied to the object with the moment of inertia J, is formed by the kinematic disturbance z:

$$\overline{M}_{\text{red}} = \frac{k \cdot (k_1l_1 - k_2l_2)}{mp^2 + k + k_1 + k_2} \cdot \bar{z}. \quad (6.252)$$

As for elements of structure in Fig. 6.64 with use of A_1, they are quasisprings whose stiffnesses in rotary motion make:

$$k_{\text{red}1} = -\frac{k_1^2l_1}{mp^2 + k + k_1 + k_2}; \quad (6.253)$$

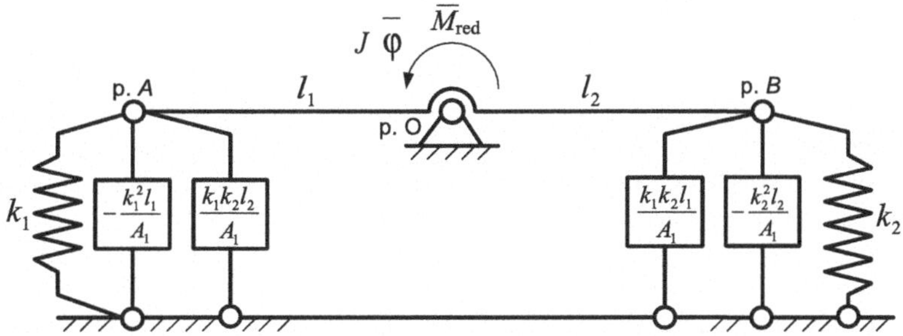

Fig. 6.64 The computational scheme of the equivalent system with one degree of freedom with an object which condition is described by the coordinate $\varphi(A_1 = mp^2 + k + k_1 + k_2)$

$$k_{\text{red2}} = -\frac{k_2^2 l_2}{A_1}; \tag{6.254}$$

$$k_{\text{red3}} = \frac{k_1 k_2 l_1}{A_1}; \tag{6.255}$$

$$k_{\text{red4}} = \frac{k_1 k_2 l_2}{A_1}. \tag{6.256}$$

Thus, the combined system (Fig. 6.57) is reduced to a lever mechanism in which the lever has the moment of inertia J. At the same time the kinematic disturbance $\bar{z}$ it will be transformed to the reduced moment of forces (Fig. 6.64). If to accept that $m = 0$, then

$$\begin{aligned} k_{\text{red1}} &= -\frac{k_1^2 l_1}{k + k_1 + k_2}; \; k_{\text{red2}} = -\frac{k_2^2 l_2}{k + k_1 + k_2}; \\ k_{\text{red3}} &= \frac{k_1 k_2 l_1}{k + k_1 + k_2}; \; k_{\text{red4}} = \frac{k_1 k_2 l_2}{k + k_1 + k_2}, \end{aligned} \tag{6.257}$$

that allows us to reduce elastic elements to points A and B, which yields the following results:

$$k_{\text{red}_A} = \frac{kk_1 + k_1 k_2 \cdot (1 + i)}{k + k_1 + k_2}; \tag{6.258}$$

$$k_{\text{red}_B} = \frac{kk_2 i + k_1 k_2 \cdot (1 + i)}{k + k_1 + k_2}. \tag{6.259}$$

Thus, consideration of features of formation of lever linkages shows that dynamic properties of mechanical oscillatory systems significantly depend on features of the partial systems that make the system basis. If partial systems are non-uniform, i.e. can make both translational and rotary motions, then lever linkages are of great importance. One can assume that lever linkages in oscillatory systems can take various forms and it depends on the choice of the generalized coordinates.

The mentioned materials demonstrate that between rotary and translational motions, while implementing structural approaches, the adequacy of ideas about transformation rules is revealed. The main thing is that lever linkages occur as a result of abstraction from some features of the rotary motion. It is characterized by the fact that constraints between standard elements in the system are separated in space, which makes the introduction of both lever linkages and lever mechanisms reasonable.

It is shown that separation of lever linkages is quite explicable if to take into account the nature of localization of the places where the constraints are fixed relative to the motionless point of the lever. If the constraints are numerous and are also located from the different sides of the lever rotation point, then the structure of the transfer ratios has to be corrected depending on the choice of attachment points of the elements in relation to the lever rotation point.

In this case, partial systems are chosen in such a way that the rotary and translational motions are physically separated and the elastic elements k_1 and k_2 act as interpartial constraints. However, there are systems in which the solid body unites separate types of motions in one (this is a flat motion). In this case partial constraints will be of other nature (they are often called inertial [11, 65]), though such systems can be also reduced to the equivalent type of the chain system that, in particular, was considered in [66–68].

6.8 Dynamic Interactions Between Solid Bodies, Having Rotation Points. Lever Linkages

Lever linkages in dynamics of machines are most often considered within the theory of plain mechanisms that found reflection in researches [13, 52, 69]. In recent years the attention to issues of the dynamic interactions caused by the vibration processes characteristic of transport problems, vibration stabilization, technologies of hardening of materials and mechatronics has significantly increased [7, 70–72]. Lever linkages have a great impact on interpartial constraints and the possibilities of implementing the modes of dynamic absorbing of external influences. Some dynamic features in the approaches, relying on use of the concept of feedback in structural interpretations of oscillatory systems, were translated into [27]. With that, many aspects of a current problem of dynamics of machines, taking into account lever linkages, have not obtained due consideration yet. In particular, this relates to the consideration of features and constructive and engineering forms of implementation of interactions in systems with several degrees of freedom.

6.8.1 Mechanical Systems with Lever Linkages (as Levers of the First Kind)

The system from two solid bodies containing flat oscillations (Fig. 6.65) concerning rotation points O_1 and O_2 is considered. Mass-and-inertia properties of solid bodies are modelled by material points m_1 and m_2 on the ends of rigid levers which are connected by elastic elements to stiffnesses k_1, k_2, k_3. Places of fixing of elastic elements (Fig. 6.65) are respectively designated by $O_1A = l_1$, $O_1B = l_1$, $O_1A_1 = l_3$, $O_2l_4 = l_4$. As the generalized coordinates in motionless basis, φ_1 and φ_2 and also y_1 and y_2 are chosen.

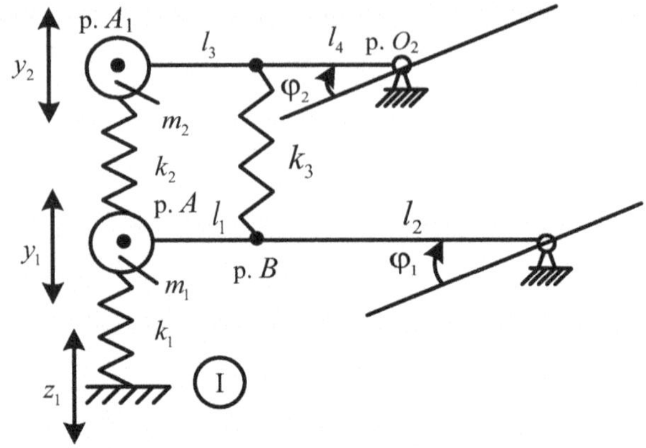

Fig. 6.65 The computational scheme of the system with two levers of the first kind

Let us write down expressions for kinetic and potential energy:

$$\mathrm{T} = \frac{1}{2} m_1 l_1^2 \dot{\varphi}_1^2 + \frac{1}{2} m_2 l_3^2 \dot{\varphi}_2^2, \tag{6.260}$$

$$\Pi = \frac{1}{2} k_1 (\varphi_1 l)^2 + \frac{1}{2} k_2 (l_3 \varphi_2 - l_1 \varphi_1)^2 + \frac{1}{2} k_3 (\varphi_2 l_4 - \varphi_1 l_2)^2, \tag{6.261}$$

Let us make a number of transformations and we will write down the equations of motion with a kinematic external influence z_1:

$$\varphi''_1 m_1 l_1^2 + \varphi_1 (k_1 l_1^2 + k_2 l_1^2 + k_3 l_2^2) - \varphi_2 (k_2 l_1 l_3 + k_3 l_2 l_4) = k l_1 z_1, \tag{6.262}$$

$$\varphi''_2 m_2 l_3^2 + \varphi_2 (k_2 l_3^2 + k_3 l_4^2) - \varphi_1 (k_2 l_1 l_3 + k_3 l_2 l_4) = 0. \tag{6.263}$$

Using Laplace's transformations, we will transform expressions (6.262) and (6.263) and we will construct the structural diagram of the original system (Fig. 6.66).

Let us transform transfer functions of partial systems:

$$\begin{aligned} & m_1 l_1^2 p^2 + k_1 l_1^2 + k_2 l_1^2 + k_3 l_2^2 - k_2 l_1 l_3 - k_3 l_2 l_4 + (k_2 l_1 l_3 + k_2 l_2 l_4) \\ & \quad = m_1 l_1^2 p^2 + k_1 l_1^2 + k_2 l_1 (l_1 - l_3) + k_3 l_2 (l_2 - l_4) + (k_2 l_1 l_3 + k_3 l_2 l_4). \end{aligned} \tag{6.264}$$

$$\begin{aligned} & m_2 l_3^2 p^2 + k_2 l_3^2 + k_3 l_4^2 - k_2 l_1 l_3 - k_3 l_2 l_4 + (k_2 l_1 l_3 + k_3 l_2 l_4) \\ & \quad = m_2 l_3^2 p^2 + l_3 (k_2 l_3 - k_2 l_1) + l_4 (k_3 l_4 - k_3 l_2) + (k_2 l_1 l_3 + k_3 l_2 l_4). \end{aligned} \tag{6.265}$$

Using (6.264) and (6.265) and the structural diagram in Fig. 6.66, we can construct an equivalent computational scheme for the original system in Fig. 6.65. This scheme will be adequate to the system which could make translational, but not rotary motions. At the same time φ_1 and φ_2 will conditionally reflect progress (Fig. 6.67).

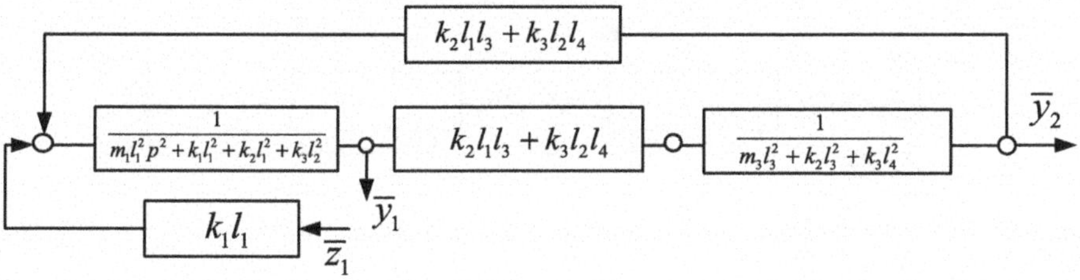

Fig. 6.66 The structural diagram of the system in Fig. 6.65

Fig. 6.67 The computational scheme in coordinates φ_1 and φ_2, the equivalent system with partial structures of translational motion

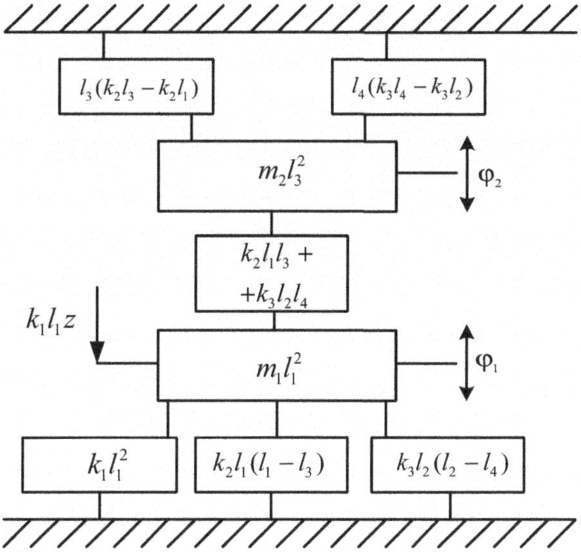

Let us note that the computational scheme according to Fig. 6.68 is display equivalent in the dynamic relation of the system having coordinates φ_1 and φ_2, but having all properties of a system executing not angular and conditionally translational motions. From this we can conclude that the system in Fig. 6.65 can be considered as the mechanical oscillatory system of chain type. The computational scheme of the chain system (prototype) is provided in Fig. 6.68.

Let us find transfer functions of the original system (see Fig. 6.65), using the structural diagram in Fig. 6.66 or the computational scheme in Fig. 6.67:

$$W_1(p) = \frac{\overline{\varphi}_1}{k_1l_1\overline{z}_1} = \frac{m_2l_3^2p^2 + k_2l_3^2 + k_3l_4^2}{(m_1l_1^2p^2 + k_1l_1^2 + k_2l_1^2 + k_3l_2^2)(m_2l_3^2p^2 + k_2l_3^2 + k_3l_4^2) - (k_2l_1l_3 + k_3l_3l_4^2) - (k_2l_1l_3 + k_3l_3l_4)^2} \quad (6.266)$$

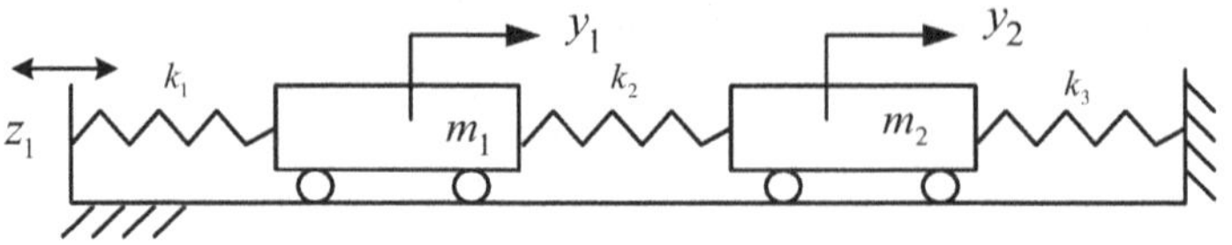

Fig. 6.68 The computational scheme of the chain mechanical system considered as a prototype for the system in Fig. 6.67

$$W_2(p) = \frac{\bar{\varphi}_2}{k_1 l_1 \bar{z}_1} = \frac{k_2 l_1 l_3 + k_3 l_2 l_4}{(m_1 l_1^2 p^2 + k_1 l_1^2 + k_2 l_1^2 + k_3 l_2^2)(m_2 l_3^2 p^2 + k_2 l_3^2 + k_3 l_4^2) - (k_2 l_1 l_3 + k_3 l_2 l_4)^2}. \quad (6.267)$$

Expressions (6.261) and (6.262) can be transformed taking into account that $y_1 = \varphi l_{11}$, $y_2 = \varphi l_{23}$, then

$$W_1'(p) = \frac{\bar{y}_1}{\bar{z}_1} = \frac{k_1 l_1^2 (m_2 l_3^2 p^2 + k_2 l_3^2 + k_3 l_4^2)}{l_1^2 l_3^2 (m_1 p^2 + k_1 + k_2 + k_3 i_1^2)(m_3 p^2 + k_2 + k_3 i_2^2) - (k_2 + k_3 i_1 i_2)^2}, \quad (6.268)$$

$$W_2'(p) = \frac{\bar{y}_2}{\bar{z}_1} = \frac{k_1 l_1 l_3 (k_2 l_1 l_3 + k_3 l_2 l_4)}{l_1^2 l_3^2 (m_1 p^2 + k_1 + k_2 + k_3 i_1^2)(m_3 p^2 + k_2 + k_3 i_2^2) - (k_2 + k_3 i_1 i_2)^2}. \quad (6.269)$$

Let us make reductions of (6.263) and (6.264) which can be written down as

$$W_1'(p) = \frac{\bar{y}_1}{\bar{z}_1} = \frac{k_1 (m_2 p^2 + k_2 + k_3 i_2^2)}{(m_1 p^2 + k_1 + k_2 + k_3 i_1^2)(m_3 p^2 + k_2 + k_3 i_2^2) - (k_2 + k_3 i_1 i_2)^2}, \quad (6.270)$$

$$W_2'(p) = \frac{\bar{y}_2}{\bar{z}_1} = \frac{k_1 (k_2 + k_3 i_1 i_2)}{(m_1 p^2 + k_1 + k_2 + k_3 i_1^2)(m_3 p^2 + k_2 + k_3 i_2^2) - (k_2 + k_3 i_1 i_2)^2}, \quad (6.271)$$

where $i_1 = l/l_{21}$, $i_2 = l/l_{43}$.

Comparing the expressions (6.261), (6.262) and (6.265), (6.266), it can be inferred that with the small angular motions the original system (see Fig. 6.65) has all properties of the usual chain system taken for an example (see Fig. 6.68). However, quite certain distinctions also appear in the system. So $k_1 + ki_{31}$ represents a quasi-elastic element which is the reduced stiffness of some structure formed by two standard elementary elastic units k_1 and k_3 and the lever of the first kind

(k_1 and k_3 are attached to the lever at pp. A and B). In turn, elastic standard elements k_2 and k_3 also form a spring $k_2 + k_3 i_1^2$ by means of the lever of the first kind (the points of fastening of the elements k_2 and k_3 are points A_1 and B_1).

If to take the expression (6.263) as a basis, then denominators of transfer functions of partial systems can be presented as

$$m_1p^2 + k_1 + k_2 + k_3 i_1^2 - k_3 i_1 i_2 - k_3 i_1 i_2 = m_1 p^2 + k_1 + k_2 + k_3 i_1(i_1 - i_2) + k_2 + k_3 i_1 i_2, \quad (6.272)$$

$$m_2p^2 + k_2 + k_3 i_2^2 - k_3 i_1 i_2 + k_3 i_1 i_2 = m_2 p^2 + k_3 i_2(i_2 - i_1) + k_2 + k_3 i_1 i_2. \quad (6.273)$$

Using (6.267) and (6.268), we will construct the computational scheme of the original system (Fig. 6.65) in the system of coordinates y_1 and y_2 (Fig. 6.69).

If to compare the computational scheme in Fig. 6.69 to the scheme in Fig. 6.68, then it can be noted that quasisprings emerged in the structure. In the lower cascade the quasispring has stiffness $k_3 i_1(i_1 - i_2)$, which is formed by the system of levers. If to compare elastic structures in Figs. 6.68 and 6.69 (the lower cascade), then in Fig. 6.68 the element k_1 corresponds to the stiffness of the elastic system системы ($k_1 + k_3 i_1(i_1 - i_2)$). Interpartial constraints in Fig. 6.68 are defined by the elastic element k_2; in Fig. 6.69 we respectively have the quasispring parameters—($k_2 + k_3 i_1 i_2$). In the upper cascade in Fig. 6.68 elastic properties are defined by the spring k_3, and in Fig. 6.69, respectively, through ($k_3 i_2(i_2 - i_1)$). Thus, the rotary systems (see Fig. 6.65), owing to existence of lever linkages, form a more complex system of elastic interactions, which predetermines introduction of such a concept as "quasispring". Structures like this are formed by the rules, the details of which are

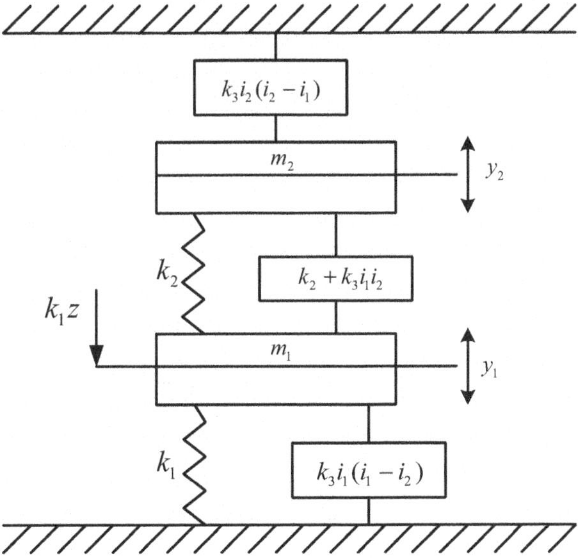

Fig. 6.69 The computational scheme of the system (Fig. 6.65) in the coordinates y_1, y_2 as the equivalent structure with partial formations of translational motion

reasonably provided in [44, 54] where they are also called compacts. Quasisprings, as a concept of dynamic synthesis within the structural theory of vibration protection systems, are connected with definitions of the generalized springs introduced in [8, 51] which, in essence, is in one conceptual space. Distinctions are in the fact that the approach can be widespread not only to static stiffnesses, but also to the dynamic ones which contain mass-and-inertia elements in their structure. However, the general properties of quasisprings remain invariable, if to take into account their opportunities to enter into connection with other elements of systems (including elastic and dissipative ones) on the basis of rules of the consecutive and parallel connection of springs [12].

6.8.2 Features of the Combined Systems

Computational scheme (Fig. 6.70) presents a combined system in which there are two partial systems.

The first system has the solid body with the moment of inertia J_1 oscillating around the point O_1 where it is accepted that $l_1 = O_1A$, $l_2 = O_1B$. The second solid body (J_2) has the center of rotation at p. O_2 and generally can consider as the lever of the second kind (the solid body J_1 is the lever of the first kind). At this $AO_{12} = A_1O_2 = l_3$, $O_2B_1 = l_4$. φ_1, φ_2 and y_1 and y_2 are considered as coordinates of state.

Believing that kinetic and potential energy are defined by expressions:

$$\mathrm{T} = \frac{1}{2}J_1\dot{\varphi}_1^2 + \frac{1}{2}J_2\dot{\varphi}_2, \tag{6.274}$$

$$\Pi = \frac{1}{2}k_1(\varphi_1 l - z_1)^2 + \frac{1}{2}k_2(l_3\varphi_2 - l_1\varphi_1)^2 + \frac{1}{2}k_3(-\varphi_2 l_4 - \varphi_1 l_2)^2. \tag{6.275}$$

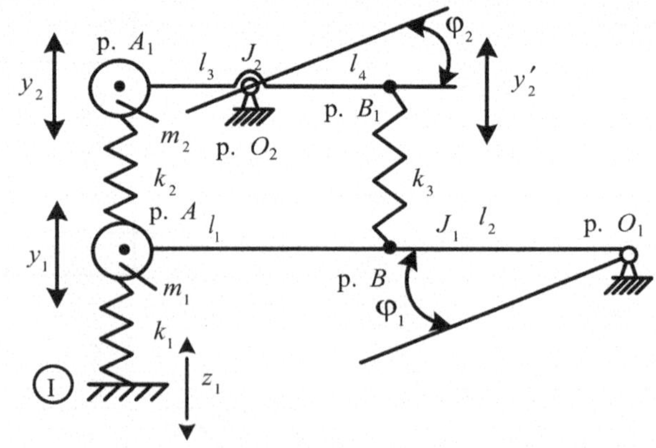

Fig. 6.70 The computational scheme of the system in the coordinates φ_1, φ_2 (the upper solid body J_2 is the lever of the second kind)

The equations of the motion in the coordinates φ_1, φ_2 are of the form of

$$\varphi_1'' J_1 + \varphi_1(k_1 l_1^2 + k_2 l_1^2 + k_3 l_2^2) - (k_2 l_3 l_1 - k_3 l_2 l_4) = k_1 l_1 z_1, \tag{6.276}$$

$$\varphi_2'' J_2 + \varphi_2(k_2 l_3^2 + k_3 l_4^2) - (k_2 l_1 l_3 - k_3 l_2 l_4) = 0. \tag{6.277}$$

The structural diagram of the system is provided in Fig. 6.71.

Let us transform transfer functions of partial systems:

$$\begin{aligned} J_1 p^2 + k_1 l_1^2 + k_2 l_3^2 + k_3 l_3^2 &= J_1 p^2 + k_1 l_1^2 + k_2 l_3^2 + k_3 l_3^2 - (k_2 l_3 l_1 - k_2 l_2 l_4) \\ &+ (k_2 l_1 l_3 - k_3 l_2 l_4) = J_1 p^2 + k_1 l_1^2 + k_2 l_3 (l_3 - l_1) \\ &+ k_3 l_2 (l_2 - l_4) + (k_2 l_3 l_1 - k_3 l_2 l_4); \end{aligned} \tag{6.278}$$

$$\begin{aligned} J_2 p^2 &+ k_2 l_3^2 + k_3 l_4^2 - (k_2 l_1 l_3 - k_3 l_2 l_4) + (k_2 l_1 l_3 - k_2 l_2 l_4) \\ &= J_2 p^2 + k_2 l_3 (l_1 - l_3) + k_3 l_4 (l_4 - l_2) + (k_2 l_1 l_3 - k_2 l_2 l_4). \end{aligned} \tag{6.279}$$

The computational scheme for the original system (see Fig. 6.70) in coordinates φ_1 and φ_2 will are of the form, as shown in Fig. 6.72.

Let us note that in the schemes of Fig. 6.72 analogies $J_1 \to m_1'$, $J_2 \to m_2'$,$(k_1 l_1^2 + k_2 l_3(l_1 - l_3) + k_3 l_2(l_2 - l_4) \to k_1'$, $(k_2 l_3(l_1 - l_3) + k_3 l_4(l_2 - l_4)) \to k_3'$, $(k_2(l_1 l_3 - k_3 l_2 l_4) \to k_2'$ are implemented.

For further researches it is important to note that the systems of rotary type, in which partial systems physically correspond to the rotating solid bodies, possess a property to act as analogs of chain translational systems. This property is illustrated by comparison of computational schemes in Fig. 6.72. Believing that $J_1 = m_1 l_1^2$, $J_2 = m_1 l_3^2$, it is possible to transform the initial mathematical model to the description in the coordinates y_1 and y_2 (at the same time $y_1 = \varphi_1 l_1$; $y_2 = \varphi_2 l_3$). Let us find expressions for the kinetic and potential energy:

$$\mathrm{T} = \frac{1}{2} m_1 \dot{y}_1^2 + \frac{1}{2} m_2 \dot{y}_2^2; \tag{6.280}$$

$$\Pi = \frac{1}{2} k_1 (y_1 - z_1)^2 + \frac{1}{2} k_2 (y_2 - y_1)^2 + \frac{1}{2} k_3 (-y_2' - \phi_1 l_2)^2,$$

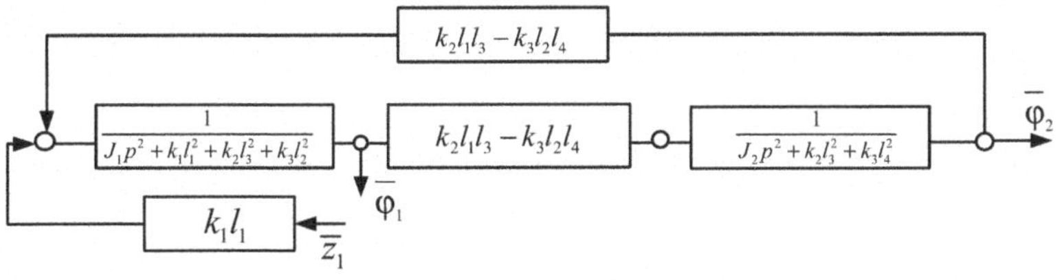

Fig. 6.71 The structural diagram of the system (Fig. 6.70) in the coordinates φ_1, φ_2

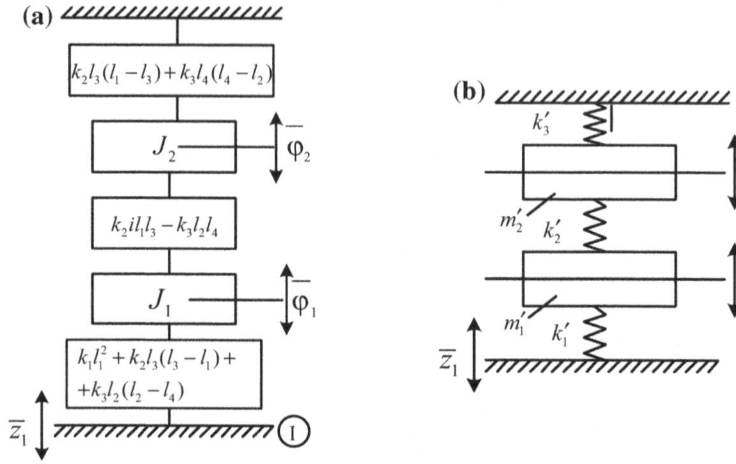

Fig. 6.72 The computational scheme of the original system (Fig. 6.70) in the coordinates φ_1 and φ_2, which corresponds to the coordinates of translational motion (**a**); computational scheme of the equivalent system of translational motion in the coordinates y'_1 and y'_2 (**b**)

where $y_2' = i_1 y_1, i = \frac{l_4}{l_2}$. As $\varphi_1 = \frac{y_1}{l_1}$,

$$\Pi = \frac{1}{2}k_1(y_1 - z_1)^2 + \frac{1}{2}k_2(y_2 - y_1)^2 + \frac{1}{2}k_3(-i_1 y_2' - y_1 l_2)^2, \tag{6.281}$$

where $i_1 = \frac{l_2}{l_4}$.

Let us make a number of intermediate calculations and we will write down the equations of motion in the system of coordinates y_1 and y_2:

$$m_1 y_1'' + y_1(k_1 + k_2 + k_3 i_2^2) - y_2(k_2 - k_3 i_1 i_2) = k_1 z, \tag{6.282}$$

$$m_2 y_2'' + y_2(k_2 + k_3 i_1^2) - y_1(k_2 - k_3 i_1 i_2) = 0. \tag{6.283}$$

The structural diagram of the system is of the form, as shown in Fig. 6.73. Let us execute transformations of transfer functions of partial systems:

$$m_1 p^2 + k_1 + k_2 + k_3 i_2^2 + k_3 i_1 i_2 - k_3 i_1 i_2 = m_1 p^2 + k_1 + k_3 i_2(i_1 + i_2) + k_2 - k_3 i_1 i_2, \tag{6.284}$$

$$\begin{aligned} m_2 p^2 + k_2 + k_3 i_1^2 &= m_2 p^2 + k_3 i_1^2 + k_2 + k_3 i_1 i_2 - k_3 i_1 i_2 \\ &= m_2 p^2 + k_2 - k_3 i_1 i_2 + k_3 i_1 (i_1 + i_2). \end{aligned} \tag{6.285}$$

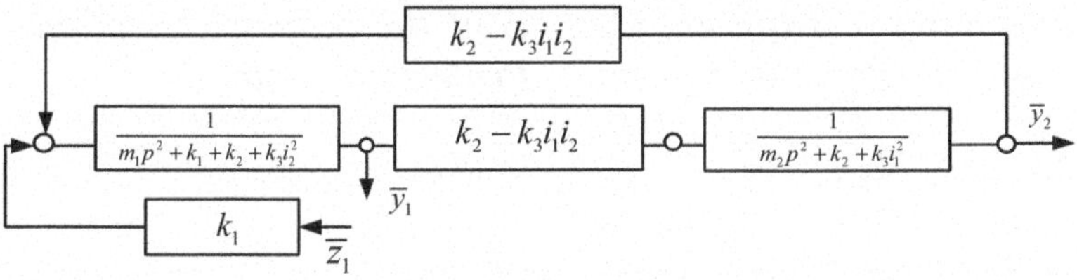

Fig. 6.73 The structural diagram of the system (Fig. 6.70) in the coordinates y_1 and y_2

Using (6.284) and (6.285), we will construct the computational scheme of the system in coordinates y_1 and y_2 with reduction of J_1 to the point A $\left(J_1 = m_1 l_1^2\right)$, and J_2 respectively to the point B $\left(J_2 = m_1 l_3^2\right)$, as shown in Fig. 6.70.

Let us construct a computational scheme of the original system in the coordinates y_1 and y_2 on the basis of the structural diagram (Fig. 6.73) and expressions (6.279) and (6.280), as it is shown in Fig. 6.74.

Such a computational scheme is also equivalent to the scheme in Fig. 6.72b. At the same time $(k_3 i_1 (i_1 + i_2))$ corresponds (is similar in its nature) k'_3; $(k_2 - k_3 i_1 i_2)$ corresponds to k'_2, and $(k_1 + k_3(i_2(i_2 + i_1))$ corresponds to k'_1; m_1 corresponds to m'_1 and m_2 corresponds to m'_2. Quasisprings give an idea of spatial forms of an arrangement of the system of elastic elements.

Similar ratios make it possible to distinguish a number of differences. For the scheme in Fig. 6.74 one can note the emergence of the quasisprings having coefficients of the reduced stiffness: in the first cascade—$k_3 i_2 (i_1 + i_2)$; in the second —$k_3 i_1 i_2$; in the third—$k_3 i_1 (i_1 + i_2)$.

Fig. 6.74 The computational scheme of the original system (Fig. 6.70) in the coordinates y_1 and y_2

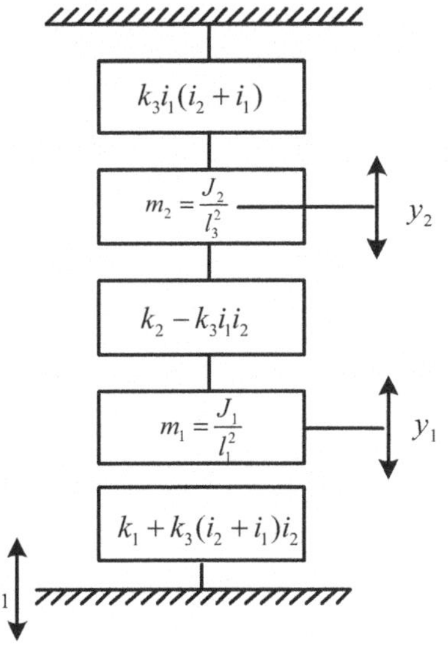

6.8.3 Systems at Different Options of the Selected Location of the Object in Vibration Protection Systems

Taking into consideration that $J_2 \to 0$, we will designate the object of protection in the form of a solid body J_2. Let us write down an expression for the transfer function at an output influence φ_1 and input influence z_1:

$$W_1(p) = \frac{\bar{\varphi}_1}{\bar{z}_1} = \frac{k_1 l_1 (J_2 p^2 + k_2 l_3^2 + k_3 l_4^2)}{A_0'}, \qquad (6.286)$$

where

$$A_0' = (J_1 p^2 + k_1 l_1^2 + k_2 l_3^2 + k_3 l_2^2)(J_1 p^2 + k_2 l_3^2 + k_3 l_4^2) - (k_2 l_1 l_3 - k_3 l_2 l_4)^2. \qquad (6.287)$$

Let us accept that $J_2 = 0$, then

$$W_1'(p) = \frac{\bar{\varphi}_1}{\bar{z}_1} = \frac{k_1 l_1}{(J_1 p^2 + k_1 l_1^2 + k_2 l_1^2 + k_3 l_2^2) - \frac{(k_2 l_1 l_3 - k_3 l_2 l_4)^2}{(k_2 l_3^2 + k_3 l_4^2)}}, \qquad (6.288)$$

$$W_1''(p) = \frac{\bar{\varphi}_1}{\bar{z}_1} = \frac{k_1 l_1}{l_1^2 (\frac{J_1 p^2}{l_1^2} + k_1 + k_2 + k_3 i_2^2) - \frac{(k_2 - k_3 i_1 i_2)^2}{(k_2 + k_3 i_2)}}. \qquad (6.289)$$

Thus, as $y_1 = \varphi_1 l_1$, (6.284) will be rearranged to a form

$$\begin{aligned} W_1'''(p) = \frac{\bar{y}_1}{\bar{z}_1} &= \frac{k_1}{(m_1 p^2 + k_1 + k_2 + k_3 i_2^2) - \frac{(k_2 - k_3 i_1 i_2)^2}{(k_2 + k_3 i^2)}} \\ &= \frac{k_1}{m_1 p^2 + k_1 + \frac{k_2^2 + k_3 i_2^2 + k_2 k_3 i_1^2 - k_2^2 + 2k_2 k_3 i_1 i_2 - k_3^2 i_1^2 i_2^2}{k_2 + k_3 i_1^2}}; \end{aligned} \qquad (6.290)$$

$$W_1'''(p) = \frac{k_1}{m_1 p^2 + k_1 + \frac{k_3 i_2^2 + k_2 k_3 i_1 (i_1 + 2 i_2)}{k_2 + k_3 i_1^2}}. \qquad (6.291)$$

According to (6.286) computational scheme of the rotary system (see Fig. 6.70) at $J_2 = 0$, i.e. with the object of protection $m_1 = J/l2_{11}$as the system of equivalent progress, is of the form, as shown in Fig. 6.75.

If to accept that the object of protection will be $m_2 = J/l2_{23}$ at $J_1 = 0$, then transfer function of the system (see Fig. 6.70) can be written as

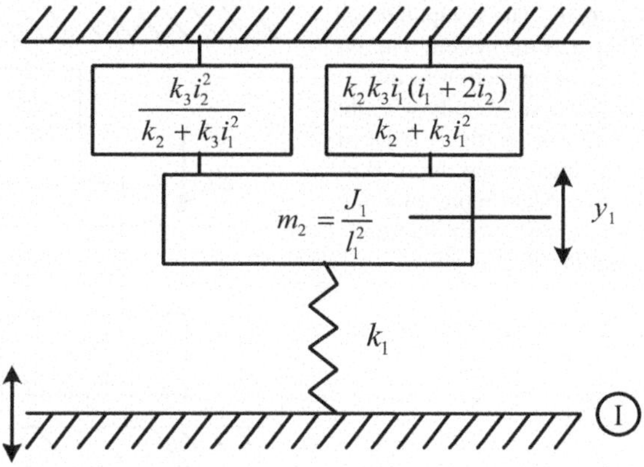

Fig. 6.75 The computational scheme of the system (see Fig. 6.70) with the object of protection $m_1 = J_1/l_1^2$ with $J_2 = 0$

$$W_2(p) = \frac{\bar{\phi}_2}{\bar{z}_1} = \frac{(k_2 l_1 l_3 - k_3 l_2 l_4) k_1 l_1}{(k_1 l_1^2 + k_2 l_1^2 + k_3 l_2^2)(J_2 p^2 + k_3 l_2^2 + k_3 l_2^2) - (k_2 l_1 l_3 - k_3 l_2 l_4)^2}. \tag{6.292}$$

Let us transform (6.287), accepting that $y_2 = \varphi l_{23}$:

$$W_2'(p) = \frac{\bar{y}_2}{\bar{z}_1} = \frac{\frac{k_2 l_1 l_3 (k_2 l_1 l_3 - k_3 l_2 l_4)}{k_1 l_1^2 + k_2 l_1^2 + k_3 l_2^2}}{(k_1 l_1^2 + k_2 l_1^2 + k_3 l_2^2)(J_2 p^2 + k_3 l_2^2 + k_3 l_2^2) - (k_2 l_1 l_3 - k_3 l_2 l_4)^2}$$

$$= \frac{\frac{k_2 l_1^2 l_3^2 (k_2 - k_3 i_1 i_2)}{l_1^2((k_1 + k_2) + k_3 i_2^2)}}{l_3^2 (m_2 p^2 + k_2 + k_3 i_1^2) - \frac{l_1^2 l_3^2 (k_2 - k_3 i_1 i_2)^2}{l_1^2 (k_1 + k_2 + k_3 i_2^2)}} = \frac{\frac{k_1 (k_2 - k_3 i_1 i_2)}{(k_1 + k_2) + k_3 i_2^2}}{m_2 p^2 + k_2 + k_3 i_1^2 - \frac{(k_2 - k_3 i_1 i_2)^2}{k_1 + k_2 + k_3 i_2^2}}$$

$$= \frac{\frac{k_1 (k_2 - k_3 i_1 i_2)}{(k_1 + k_2) + k_3 i_2^2}}{m_2 p^2 + \frac{k_1 k_2 + k_2^2 + k_2 k_3 i_2^2 + k_1 k_3 i_1^2 + k_1 k_2 i_1^2 + k_3^2 i_1^2 i_1^2 - k_2^2 + 2 k_2 k_3 i_1 i_2 - k_3^2 i_1^2 i_2^2}{k_1 + k_2 + k_3 i_2^2}}$$

$$= \frac{\frac{k_1 (k_2 - k_3 i_1 i_2)}{(k_1 + k_2) + k_3 i_2^2}}{m_2 p^2 + \frac{k_1 k_2 + k_3 (k_1 + k_2) i_1^2 + k_2 k_3 i_2 (i_2 + 2 i_1)}{k_1 + k_2 + k_3 i_2^2}};$$

$$W_2'(p) = \frac{\bar{y}_2}{\bar{z}_1} = \frac{\frac{k_1 (k_2 - k_3 i_1 i_2)}{(k_1 + k_2) + k_3 i_2^2}}{m_2 p^2 + \frac{k_1 k_2 + k_3 (k_1 + k_2) i_1^2 + k_2 k_3 i_2 (i_2 + 2 i_1)}{k_1 + k_2 + k_3 i_2^2}}. \tag{6.293}$$

Using (6.288) we will construct the computational scheme of the system in which the solid body J acts as the object of protection $_2(m_2 = J/l2_{23})$. The scheme is provided in Fig. 6.76.

In the scheme of Fig. 6.76 Q'_1 reflects the external force (the reduced kinematic disturbance z_1) which is defined by expression.

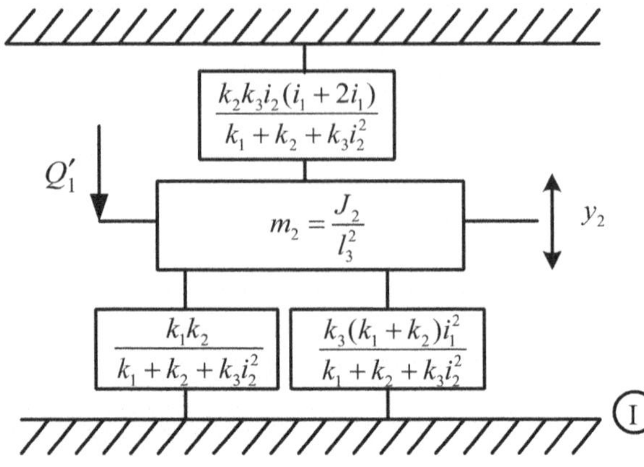

Fig. 6.76 The computational scheme of the system with the object of protection in the form of a solid body with the moment of inertia J_2 reduced to the mass m_2 at the p. B in the equivalent representation with a translational motion along the coordinate y_2

$$Q_1' = \frac{k_1(k_2 - k_3 i_1 i_2)}{k_1 + k_2 + k_3 i_2^2} z_1. \tag{6.294}$$

Let us note that in this case the equivalent representation of vibration protection systems, obtained at $J_1 = 0$, elastic properties of the system are determined by three quasisprings with stiffnesses:

$$k_{\text{red}1} = \frac{k_1 k_2}{k_1 + k_2 + k_3 i_2^2}; \tag{6.295}$$

$$k_{\text{red}2} = \frac{k_3(k_1 + k_2) i_1^2}{k_1 + k_2 + k_3 i_2^2}; \tag{6.296}$$

$$k_{\text{red}3} = \frac{k_2 k_3 i_2 (i_2 + 2i_1)}{k_1 + k_2 + k_3 i_2^2} z_1. \tag{6.297}$$

The developed approach allows us to consider mass-and-inertia properties of levers at the selected object of vibration protection. So, for the scheme in Fig. 6.66 at $J_1 \neq 0$, the stiffnesses of quasisprings will depend on the frequency of external disturbance.

$$k'_{\text{red}1} = \frac{k_1 k_2}{m_1 p^2 + k_1 + k_2 + k_3 i_2^2}; \; k''_{\text{red}1} = \frac{k_3 (k_1 + k_2) i_1^2}{m_1 p^2 + k_1 + k_2 + k_3 i_2^2};$$
$$k'''_{\text{red}1} = \frac{k_2 k_3 i_2 (i_2 + 2i_1)}{m_1 p^2 + k_1 + k_2 + k_3 i_2^2}.$$

The expression for the reduced external force will also change in the same way:

$$Q_1' = \frac{k_1(k_2 - k_3 i_1 i_2)}{m_1 p^2 + k_1 + k_2 + k_3 i_{21}^2}.$$

In its turn, the accounting of inertial properties of the lever in the form of a solid body with the moment of inertia $J_2 \neq 0$ is carried out according to the same scheme. So, for the scheme in Fig. 6.75, the dynamic stiffness of the quasispring

$$k_{red} = \frac{k_3 i_2^2 + k_2 k_3 i_1 (i_1 + 2 i_2)}{m_2 p^2 + k_2 + k_3 i_1^2}. \quad (6.298)$$

Issues of taking into consideration of mass-and-inertia properties of lever linkages have independent significance, some important features are considered in [63].

Dynamic properties of mechanical oscillatory systems depend on structure of elementary units, structure of the formed constraints defining a spatial configuration of systems in general. With that, the form and features of the structure of partial systems are of great importance. With the existing research methods, mechanical oscillatory systems seldom become an object of the detailed studying. At the same time, with two types of the motion of partial systems (rotary and translational motion) a large variety of constraints between partial systems is possible, even for systems with two degrees of freedom. Prospects of the data of all possible combinations of interaction of partial systems to the equivalent schemes of translational motion are of interest to further researches.

Properties of equivalence of the representations of systems with different types of motions at the partial level are quite explainable if to take into account properties of small oscillations. Lever linkages play an essential role in mechanical oscillating motions, creating an opportunity to estimate spatial forms of interaction that is important for development of theoretical ideas of laws of formation of dynamic states of the multidimensional systems distributed in space.

References

1. Weits, V.L., Kochura, A.E., Martynenko, A.M.: Dynamic Calculations of Drive Machines, 352 pp. Mashinostroenie Publ, Leningrad (1971)
2. Wolfson, I., Kolovsky, M.Z.: Nonlinear Problems of Machine Dynamics, 284 pp. Mashinostroenie Publication, Leningrad (1968)
3. Belokobylsky, S.V., Eliseev, S.V., Kashuba, V.B.: Applied Problems of the Structural Theory of Vibration Protection Systems, 363 pp. Polytekhnika Publication, St.Petersburg (2013)
4. Briskin, E.S., Zhoga, V.V., Chernyshev, V.V., et al.: Dynamics and motion control of walking machines with cyclic propellers. In: Briskin E.S. (ed.), 191 pp. Mashinostroenie Publication, Moscow (2009)
5. Kreinin, G.V., Bessonov A.P., Pavlov B.I., et al.: Kinematics, dynamics and accuracy of mechanisms. In: Kreinin, G.V., (ed.), 216 pp. Mashinostroenie Publication, Moscow (1984)
6. Shchepetilnikov, V.A.: The Balancing of Mechanisms, 256 pp. Mashinostroenie Publication, Moscow (1982)

7. Eliseev, S.V., Reznik, YuN, Khomenko, A.P.: Mechatronic Approaches in the Dynamics of Mechanical Oscillatory Systems, 384 pp. Nauka Publication, Novosibirsk (2011)
8. Eliseev, S.V., Reznik, YuN, Khomenko, A.P., et al.: Dynamic Synthesis in Generalized Problems of Vibration Protection and Vibration Isolation of Technical Objects, 523 pp. ISU Publication, Irkutsk (2008)
9. Eliseev, S.V., Moskovskikh, A.O., Bolshakov, R.S., Savchenko, A.A.: The possibilities of integrating the methods of the theory of circuits and the theory of automatic control in problems of the dynamics of machines (Electronic resource). technomag.edu.ru. In: Science and Education: Electronic Science and Technical ed., vol. 6. Access mode: http://technomag.edu.ru/doc/378699.html. Access date 10 June 2012 (2012)
10. Eliseev, S.V. Trofimov, A.N., Bolshakov, R.S., Savchenko, A.A.: The concept of feedback in the dynamics of mechanical systems and dynamic oscillation damping (Electronic resource). technomag.edu.ru. In: Science and Education: Electronic Science and Technical ed., vol. 5. Access mode http://technomag.edu.ru/doc/378353.html. Access date 10 May 2012 (2012)
11. Kolovsky, M.Z.: Automatic Control of Vibration Protection Systems, 320 pp. Nauka Publication, Moscow (1976)
12. Khomenko, A.P., Eliseev, S.V., Ermoshenko, YuV: System Analysis and Mathematical Modeling in Mechatronics of Vibration Protection Systems, 288 pp. IrGUPS Publication, Irkutsk (2012)
13. Levitsky, N.I.: Fluctuations in the Mechanisms, 358 pp. Nauka Publcation, Moscow (1988)
14. Eliseev, S.V., Volkov, L.N., Kukharenko, V.P.: Dynamics of Mechanical Systems with Additional Constraints, 214 pp. Nauka Publication, Novosibirsk (1990)
15. Bessonov, A.P.: Fundamentals of the Dynamics of Mechanisms with a Variable mass of Units, 282 pp. Nauka Publication, Moscow (1967)
16. Lurie, A.I.: Operational Calculus and Application in Engineering, 368 pp. Nauka Publication, Moscow (1959)
17. Druzhinsky, I.A.: Mechanical Chains, 240 pp. Mashinostroenie Publication, Moscow (1977)
18. Harris, C.M., Allan, G.: Shock and Vibration Handbook, 877 pp. Mc Graw-Hill, N.–Y (2002)
19. Eliseev, S.V.: Impedance methods in the study of mechanical systems. In: Fundamentals of the Theory, 86 pp. IPI Publication, Irkutsk (1979)
20. Parshuta, E.A.: Evaluation of the dynamic properties of vibration protection systems in the study and design of engineering objects: author's abstract of Ph.D. (Engineering) Thesis, 17 pp. BrSU Publication, Bratsk (2014)
21. Khomenko, A.P., Eliseev, S.V., Artyunin, A.I., et al.: Mechanisms in elastic oscillatory systems: features of taking into account dynamic properties, problems of vibration protection of machines, instruments and equipment, 187 pp. IrGUPS Publication, Irkutsk. Dep. in VINITI 08.15.13, No. 243-B (2013)
22. Eliseev, S.V., Kaimov, E.V., Kinash, N. Zh.: Features of taking into account slip in kinematic pairs of dynamic vibration dampers. Bull. Irkutsk State Tech. Univ. **10**(93), 43–49 (2014)
23. Eliseev, S.V., Trofimov, A.N., Kaimov, E.V.: Constructive-technical forms of the use of linkages in problems of the dynamics of vehicle suspensions. Internat. J. Appl. Basic Res. 5–2, 14–23 (2014)
24. Ragulskis, K.M.: Mechanisms on a Vibrating Basis (Questions of Dynamics and Stability), 232 pp. Institute of Energy and Electrical Engineering, Academy of Sciences of the Lithuanian SSR, Kaunas (1963)
25. Eliseev, S.V., Nerubenko, G.P.: Dynamic Vibration Dampers, 182 pp. Nauka Publication, Novosibirsk (1982)
26. Korenev, B.G., Reznikov, P.M.: Dynamic Vibration Dampers. Theory and Technical Applications, 535 pp. Nauka Publication, Moscow (1978)
27. Eliseev, S.V., Khomenko, A.P.: Dynamic Damping. Feedback Concept and Structural Methods of Mathematical Modeling, 357 pp. Nauka Publication, Novosibirsk (2014)
28. Eliseev, S.V.: Structural Theory of Vibration Protection Systems, 224 pp. Nauka Publication, Novosibirsk (1978)

29. Eliseev, S.V., Kovyrshin, S.V., Bolshakov, R.S.: Design features of compacts of elastic elements in mechanical oscillatory systems. Interactions with the elements of the system and forms of connections. Mod. Technol. Syst Anal. Model. **4**(36), 61–70 (2012)
30. Mandelshtam, L.I.: Lectures on the Theory of Vibrations, 466 pp. Nauka Publication, Moscow (1972)
31. Eliseev, S.V., Ol'kov, V.V., Zasyadko, A.A.: On the possibilities of active vibration protection using devices with motion conversion. In: Proceedings of Jubilee Science-Technical Conference of the Department of mechanics "Mechanics and Control Processes", pp. 89–97. Irkutsk(1971)
32. Frolov, K.V., Furman, F.A.: Applied Theory of Vibration Protection Systems, 286 pp. Mashinostroenie Publication, Moscow (1985)
33. Alabuzhev, P.M., Gritchin, A.A., Kim, L.I., et al.: Vibration systems with quasi-zero stiffness. In: Ragulskis, K.M. (ed.), 96 pp. Mashinostroenie Publication, (1986)
34. Blekhman, I.I.: Vibration Mechanics, 394 pp. Nauka Publication, Moscow (1994)
35. Stoker, J.J.: Nonlinear Oscillations in Mechanical and Electrical Systems, 264 pp. Publishing House of Foreign Literature, Moscow (1952)
36. Dembaremdiker, A.D.: Shock Absorbers for Vehicles, 199 pp. Mashinostroenie Publication, Moscow (1985)
37. Kolchin, N.N.: Machine Mechanics, vol. 2, 486 pp. Mashinostroenie Publication, Leningrad (1965)
38. Khachaturov, A.A., Afanasyev, V.L., Vasilyev, V.S., et al.: Dynamics of the Road—Bus—Car—Driver System. In: Khachaturov, A.A. (ed.), 536 pp. Mashinostroenie Publication, Moscow (1976)
39. Eliseev, S.V., Khomenko, A.P., Upyr, R.Yu.: Mechatronics of Vibration Protection Systems with Lever Linkages. Mod. Technol. Syst. Anal. Model. **3**(23), 104–119 (2009)
40. Rabinovich, M.I., Trubetskov, D.I.: Introduction to the Theory of Oscillations and Waves, 455 pp. Nauka Publication, Moscow (1992)
41. Eliseev, S.V., Khomenko, A.P., Barsukov, S.V.: Modern Problems of the Dynamics of Machines. Protection Against Vibrations and Shocks, 165 pp. IrGUPS Publication, Irkutsk. Dep. in VINITI 03/21/11, No. 135-B (2011)
42. Atabekov, G.I.: Linear Electrical Circuits, 592 pp. Energia Publication, Moscow (1978)
43. Goverdovsky, V.N.: Development of the theory and methods of designing machines with infra-frequency vibration protection systems: author's abstract of Dr. Sc. (Engineering) Thesis, 42 pp. SGUPS Publication, Novosibirsk (2006)
44. Kiryukhin, A.V., Tikhonov, V.A., Chistyakov, A.G., Yablonsky, V.V.: Active vibration protection: purpose, principles, condition. 1. Purpose and principles of development. In: Problems of Mechanical Engineering and Automation, vol. 2, pp. 108–111. OJSC "National Institute of Aviation Technologies" Publication, Moscow (2011)
45. Hendricks, S.L.: Identification of mass, damping and stiffness matrices for large linear vibratory system. In: Guidance, Control and Dynamics, vol. 7, pp. 244–245 (1984)
46. Sinev, A.V.: Determination of the zeros of the transfer functions of mechanical oscillatory systems. In: Controlled Mechanical Systems, pp. 25–32. Polytechnic Institute, Irkutsk (1984)
47. Khomenko, A.P., Eliseev, S.V., Artyunin, A.I.: Dynamic damping of vibrations in vibration protection systems with joints. Bull. BelSUT Sci. Transp. **1**, 83–90 (2011)
48. Eliseev, S.V., Ermoshenko, Yu.V.: Dynamic Properties of Vibration Protection Systems. Limit transitions. Mod. Technol. Syst. Anal. Model. **4**(28), 23–30 (2010)
49. Sinev, A.V.: The choice of parameters of vibration isolation systems and dynamic dampers based on the methods of synthesis of chains. Mashinovedenie **1**, 28–34 (1972)
50. Gardner, M.F., Burns, J.L.: Transients in Linear Systems: Translation from English, 3rd edn, 551 pp. Fizmatgiz Publication, Moscow (1961)
51. Eliseev, S.V., Belokobylsky, S.V., Upyr, R.Yu.: The generalized spring in the tasks of machines and equipment. In: The Assistant of Scientific Works (Branch Mechanical Engineering, Construction), vol. 1, issue 3(25), pp.79–89. PoltSTU Publication, Poltava (2009)

52. Belokobylsky, S.V., Eliseev, S.V., Sitov, I.S.: Dynamics of Mechanical Systems. Lever Linkages and Inertial-Elastic Constraints, 319 pp. Polytekhnika Publication, St.Petersburg (2013)
53. Vargunin, V.N., Gusarov, V.N., Ivanov, B.G., et al.: Designing and Calculation of Articulation Devices and Assemblies. In: Mulyukin, O.P. (ed.), 86 pp. SamGAPS Publication, Samara (2006)
54. Eliseev, S.V., Artyunin, A.I., Kaimov, E.V.: Equivalent transformations in the structures of mechanical oscillatory systems containing material points and solids. In: Proceedings of XVIII Baikal All-Russia conference "Information and Mathematical Technologies in Science and Management", Part 1, pp. 173–182 Irkutsk (2013)
55. Eliseev, S.V., Piskunova, V.A., Savchenko, A.A.: The interaction of solids in oscillatory systems with elastic constraints and joints under the action of external vibration perturbations. In: Science and Education: Electron Scientific and Technical ed. vol. 1. Access mode http://technomag.edu.ru/doc/486817.html. Access date is 25 Apr 2014 (2013)
56. Khomenko, A.P., Eliseev, S.V.: Mechatronic vibration protection systems. Some issues of ensuring the sufficiency of computational schemes and structural interpretations. Mod. Technol. Syst. Anal. Model. **1**(33), 8–13 (2012)
57. Khomenko, A.P., Eliseev, S.V.: Unconventional approaches to the construction of mathematical models of mechanical oscillatory systems with lever linkages. News of Transsib **4**(12), 75–86 (2012)
58. Wiercigroch, M.: Modelling of dynamical systems with motion dependent discontinuities. Chaos, Solitons Fractals **11**(15), 2429–2442 (2000)
59. Makhutov, N.A., Petrov V.P., Kuksova V.I., Moskvitin G.V.: Modern trends in the development of scientific research on problems of machine science and mechanical engineering. In: Problems of Mechanical Engineering and Automation, vol. 3, pp. 16–37. OJSC National Institute of Aviation Technologies, Moscow (2008)
60. Eliseev, S.V., Artyunin, A.I., Bolshakov, R.S.: Some questions of the interaction dynamics in mechanical oscillatory systems with lever linkages. In: Mechanical Engineering and Life Safety, vol. 4, issue 14, pp. 36–45. Murom Institute (branch) of the Alexander and Nikolai Stoletovs Vladimir state unity, Murom (2012)
61. Kauderer, G.: Nonlinear Mechanics, 778 pp. Publishing House of Foreign Literature, Moscow (1961)
62. Andreychikov, A.V.: Development of pneumatic systems for vibration isolation of a locomotive driver's seat using automated methods of search design Ph.D. (Engineering) Thesis, 295 pp. Bryansk (1984)
63. Eliseev, S.V., Kovyrshin, S.V., Parshuta, E.A.: Some questions of the theory of vibration isolation. Justification of structural approaches. In: News of ISEA, vol. 3, pp. 121–127. BSUEL Publication, Irkutsk (2013)
64. Eliseev, S.V., Khomenko, A.P., Logunov, A.S.: Use of pneumatic protection schemes for mechanisms or devices with movement conversion. Mod. Technol. Syst. Anal. Model. 1(25), 8–13 (2010)
65. Eliseev, S.V.: The theory of active vibration protection systems and its applications: systems: author's abstract of Dr. Sc. (Engineering) Thesis, 35 pp. Tymoshenko Institute of Mechanics of AS Ukr. SSR, Kiev (1973)
66. Genkin, M.D., Eliseev, S.V., Migirenko, G.S., Frolov, K.V.: The principles of modern vibration protection. In: Collection of Scientific Papers Vibration Isolation Mechanisms and Machines, pp. 3–13. Novosibirsk (1984)
67. Kernop, D.: Principles of designing oscillation control systems using semi-active dampers. In: Dynamics of Systems, Mechanics and Control: Proceedings of ASME 1990, vol. 112, issue 3, pp. 448–453 (1990)
68. Levitskij, N.I. (ed.): Theory of Mechanisms and Machines. Terminology, vol. 93, pp. 34. Nauka Publication, Moscow (1978)

69. Eliseev, S.V., Belokobylsky, S.V., Upyr, R.Yu., Gozbenko V.E.: Lever Linkages in the Problems of the Dynamics of Mechanical Oscillatory Systems. Theoretical Aspects, 159 pp. IrGUPS Publication, Irkutsk. Dep. in VINITI 27.11.09, No. 737-B (2009)
70. Ivanov, B.G.: Development of methods for calculating the dynamics and strength of units of transport equipment with articulation constraints: Dr. Sc. (Engineering) Thesis, 48 pp. Korolev Samara Aerospace University Publication, Samara (2007)
71. Lavrus, V.V.: Improvement of pneumatic articulation railway transport systems: author's abstract of Ph.D. (Engineering) Thesis, 20 pp. Oryol State Technical University, Oryol (2006)
72. Panovko, Ya.G.: Introduction to the Theory of Mechanical Vibrations, 255 pp. Nauka Publication, Moscow (1991)

Chapter 7
Lever Linkages in Mechanical Oscillatory Systems

7.1 Some Questions of the Theory of Lever Linkages in Dynamics Mechanical Oscillatory Systems

In preliminary studies, the evaluation of dynamic properties, as a rule, is carried out on the computational schemes of mechanical oscillatory systems with several degrees of freedom, taking into account the structural and technical features of technical objects. At the same time, accounting and forms of interaction of elements of systems in which various dynamic constraints are implemented are of great importance.

In this regard, of particular interest are lever linkages in complex motions of nodes and elements that perform, in particular, combined angular and translational oscillatory motions [1].

The issues of rational distribution of mass-and-inertia, elastic and dissipative elements, which in general form some spatial structure of dynamic interactions, are of great importance when drawing up the computational schemes of transportation objects. This is due to the expansion of ideas about the forms of manifestation of dynamic effects and transformation of the motions of the interacting elements. A number of developments in problems of dynamic synthesis of vibration protection systems is given in [2–4].

Further, it is proposed to substantiate the existence of lever linkages, which are a form of manifestation of stable relations between the motion parameters of mechanical oscillatory systems with different numbers of degrees of freedom.

7.1.1 Properties of Lever Linkages

In the following, the methodological basis of the structural theory of vibration protection systems is used, within which a linear mechanical oscillatory system can

© Springer Nature Switzerland AG 2020

S. V. Eliseev and A. V. Eliseev, *Theory of Oscillations*, Studies in Systems, Decision and Control 252, https://doi.org/10.1007/978-3-030-31295-4_7

be reduced to a structural diagram of the dynamically equivalent automatic control system. In this case, the structural diagram (as a structural analogue) displays the properties of an ordinary mathematical model in the form of a system of linear ordinary differential equations with constant coefficients. If a mechanical oscillatory system has several degrees of freedom, then its structure is formed on the ideas of partial systems and the connections between them. The transition to structural mathematical models is based on Laplace transformations.

The concept of partial systems is connected with the concept of the simplest systems with one degree of freedom, which can be obtained by simplifying a more complex initial mechanical oscillatory system. If the initial mechanical system is described by several generalized coordinates, then if the motion stops along all generalized coordinates, except for one (their zeroing), it is possible to obtain a partial system. When considering many tasks of the dynamics of machines and equipment, in particular, the tasks of vibration protection, they are limited to ideas about the motion of an object and system elements in a plane. In this case, partial systems usually have the form of a material point, which executes rectilinear oscillations on an elastic element supported by a bearing surface, or a material point located on a weightless rigid rod, which makes angular oscillations relative to a fixed point of rotation. With a vertical suspension of a rod, a model of a mathematical pendulum can be obtained. In the general case, one can consider two physical models of partial systems (with translational and angular motions), which are analogs; here, the equations of motion in linear and angular coordinates have the same form.

The partial system of angular oscillations is usually associated with spatial forms of the manifestation of motions, and the parameters of the motion depend on the location of the typical elements of the system, the type of their connection and the places of application of external disturbances. The axis of rotation of the mass-and-inertia element of the partial system of angular motion is usually perpendicular to the plane of the considered motion. But this is not always the case, since there are other forms of rotation when, for example, the axis of rotation lies in the plane of motion, and not perpendicular to it. There are also possible generalized ideas about the forms of partial systems, if we take into account that a generalized coordinate can characterize a helical motion. The physical forms of such partial systems can be implemented, for example, in the form of a bifilar suspension or a screw non-locking mechanism. In this case, a more complex system will consist of several partial systems. Considering combinations of partial systems of various types, one can “assemble” more complex structures or systems that can become models of various technical objects.

If the above considerations can exist then it is possible to build complex systems from some elementary structures. This approach is based both on ideas about more complex partial systems (for example, systems with two degrees of freedom) and on expanding ideas about the forms and essence of interpartial constraints.

In the theory of mechanical oscillations, springs and devices of a dissipative nature (or dampers, in which there is viscous resistance) are used as the connecting units of mass-and-inertia elements.

In the theory of mechanical oscillations, springs and devices of a dissipative nature (or dampers, in which viscous resistance occurs) are used as connecting units of mass-and-inertia elements. More complex structures are built from elementary standard units with the help of certain rules for connecting elements and observing the laws of mechanics or the theory of chains. These approaches are reflected in many papers on the theory of oscillations and its various applications.

Some examples. Consider several options for the formation of structures using partial systems (Fig. 7.1); abstractions of physical properties of material objects in the form of "bundles" are used: the material body of translational motion is a material point; a solid body of rotational (or angular) motion is a material point on a weightless rod fixed with the possibility of rotation in the plane. Elastic elements are considered in the framework of the usual ideas about abstractions of properties.

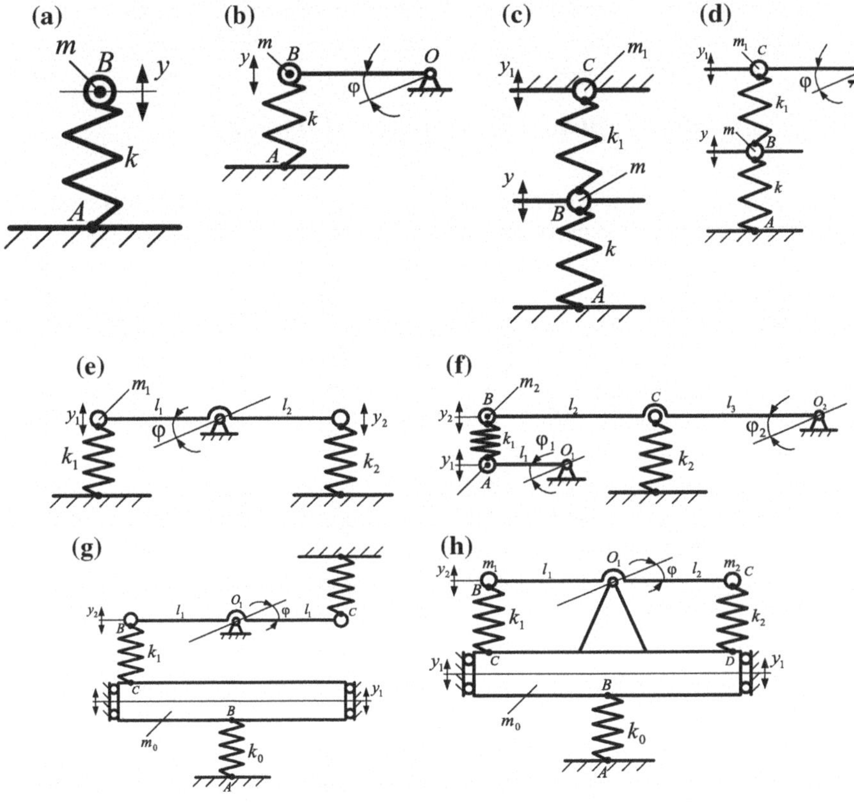

Fig. 7.1 Options for the representation of partial systems and types of constraints between partial systems. **a** Translational; **b** rotational; **c** the consecutive elastic constraint of two translational partial systems; **d** the connection of the translational and rotational systems; **e** is the rotational system of lever type; **f** the connection of two rotational systems; **g** the connection of the rotational system of the lever type and the system of the translational type with a solid body (and not a material point); **h** the connection of the rotational system of the lever type with a point of rotation on a solid body that executes translational motion

A characteristic feature in the choice of forms, the construction of partial systems and their simplest interactions is that the motion of system elements is implemented in a certain space. Such properties are accompanied by the need to fix ideas about the places of fastening elements, arm lengths of arms, the location of points of rotation, etc. Even with the use of the simplest typical elements in the form of ideal springs, two partial systems can form rather complex structures that disturbances create different dynamic modes and effects. With all the conventions of the simplicity of partial systems, it follows that the interactions of partial systems are not so simple, which requires the development of some methodological provisions. As such, the possibilities of expanding the set of elementary units of mechanical oscillatory systems by introducing elementary units that supplement the properties of elastic and dissipative elements through the possibilities introduced by lever linkages and motion transformation devices can be considered. Such approaches were developed in the structural theory of vibration protection systems.

7.1.2 Combining Typical Elements into Compacts

Another direction in the development of the concept of using the interaction of partial systems could be the introduction into the structure of interpartial constraints of elements of a generalized form, formed according to certain rules from typical compounds of elements of an extended set. The capabilities of such elements in recent years have been discussed as forms of generalized springs with reduced stiffnesses, as well as quasi-elastic elements, compacts, or quasi-springs. In the elaboration of ideas about interactions of this kind, the approaches developed in [5] have been applied. Possible forms of interaction of partial systems are represented by the structures shown in Fig. 7.2.

Analysis of the schematic diagrams Fig. 7.2 gives an idea of the possibilities of transforming the initial positions, which can be implemented on the basis of using the methods of structural mathematical modeling.

Consider as an example the mechanical oscillatory system in Fig. 7.2d. If we assume that $m_2 = 0$, then the system is transformed into a structure with one degree of freedom with an elastic block, which will consist of three springs with stiffnesses k_1, k_2, $\frac{k_3 k_4}{k_3+k_4}$, connected by a lever of the second kind. After a series of transformations, the system can be reduced to the form of a conventional system with one degree of freedom with mass m_1 and some generalized spring with reduced stiffness. The presence of the lever linkage l_1/l_2 substantially affects the values of the reduced stiffness. In Fig. 7.2d the lever of the second kind can be considered as the limiting form of the state of a certain solid body, which with the existing points of attachment of elastic elements k_2 and k_3 and the point of rotation O has a moment of inertia $\to 0$.

That formal technique under certain conditions can be considered as an illustration of the formation of a lever linkage. In the general case, the properties under consideration are more complex, and the characteristics of the system are dependent

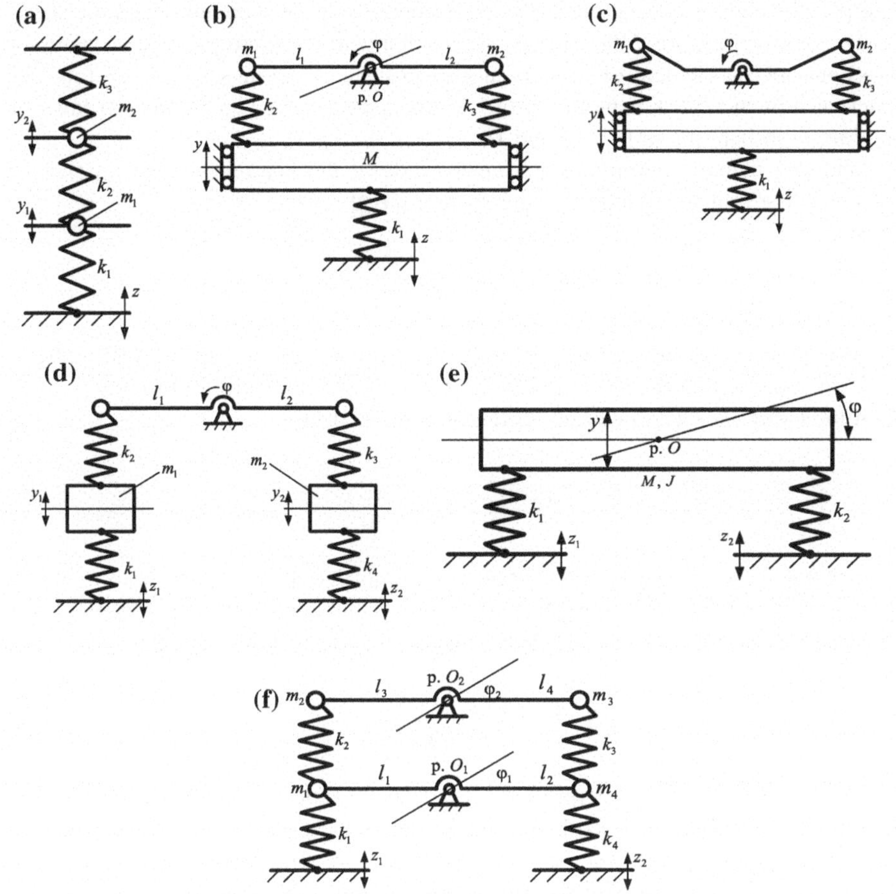

Fig. 7.2 Variants of the construction of systems with two degrees of freedom. **a** The translational system with elastic constraints; **b** the system with a rotational body interacting with the object of translational motion; **c** the diagram of the lever dynamic absorber on the object of translational motion; **d** the lever type system with spaced-apart translational motions; **e** the solid body on elastic supports; **f** two rotational solid bodies with elastic constraints

on the frequency of external influences. It is appropriate to note the fact that only linear systems are considered in the developed approaches.

7.1.3 Lever Linkages in Mechanical Oscillatory Systems with One Degree of Freedom

The presence of lever linkages formed in mechanical oscillatory systems in which solid bodies with angular oscillations are used, makes it possible to identify and interpret the necessary relations between the parameters of static and dynamic states fairly simply. Figure 7.3a shows a mechanical oscillatory system with a mass-and-inertia element of mass m and coordinate y_1, springs k_1 and k_2, force of external influence Q, kinematic perturbations $z_1(t)$, $z_2(t)$, given by known laws of motion, with bearing surfaces I and II.

With a kinematic perturbation from the side of the bearing surface ($z_1(t) \neq 0$, $z_2(t) = 0$) the transfer function of the system has the form

$$W(p) = \frac{\bar{y}}{\bar{z}} = \frac{k_1}{mp^2 + k_1 + k_2}, \tag{7.1}$$

where $p = j\omega$ is a complex variable (ω is the frequency of an external disturbance).

If we assume that the system is in the position of static equilibrium and $p \to 0$, then the expression (7.1) can be represented in the form

$$i_1 = \left|\frac{\bar{y}}{\bar{z}_1}\right| = \frac{k_1}{k_1 + k_2}. \tag{7.2}$$

The relation (7.2) can be considered as a manifestation of a lever linkage. In this case, i_1 is interpreted as a transfer ratio of a lever linkage formed by a virtual lever of the first kind ($z_2 = 0$, $Q = 0$). If we take $z_1 = 0$, $Q = 0$, then we get

$$i_2 = \left|\frac{\bar{y}}{\bar{z}_2}\right| = \frac{k_2}{k_1 + k_2}. \tag{7.3}$$

In this case, i_2 can be interpreted as the transfer ratio of the virtual lever of the first kind. The difference is explained by the fact that different input-output pairs are used. In the static version ($Q = 0$, z_1 = const or z_2 = const alternately) the transfer ratio i1 and i2 reflects the inherent oscillatory systems of lever linkages.

Under the action of harmonic kinematic perturbations, the transfer ratio of the lever linkage, determined from (7.1), will depend on the perturbation frequency (p = jω 0); then

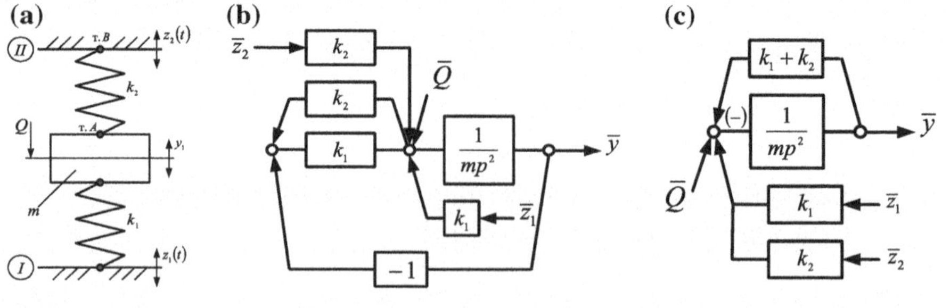

Fig. 7.3 The computational scheme of the system with one degree of freedom (**a**); structural model of the system (**b**); the same with the allocated object of protection m (**c**)

$$\underset{Q=0,\,z_2=0}{i_1(p)} = \left|\frac{\bar{y}_1}{\bar{z}_1}\right| = \left|\frac{k_1}{mp^2 + k_1 + k_2}\right|; \tag{7.4}$$

$$\underset{Q=0,\,z_1=0}{i_2(p)} = \left|\frac{\bar{y}_1}{\bar{z}_2}\right| = \left|\frac{k_2}{mp^2 + k_1 + k_2}\right|. \tag{7.5}$$

It should be noted that when going through resonance with increasing frequency ω, the nature of the constraints changes from a lever of the first kind to a lever of the second kind. In this case (Fig. 7.4), the graphs of dependences for the transfer ratios of lever linkages coincide with the amplitude-frequency characteristics of the system under kinematic perturbation. An important circumstance is that the transfer ratio of the lever linkage depends on, and therefore, can be regulated by the frequency of external kinematic perturbation.

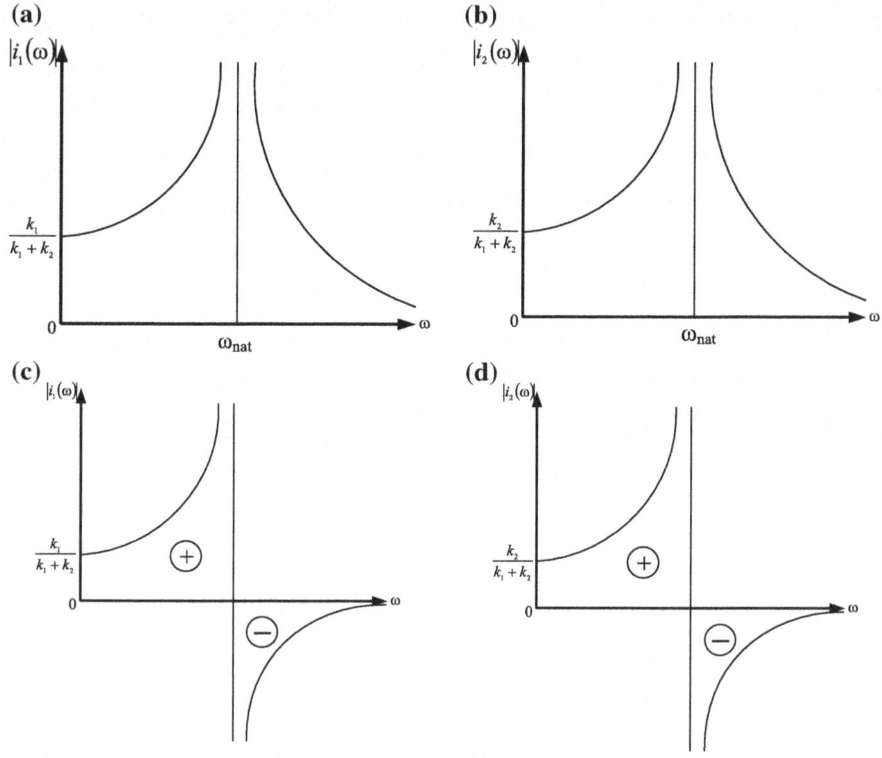

Fig. 7.4 The graphs of $i_1(\omega)$ and $i_2(\omega)$, the characteristic points for which are the resonance frequencies (**a**) input action z1 ≠ 0 (z2 = 0, Q = 0); (**b**) input action z2 ≠ 0 (z1 = 0, Q = 0); input action $z_1 \neq 0$ (**c**) and $z_2 \neq 0$ (**d**) with indication of the area of correspondence to lever linkages according to the type of lever of the first kind ⊕ and the second kind ⊖

7.1.4 Force Perturbation

In the case of a force perturbation $Q \neq 0$ ($z_1 = 0, z_2 = 0$), the transfer function takes the form

$$W_2(p) = \frac{\bar{y}}{\bar{Q}} = \frac{1}{mp^2 + k_1 + k_2}. \tag{7.6}$$

In this case, dynamic responses $\bar{R}$ at points A and B can be found (Fig. 7.3a).

$$|\overline{R}_A| = |k_1 \cdot \bar{y}| = \frac{k_1 \cdot \overline{Q}}{mp^2 + k_1 + k_2}, \tag{7.7}$$

$$|\overline{R}_B| = |k_2 \cdot \bar{y}| = \frac{k_2 \cdot \overline{Q}}{mp^2 + k_1 + k_2}. \tag{7.8}$$

From (7.7) and (7.8) it is possible to find transfer ratios of lever linkages under force perturbations, using the relationship between external influence and dynamic constraint reaction.

Thus

$$i_1(p) = \frac{\bar{y}}{\bar{z}_1} = \frac{k_1}{mp^2 + k_1 + k_2} = \frac{\overline{R}_A}{\overline{Q}} = \frac{k_1}{mp^2 + k_1 + k_2}; \tag{7.9}$$

$$i_2(p) = \frac{\bar{y}}{\bar{z}_2} = \frac{k_2}{mp^2 + k_1 + k_2} = \frac{\overline{R}_B}{\overline{Q}} = \frac{k_2}{mp^2 + k_1 + k_2}. \tag{7.10}$$

From comparison (7.9), (7.10) and (7.4), (7.5), it follows that the lever linkages are preserved under various types of effects on the lever "mechanisms".

Thus, if there are solid bodies in the systems that perform angular oscillations, then lever linkages are associated with the use of geometric relations characterizing the "metric" of the system. In systems where elements make translational motion, transfer ratios of lever linkages are displayed by the ratio of stiffness parameters. In both cases, the transfer ratio lever linkages are dimensionless. The developed approach can be extended to mechanical systems in the form of toothed gearings. Mechanical oscillatory systems, the mass-and-inertia elements of which perform rotational motions, can be reduced to equivalent computational schemes of systems where the mass-and-inertia elements make translational motions. The exception is made by mechanical systems in which partial blocks (or systems) implement different types of motion and are a complex of solids making different types of motions. In this case, the transfer ratios of lever linkages can reflect the helical nature of the motion, when there is a constant relationship between the parameters of rotation and translational motion. Thus, the screw pair implements a special form of constant relations between the motion parameters. That lever linkage will also depend on the frequency, but the transfer ratio of this lever linkage will have the dimension of an angle or length, which is typical for a screw pair.

7.1.5 Interpretation of Lever Linkages in Systems with Two Degrees of Freedom

We also note that all the above-mentioned transfer ratios of lever linkages should be considered taking into account the forces applied to the system. With the simultaneous action of several disturbances, it is necessary to determine the functional relationship between them. Consider the mechanical system (Fig. 7.5) in the form of a solid body with mass-and-inertia parameters M, J on the elastic supports k_1 and k_2.

The bearing surfaces I and II have periodic oscillations $z_1(t)$ and $z_2(t)$. The position of the center of gravity (p. O) is determined by the distances l_1 and l_2. The external force is represented by the harmonic force action Q, applied at p. O.

Figure 7.5a shows two coordinate systems: y_1, y_2 and y_0, φ, which determine small oscillations of the system relative to the position of static equilibrium. Figure 7.5b shows the structural diagram (or structural mathematical model) of the system in coordinates $\bar{y}_0$, $\bar{\varphi}$. The transfer functions of the system are determined from the structural diagram in Fig. 7.5b

$$W_1(p) = \frac{\bar{y}_0}{\bar{Q}} = \frac{Jp^2 + k_1 l_1^2 + k_2 l_2^2}{A_0}; \tag{7.11}$$

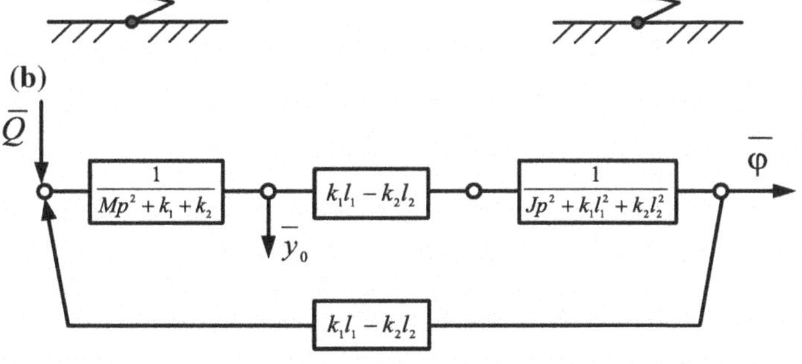

Fig. 7.5 The computational scheme (**a**) and structural diagram (**b**) of the system with a solid body

$$W_2(p) = \frac{\bar{\varphi}}{\bar{Q}} = \frac{Jp^2 + k_1 l_1^2 + k_2 l_2^2}{A_0}, \tag{7.12}$$

where $A_0 = (Mp^2 + k_1 + k_2) \cdot (Jp^2 + k_1 l_1^2 + k_2 l_2^2) - (k_1 l_1 - k_2 l_2)^2$ is the characteristic equation. From expressions (7.11) and (7.12), we can find the transfer ratio, which characterizes the lever linkages between the coordinates of motion $\bar{\varphi}$ and $\bar{y}_0$

$$i_0(p) = \frac{\bar{\varphi}}{\bar{y}_0} = \frac{(k_1 l_1 - k_2 l_2)^2}{Jp^2 + k_1 l_1^2 + k_2 l_2^2}. \tag{7.13}$$

As $p \to 0$, we find that

$$\underset{p \to 0}{i_0(p)} = \frac{(k_1 l_1 - k_2 l_2)^2}{k_1 l_1^2 + k_2 l_2^2} = \left(\frac{k_1 l - k_2 i_1}{k_1 + k_2 i_1^2}\right) \cdot \frac{1}{l_1}, \tag{7.14}$$

where $i_1 = l_2/l_1$ is the transfer ratio.

From (7.14) it follows that the dimension $i_0(p)$ corresponds to the reciprocal magnitude of the unit of length (m^{-1}). Note that in this case $\bar{\varphi}$ and $\bar{y}_0$ are interconnected as elements of a screw pair. When the screw pair is rotated through an angle $\bar{\varphi}$, there is a translational motion by a certain amount (pitch). The transfer ratio (7.14) can have positive and negative values, which corresponds, if we use the screw–nut analogy, right and left threads.

Thus, in the static mode, the transfer ratio (7.14) reflects the properties of a specific lever, the physical form of which is implemented in the form of a screw pair. If in the preceding considerations there were ideas about levers of the first and second kind, in this case the virtual lever takes the form of a screw mechanism.

Under the action of the harmonic force, the transfer ratio of the lever linkage $i_0(p)$ depends on the frequency ω; while at frequency

$$\omega_{nat}^2 = \frac{k_1 l_1^2 + k_2 l_2^2}{J} \tag{7.15}$$

a resonance occurs, and with a further increase in frequency, the transfer ratio will change its sign, which is typical of systems with other types of motion. The change of the sign of $i_0(p)$ corresponds to the transition of the virtual "screw" lever (conventionally) from the "left" thread to the "right" one. The issues of dynamic interactions are considered in more detail in [6, 7].

The given considerations about the forms and properties of lever linkages, manifested in mechanical oscillatory systems, can be considered as a certain scientific concept of building and developing new principles for controlling the dynamic state of systems, in which adjustment options will be determined not only by the parameters of the system as such, but also and have adaptive properties when taking into account the frequency of external influences and features of its type.

Note that in mechanics the concepts of power and kinematic screws are already used.

Thus, the lever linkages, the manifestations of which are characteristic of the lever mechanisms of the first and second kind, play a significant role in the dynamic interactions of the elements of mechanical oscillatory systems, including vibration protection systems. The manifestation of lever linkages is characteristic of elements in the form of solid bodies, which have finite dimensions and execute angular motions. Lever linkages are determined by the geometrical parameters of the relative position of the attachment points of the system elements with mass-and-inertia bodies and bearing surfaces.

When executing translational oscillatory motions in systems with one degree of freedom, lever linkages also arise, whose transfer ratios are determined through the ratio of the stiffness coefficients of the elastic elements in statics. With harmonic effects, the transfer ratios are determined by the coefficients of the dynamic stiffness. In this case, the transfer ratios of the lever linkages become dependent on the frequency of the harmonic force. The dependence of the transfer ratio on the frequency reflects the properties of lever linkages in the formation of interactions between elements of mechanical oscillatory systems.

The forms of interaction depend on the choice of coordinate systems, as well as on the forms of partial systems and their constraints. During interactions of partial systems of translational and angular motions, manifestations of the screw lever are possible as some generalized idea of the combination of two types of motion. Taking into account lever linkages predetermines spatial ideas about the forms of dynamic interactions of elements of mechanical oscillatory systems.

7.2 Lever Linkages: Virtual Lever Mechanisms, the Features of Oscillatory Processes

Lever interactions in mechanical oscillatory systems are quite diverse in their manifestations and are reflected in the formation of reduced mass-and-inertia and elastic-dissipative characteristics of technical systems, which expands ideas about the features of the vibrational states of mechanical systems, the formation of various constraints obtained through the connection of typical elements with each other and with the structure of a more complex type. In this regard, of interest are the ideas of constructing mathematical models that reflect the properties of quasi-springs or compacts, forming structures, the interactions of which, despite the complexity of the forms, obey the rules of consecutive and parallel connection of springs and the rules of structural transformations characteristic of the theory of automatic control. Note that the mechanical oscillatory system with several degrees of freedom is considered in terms of identifying characteristic constraints between system state parameters that take the form of lever linkages.

The systems in Fig. 7.6 differ in the number of degrees of freedom and the possibilities of forming interactions between mass-and-inertia elements. Masses m_1 are selected as an object of protection in both schemes Fig. 7.6a, b.

For the system in Fig. 7.6a the equations of motion can be set-up using the Lagrange equation of the second kind and the subsequent Laplace transformations. The coefficients of the equations of motion for the system in Fig. 7.6a are given in Table 7.1, where $p = j\omega$ is a complex variable $\left(j = \sqrt{-1}\right)$.

External disturbances in the system are represented by the vibrations of the bearing surfaces $\bar{z}_1(t)$ and $\bar{z}_2(t)$, as well as by force perturbations $\overline{Q}_1$ and $\overline{Q}_2$, directly applied to the mass-and-inertia elements m_1 and m_2, respectively.

Figure 7.7 shows the variants of structural diagrams in which, by means of transformations, features of structural interpretations of the mathematical model are highlighted. In particular, Fig. 7.7b and c, show, respectively, that force perturbation $\overline{Q}_2$ and kinematic perturbation $\bar{z}_2$ are possible to the input of the partial system $(m_1p^2,\ k_1)$, which indicates the possibilities of equivalent transfer of force factors from one point of the system to another.

Thus, a feature of the computational scheme Fig. 7.7c is that when converting the original system (Fig. 7.6a), external disturbances Q_2 and z_2 are reduced to an equivalent form. In this case at p. 1 Fig. 7.6a all forces are summed up. The principle of superposition can be used to obtain a system reaction. In particular, when considering generalized external actions, the conditions $z_1 = \alpha z_2$ or $z_1 = z_2 = z$; $Q_1 = \alpha Q_2$ ($Q_1 = \beta k_1 z_1$, $Q_2 = k_3 z_2$), where α and β are the coupling coefficients between external influences.

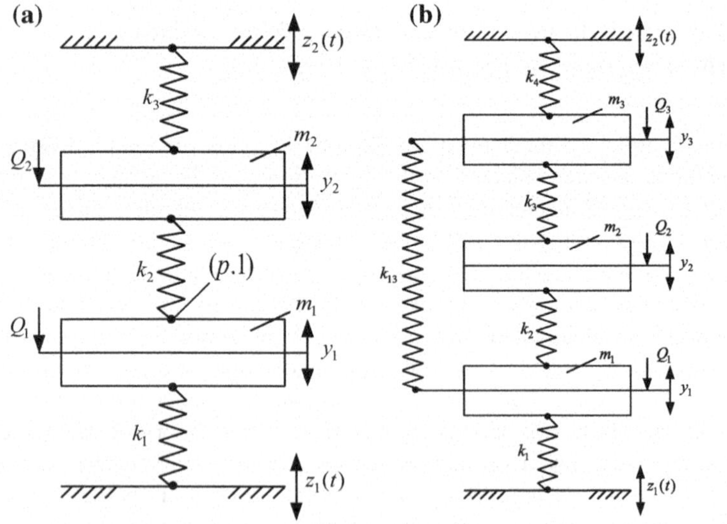

Fig. 7.6 The computational scheme of vibration protection systems with two (**a**) and three (**b**) degrees of freedom

Table 7.1 The coefficients of the equations of motion of the system in the coordinates y_1, y_2

a_{11}	a_{12}
$m_1p^2 + k_1 + k_2$	$-k_2$
a_{21}	a_{22}
$-k_2$	$m_2p^2 + k_3 + k_3$
Generalized forces	
$\overline{Q}'_1 = \overline{Q}_1 + k_1\overline{z}_1$	$\overline{Q}'_2 = \overline{Q}_2 + k_3\overline{z}_2$

Note $\overline{Q}'_1$, $\overline{Q}'_2$ are the generalized forces, referred to the corresponding generalized coordinates y_1 and y_2

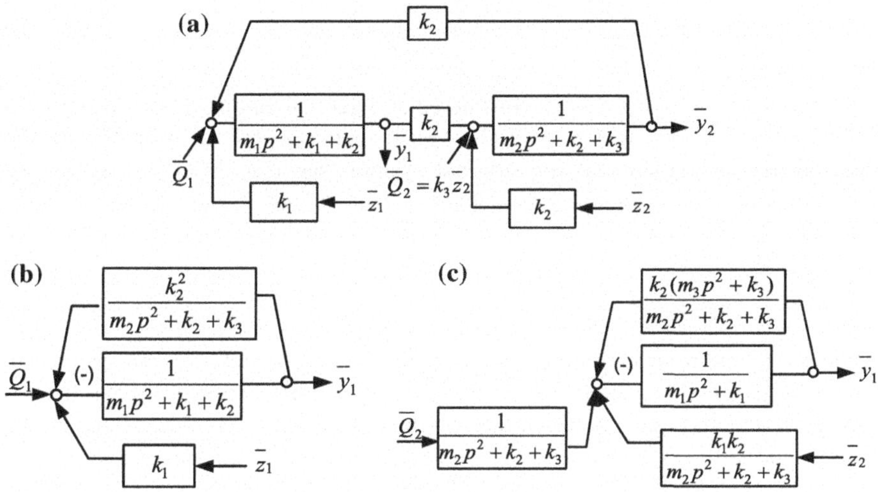

Fig. 7.7 Structural diagrams of the vibration protection system (Fig. 7.6a). **a** The detailed structural diagram; **b** the structural diagram with the excluded coordinate $\overline{y}_2$ at $z_2 = 0$, $Q_2 = 0$; **c** the scheme with excluded coordinate $\overline{y}_2$ ($\overline{Q}_1 = 0$, $\overline{z}_1 = 0$, $\overline{Q}_2 \neq 0$, $\overline{z}_2 \neq 0$)

7.2.1 *Features of Mathematical Models of Dynamic Interactions Between Partial Systems*

1. The case of kinematic perturbation from the side of the lower bearing surface is considered (Fig. 7.6a); in this case, $z_1 \neq 0$; $z_2 = 0$; $Q_1 = 0$; $Q_2 = 0$. Note that when $Q_1 = 0$; $\overline{Q}'_1 = k_1\tilde{z}_1$, because $\overline{Q}'_1 = \overline{Q}_1 + k_1\tilde{z}_1$ in a reduced form. Using the structural diagram in Fig. 7.7a, we define the transfer function with kinematic perturbation z_1,

$$W_{1\mathrm{add}}(p) = \frac{\overline{y}_1}{\overline{z}_1} = \frac{k_1(m_2p^2 + k_2 + k_3)}{A_0}, \tag{7.16}$$

$$W_{2\text{add}}(p) = \frac{\bar{y}_2}{\bar{z}_1} = \frac{k_1 k_2}{A_0}, \tag{7.17}$$

where A_0 is the characteristic frequency equation:

$$A_0 = (m_1 p^2 + k_1 + k_2)(m_2 p^2 + k_2 + k_3) - k_2^2. \tag{7.18}$$

Using (7.17) and (7.18), we find the ratio of coordinates in Laplace transformations:

$$i_{\text{gen}}(p) = \frac{\bar{y}_2}{\bar{y}_1} = \frac{k_1 k_2}{(m_2 p^2 + k_2 + k_3)k_1} = \frac{k_2}{m_2 p^2 + k_2 + k_3}. \tag{7.19}$$

We suppose that $i_{\text{gen}}(p)$ can be considered as a generalization of the transfer ratio of the lever linkage, which manifests itself between the coordinates of the mass-and-inertia elements of the original system (see Fig. 7.6a). When $p \to 0$, which corresponds to proximity to a static external effect

$$\underset{p \to 0}{i_{\text{gen}}(p)} = \frac{k_2}{k_2 + k_3}. \tag{7.20}$$

For $p = 0$, the transfer ratio of the lever linkage $i_{\text{gen}}(p)$ takes a constant value, defined by expression (7.20), i.e. in the case of a static action from the bottom bearing surface, the ratios between the coordinates y_1 and y_2 are determined by the ratios by the stiffness coefficients of the elastic elements, and the system is, in a sense, analogous to the lever. Relations between the parameters of the state of the system in this case can be called leverage. In systems of translational motion with given forms of external influences (displacements of the lower base), that sort of a virtual lever that creates relationships (7.20) can be referred to as levers of the first kind.

The peculiarity of the virtual lever is that the transfer ratio (7.20) depends on the frequency of external influence. In particular, with

$$\omega^2 = \frac{k_2 + k_3}{m_2}. \tag{7.21}$$

the transfer ratio $i_{\text{gen}}(p) \to \infty$, which corresponds to the particular forms of the location of the pivot point of the virtual-lever. With increasing $p = j\omega$ (ω is the frequency of external influence), the transfer ratio of the virtual-lever changes its sign, i.e. becomes negative. In this case, the virtual lever of the first kind is transformed into a lever of the second kind. In the physical sense, the above refers to the forms of motion of elements with masses m_1 and m_2. They either move in phase, or move in opposite directions. If $p \to \infty$, then $i_{\text{gen}}(p) \to 0$, which corresponds to the manifestations of small motions from the element m_2 (coordinate y_2) with the kinematic perturbation z_1.

2. At external disturbance $z_2 \neq 0$ ($z_1 = 0$, $Q_1 = 0$, $Q_2 = 0$) the transfer functions of the system take the form

$$W_1'(p) = \frac{\bar{y}_1}{\bar{z}_2} = \frac{k_2 k_3}{A_0}; \tag{7.22}$$

$$W_2'(p) = \frac{\bar{y}_2}{\bar{z}_2} = \frac{k_2(m_1 p^2 + k_1 + k_2)}{A_0}, \tag{7.23}$$

then

$$\underset{z_2 \neq 0}{i'_{\text{gen}}(p)} = \frac{\bar{y}_2}{\bar{y}_1} = \frac{(m_1 p^2 + k_1 + k_2)k_3}{k_2 k_3} = \frac{m_1 p^2 + k_1 + k_2}{k_2}. \tag{7.24}$$

When $p \to 0$

$$\underset{z_2 \neq 0, p \to 0}{i'_{\text{gen}}(p)} = \frac{\bar{y}_2}{\bar{y}_1} = 1 + \frac{k_3}{k_2}, \tag{7.25}$$

which is different from expression (7.20). In this case, the condition

$$i_{\text{gen}}(p) \cdot i'_{\text{gen}}(p) = 1. \tag{7.26}$$

3. With a total kinematic effect from the side of the bearing surfaces, $z_1 = z_2 = z$ for $Q_1 = 0$; $Q_2 = 0$ the transfer functions of the system (Fig. 7.6a) take the form

$$W_1''(p) = \frac{\bar{y}_1}{\bar{z}} = \frac{k_1(m_2 p^2 + k_2 + k_3) + k_2 k_3}{A_0}; \tag{7.27}$$

$$W_2'(p) = \frac{\bar{y}_2}{\bar{z}} = \frac{k_3(m_1 p^2 + k_1 + k_2) + k_2 k_1}{A_0}. \tag{7.28}$$

The system's response to a common kinematic perturbation consists of two components: the action from the lower bearing surface, and also from the upper bearing surface z_2 (in this case they coincide—$z_1 = z_2 = z$). Relation of coordinates $\bar{y}_1$ and $\bar{y}_2$ has the form

$$i'_{\text{gen}}(p)\underset{z_1=z_2=z}{} = \frac{\bar{y}_2}{\bar{y}_1} = \frac{k_2(m_1p^2+k_1+k_2)+k_1k_2}{k_1(m_2p^2+k_2+k_3)+k_2k_3} = \frac{k_3(m_1p^2)+k_1k_3+k_2k_3+k_1k_2}{k_1(m_3p^2)+k_1k_2+k_1k_3+k_2k_3}. \tag{7.29}$$

When $p \to \infty$, i.e. in a static state

$$\underset{p\to 0}{i_{\text{gen}}(p)} = 1. \tag{7.30}$$

In turn, as p $\to \infty$, expression (7.28) is reduced to the form

$$\underset{p\to\infty}{i_{\text{gen}}(p)} = \frac{k_3m_1}{k_1m_3}. \tag{7.31}$$

For comparison, when $z_1 = z$, $z_2 = 0$ $\underset{p\to\infty}{i_{\text{gen}}(p)} \to 0$; with $z_2 = z$, $z_1 = 0$ ($Q_1 = 0$, $Q_2 = 0$)—$\underset{p\to\infty}{i_{\text{gen}}(p)} \to \infty$.

The ratio $\left|\frac{\bar{y}_2}{\bar{y}_1}\right| = |i_{\text{gen}}(p)|$ modulus, defined by the expression (7.29), depends on the frequency of external influence, which is shown in Fig. 7.8.

Thus, the lever linkages are quite diverse, determined by the relationships of the stiffnesses of the elastic elements and depend on the frequency of external influence. The "fracture" points of the frequency characteristics in the region of frequency changes on the x-axis determine the sign of the transfer ratio, which pre-determines the type of lever linkages (levers of the first or second kind), as well as the forms of mutual motions of the elements m_1 and m_2 in the coordinates y_1 and y_2.

If $z_2 = \alpha z_1$ is adopted, which is quite feasible with consistent motions of the lower and upper bearing surfaces, the transfer ratio (7.29) transforms to the form

$$i'_{\text{gen}}(p) = \frac{\alpha k_3(m_1p^2+k_1+k_2)+k_1k_2}{k_1(m_2p^2+k_2+k_3)+\alpha k_2k_3}, \tag{7.32}$$

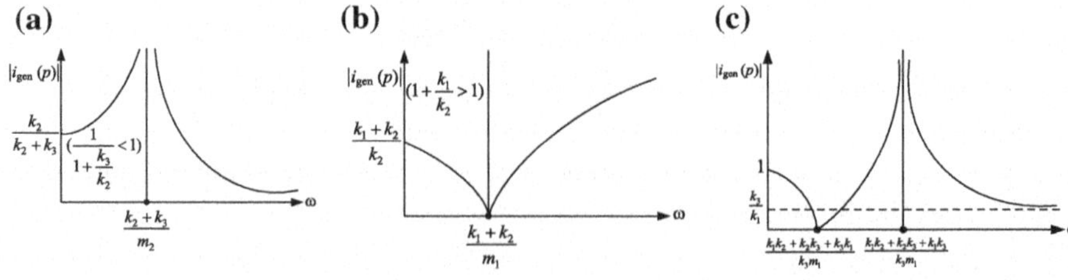

Fig. 7.8 Dependencies of transfer ratio $i_{\text{gen}}(p)$ on conditions of kinematic disturbance with increasing frequency of external influence. **a** Kinematic perturbation $z_1 \neq 0$ ($z_2 = 0$); **b** $z_2 \neq 0$ ($z_1 = 0$); **c** $z = z_1 + z_2$ is the total action

where α, as the coupling coefficient of the motions of the bearing surfaces can vary in the range $-\infty < \alpha < \infty$. For α = 1, z1 = z2 = z, which is taken into account, for example, by the expression (7.29).

If α is negative, the upper and lower bearing surfaces oscillate in antiphase; when α > 0, these oscillations occur in the phase, but the amplitudes of oscillations can be different, including equal, which predetermines quite wide possibilities for changing the dynamic properties of lever linkages. The results of the study and the comparative analysis of the interaction peculiarities for various types of disturbance were further reflected, which is associated with the formation of a specific system of dynamic connections between the elements of a mechanical oscillatory system.

7.2.2 *Taking Account of the Features of Force Perturbations*

Cases are considered when $z_1 = 0$, $z_2 = 0$, and force perturbations assigned to the corresponding generalized coordinates y_1 and y_2 are not zero (Q1 ≠ 0, Q2 ≠ 0).

1. Provided that $Q_2 = \beta Q_1$, where β as the coupling coefficient between external influences Q_1 and Q_2 can vary in the range $-\infty < \beta < \infty$. In particular, when β = 1, the forces are equal and have the same direction; when β = −1, the forces are equal, but the directions are different. When β = 0, the force is $Q_2 = 0$. If we accept that $Q_1 \neq 0$ ($Q_2 = 0$), then from the structural diagram in Fig. 7.7 you can define transfer functions

$$W_1''(p) = \frac{\bar{y}_1}{\bar{Q}_1} = \frac{m_2p^2 + k_2 + k_3}{A_0}; \tag{7.33}$$

$$W_2''(p) = \frac{\bar{y}_2}{\bar{Q}_1} = \frac{k_2}{A_0}. \tag{7.34}$$

Then the transfer ratio between the coordinates $\tilde{y}_1$ and $\tilde{y}_2$ is determined:

$$\underset{\substack{Q_1 \neq 0 \\ Q_2 = 0}}{i_{\text{gen}}(p)} = \frac{\bar{y}_2}{\bar{y}_1} = \frac{k_2}{m_2p^2 + k_2 + k_3}. \tag{7.35}$$

In turn, when $Q_1 = 0$, $Q_2 \neq 0$, we obtain that

$$W_1'''(p) = \frac{\bar{y}_1}{\bar{Q}_2} = \frac{k_2}{A_0}; \tag{7.36}$$

$$W_2'''(p) = \frac{\bar{y}_2}{\bar{Q}_2} = \frac{m_1p^2 + k_1 + k_2}{A_0}, \tag{7.37}$$

from where it follows that

$$\underset{\substack{Q_1 \neq 0 \\ Q_2 = 0}}{i_{\text{gen}}(p)} = \frac{m_1p^2 + k_1 + k_2}{A_0}. \tag{7.38}$$

Note that expression (7.35) coincides with expression (7.19) corresponding to the external effect $z_1 \neq 0$ ($z_2 = 0$, $Q_1 = 0$, $Q_2 = 0$). In turn, the expression (7.37) corresponds to the expression (7.23) (for $z_1 \neq 0$, $z_2 = 0$, $Q_1 = 0$, $Q_2 = 0$). If we assume that there is a relationship between external forces $Q_2 = \beta Q_1$, then with the total effect of such disturbances, we can obtain that

$$W_1^{\text{IV}}(p) = \frac{\bar{y}_1}{\bar{Q}_2} = \frac{\beta \cdot (m_2p^2 + k_2 + k_3) + k_2}{A_0}; \tag{7.39}$$

$$W_2^{\text{IV}}(p) = \frac{\bar{y}_2}{\bar{Q}_2} = \frac{\beta k_2 + (m_1p^2 + k_1 + k_2)}{A_0}. \tag{7.40}$$

Using (7.39) and (7.40), we find

$$\underset{\bar{Q}_2 = \bar{Q}_1\beta}{i_{\text{gen}}(p)} = \frac{(m_1p^2 + k_1 + k_2) + \beta k_2}{\beta k_2(m_2p^2 + k_2 + k_3) + k_2}. \tag{7.41}$$

Expression (7.40) with $\beta = 1$ determines the case when two force ($Q_1 = Q_2$) (and not kinematic) factors act on the system simultaneously. Note that (7.40) does not coincide with the expression (7.28).

When $p \to 0$

$$\underset{p \to 0}{i_{\text{gen}}(p)} = \frac{k_1 + k_2(1 + \beta)}{\beta k_3 + k_2(1 + \beta)}; \tag{7.42}$$

when $p \to \infty$

$$\underset{p \to \infty}{i_{\text{gen}}(p)} = \frac{m_1}{\beta m_2}. \tag{7.43}$$

Thus, external influences affect the formation of lever linkages not only through changes in the frequency of external influences and the elastic parameters of the system, but also through the connection parameters between external influences.

Such approaches were considered in works on the dynamics of transport systems and the vibration protection of technical systems. In general, the developed approach can be extended to variants of combined interactions (force and kinematic).

7.2.3 *Lever Linkages in Systems with Three Degrees of Freedom*

The computational scheme of a system with three degrees of freedom can be transformed into a form, as shown in Fig. 7.9, where the mass-and-inertia element m_3 is chosen as the object of protection. The structural diagram (Fig. 7.9b) can be constructed on the basis of a mathematical model in the form of a system of three differential equations, obtained using the Lagrange equation of the second kind and the subsequent Laplace transformations. The system (Fig. 7.9b) consists of three partial blocks interacting with each other. Kinematic perturbations $z_1(t)$ and $z_2(t)$ from the side of the bearing surfaces are chosen as external influences.

The computational scheme Fig. 7.9a is transformation of the circuit in Fig. 7.6b. Differences between Fig. 7.9a, b consists in allocating the object of protection and determining its position (in this case, the element m_3). It can be shown that in the system in Fig. 7.9a the connections between the motions along the coordinates y_1 and y_2 as a whole will be the same as in the system with two degrees of freedom (Fig. 7.6a). In the relationship between y_2 and y_1 certain local properties of dynamic interactions between elements arise. Below, based on the use of Kramer's formulas [8], it is shown that in a system with two (Fig. 7.6a) and three (Fig. 7.6a, b) degrees of freedom, the basic interaction scheme is preserved. The latter is explained by the peculiarities of intercoordinate constraints, which, in essence, ensure the formation of a closed chain or block through the elastic elements k_2, k_3, k_{13}. In this case, the motion of the element m_3 with the coordinate y3 in the presence of the elastic support of the elements m_1 and m_2 through the springs k_3 and k_{13} creates the same scheme of dynamic interactions as Fig. 7.6, and provided that both bearing surfaces oscillate synchronously and the condition ($z_1 = z_2 = z$) is fulfilled.

With that, the features of external disturbances, the consideration of which was presented in the previous section, are also important.

The elements m_1 and m_2 form a dynamic absorber. Consider the features of the lever linkages between the coordinates y_1 and y_2 (Table 7.2). Using Kramer's formulas, we write that

$$\bar{y}_1 = \frac{k_1\bar{z}_1(a_{22}a_{33} - a_{23}^2) + Q_2(a_{12}a_{32} - a_{12}a_{33}) + k_3\bar{z}_2(a_{12}a_{23} - a_{13}a_{22})}{A_{10}}; \quad (7.44)$$

$$\bar{y}_2 = \frac{k_1\bar{z}_1(a_{23}a_{31} - a_{21}a_{33}) + Q_2(a_{11}a_{33} - a_{23}^2) + k_3\bar{z}_2(a_{13}a_{12} - a_{11}a_{23})}{A_{10}}; \quad (7.45)$$

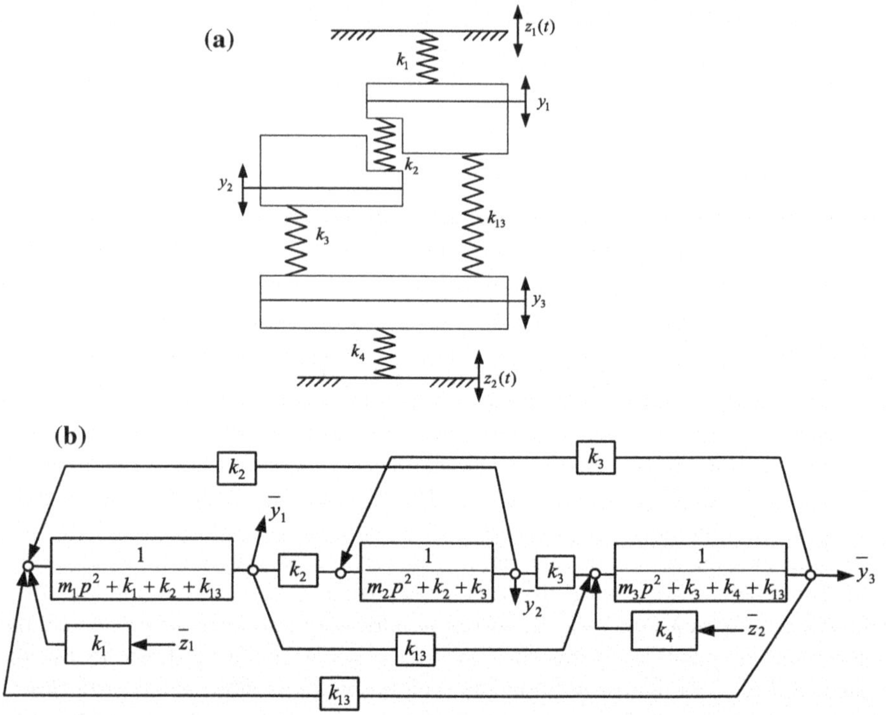

Fig. 7.9 Computational scheme (**a**) and structural diagram (**b**) of a vibration protection system with three degrees of freedom

Table 7.2 The coefficients of the equations of motion in the coordinates $\bar{y}_1$–$\bar{y}_3$

a_{11}	a_{12}	a_{13}
$m_1p^2 + k_1 + k_2 + k_{13}$	$-k_2$	$-k_{13}$
a_{21}	a_{22}	a_{23}
$-k_2$	$m_2p^2 + k_2 + k_3$	$-k_3$
a_{31}	a_{32}	a_{33}
$-k_{13}$	$-k_3$	$M_3p^2 + k_4 + k_4 + k_{13}$
Generalized forces		
k_1z_1	0	k_2z_2

$$\bar{y}_3 = \frac{k_1\bar{z}_1(a_{21}a_{32} - a_{22}a_{31}) + Q_2(a_{12}a_{31} - a_{11}a_{32}) + k_3\bar{z}_2(a_{11}a_{22} - a_{12}^2)}{A_{10}}, \quad (7.46)$$

where $A_{10} = a_{11}a_{22}a_{33} - a_{11}a_{32}^2 - a_{22}a_{13}^2 - a_{33}a_{12}^2 + 2a_{12}a_{13}a_{23}$ is the characteristic equation.

1. If we accept that $\overline{z_2} = 0$, then at the input action $\overline{z_1} \neq 0$, we obtain the transfer functions with kinematic perturbation

$$W_{10}(p) = \frac{\bar{y}_1}{\bar{z}_1} = \frac{k_1(a_{22}a_{33} - a_{23}^2)}{A_{10}}; \tag{7.47}$$

$$W_{20}(p) = \frac{\bar{y}_2}{\bar{z}_1} = \frac{k_1(a_{23}a_{31} - a_{21}a_{32}) + k_2}{A_{10}}; \tag{7.48}$$

$$\underset{z_2=0}{i_{gen}(p)} = \frac{\bar{y}_2}{\bar{y}_1} = \frac{k_1(a_{23}a_{31} - a_{21}a_{33})}{k_1(a_{22}a_{33} - a_{23}^2)} = \frac{a_{23}a_{31} - a_{21}a_{33}}{a_{22}a_{33} - a_{23}^2}; \tag{7.49}$$

or

$$\underset{z_2=0}{i_{gen}(p)} = \frac{k_3k_{13} + k_2(m_3p^2 + k_3 + k_4 + k_{13})}{(m_3p^2 + k_3 + k_4 + k_{13})(m_2p^2 + k_2) + (m_3p^2 + k_4 + k_{13})k_3}. \tag{7.50}$$

When $p \to 0$, the transfer ratio of the lever linkage will be determined

$$\begin{aligned}\underset{p\to 0}{i_{gen}(p)} &= \frac{k_3k_{13} + k_2k_3 + k_2k_4 + k_2k_{13}}{(k_3 + k_4 + k_{13})k_2 + (k_4 + k_{13})k_3} \\ &= \frac{k_2(k_3 + k_4 + k_{13}) + k_3k_{13}}{k_2(k_3 + k_4 + k_{13}) + k_3(k_4 + k_{13})}.\end{aligned} \tag{7.51}$$

$$\underset{p\to\infty}{i_{gen}(p)} = 0.$$

Note that with frequency

$$\omega^2 = \frac{k_2(k_3 + k_4 + k_{13}) + k_3k_{13}}{k_2m_3} \tag{7.52}$$

the value of $\underset{z_2=0}{i_{gen}(p)}$ becomes zero. Since the denominator (7.50) is a biquadratic equation, with two values of the frequency of external influence $\underset{z_2=0}{i_{gen}(p)} \to \infty$. Note that the form of the frequency characteristic $\left|\frac{\bar{y}_2}{\bar{y}_1}\right|$ corresponds to the amplitude-frequency characteristic of a system of elements m_1, m_2, k_1, k_2, k_3, k_4, k_{13} with kinematic perturbation $\bar{y}_3$.

2. When considering the case of $z_1 = 0$, $z_2 \neq 0$, the corresponding transfer functions take the form

$$W''_{10}(p) = \frac{\bar{y}_1}{\bar{z}_2} = \frac{k_3(a_{12}a_{23} - a_{13}a_{22})}{A_{10}}; \tag{7.53}$$

$$W_{20}(p) = \frac{\bar{y}_2}{\bar{z}_2} = \frac{k_3(a_{13}a_{12} - a_{11}a_{23})}{A_{10}}. \tag{7.54}$$

Taking into account (7.53) and (7.54)

$$\underset{z_1=0}{i_{\text{gen}}(p)} = \frac{k_2k_3 + k_{13}(m_2p^2 + k_2 + k_3)}{k_2k_{13} + k_3(m_1p^2 + k_1 + k_2 + k_{13})}. \tag{7.55}$$

When $p \to 0$, expression (7.55) transforms to the form

$$\underset{z_1=0}{i_{\text{gen}}(p)} = \frac{k_2k_3 + k_{13}(k_2 + k_3)}{k_2k_{13} + k_3(k_1 + k_2 + k_{13})}, \tag{7.56}$$

and with $p \to \infty$, respectively,

$$\underset{z_1=0}{i_{\text{gen}}(p)} = \frac{k_{13}m_2}{k_3m_1}. \tag{7.57}$$

Note that with an external disturbance $z_2 \neq 0$, the dependence of the transfer ratio $\underset{z_1=0}{i_{\text{gen}}(p)}$ on the frequency ω has the form, as shown in Fig. 7.10.

Theoretically, it is interesting to see the ratio of parameters in which the resonance and dynamic absorbing modes coincide. In this case, from the condition

$$(k_2 + k_3)k_{13}^2 + k_{13}k_3(k_1 + k_2) = k_{13}k_3(k_2 + k_3) + k_2k_3^2 \tag{7.58}$$

ratios between stiffnesses can be found. The transfer ratio $i_{\text{gen}}(p)$ becomes undetermined and has the form $\frac{0}{0}$ that can be resolved based on the L'Hôpital rule [9], within the framework of which in this situation $i_{\text{gen}}(p)$ will have a final value, $z_1 = 0$, and the lever linkage will correspond to ideas about the virtual lever of the first kind. The location of zones corresponding to certain forms of lever linkages can be determined from the graphs in Fig. 7.11, obtained on the basis of the dependences shown in Fig. 7.10.

3. With the total effect of two force factors ($z_1 = 0$, $z_2 = 0$), the interaction parameters can be estimated in the same way as before, based on the assumption that $z_2 = \alpha z_1$, then

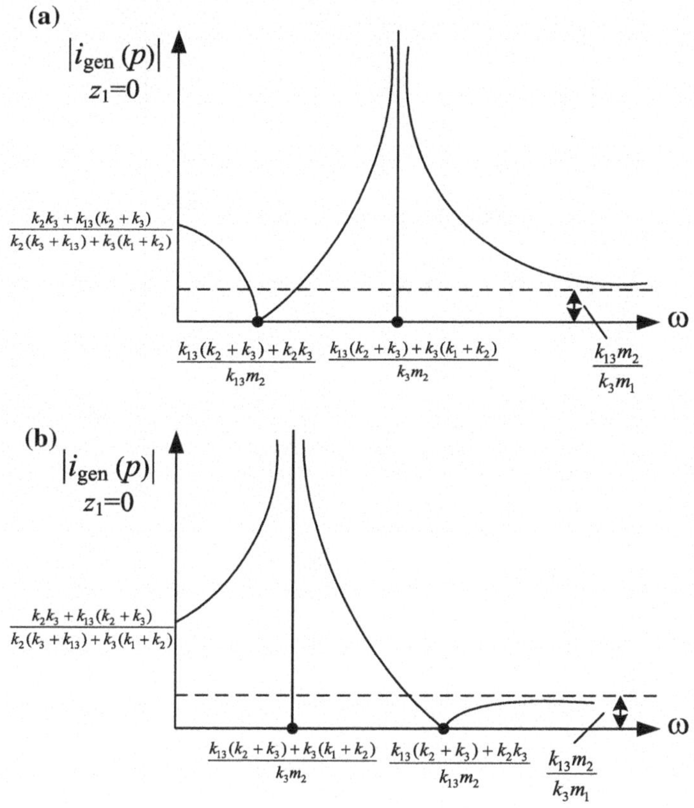

Fig. 7.10 Frequency characteristics of the transfer ratio of lever linkages $i_{\text{gen}}(p)$ with various combinations of parameters, depending on the frequency of external influence. **a** Dynamic absorbing mode in the preresonance region; **b** dynamic absorbing mode in the superresonance region

$$
\begin{aligned}
\underset{z_1=0}{i_{\text{gen}}(p)} &= \frac{k_1(a_{23}a_{31} - a_{21}a_{33}) + k_3\alpha(a_{13}a_{12} - a_{11}a_{23})}{k_1(a_{22}a_{33} - a_{23}^2) + k_3\alpha(a_{12}a_{23} - a_{13}a_{22})} \\
&= \frac{k_1k_3k_{13} + k_1k_2(m_3p^2 + k_3 + k_4 + k_{13}) + k_3\alpha(k_{13}k_2 + k_3(m_1p^2 + k_1 + k_2 + k_{13}))}{k_1a_{22}a_{33} - k_1k_3^2 + k_3k_3k_2\alpha + k_3k_{13}a_{22}} \\
&= \frac{p^2(m_3k_1k_2 + k_3^2\alpha m_1) + k_1k_3k_{13} + k_1k_2(k_3 + k_4) + k_3^2k_{13}\alpha + k_3(k_1 + k_2 + k_{13})}{a_{22}(k_1a_{33} - k_3k_{13}\alpha) - k_1k_3^2 + k_3^2k_2\alpha}.
\end{aligned}
\tag{7.59}
$$

In this case, the graph of dependence of $i_{\text{gen}}(p)$ on frequency will have quite diverse forms, which are determined not only by the ratios of the system parameters m_1, m_2, k_1–k_4, k_{13}, but also by the coordinate values of the relationship between

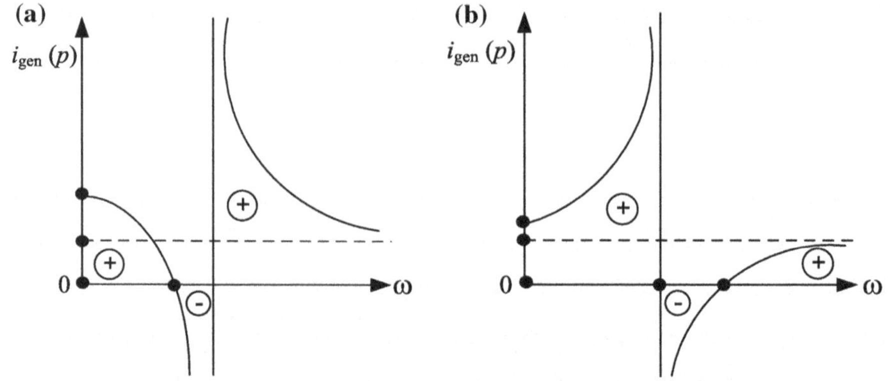

Fig. 7.11 Dependencies of transfer ratio lever linkage from frequency. The mode of dynamic oscillation damping in the preresonance (**a**) region; and in the superresonance (**b**) regions. The plus sign indicates a lever linkage of the first kind, and a minus sign indicates that of the second kind

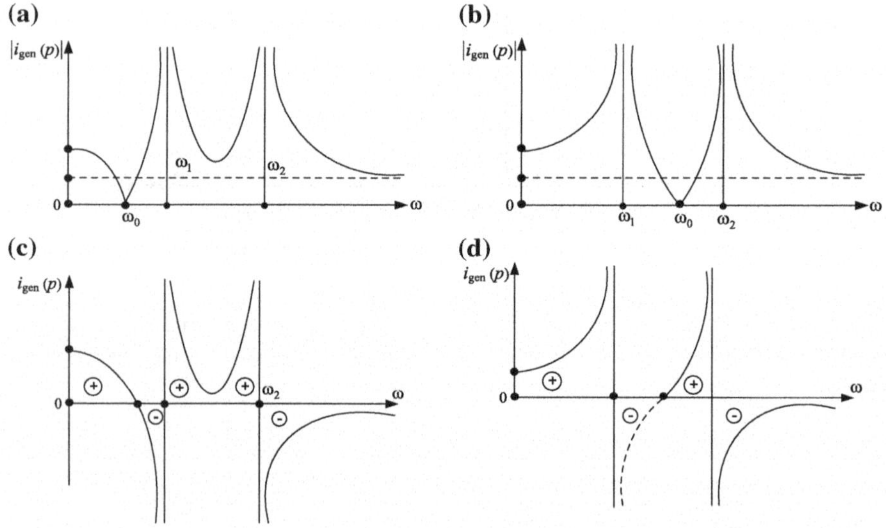

Fig. 7.12 Options for the location of frequency zones corresponding to certain types of lever linkages. **a** Dynamic oscillation absorbing in the preresonance zone; **b** the same in the interresonance zone (the signs (+) and (−) correspond to the lever linkages characteristic of the levers of the second and first kinds); **c** the ratio of amplitudes with the force reduced to m_1; **d** the same with a force reduced to m_2

external influences α (Fig. 7.12). There are options for the location of the graphs of dependencies $|i_{gen}(p)|$ on the frequency ω.

Dynamic interactions in mechanical oscillatory systems for various purposes, including vibration protection, can manifest themselves through ratios of the amplitudes of oscillations along the coordinates of the coupled motions. Such types of relations are characteristic of static states, when the applied perturbations are not periodic, but their action is expressed in establishing a certain distribution of displacements over coordinates. The relations between the displacements of coordinates depend not only on the configuration of the system, but also on the ratio of the stiffnesses. A generalized approach is being developed in the evaluation of lever linkages in mechanical oscillatory systems, perceived as some relationships between two coordinates in the form characteristic of relationships in levers of the first and second kind. In relation to the systems of translational oscillatory motions, it is proposed to correlate the lever linkages with the virtual levers. Note that the transfer ratios of virtual levers depend on the parameters of the system, which can be affected by both static and dynamic force and kinematic perturbations.

The implementation of lever linkages correlates with the ideas about the choice of certain frequency ranges in which a certain form of lever linkage dominates. A change in the frequency of a harmonic external disturbance affects the parameters of the transfer ratio. At the same time, the total interaction of external factors has a significant effect on the parameters of lever linkages. Instructional techniques for evaluating the possibilities of changing dynamic properties are based on the allocation of coefficients of external relations. The practical implementation of that change in the dynamic state can be achieved by introducing a special generator of disturbing forces. The concepts of lever linkages form the basis for determining the forms of joint interrelated motions of individual elements in systems with several degrees of freedom and the role of controlling the dynamic state of factors, one of which is the frequency of external disturbance.

7.3 Dynamics of Interaction in Mechanical Systems with Lever Linkages

Lever mechanisms as part of mechanical oscillatory systems have features that are manifested in changes in dynamic properties with respect to systems of the usual type and must be taken into account when determining dynamic responses. Note that the lever mechanisms in the structure of the mechanical system affect the conditions for the formation of relations between the coordinates of the motion of the system elements, as well as the forms of elastic constraints. In particular, the task of creating mathematical models of systems with lever linkages in different coordinates seems to be understudied, which is typical for systems that include units in the form of a solid body.

Consider a mechanical system with two degrees of freedom, which includes the use of levers of the first kind (Fig. 7.13a) and a bundle of levers through a gearing (Fig. 7.13b).

The introduction of lever linkages into mechanical chains causes certain difficulties in using the tools of the theory of circuits, which was considered in [6]. In

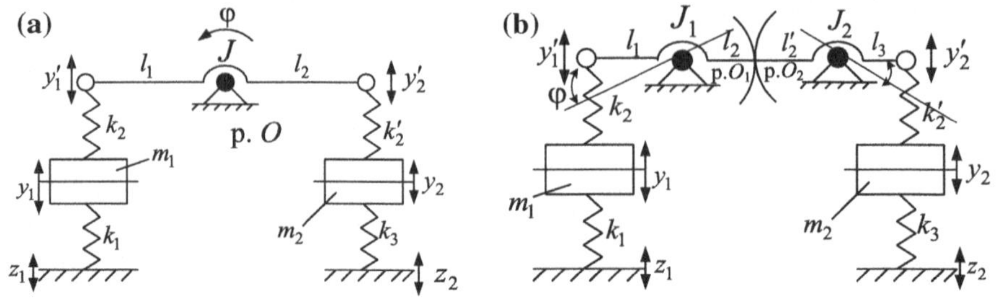

Fig. 7.13 Computational schemes of mechanical oscillatory systems. **a** With a lever of the first kind; **b** with a toothed gearing of the motion between the units

this regard, of interest is a comparative analysis of systems with levers in their comparison with the known circuit solutions in the form of a chain of two bodies and a solid body on elastic supports.

The system under consideration (Fig. 7.13a) consists of two elements with masses m_1 and m_2, which interact with each other through a lever of the first kind. The lever has a center of rotation in the form of a fixed point O and arms, respectively, l_1 and l_2. External disturbance is represented by the motion of the base $z_1(t)$ and $z_2(t)$ of a harmonic form; spring stiffnesses are denoted by k_1, k_2, k_2', k_3, respectively. To derive the equations of motion, it is assumed that the lever has a moment of inertia J relative to the point of rotation: the transfer ratio $i_1 = l_2/l_1$ and is characterized in this case by changing the direction of the speed of motion along the ends of the lever; the properties of the system are assumed to be linear, the friction is absent, and the motions of the elements of the system are considered small.

7.3.1 Mathematical Models and Their Features

To build a mathematical model of the system (see Fig. 7.13a), we introduce the coordinate system y_1 and y_2 (relative to the fixed base), as well as the angle of rotation of the lever with the moment of inertia J around the point O in the form of φ. We write the expressions for the kinetic and potential energies of the system:

$$T = \frac{1}{2}m_1\dot{y}_1^2 + \frac{1}{2}m_2\dot{y}_2^2 + \frac{1}{2}J\dot{\varphi}^2; \tag{7.60}$$

$$\Pi = \frac{1}{2}k_1(y_1 - z_1)^2 + \frac{1}{2}k_2(y_1' - y_1)^2 + \frac{1}{2}k_2'(y_2' - y_2)^2 + \frac{1}{2}k_3y_2^2, \tag{7.61}$$

where y_1', y_2' are coordinates of the ends of the lever: $y_1' = l_1\varphi$; $y_2' = l_2\varphi$ in accordance with Fig. 7.13a; $z_2(t) = 0$. Between y_1' and y_2' there is a $y_1'/l_1 = y_2'/l_2$, whence it follows that $y_2' = iy_1'$ ($i = -l_2/l_1$). We rewrite (7.60) and (7.61) with the condition of the connection of coordinates, then

$$\mathrm{T} = \frac{1}{2}m_1\dot{y}_1^2 + \frac{1}{2}m_2\dot{y}_2^2 + \frac{1}{2}J\left(\frac{\dot{y}_1'}{l_1}\right)^2; \tag{7.62}$$

$$\Pi = \frac{1}{2}k_1(y_1 - z_1)^2 + \frac{1}{2}k_2(y_1' - y_1)^2 + \frac{1}{2}k_2'(iy_1' - y_2)^2 + \frac{1}{2}k_3y_2^2. \tag{7.63}$$

After a series of transformations, we obtain the system of equations of motion:

$$m_1\ddot{y}_1 + y_1(k_1 + k_2) - k_2y_1' = k_1z_1; \tag{7.64}$$

$$m_2\ddot{y}_2 + y_2(k_3 + k_2') - k_2'iy_1' = 0; \tag{7.65}$$

$$(I/l_1^2)\ddot{y}_1' + y_1'(k_2 + k_2'i^2) - k_2y_1 - k_2'iy_2 = 0. \tag{7.66}$$

The coefficients of the system of equations of motion (7.64)–(7.66) are presented in Table 7.3.

Using (7.65) with $J = 0$, we find that $y_1'(k_2 + k_2'i^2) - k_2y_1 - k_2'iy_2 = 0$, whence

$$y_1' = \frac{k_2y_1 + k_2'iy_2}{k_2 + k_2'i^2} = ay_1 + by_2, \tag{7.67}$$

where

$$a = k_2/(k_2 + k_2'i^2); \; b = k_2'i/(k_2 + k_2'i^2). \tag{7.68}$$

Perform transformations of Eqs. (7.64) and (7.65) and get:

$$m_1\ddot{y}_1 + y_1(k_1 + k_2) - k_2(ay_1 + by_2) = k_1z_1; \tag{7.69}$$

$$m_2\ddot{y}_2 + y_2(k_2' + k_3) - k_2'i(ay_1 + by_2) = 0. \tag{7.70}$$

Table 7.3 The coefficients of the equations of motion (7.64)–(7.66)

a_{11}	a_{12}	a_{13}
$m_1p^2 + k_1 + k_2$	0	$-k_2$
a_{21}	a_{22}	a_{23}
0	$m_1p^2 + k_2' + k_3$	$-k_2'i$
a_{31}	a_{32}	a_{33}
$-k_2$	$-k_2'i$	$(J/l_1^2)p^2 + k_2 + k_2'i^2$
Q_1	Q_2	Q_3
k_1z_1	0	0

Note Q_1–Q_3 are the generalized forces corresponding to the generalized coordinates y_1, y_2, y_3'

Table 7.4 The coefficients of the equations of motion of the system in the coordinates y_1 and y_2

a_{11}	a_{12}
$m_1p^2 + k_1 + k_2 - k_2b$	$-k_2b$
a_{21}	a_{22}
$-k_2'ia$	$m_1p^2 + k_2' + k_3 - k_2'ib$
Q_1	Q_2
k_1z_1	0

After reduction to the unified form we will obtain Table 7.4 coefficients of equations (7.69) and (7.70), in which the coordinate of the motion y_1' is excluded using the relation (7.67).

Check the ratio between the coefficients a_{21} and a_{21}:

$$-k_2b = a_{12} = -\frac{k_2k_2'i}{k_2 + k_2'i^2} = \frac{-k_2k_2'i}{k_2 + k_2'i^2}, \tag{7.71}$$

$$-k_2'ia = a_{21} = -\frac{k_2k_2'i}{k_2 + k_2'i^2}. \tag{7.72}$$

Thus, from (7.71) and (7.72) it follows that the symmetry of the matrix of coefficients of the equations (Table 7.4) is preserved. The structural diagram of the system at $J = 0$ has the form, as shown in Fig. 7.14.

Expand expressions for reduced stiffness:

$$k_{\text{red}_1} = k_1 + k_2 - k_2a = k_1 + \frac{k_2k_2'i^2}{k_2 + k_2'i^2}. \tag{7.73}$$

In turn,

$$k_{\text{red}_2} = k_3 + k_2' - k_2'ib = k_3 + \frac{k_2k_2'}{k_2 + k_2'i^2}. \tag{7.74}$$

If $J \neq 0$, then the structural diagram of the system takes the form, as shown in Fig. 7.15 in view of (7.73) and (7.74).

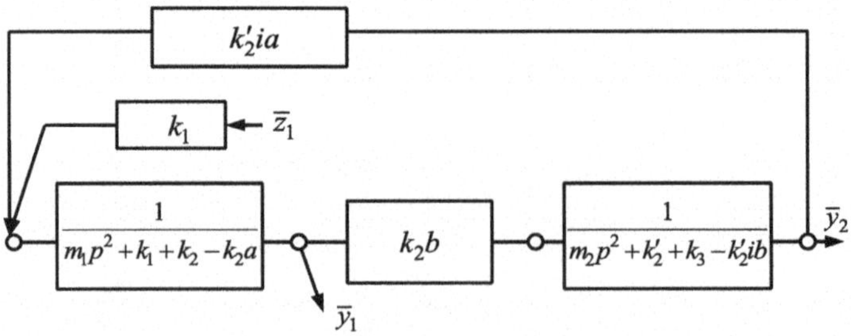

Fig. 7.14 The structural diagram of a mechanical system with lever linkages at $J = 0$

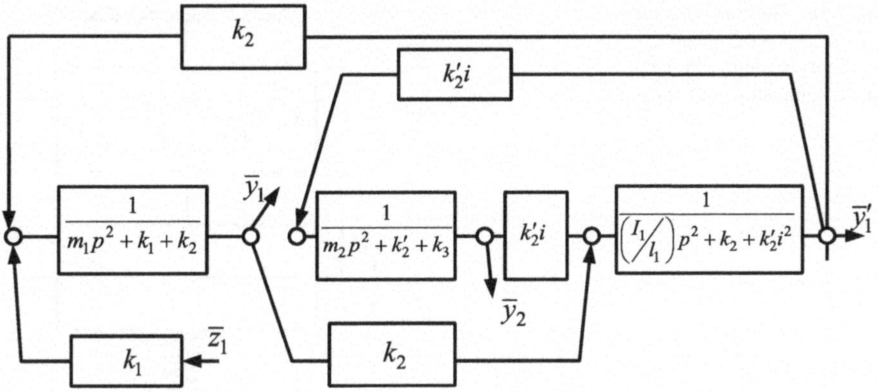

Fig. 7.15 The structural diagram of the system according to Fig. 7.13a with the inertial lever

The peculiarity of the system is that there are no connections between the partial systems in the coordinates y_1 and y_2. Assuming that $k_1z_1 = Q_1$, you can find the transfer function of the system:

$$W(p) = \frac{\bar{y}_1}{\bar{Q}} = \frac{\left[(J/l_1^2)p^2 + k_2 + k_2' i^2\right](m_2p^2 + k_2' + k_3) - (k_2' i)^2}{A}, \tag{7.75}$$

where A is the characteristic equation:

$$\begin{aligned} A = {} & (m_1p^2 + k_1 + k_2) \cdot (J/l_1^2p^2 + k_2 + k_2' i^2)(m_2p^2 + k_2' + k_3) \\ & - (k_2' i)^2(m_1p^2 + k_1 + k_2) - k_2^2(m_2p^2 + k_2' + k_3). \end{aligned} \tag{7.76}$$

7.3.2 Reduced System Stiffnesses

From (7.75) it follows that the reduced elasticity, that is, the stiffness of the elastic compact in the system in Fig. 7.13 when applying a static force Q_1 to an element with mass m_1, will be determined

$$k_{\text{red}} = \frac{i^2 k_2 k_2' k_3}{k_2' k_2 + k_3 k_2 + k_2' k_3 i^2}. \tag{7.77}$$

Similar expression, using (7.76) and (7.77), can be obtained directly from the computational scheme Fig. 7.14, assuming $J = 0$, $m_2 = 0$, then

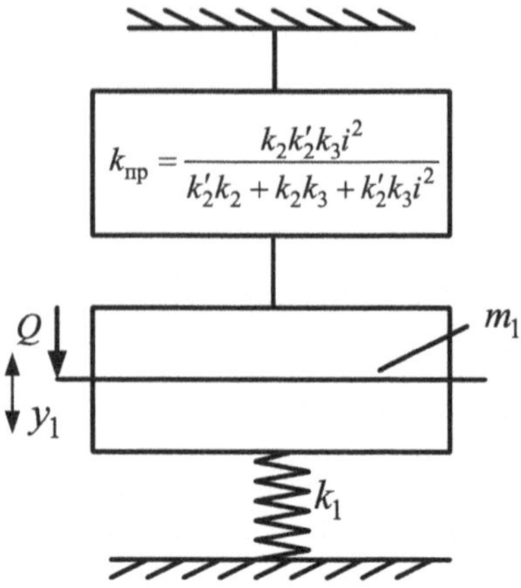

Fig. 7.16 The computational scheme of a mechanical oscillatory system with an elastic compact (quasi-spring)

$$k_{red} = \frac{\left(\frac{k_2' k_3}{k_2' + k_3}\right) i^2 k_2}{\left(\frac{k_2' k_3}{k_2' + k_3}\right) i^2 + k_2} = \frac{i^2 k_2 k_2' k_3}{k_2 k' + k_2 k_3 + k_2' k_3 i^2}. \tag{7.78}$$

The algorithm for obtaining (7.78) consists in isolating two stages from successively connected elastic elements and taking into account the properties of the lever joint. To obtain the expression (7.77), if we use (7.75) and (7.76), it is also necessary to accept $p = 0$ in partial systems $m_2 p^2 + k_2' + k_3$ and $J/l_1^2 p^2 + k_2 + k_2' i^2$, then bring the system to the form $m_1 p^2 + k_1 + k_{red}$. Figure 7.16 shows that, taking into account the transformations, the compact (quasi-spring) of elastic elements in the case of static loads takes the place of the usual elastic element.

When $i = 1$, expression (7.78) takes the form of consecutively connected elastic elements. In turn, when $i = 0$, which corresponds to $l_2 = 0$, the system takes a simplified form, at which $k_{red} = 0$. In this case, the interaction through the lever with the other elements does not occur. Assuming $k_2' = \infty$, i.e. count the mass m_2 directly connected with the lever, then

$$k_{red} = \frac{k_2 k_3 i^2}{k_2 + k_3 i^2}, \tag{7.79}$$

which coincides with the results, for example, given in [1, 10].

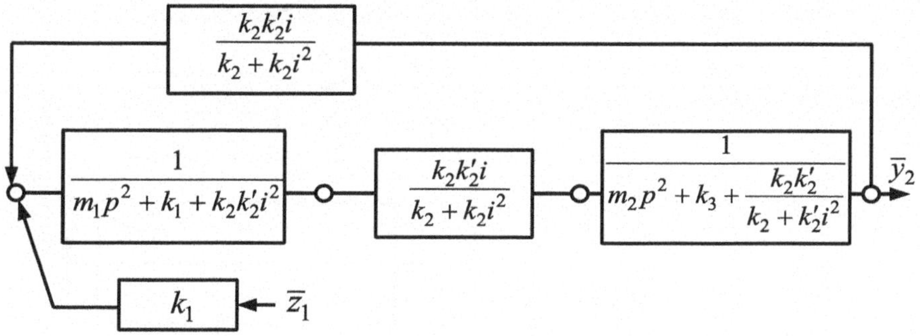

Fig. 7.17 The structural diagram of the system with lever linkages

7.3.3 *The Mode of Dynamic Interactions with a Lever Mechanism*

Accepting that $J = 0$, you can convert the structural diagram in Fig. 7.15 to the form, as shown in Fig. 7.17. In this case, the interaction between the partial systems will be carried out through an elastic compact (quasi-spring) with stiffness

$$W'(p) = \frac{k_2k_2'i}{k_2 + k_2'i^2}. \tag{7.80}$$

Using the structural diagram in Fig. 7.17, it is possible to carry out a transformation, which gives the system under consideration the form of a conventional chain structure containing compact sets of elastic elements connected by lever linkages. The structural diagram with the transformed elements is presented in Fig. 7.18.

The corresponding computational scheme at the level of using individual units, taking into account their physical nature, is shown in Fig. 7.19.

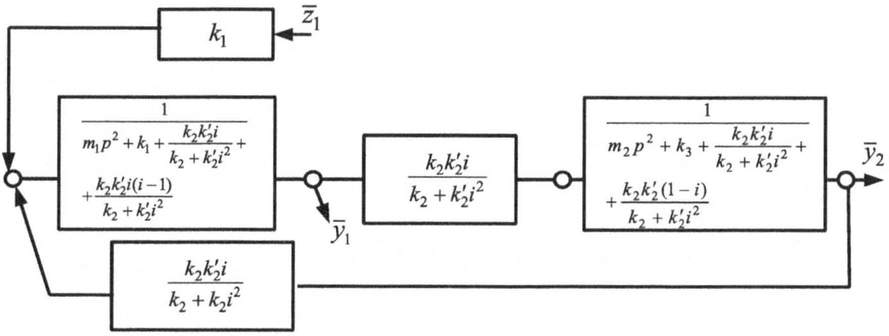

Fig. 7.18 The structural diagram of the system, reduced to a chain view

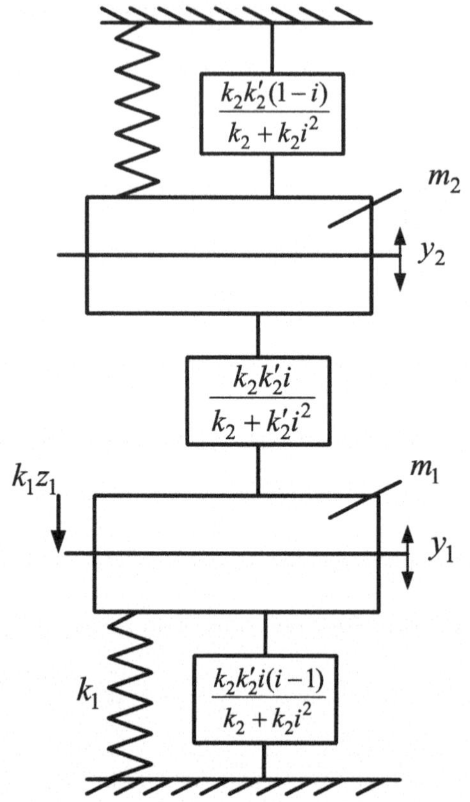

Fig. 7.19 The computational scheme of the system, reduced to a chain view and containing lever linkages

A feature of the system in Fig. 7.19 is that the lever linkages can be introduced into the structure of the compact of elastic elements (or quasi-springs), and this has not been considered in the scientific literature before in this perspective. In the theory of chains, special techniques are used to account for lever linkages that do not reflect the common nature of dynamic connections. We also note that the relationship between the partial systems in the physical sense is implemented through a lever mechanism, which turns the rotational force into force factors of interaction between the masses m_1 and m_2. Thus, a system with lever linkages can be represented by a chain system with elastic elements forming some compacts (quasi-springs); between the partial systems, elastic constraints are implemented (and the mass-and-inertia properties of the lever for static calculations are assumed to be small). The type of lever device (levers of the first or second kind) is chosen to build the transfer functions of the system, since the levers of the first kind have a transfer ratio with a minus sign. This is important for determining the sign of the additional feedback introduced by the levers with the corresponding changes in the characteristic equation. According to the structural diagram in Fig. 7.19, the frequencies of partial systems and the frequencies of natural oscillations of the system, as well as the dynamic properties, will depend on the type of lever linkages, which, in particular, was reflected in [11]. To verify the correctness of the approach, we define the static stiffness of the system in Fig. 7.19:

$$\begin{aligned} k_{\text{red}} &= \frac{\dfrac{\left[k_3(k_2+k_2'i^2)+k_2k_2'(1-i)\right]}{(k_2+k_2i^2)}\dfrac{k_2k_2'i}{k_2+k_2'i^2}}{\dfrac{k_3(k_2+k_2'i^2)+k_2k_2'(1-i)}{(k_2+k_2'i^2)}+\dfrac{k_2k_2'i}{(k_2+k_2'i^2)}}+\frac{k_2k_2'i}{k_2+k_2'i^2} \\ &= \frac{k_2k_2'k_3i}{k_2k_2'+k_3k_2+k_3k_2'i^2}, \end{aligned} \tag{7.81}$$

which coincides with the expression (7.78).

Accounting for the dynamic properties of a system with an inertial lever requires independent consideration, but it is done in the same way. When using the lever of the first kind, the speeds of the ends of the leech have different directions, therefore, if we take i < 0, then in accordance with the initial Eqs. (7.64)–(7.66) the feedbacks in the structural diagrams will be negative. This implies in the characteristic equation (Fig. 7.18) the emergence of the plus sign in front of the last member. In this case, it is necessary to take into account that the minus sign of the transfer ratio does not change the parameters of the partial systems, i.e. the lever of the first kind in this case provides negative feedback for the convolutions of the system. If we use a lever of the second kind, in which $i > 0$, then the feedback in the system will be positive, which changes sign in the characteristic equation of the system—it becomes negative. Taking into account the characteristics of lever linkages leads to the fact that the dynamic properties of systems will vary for different types of lever linkages.

7.3.4 Properties of Systems with Complex Lever Linkages

Consider a more complex mechanical system with lever linkages, including gearing (Fig. 7.13b). We write the expressions for the kinetic and potential energies, assuming that the components of the lever linkages have moments of inertia J_1 and J_2, respectively, and have fixed points of rotation O_1 and O_2 (Fig. 7.13b):

$$\mathrm{T} = \frac{1}{2}m_1\dot{y}_1^2 + \frac{1}{2}m_2\dot{y}_2^2 + \frac{1}{2}J_1\dot{\varphi}^2 + \frac{1}{2}J_2\dot{\varphi}_1^2, \tag{7.82}$$

$$\Pi = \frac{1}{2}k_1(y_1 - z_1)^2 + \frac{1}{2}k_2(y_1' - y_1)^2 + \frac{1}{2}k_2'(y_2' - y_2) + \frac{1}{2}k_3(y_2 - z_2)^2, \tag{7.83}$$

where φ and φ_1 are the angular coordinates of the rotation of the gearing elements.

We introduce a number of necessary ratios for further calculations:

$$\varphi = \frac{y_1'}{l_1};\ \varphi_1 = \frac{y_2'}{l_3};\ l_2\varphi = \varphi_1 l_2';\ \frac{y_1'}{l_1} = \frac{y_2'}{l_3}\cdot\frac{l_2'}{l_2};$$
$$y_2' = \frac{y_1' l_3 l_2}{l_1 l_2'},\ i = \frac{l_2}{l_1};\ i_1 = \frac{l_3}{l_2'};\ \varphi_1 = \varphi\frac{l_2}{l_2'}. \tag{7.84}$$

Taking into account relations (7.84), expressions (7.82) and (7.83) will take the form

$$\mathrm{T} = \frac{1}{2}m_1\dot{y}_1^2 + \frac{1}{2}m_2\dot{y}_2^2 + \frac{1}{2}I_1\dot{\varphi}^2 + \frac{1}{2}I_2\dot{\varphi}^2\left(\frac{l_2}{l_2'}\right), \tag{7.85}$$

$$\Pi = \frac{1}{2}k_1(y_1 - z_1)^2 + \frac{1}{2}k_2(\varphi l_1 - y_1)^2 + \frac{1}{2}k_2'(\varphi l_1 i i_1 - y_2)^2 + \frac{1}{2}k_3(y_2 - z_2)^2. \tag{7.86}$$

Since $l_2 = il_1$, and $l_2' = l_3/i_1$, we can introduce the following relation:

$$l_2/l_2' = i_1 i * l_1/l_3 = i_2. \tag{7.87}$$

The system of equations with (7.84)–(7.86) takes the form

$$m_1\ddot{y}_1 + y_1(k_1 + k_2) - k_2\varphi l_1 = k_1 z_1, \tag{7.88}$$

$$m_2\ddot{y}_2 + y_2(k_2' + k_3) - k_2' i i_1 l_1\varphi = k_3 z_2, \tag{7.89}$$

$$\ddot{\varphi}(J_1 + J_2 i_2^2) + \varphi\left[k_2 l_1^2 + k_2'(l_1 i i_1)^2\right] - k_2 l_1 y_1 - k_2' l_1 i i_1 y_2 = 0. \tag{7.90}$$

Let $J_1 = 0$, $J_2 = 0$, then

$$\varphi = a y_1 + b y_2, \tag{7.91}$$

where

$$a = \frac{k_2 l_1}{k_2 l_1^2 + k_2'(l_1 i i_1)^2}, \tag{7.92}$$

$$b = \frac{k_2' l_1 i i_1}{k_2 l_1^2 + k_2'(l_1 i i_1)^2}. \tag{7.93}$$

Taking into account (7.92) and (7.93), the system of equations (7.88) and (7.89) is converted to the form:

Table 7.5 The coefficients of the equations in the coordinates y_1 and y_2

a_{11}	a_{12}
$m_1p^2 + k_1 + \frac{k_2k_2'(l_1 i i_1)^2}{k_2 l_1^2 + k_2'(l_1 i i_1)^2}$	$-\frac{k_2k_2' l_1(l_1 i i_1)}{k_2 l_1^2 + k_2'(l_1 i i_1)^2}$
a_{21}	a_{22}
$-\frac{k_2k_2' l_1(l_1 i i_1)}{k_2 l_1^2 + k_2'(l_1 i i_1)^2}$	$m_2p^2 + k_3 + \frac{k_2k_2' l_1^2}{k_2 l_1^2 + k_2'(l_1 i i_1)^2}$
Q_1	Q_2
$k_1 z_1$	$k_2 z_2$

Note Q_1, Q_2 are the generalized forces corresponding to the generalized coordinates y_1 and y_2

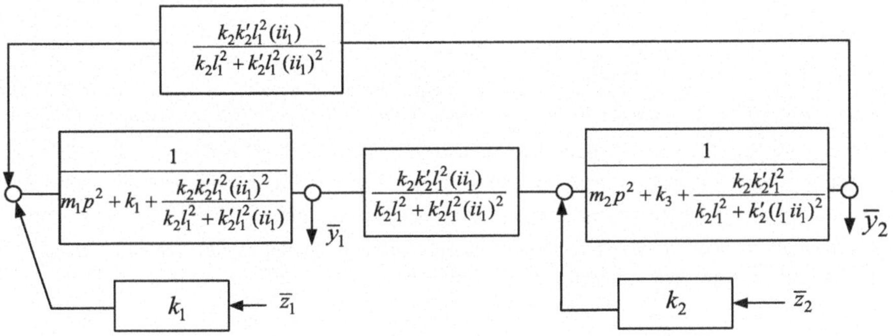

Fig. 7.20 The structural diagram of the system according to Fig. 7.13b

$$m_1\ddot{y}_1 + (k_1 + k_2)y_1 - \frac{k_2^2 l_1^2}{k_2 l_1^2 + k_2'(l_1 i i_1)^2} - \frac{k_2 l_1 k_2'(l_1 i i_1)y_2}{k_2 l_1^2 + k_2'(l_1 i i_1)^2} = k_1 z_1, \tag{7.94}$$

$$m_2\ddot{y}_2 + (k_2' + k_3)y_2 - \frac{(k_2')^2(l_1 i i_1)^2 y_2}{k_2 l_1^2 + k_2'(l_1 i i_1)^2} - \frac{k_2 l_1 k_2'(l_1 i i_1)}{k_2 l_1^2 + k_2'(l_1 i i_1)^2} = k_3 z_2. \tag{7.95}$$

Table 7.5 shows the coefficients of equations (7.94) and (7.95).

The structural diagram of the system according to Fig. 7.13b is shown respectively in Fig. 7.20.

From the structural diagram it follows that the introduction of lever and elastic elements forms a system in which the partial blocks have connections determined by the transfer function

$$W'(p) = \frac{k_2 k_2'(i i_1)}{k_2 + k_2'(i i_1)^2}. \tag{7.96}$$

In the structural diagram in Fig. 7.20 it is possible to produce a number of equivalent transformations (Fig. 7.21), assuming that

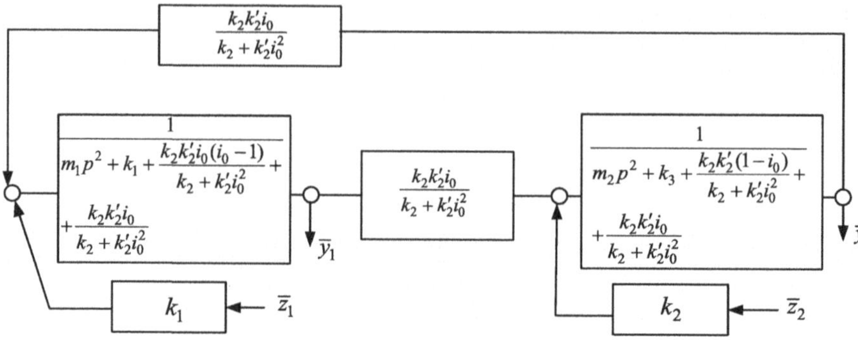

Fig. 7.21 The structural diagram of the system according to Fig. 7.13b, reduced to a chain view

$$ii_1 = i_0. \tag{7.97}$$

Note that the above transformations for the computational scheme Fig. 7.13b are similar to the transformations for the scheme Fig. 7.13a. If you do not take into account the signs of the transfer ratio, then for the scheme in Fig. 7.21 a system is presented in which positive feedback is introduced between the partial blocks. In this case, the characteristic equation will be:

$$\left(m_1p^2 + k_1 + \frac{k_2k_2'i_0^2}{k_2 + k_2'i_0^2}\right) \cdot \left(m_2p^2 + k_3 + \frac{k_2k_2'}{k_2 + k_2'i_0^2}\right) - \frac{\left(k_2k_2'i_0\right)^2}{\left(k_2 + k_2'i_0^2\right)^2} = 0. \tag{7.98}$$

Let us consider in more detail the structure with the transfer ratio $i_0 = ii_1$. Note that $i = l_2/l_1$ is the transfer ratio of the lever of the first kind (see Fig. 7.13b), for which $i = -l_2/l_1$. In turn, $i_1 = l_2/l_2'$ is the transfer ratio of a toothed gearing drive with external gearing ensuring rotation of the transmission elements in opposite directions; therefore $i_1 = -l_2/l_2'$. The combination of two transfer relations will thus always be positive. The above allows us to conclude that the connection of two levers of the first kind through an external gearing turns the entire unit into a lever of the second kind, which in itself is an unconventional representation of the characteristics of lever linkages.

Thus, lever linkages in the structures of mechanical systems, containing elastic and mass-and-inertia elements, create spatial (in this case, two-dimensional) interactions. With the analysis of static equilibrium, the system can be considered at the levels of allocation of structural formations from elastic elements and levers, in turn, interconnected by a toothed mechanism. Such structural formations can also be called compacts, or quasi-springs, which can have quite complex schemes and consist of various combinations of levers and springs. It is noteworthy that the quasi-spring in static transformations, for example, when determining the reduced stiffness of a mechanical system (stiffness at the point of application of force), behaves as a conventional spring in the form of a typical elementary unit.

These structural representations allow us to propose a method for determining reduced stiffness based on the use of transfer functions for these purposes. The transfer functions calculated by simple transformations make it possible to obtain the necessary data on the reduced stiffnesses of the system and thereby take into account the constructive and engineering features of the system. That method for solving problems of dynamics opens up the opportunity of introducing and accounting for lever linkages in structural interpretations of mechanical oscillatory systems with the subsequent determination of dynamic responses in the established order.

7.4 Features of Mechanical Oscillatory Systems Containing Units in the Form of Solid Bodies

The introduction of motion transformation devices can be viewed from the standpoint of the possibilities of transforming the structures of mechanical oscillatory systems based on the joint of units as an approach to simplify models and the method of constructing non-traditional forms of mechanical oscillatory systems.

Figure 7.22 shows the computational scheme of the vibration protection system (VPS), in which there are two blocks marked with contours I and II. At the bottom of

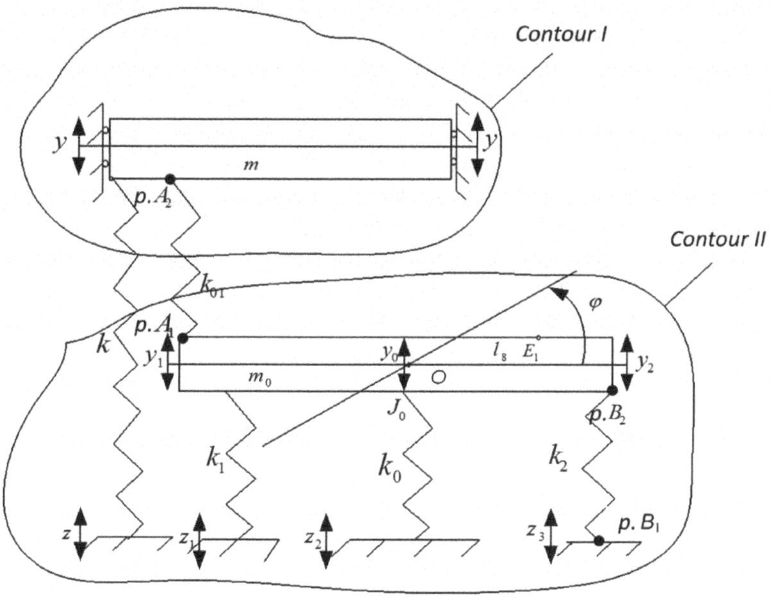

Fig. 7.22 The computational scheme of the vibration protection system, which has two interaction circuits. m_0 is the mass of the intermediate body (can turn into m_1 and m_2, connected by a lever); m is the mass of the object; k, k_1, k_0, k_2 are the elastic elements of the intermediate body resting on the base; z to z_3—kinematic perturbations

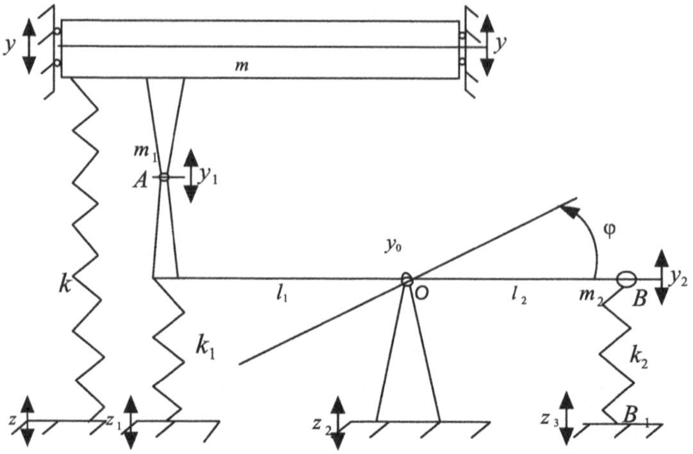

Fig. 7.23 The transformed computational scheme containing joints

the block II is a solid body with a mass and moment of inertia; in addition, the system includes elastic (k_{01}, k_0, k, k_1, k_2,) elements and the object of protection. It is assumed that the forces of resistance to motion along the coordinates are rather small.

The contours I and II (Fig. 7.22) are in interaction through the elastic connecting element k_{01}. In turn, a solid body rests against not only on the elastic supports k_1 and k_2, but has an elastic constraint k_0, the line of action of which passes through the center of gravity of the solid body at point O. We assume that $A_1O = l_1$, $B_2O = l_2$, and the added mass m does not cause significant changes in the mass-and-inertia parameters of the system. Developing the provisions on joints of solid bodies as connections that take the form of a rotary hinge, we note that, by setting the stiffnesses k_{01} and k_0 to be large enough, we can convert the computational scheme to the form, as shown in Fig. 7.23.

Obtained as a result of simplification of the scheme in Fig. 7.22, the computational scheme in Fig. 7.23 is given in [12], where it was used to justify an increase in the efficiency of oscillations damping of transportation device systems. However, the issue of the features of the joints has not been studied.

7.4.1 The Method of Constructing a Mathematical Model

The expressions for the kinetic and potential energies for the VPS computational scheme shown in Fig. 7.23, have the form

$$\mathrm{T} = \frac{1}{2}(m + m_1)\dot{y}^2 + \frac{1}{2}m_2\dot{y}_2, \tag{7.99}$$

$$\Pi = \frac{1}{2}k(y-z)^2 + \frac{1}{2}k_2(y-z_1)^2 + \frac{1}{2}k_2(y_2-z_3)^2, \tag{7.100}$$

where $\dot{y}_2$ is the speed of the element mass m_2 in absolute motion, determined by the expression

$$\dot{y}_2 = -\frac{l_2}{l_1}\dot{y} + \dot{z}_2. \tag{7.101}$$

Understanding the transfer ratio of the lever in the form of $i = \frac{l_2}{l_1}$, we transform (7.99) and (7.100) to the form

$$\mathrm{T} = \frac{1}{2}(m+m_1)\dot{y}^2 + \frac{1}{2}m_2[-\dot{y}i + \dot{z}_2(1+i)]^2, \tag{7.102}$$

$$\Pi = \frac{1}{2}k(y-z)^2 + \frac{1}{2}k_1(y-z_1)^2 + \frac{1}{2}k_2(y_{2\text{аб}}-z_3)^2 = \frac{1}{2}k(y-z)^2 + \frac{1}{2}k_1(y-z_1)^2 + \frac{1}{2}k_2[-yi + z_2(1+i) - z_3]^2. \tag{7.103}$$

We obtain for the system (Fig. 7.23) the differential equation of motion

$$\begin{aligned} &\ddot{y}\left(m+m_1+m_2i^2\right) + y\left(k+k_1+k_2i^2\right) \\ &= m_2i(1+i)\ddot{z}_2 + z_2k_2(1+i) + k_1z_1 + kz - k_2iz_3. \end{aligned} \tag{7.104}$$

To simplify the calculations, we assume that $z_1 = z_2 = z_3$; $k_1 = 0$ and $k_2 = 0$ and find the transfer function of the system

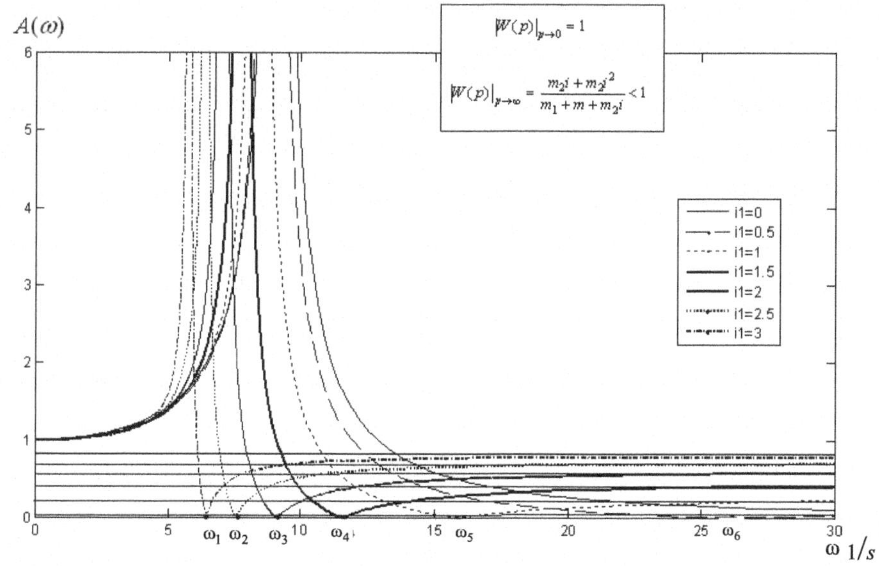

Fig. 7.24 The family of amplitude-frequency characteristics of the system with the transfer function (7.104)

Table 7.6 Natural frequencies in comparison with frequencies of dynamic absorbing of oscillations ω_1–ω_6 for the system in Fig. 7.24

i (transfer ratio)	0.5	1	1.5	2	2.5	3
Frequencies of dynamic absorbing, 1/s	ω_{dyn1}	ω_{dyn2}	ω_{dyn3}	ω_{dyn4}	Ω_{dyn5}	Ω_{dyn6}
	25.81	15.81	11.54	9.12	7.55	6.45
Natural frequencies oscillations, 1/s	Ω_{nat1}	Ω_{nat2}	Ω_{nat3}	Ω_{nat4}	Ω_{nat5}	ω_{nat6}
	8.94	8.45	7.78	7.07	6.38	5.77

$$W(p) = \frac{\bar{y}(p)}{\bar{z}(p)} = \frac{m_2(1+i)ip^2 + k}{(m_1 + m + m_2 i^2)p^2 + k}. \tag{7.105}$$

As an example, Fig. 7.24 shows a family of amplitude-frequency characteristics, built on the basis of (7.104) when the parameter i varies within limits from 0 to 3 in increments of 0.5. The following system parameters are taken as initial values: m = 100 kg; $m_1 = m_2$ = 20 kg; k = 10,000 N/m. In Fig. 7.24 dynamic absorbing frequencies are denoted through ω_1–ω_6.

The corresponding values of natural frequencies in comparison with the frequencies are given in Table 7.6.

The feature of the amplitude-frequency characteristics is that with an increase in i, the frequencies of the dynamic absorbing shift to the left, i.e. the frequency difference $\omega_{nat} - \omega_{dyn}$ decreases. At the same time, the natural oscillation frequency also decreases, but much more slowly. At high frequencies, the transfer coefficient of the amplitude of oscillations after the dynamic absorbing mode tends to the limiting value (the more i is, the greater the value of $|W(p)|_{p\to\infty}$ will be).

If $i \to \infty$, we obtain that $|W(p)|_{p\to 0} = \frac{m_2 i + m_2 i^2}{m_1 + m + m_2 i} = 1$.

The mathematical model of the system (7.103) can also be obtained by simplifying a more complex model (Fig. 7.23), described not by two (y, y_2), but by three coordinates: y_0, y, φ; y, y_1, y_0; y_1, y_2, y.

7.4.2 Building a Complete Mathematical Model

Let us return to the computational scheme shown in Fig. 7.25. Then the expressions (7.98) and (7.99) for kinetic and potential energy can be converted to

$$\mathrm{T} = \frac{1}{2} m_0 \dot{y}_0^2 + \frac{1}{2} J_0 \dot{\phi} + \frac{1}{2} m \dot{y}^2, \tag{7.106}$$

$$\begin{aligned} \Pi = {} & \frac{1}{2} k (y - z)^2 + \frac{1}{2} k_1 (y_1 - z_1)^2 + \frac{1}{2} k_0 (y_0 - z_2)^2 \\ & + \frac{1}{2} k_2 (y_2 - z_3)^2 + \frac{1}{2} k_{01} (y - y_1)^2. \end{aligned} \tag{7.107}$$

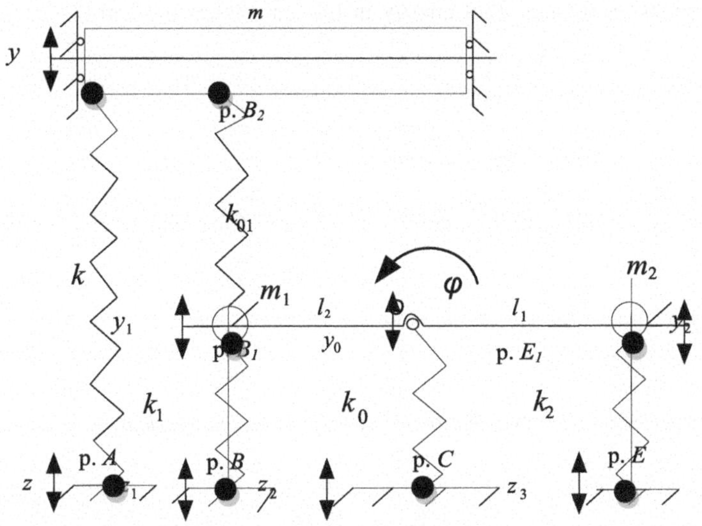

Fig. 7.25 The computational scheme of the VPS as in Fig. 7.23, but with spaced-apart masses m_1 and m_2

while $y_0 = ay_1 + by_2$, $\varphi = c(y_2 - y_1)$.

Coordinates of pp. A_1 and A_2 (see Fig. 7.22) are defined as follows: $y_{A1} = y_1$ and $y_{A2} = y$. To obtain joints between elements, the following conditions must be met:

$$y_1 - y = 0, y_{10} = 0. \tag{7.108}$$

Take the coordinate values

$$y_{10} = y_0 - z_2. \tag{7.109}$$

Consider the motion of the system in the coordinates y_0, φ. We take the notation

$$a = \frac{l_2}{l_1 + l_2}; \quad b = \frac{l_1}{l_1 + l_2}; \quad c = \frac{1}{l_1 + l_2}. \tag{7.110}$$

When moving the object of protection in the coordinate system y_0, y_2 and y_1, the expression for the kinetic energy can be reduced to

$$\mathrm{T} = \frac{1}{2} m_0 (a\dot{y}_1 + b\dot{y}_2)^2 + \frac{1}{2} J_0 c^2 (\dot{y}_2 - \dot{y}_1)^2 + \frac{1}{2} m\dot{y}^2. \tag{7.111}$$

The expression for potential energy in this case takes the form

$$\begin{aligned}\Pi &= \frac{1}{2}k(y-z)^2 + \frac{1}{2}k_1(y_1-z_1)^2 + \frac{1}{2}k_0(ay_1+by_2-z_2)^2 \\ &+ \frac{1}{2}k_2(y_2-z_3)^2 + \frac{1}{2}k_{01}(y-y_1)^2.\end{aligned} \tag{7.112}$$

The system of differential equations of motion of the system can be represented in the form

$$\begin{aligned}&\ddot{y}_1\left(m_0a^2+J_0c^2\right)+y_1\left(k_1+k_{01}k_0a^2\right)+\ddot{y}_2\left(m_0ab-J_0c^2\right)+y_2k_0ab \\ &= -k_{01}yk_1z_1+k_0az_2,\end{aligned} \tag{7.113}$$

$$\begin{aligned}&\ddot{y}_1\left(m_0ab-J_0c^2\right)+y_1(-k_0ab)+\ddot{y}_2(m_0b^2+J_0c^2)+y_2(k_0b_2+k_2) \\ &= k_0bz_2+k_2z_3,\end{aligned} \tag{7.114}$$

$$y_1k_{01}+\ddot{y}m+y(k+k_{01})=kz. \tag{7.115}$$

Table 7.7 shows the values of the coefficients of equations in the coordinates y_1, y_2 and y.

The generalized forces on the coordinates y, y_2 and y_1 are

$$Q_{y_1}=k_1z_1+k_0az_2; \quad Q_{y_2}=k_0bz_2+k_2z_3; \quad Q_y=kz. \tag{7.116}$$

Equations (7.112)–(7.114) describe the motions in the coordinate system reflecting the vertical motions of the mass-and-inertia elements of the vibration protection systems. This mathematical model can be called basic. A feature of the matrix is that $a_{13} = -k_{01}$, $a_{23} = a_{32} = 0$; it depends on the features of the dynamic interactions, determined by the structure of the elements of the vibration protection systems (see Fig. 7.22), and the choice of the system of generalized coordinates.

Table 7.7 The values of the coefficients of equations (7.112)–(7.114) in the coordinates y_1, y_2 and y

a_{11}	a_{12}	a_{13}
$(m_0a^2 + J_0c^2) \cdot p^2 + k_1 + k_0a^2$	$(m_0ab - J_0c^2) \cdot p^2 + k_0ab$	$-k_{01}$
a_{21}	a_{22}	a_{23}
$(m_0ab - J_0c^2) \cdot p^2 + k_0ab$	$(m_0b^2 + J_0c^2) \cdot p^2 + k_2 + k_0b^2$	0
a_{31}	a_{32}	a_{33}
$-k_{01}$	0	$mp^2 + k + k_{01}$

1. We turn to the coordinate system $y_{11} = y - y_1$, y_0 and φ; we write the series of relations $y = y_{11} + y_1 = y_{11} + y_0 - l_1\varphi$ and get the expression for the kinetic and potential energy:

$$T = \frac{1}{2}m_0(\dot{y}^0)^2 + \frac{1}{2}J_0\dot{\varphi}^2 + \frac{1}{2}m(\dot{y}_{11} + \dot{y}_0 - l_1\dot{\varphi})^2; \tag{7.117}$$

$$\begin{aligned} \Pi = &\frac{1}{2}k(y_{11} + y_0 - l_1\varphi - z)^2 + \frac{1}{2}k_1(y_0 - l_1\varphi - z_1)^2 \\ &+ \frac{1}{2}k_0(y_0 - z_2)^2 + \frac{1}{2}k_2(y_0 + l_2\varphi - z_3)^2 + \frac{1}{2}k_{01}(y_{11})^2. \end{aligned} \tag{7.118}$$

We write the equations of motion of the system in the coordinates φ, y_0, y_{11}:

$$\begin{aligned} &\ddot{\varphi}\cdot(J_0 + ml_1^2) + \ddot{\varphi}\cdot(kl_1^2 + k_1l_1^2 + k_2l_2^2) + \ddot{y}_0(-ml_1) + y_0(-kl_1 + k_1l_1 + k_2l_2) \\ &+ \ddot{y}_{11}(-ml_1) + y_{11}(-kl_1) = -kl_1z + k_1l_1z_1 + k_2l_2z_3; \end{aligned} \tag{7.119}$$

$$\begin{aligned} &\ddot{\varphi}(-ml_1) + \varphi(-kl_1 + k_1l_1 + k_2l_2) + \ddot{y}_0(m + m_0) + y_0(k + k_1 + k_0 + k_2) \\ &+ \ddot{y}_{11}(m) + y_{11}(k) = kz - k_1z_1 + k_0z_2 - k_2z_3; \end{aligned} \tag{7.120}$$

$$\ddot{\varphi}\cdot(-ml_1) + \varphi(-k_1l_1) + \ddot{y}_0(m) + y_0(k) + \ddot{y}_{11}(m) + y_{11}(k + k_{01}) = kz. \tag{7.121}$$

The values of the coefficients of the equations in the coordinates φ, y0 and y11 are respectively presented in Table 7.8.

The generalized forces on the coordinates φ, y_0 and y_{11} are

$$Q_\varphi = -kl_1z + k_1l_1z_1 + k_2l_2z_3;\ Q_{y_0} = kz + k_1z_1 + k_0z_2 - k_2z_3;\ Q_{y_{11}} = kz. \tag{7.122}$$

The mathematical model in the coordinate system φ, y_0 and y_{11} differs from the previous model (coordinates y_1, y_2 and y) in that the zero cells (Table 7.8) are missing, i.e. connections exist between all partial systems. As for the coordinate $y_{11} = y - y_1$, then it can be "zeroed out" by the assumption $k_{01} \to \infty$ with the formation of a joint or kinematic pair. In these conditions it is possible to connect

Table 7.8 The values of the coefficients of equations (7.119)–(7.121) in the coordinates φ, y_0 and y_{11}

a_{11}	a_{12}	a_{13}
$(ml_1^2 + J_0)p^2 + kl_1^2 + k_1l_1^2 + k_2l_2^2$	$(-ml_1)p^2 - kl_1 + k_1l_1 + k_2l_2$	$(-ml_1)p^2 - kl_1$
a_{21}	a_{22}	a_{23}
$(-ml_1)p^2 - kl_1 + k_1l_1 + k_2l_2$	$(m_0 + m)p^2 + k + k_1 + k_0 + k_2$	$(m)p^2 + k$
a_{31}	a_{32}	a_{33}
$(-ml_1)p^2 - k_1l_1$	$(m)p^2 + k$	$(m)p^2 + k + k_{01}$

solid bodies m and m_1. Physically, this means that the mass m is attached to the VPS element at p. A with mass-and-inertia parameters m_0, J_0 and changes them (both the total mass and the moment of inertia). The motion of the system will be described in this case by the coordinates y_0 and φ. The necessary data for obtaining transfer functions can be obtained by excluding the column and row containing y_{11} (in fact, the variable is eliminated, and the matrix order is reduced by one).

2. For further calculations, we introduce the coordinate system y_0, y_1 and y_{11}. Then the expression for the kinetic energy in the coordinates y_0, y_1 and y_{11} is converted to

$$\mathrm{T} = \frac{1}{2}m_0(\dot{y}_0)^2 + \frac{1}{2}J_0\dot{\varphi}^2 + \frac{1}{2}m(\dot{y}_1 + \dot{y})^2. \tag{7.123}$$

We will replace the coordinate system with respect to the expression for the kinetic energy, assuming that

$$y = y_1 + y_{11}; y_{11} = y - y_1, \varphi = c(y_2 - y_1); y_0 = ay_1 + by_2; y_2 = \frac{y_0 - ay_1}{b}.$$

We write

$$\varphi = cy_2 - cy_1 = \frac{c \cdot (y_0 - ay_1)}{b} - cy_1 = \frac{cy_0 - cay_1 - bcy_1}{b} = \frac{c}{b}(y_0 - y_1)$$

and find $\varphi = a_0(y_0 - y_1)$ where $a_0 = \frac{c}{b}$; $y_0 = ay_1 + by_2$;

$$y_2 = \frac{y_0 - ay_1}{b} = \frac{1}{b}y_0 - \frac{a}{b}y_1 = a_{10}y_0 - a_{01}y, \quad a_{10} = \frac{1}{b}, \quad a_{01} = \frac{a}{b}.$$

We write the expression for the potential energy in expanded form in the coordinates y_0, y_1 and y_{11}:

$$\begin{aligned}\Pi &= \frac{1}{2}k(y_1 + y_{11} - z)^2 + \frac{1}{2}k_1(y_1 - z_1)^2 + \frac{1}{2}k_0(y_0 - z_2)^2 \\ &+ \frac{1}{2}k_2(a_{10}y_0 - a_{01}y_1 - z_3)^2 + \frac{1}{2}k_{01}(y_{11})^2,\end{aligned} \tag{7.124}$$

The equations of motion of the system in the coordinates y_0, y_1 and y_{11} have the form

$$\begin{aligned}&\ddot{y}_1\left(Ja_0^2 + m\right) + y_1\left(k + k_1 + k_2a_{01}^2\right) + \ddot{y}_0\left(-Ja_0^2\right) + y_0(-k_2a_{01}a_{10}) + \ddot{y}_{11}(-m) + y_{11}(k) \\ &= kz - k_2a_{01}z_3 + k_1z_1;\end{aligned}$$

$$\tag{7.125}$$

Table 7.9 The values of the coefficients of equations (7.124) and (7.125) in the coordinates y_1, y_0 and y_{11}

a_{11}	a_{12}	a_{13}
$(m+J_0a_0^2)p^2+k+k_1+k_2la_{01}^2$	$(-J_0a_0^2)p^2-k_2a_{01}a_{10}$	$(-m)p^2+k$
a_{21}	a_{22}	a_{23}
$(-J_0a_0^2)p^2-k_2a_{01}a_{10}$	$(m+J_0a_0^2)p^2+k+k_2a_{10}^2$	0
a_{31}	a_{32}	a_{33}
$(-m)p^2+k$	0	$(m)p^2+k+k_{01}$

$$\ddot{y}_1\left(-Ja_0^2\right)+y_1\left(-k_2a_{01}a_{10}\right)+\ddot{y}_0\left(m_0+J_0a_0^2\right)+y_0\left(k+k_2a_{10}^2\right)=k_0z_2+k_2a_{01}z_3; \tag{7.126}$$

$$\ddot{y}_1(-m)+y_1(k)+\ddot{y}_{11}(m)+y_{11}(k+k_{01})=-kz. \tag{7.127}$$

Table 7.9 presents the values of the coefficients of equations (7.125)–(7.127) in the coordinates y_0, y_1 and y_{11}.

The generalized forces along the coordinates y_{11}, y_0, y_1 and have the form

$$Q_{y11}=-kz;\;\; Q_{y_0}=k_0z_2+k_2a_{01}z_3;\;\; Q_{y_1}=kz-k_2a_{01}z_3+k_1z_1. \tag{7.128}$$

In this coordinate system y_{11}, y_0, y_1 the introduction of a joint is also possible along the coordinate y_{11}. To obtain the transfer function of a system that has two degrees of freedom y_0 and y_1, it is necessary to exclude the corresponding columns and row. Note that if $y_{11}=0$, then $y_1=y$, the motion of the system with one hinge at point A will be described by the coordinates y_0 and y_1. In this case, the lever has an elastic support for its center of rotation. The greatest interest is still the case with two joints.

3. Consider the coordinate system y_{10}, y and y_{11}. Entering a series of relations ($y_{10}=y-z_2$), we write the expressions for the kinetic energy of the system:

$$\mathrm{T}=\frac{1}{2}m_0(\dot{y}_{10}+\dot{z}_2)^2+\frac{1}{2}J_0a_0^2(\dot{y}_{10}+\dot{z}_2-\dot{y}+y_{11})^2+\frac{1}{2}m(\dot{y})^2, \tag{7.129}$$

where $a_0=\frac{c}{b}$.

The expression for the potential energy of the system in the coordinates y_{10}, y and y_{11} is

$$\Pi=\Pi_1+\Pi_2+\Pi_3+\Pi_4+\Pi_5, \tag{7.130}$$

where

$$\begin{aligned}\Pi_1 &= \frac{1}{2}k(y-z)^2 = \frac{1}{2}k\left(y^2 - 2yz + z^2\right);\\ \Pi_2 &= \frac{1}{2}k_1\left(y^2 - 2y_{11} + y_{11}^2 - 2z_1 y + 2z_1 y_{11} + z_1^2\right);\\ \Pi_3 &= \frac{1}{2}k_0 (y_{10})^2;\\ \Pi_4 &= \frac{1}{2}k_2 (a_{10}y_{10} - a_{01}y + a_{01}y_{11} + z_0)^2,\end{aligned}$$

where $z_0 = a_{10}z_2 - z_3$. We make a series of auxiliary calculations and write down the equations of motion of the system in the coordinates y_{10}, y and y_{11}:

$$\begin{aligned}&\ddot{y}_{10}\left(m_0 + J_0 a^2\right) + y_{10}\left(k_0 + k_2 a_{10}^2\right) + \ddot{y}\left(-Ja_0^2\right) + y\left(-k_2 a_{10}a_{01}\right)\\ &\quad + \ddot{y}_{11}\left(J_0 a_0^2\right) + y_{11}\left(k_2 a_{10}a_{01}\right) = -k_2 a_{10}z_0 - m_0\ddot{z}_2 - J_0 a^2 \ddot{z}_2;\end{aligned} \tag{7.131}$$

$$\begin{aligned}&\ddot{y}_{10}\left(-J_0 a_0^2\right) + y_{10}\left(-k_2 a_{10}a_{01}\right) + \ddot{y}\left(m + Ja_0^2\right) + y\left(k + k_1 + k_2 a_{01}^2\right) + \ddot{y}_{11}\left(-J_0 a_0^2\right)\\ &\quad + y_{11}\left(-k_1 - k_2 a_{01}^2\right) = J_0 a_0^2 \ddot{z}_2 + kz + k_1 z_1 + k_2 a_{01}z_0;\end{aligned} \tag{7.132}$$

$$\begin{aligned}&\ddot{y}_{10}\left(J_0 a_0^2\right) + y_{10}\left(k_2 a_{10}a_{01}\right) + \ddot{y}\left(-Ja_0^2\right) + y\left(-k_1 - k_2 a_{01}^2\right) + \ddot{y}_{11}\left(-J_0 a_0^2\right)\\ &\quad + y_{11}\left(k_1 + k_2 a_{01}^2 + k_{01}\right) = -J_0 a_0^2 \ddot{z}_2 - k_1 z_1 - k_2 a_{01}z_0.\end{aligned} \tag{7.133}$$

Table 7.10 shows the coefficients of equations (7.131)–(7.133).

The generalized forces on the coordinates y_{10}, y and y_{11} are

$$\begin{aligned}Q_{y_{10}} &= -\left(m_0 + J_0 a_0^2\right)\ddot{z}_2 - k_2 a_{10}z_0;\\ Q_y &= \left(J_0 a_0^2\right)\ddot{z}_2 + kz + k_1 z_1 + k_2 a_{01}z_0;\\ Qy_{11} &= -\left(J_0 a_0^2\right)\ddot{z}_2 - k_1 z_1 - k_2 a_{01}z_0.\end{aligned} \tag{7.134}$$

Table 7.10 The values of the coefficients of equations (7.129)–(7.131) in the coordinates y_{10}, y and y_{11}

a_{11}	a_{12}	a_{13}
$(m_0 + J_0 a_0^2)p^2 + k_0 + k_2 a_{10}^2$	$(-J_0 a_0^2)p^2 - k_2 a_{10} a_{01}$	$(J_0 a^2)p^2 + k_2 a_{10} a_{01}$
a_{21}	a_{22}	a_{23}
$(-J_0 a_0^2)p^2 - k_2 a_{10} a_{01}$	$(m + J_0 a_0^2)p^2 + k + k_2 a_{01}^2 + k_1$	$(-J_0 a_0^2)p^2 - k_1 - k_2 a_{01}^2$
a_{31}	a_{32}	a_{33}
$(J_0 a_0^2)p^2 + k_2 a_{10} a_{01}$	$(-J_0 a_0^2)p^2 - k_1 - k_2 a_{01}^2$	$(J_0 a_0^2)p^2 + k_1 + k_2 a_{01}^2 + k_{01}$

When $z_0 = a_{10}z_2 - z_3$, we obtain the following expressions for the generalized forces:

$$\begin{aligned} Q_{y_{10}} &= -\left(m_0 + J_0 a_0^2\right)\ddot{z}_2 - k_2 a_{10} z_0 \\ &= -(m_0 + J_0 a_0^2)p^2 - k_2 a_{10}^2 z_2 - k_2 a_{10} z_3, \end{aligned} \tag{7.135}$$

$$\begin{aligned} Q_y &= -\left(J_0 a_0^2\right)\ddot{z}_2 + kz + k_1 z_1 + k_2 a_{01} z_0 \\ &= J_0 a_0^2 p^2 + kz + k_1 z_1 + k_2 a_{10} a_{01} z_2 - k_2 a_{01} z_3, \end{aligned} \tag{7.136}$$

$$\begin{aligned} Q_{y11} &= -\left(J_0 a_0^2\right)\ddot{z}_2 - k_1 z_1 - k_2 a_{10} z_0 \\ &= -J_0 a_0^2 p^2 - k_1 z_1 - k_2 a_{10} a_{01} z_2 + k_2 a_{10} z_3. \end{aligned} \tag{7.137}$$

In this coordinate system, it is possible to reach two joints: along the y_{10} coordinate and along the y_{11} coordinate. Using the matrix (Table 7.10) and excluding the corresponding rows and columns, we obtain the equation of motion for a system with two joints.

$$(m_0 + J_0 a_0^2)p^2 + k + k_2 a_{01}^2 + k_1 = J_0 a_0^2 \ddot{z}_2 + kz + k_1 z_1 + k_2 a_{01} z_0. \tag{7.138}$$

For simplicity, we take $k_2 = 0$, $k_1 = 0$, $z_1 = 0$, $z_3 = 0$; at the same time $z = z_2$, then the Eq. (7.136) rearranges to the following form

$$\bar{y}(m_0 + J_0 a_0^2)p^2 + k = (J_0 a_0^2 p^2 + k)\bar{z}, \tag{7.139}$$

where the transfer function of the system can be found

$$W(p) = \frac{\bar{y}}{\bar{z}_2} = \frac{J_0 a_0^2 p^2 + k}{m + J_0 a_0^2 + k}. \tag{7.140}$$

A comparison of (7.138) and (7.104) shows that the structures of transfer functions coincide, but the moment of inertia of a solid body $J_0\, a_0^2$ does not yet reflect the features of the computational scheme, if it consists of two material points m_1 and m_2 connected by a rod.

4. For complete coincidence of the results, it is necessary to present a detailed the computational scheme. For the derivation of (7.105), the scheme shown in Fig. 7.23 was used. A feature of this computational scheme is to take into account the mass-and-inertia properties of lever linkages. In the computational scheme (Fig. 7.25), the masses m_1 and m_2 are indicated; taking into account the peculiarities of their motion is an essential factor for the coincidence of expressions (7.140) and (7.105).

We choose for further calculations the system of the coordinates y_{10}, y and y_{11}, assuming that $y_{10} = y_0 + z$, $y_{11} = y - y_1$, where $y_0 = ay_1 + by_2$.

We write the expressions for the kinetic and potential energies of the system, taking into account the rotation of the lever B_1E_1 relative to the point O:

$$\mathrm{T} = \frac{1}{2}m\dot{y}^2 + \frac{1}{2}m_1\dot{y}_1^2 + \frac{1}{2}m_2\dot{y}_2^2, \tag{7.141}$$

$$\begin{aligned}\Pi = &\frac{1}{2}k(y - z)^2 + \frac{1}{2}k_1(y_1 - z_1)^2 + \frac{1}{2}k_2(y_2 - z_3)^2\\ &+ \frac{1}{2}k_{01}(y - y_1)^2 + \frac{1}{2}k_0(y_0 - z_2)^2.\end{aligned} \tag{7.142}$$

Let us make a number of transformations:

$$\begin{aligned}&y_2 = \frac{y_0 - ay_1}{b} = \frac{a_0y_0 - iy_1}{b}; \quad i = \frac{a}{b} = \frac{l_2}{l_1}; \quad a_0 = \frac{1}{b};\\ &y_2 = a_0y_{10} + a_0z_2 - iy + iy_{11}.y_0 = y_{10} + y_{11}\end{aligned}$$

and we obtain (7.141) and (7.142) in the form

$$\mathrm{T} = \frac{1}{2}m\dot{y}^2 + \frac{1}{2}m_1(\dot{y} - \dot{y}_1)^2 + \frac{1}{2}m_2(a_0\dot{y}_{10} + a_0\dot{z}_2 - i\dot{y} + i\dot{y}_{11})^2, \tag{7.143}$$

$$\begin{aligned}\Pi = &\frac{1}{2}k(y - z)^2 + \frac{1}{2}k_1(y - y_{11} - z_1)^2 + \frac{1}{2}k_2(a_0y_{10} - iy + iy_{11} + z_0)^2\\ &+ \frac{1}{2}k_{01}(y_{11})^2 + \frac{1}{2}k_0(y_{10})^2,\end{aligned} \tag{7.144}$$

where $z_0 = a_{10}z_2 - z_3$. Making a number of auxiliary calculations, similar to the above in the derivation of the Lagrange equations of the second kind, we obtain the equations of motion

$$\begin{aligned}&\ddot{y}(m_1 + m_2i^2 + m) + y(k + k_1 + k_2i^2) + \ddot{y}_{11}(-m_1 - m_2i^2) + y_{11}(-k_1 - k_2i^2)\\ &\quad + \ddot{y}_{10}(-m_2ia_0) + y_{10}(-k_2a_0i) = m_2a_0iz_2 + kz + k_1z_1 + k_2z_0i;\end{aligned} \tag{7.145}$$

$$\begin{aligned}&\ddot{y}(-m_1 - m_2i^2) + y(-k_1 - k_2i^2) + \ddot{y}_{11}(m_1 + m_2i^2) + y_{11}(k_1 + k_{01} + k_2i^2)\\ &\quad + \ddot{y}_{10}(m_2ia_0) + y_{10}(k_2a_0i) = -m_2a_0i\ddot{z}_2 - k_1z_1 - k_2z_0i;\end{aligned} \tag{7.146}$$

$$\begin{aligned}&\ddot{y}(-m_2ia_0) + y(-k_2i^2a_0) + \ddot{y}_{11}(m_2ia_0) + y_{11}(k_2ia_0)\\ &\quad + \ddot{y}_{10}(m_2a_0^2) + y_{10}(k_2a_0^2 + k_0) = -m_2a_0^2\ddot{z}_2 - k_2a_0z_0.\end{aligned} \tag{7.147}$$

Table 7.11 shows the coefficients of equations (7.143)–(7.145).

Table 7.11 The values of the coefficients of equations (7.143)–(7.145) in the coordinates y, y_{11} and y_{10}

a_{11}	a_{12}	a_{13}
$(m_1+m+m_2i^2)p^2+k+k_1+k_2i^2$	$(-m_1-m_2i^2)p^2-k_1-k_2i^2$	$-m_2ia_0p^2-k_2a_9i$
a_{21}	a_{22}	a_{23}
$(-m_1-m_2i^2)p^2-k_1-k_2i^2$	$(m_1+m_2i^2)p^2+k_1+k_{01}+k_2i^2$	$m_2ia_0p^2+k_2a_0i$
a_{31}	a_{32}	a_{33}
$-m_2ia_0p^2-k_2a_0i$	$m_2ia_0p^2+k_2a_0i$	$m_2a_0^2p^2+k_2a_0^2+k_0$

The generalized forces in this case are

$$\begin{aligned} Q_y &= m_2a_0i\ddot{z}_2+kz+k_1z_1+k_2z_0i,\\ Q_x &= -a_0im_2\ddot{z}_2-k_1z_1-k_2iz_0,\\ Q_{y_{10}} &= -m_2a_0^2\ddot{z}_2-k_2a_0z_0. \end{aligned} \tag{7.148}$$

Excluding columns and rows from the matrix along the coordinates y_{11} and y_{10}, we find the equation of motion for the system with coordinate y

$$\ddot{y}\left(m_1+m+m_2i^2\right)+y\left(k+k_1+k_2i^2\right)=m_2a_0i\ddot{z}_2+kz+k_1z_1+k_2z_0i. \tag{7.149}$$

To obtain the transfer function "the displacement y at the input z_2", we assume that $z = z_2$, $z_1 = 0$, $z_3 = z_2$, $k_1 = 0$, $k_2 = 0$.

$$W(p)=\frac{\bar{y}}{\bar{z}_2}=\frac{m_2(i+1)a_0ip^2+k}{(m+m_1+m_2i^2)p^2+k}. \tag{7.150}$$

Note that, then

$$W(p)=\frac{\bar{y}}{\bar{z}_2}=\frac{m_2i(i+1)p^2+k}{(m+m_1+m_2i^2)p^2+k}. \tag{7.151}$$

Expressions (7.105) and (7.150) are the same. A similar result in determining the transfer function (7.148) can be obtained with $k_{01}\to\infty$ and $k_0\to\infty$.

Thus, by choosing a system of generalized coordinates accordingly, one can construct a mathematical model of a mechanical system with lever linkages, simplifying the initials of high order. In this case, the system obtains a smaller number of degrees of freedom than in the initial situation. The joint is possible between two bodies when two bodies are connected into a kinematic pair of rotational type (class V). However, it is possible to connect the solid body with another body with the loss of the possibility of relative motion. Obtaining a mathematical model of a system with joints is carried out on the basis of using the matrices of the coefficients of the equations of motion in coordinates, the choice of which corresponds to the chosen type of joints. The formation of a coordinate system involves the introduction of the coordinates of relative motion, which characterizes a possible joint, with the subsequent "zeroing" of the generalized coordinate.

A similar result can be obtained not only by excluding (or "zeroing") the coordinates of relative motion, but also by increasing the stiffnesses of the springs that are in relative connections (in our case, $k_{01} \to \infty$, $k_0 \to \infty$). The attention to detail of the transformations discussed above is due to the fact that the separation of a joint, or a kinematic pair, makes it possible to connect two mating points with an elastic element.

References

1. Belokobylsky, S.V., Eliseev, S.V., Sitov, I.S.: Dynamics of Mechanical Systems. Lever Linkages and Inertial-Elastic Constraints, 319 pp. Polytekhnika Publication, St. Petersburg (2013)
2. Eliseev, S.V., Reznik, Y.N., Khomenko, A.P.: Mechatronic Approaches in the Dynamics of Mechanical Oscillatory Systems, 384 pp. Nauka Publication, Novosibirsk (2011)
3. Eliseev, S.V., Khomenko, A.P.: Dynamic Damping. Feedback Concept and Structural Methods of Mathematical Modeling, 357 pp. Nauka Publication, Novosibirsk (2014)
4. Khomenko, A.P., Eliseev, S.V., Ermoshenko, Y.V.: System Analysis and Mathematical Modeling in Mechatronics of Vibration Protection Systems, 288 pp. IrGUPS Publication, Irkutsk (2012)
5. Belokobylsky, S.V., Eliseev, S.V., Kashuba, V.B.: Applied Problems of the Structural Theory of Vibration Protection Systems, 363 pp. Polytekhnika Publication, St. Petersburg (2013)
6. Eliseev, S.V., Bolshakov, R.S., Kinash, N.Z.: Lever linkages: virtual lever mechanisms, oscillatory process features. News SWSU **1**(58), part 1, 8–17 (2015)
7. Eliseev, S.V., Bolshakov, R.S., Kinash, N.Z.: Lever linkages: virtual lever mechanisms, oscillatory process features. News SWSU **2**(59), part 2, 8–15 (2015)
8. Druzhinsky, I.A.: Mechanical Chains, 240 pp. Mashinostroenie Publication, Moscow (1977)
9. Vygodsky, M.Y.: Handbook of Higher Mathematics, 991 pp. AST, Astrel Publication, Moscow (2006)
10. Khomenko, A.P., Eliseev, S.V., Artyunin, A.I., et al.: Structural Mathematical Models in Problems of the Dynamics of Mechanical Oscillatory Systems, 231 pp. IrGUPS Publication, Irkutsk. Dep. in VINITI 30 Mar 15, No. 62-B (2015)
11. Belokobylsky, S.V., Eliseev, S.V.: Generalized ideas about the tasks of vibration protection. Syst. Meth. Technol. **1**(17), 7–15 (2013)
12. Fomina, I.V.: Development of methods for constructing mathematical models with joints of units. Author's Abstract of Ph.D. (Engineering) Thesis, 18 pp. IrGUPS Publication, Irkutsk (2011)

Chapter 8
Some Applications of the Methods Structural Mathematical Modeling

As was shown earlier, lever linkages and mechanisms are widely represented in various constructive and engineering forms, characteristic for solving many problems of machine dynamics, including the tasks of vibration isolation and vibration protection of technical objects. In mechanical oscillatory systems of the usual type, lever linkages are manifested in specific forms, which is quite simply revealed when using structural mathematical models of the system and the corresponding tools of frequency analysis of dynamic systems.

In a more illustrative form, lever linkages arise when considering dynamic interactions in mechanical oscillatory systems, where partial systems reflect the properties of solids making angular oscillations. In such cases, lever linkages are implemented in the form of levers of the first and second kind, as well as their various forms of simple and complex connections. Mechanical oscillatory systems as computational schemes of technical systems and vibration protection systems in particular, can incorporate various structural formations from typical elementary units, including mechanisms consisting of solid bodies connected by certain kinematic pairs. Most often, such formations are treated as separate blocks with zero degrees of freedom, or so-called Assur groups.

In problems of the dynamics of planar mechanisms, two-arm Assur groups are widely used (one example of this approach is given in Chap. 4 of this monograph to illustrate the definition of the above parameters). The studies [1, 2] reflected the questions of estimating the dynamic properties of vibration protection systems, in which the lever linkages are implemented by articulation linkages. As for the display of dynamic properties of this kind, it is necessary to take into account the peculiarities of the motion of such structural formations, since the vibration protection system is considered, as a rule, in a state of small oscillations relative to the position of static equilibrium. Similar examples are typical for the dynamics of transportation vehicles and technological machines of the vibration type.

Although, as shown in previous chapters, lever linkages are also used for conventional mechanical oscillatory systems, a large variety of lever linkages are characterized by mechanical systems that have motion transformation devices, as

© Springer Nature Switzerland AG 2020

S. V. Eliseev and A. V. Eliseev, *Theory of Oscillations*, Studies in Systems, Decision and Control 252, https://doi.org/10.1007/978-3-030-31295-4_8

well as mechanical chains, etc. Such approaches are often implemented in the construction of suspensions of transportation vehicles, when it is necessary to create which solves the problems of creating certain spatial structures of the vibrational field.

Taking consideration of the features of lever linkages introduced by mechanical chains and mechanisms requires the development of quite specific methods for constructing mathematical models of technical objects and analyzing their dynamic properties. The following materials use some constructive and engineering solutions protected by patents for utility models [3–5].

8.1 Additional Masses in the Structure of Lever Mechanisms

Mechanisms in the structures of vibration protection systems are most often considered in the mode of small oscillations relative to the position of static equilibrium or steady motion. In such cases, the mechanisms implement the function of imposing additional constraints that create certain conditions in solving specific problems of the dynamics of machines and equipment. In the future, the methodological foundations of approaches to the construction of mathematical models of mechanical oscillatory systems, in particular, vibration protection systems, are developed, the structure of which has additional mechanisms.

8.1.1 Options of Location of Additional Masses

The generalized approach associated with the structural interpretations of mechanisms as a part of mechanical oscillatory systems is considered using the concepts of motion transformation devices. In essence, that kind of definition can be attributed to any mechanism, since the transfer of motion in a mechanical chain is somehow connected with the transformation of motion. A motion transformation device (MTD), in particular, as shown in a chapter of this monograph, may be a non-locking helical mechanism with functions of a typical second-order differentiating unit. The screw mechanism in a specific statement of the problem can be interpreted as a kind of lever in which a constant relation connects the parameters of the rotational and translational motion of the elements of the kinematic pair. Details of such representations are given in [6]. Figure 8.1 shows options for the arrangement of mechanisms in the basic computational scheme (Fig. 8.1a), where the object of protection with mass M relies on an elastic element with stiffness k and a generalized mechanism designated as MTD. In turn, Fig. 8.1b presents one of the interpretations of the MTD in the form of a screw mechanism. In subsequent studies, pivot-lever mechanisms are also used.

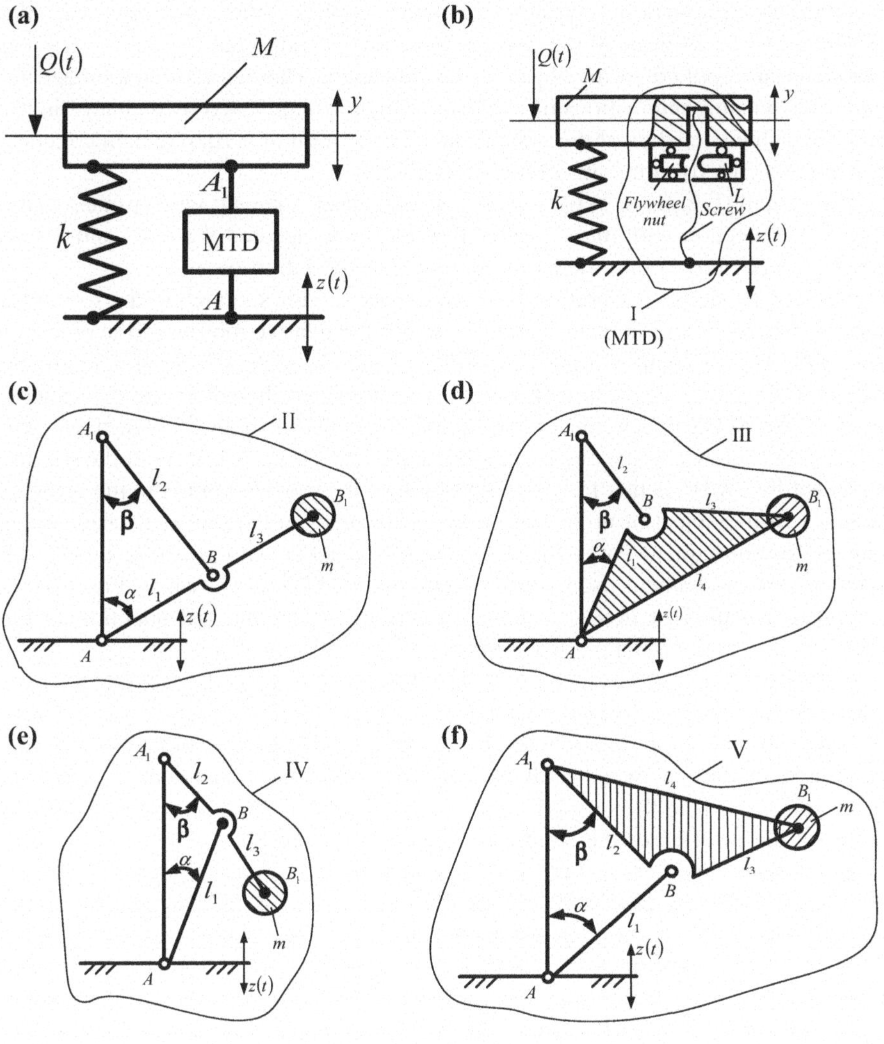

Fig. 8.1 Vibration protection systems with motion transformation devices. **a** The basic computational scheme with a supporting elastic element and a motion transformation device (MTD); **b** the MTD in the form of a non-locking mechanism; **c** the MTD in the form of the articulation linkage with an additional mass and lengthening of the lower lever; **d** the MTD with a curved lower arm; **e** The MTD with additional mass on the extended upper arm; **f** The MTD with curved upper arm

In this case, the following harmonic perturbations are considered: force $Q(t)$ and kinematic $z(t)$; the coordinate system associated with the fixed base is used. Small motions of the system are assumed with respect to the position of static equilibrium, there are no resistance forces, and the system has linear properties.

Vibration protection systems variants reflect the features of the structural-technical forms of the devices, which form additional inertia forces when vibrations are applied to an object to be protected. These forces implement the effects of dynamic interactions of elements, which is manifested by a change in the frequency characteristics of the systems and the creation of certain dynamic modes, in particular, the dynamic absorbing of oscillations.

The system (Fig. 8.1a, b) uses the supporting elastic element with stiffness k, the object of protection with mass M, and also takes into account the mass-and-inertia parameters (L and m) introduced by motion transformation devices.

Figure 8.1b shows a vibration protection system with a device in the form of a non-locking screw mechanism (circuit I). When the relative motion of the object of protection M as a result of relative motions of the elements, elastic and additional inertial forces are formed. In this case, L is a parameter characterizing the reduced mass of the MTD depending on the moment of inertia of the flywheel nut, the radius of the screw and the angle of inclination of the helix. A feature of the scheme presented Fig. 8.1b, is that the axis of rotation of the flywheel nut is in the plane of motion y of the object of protection. Resistance forces in the screw mechanism (as well as for subsequent cases Fig. 8.1c–f) are not taken into account. In [7], the dynamics of mechanical oscillatory systems with MTD as shown in Fig. 8.1b is described in sufficient detail, which to a certain extent initiated attention to the consideration of the dynamic features created by other mechanisms.

Flat articulation linkages (Fig. 8.1c–f) are characterized by the fact that the axes of the rotary hinges are located perpendicular to the axis of the plane of motion y of the object of protection. In contours II–V, (Fig. 8.1c–f) there are the options for the location of additional weights of mass m to ensure the adjustment of vibration protection systems in the simplest ways, which is achieved, for example, by simply changing the position of the additional mass m (l_3 changes).

In addition, the diagram (Fig. 8.1d, f) shows the use of curved levers. The latter is associated with the possibility of bringing the adjustment mechanisms out to the service area or with moving additional masses. The upper and lower arms (Fig. 8.1c–f) are considered as weightless rigid rods. It is assumed that external disturbances can be force (they are attached directly to the object of protection M) or kinematic, which is associated with the vibrations of the base. We consider the harmonic forms of vibration and small motions of all elements of vibration protection systems relative to the position of static equilibrium. Parameters of vibration protection systems (angles of inclination of the levers α and β, lengths of units l_1–l_4, additional mass m), shown in Fig. 8.1c–f, determine the configuration of the vibration protection system and, therefore, its dynamic properties. When adjusting a vibration protection system, the mentioned parameters can vary in one way or another, which, in particular, relates to the study of effects arising from various types of external disturbances.

8.1.2 *Mathematical Models of the System. Forced Vibrations*

When analyzing the systems in Fig. 8.1, it is necessary to take into account the features of force and kinematic external disturbances.

Force perturbation ($Q \neq 0, z = 0$). Consider the scheme in Fig. 8.1c. Point O_1 in Fig. 8.2 is an instantaneous velocity center. The speed of point B can be found from the relations:

$$\dot{y}_B = \dot{y} \cdot a; \tag{8.1}$$

$$a = \frac{i \cdot \cos \beta}{\sin \alpha \cdot (\cos \alpha + i \cdot \cos \beta)}. \tag{8.2}$$

In this case $i = \frac{l_2}{l_1}$ is the transfer ratio of the lengths of the upper and lower rods.

Determination of kinematic parameters. The angular velocity of the lower arm ω_1 relative to point A can be found

$$\omega_1 = \frac{\dot{y}_B}{AB} = \frac{\dot{y}_B}{l_1} = \frac{y \cdot a}{l_1} = \dot{y} \cdot \frac{i \cdot \cos \beta}{l_1 \cdot \sin \alpha \cdot (\cos \alpha + i \cdot \cos \beta)}. \tag{8.3}$$

The speed of point B_1 in this case is determined by the expression

$$\dot{y}_{B_1} = \omega_1 \cdot (l_1 + l_3) = \dot{y} \cdot \frac{i \cdot (l_1 + l_3) \cdot \cos \beta}{l_1 \cdot \sin \alpha \cdot (\cos \alpha + i \cdot \cos \beta)} = \dot{y} \cdot b; \tag{8.4}$$

$$b = \frac{i \cdot (l_1 + l_3) \cdot \cos \beta}{l_1 \cdot \sin \alpha \cdot (\cos \alpha + i \cdot \cos \beta)}. \tag{8.5}$$

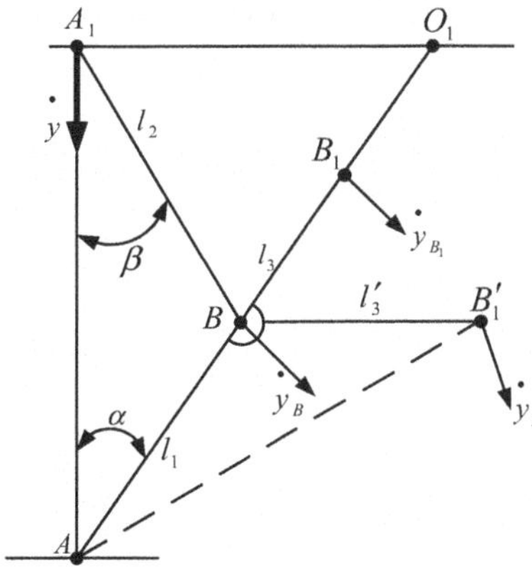

Fig. 8.2 The kinematic diagram of the system presented in Fig. 8.1c

If a curved **lever** is used in the diagram in connection with the position of p. B_1', then the velocity vector of p. B_1' will be directed perpendicular to AB_1. As for the length of AB_1', it can be defined as the length l_4 from the geometric relations

$$\left|\dot{y}_{B_1'}\right| = \dot{y} \cdot \frac{i \cdot l_4 \cdot \cos \boldsymbol{\beta}}{l_1 \cdot \sin \boldsymbol{\alpha} \cdot (\cos \boldsymbol{\alpha} + i \cdot \cos \boldsymbol{\beta})}. \tag{8.6}$$

In the presence of a curved lever, you can introduce a parameter

$$b_1 = \frac{i \cdot l_4 \cdot \cos \boldsymbol{\beta}}{l_1 \cdot \sin \boldsymbol{\alpha} \cdot (\cos \boldsymbol{\alpha} + i \cdot \cos \boldsymbol{\beta})}. \tag{8.7}$$

Increasing the distance l_4 (or l_3) increases the parameter b. To compose the equation of motion of the system whose scheme is shown in Fig. 8.1c, we write expressions for the kinetic and potential energy

$$\mathrm{T} = \frac{1}{2} M (\dot{y})^2 + \frac{1}{2} m (\dot{y}_{B_1'})^2; \tag{8.8}$$

$$\Pi = \frac{1}{2} \cdot k (y - z)^2. \tag{8.8'}$$

The equation of motion of the system will take the form

$$\ddot{y} \cdot \left(M + m b_1^2\right) + k y = Q. \tag{8.9}$$

From (8.9), one can proceed to the structural diagram of the system (Fig. 8.3a–d), using the methodological techniques described in the previous sections for the transformation.

Thus, the MTD under force perturbation is transformed into a unit with a transfer function of a typical second order differentiating element. The methods of introducing such an additional constraint correspond to the transformation rules in the structural theory of vibration protection systems. The transfer function of the system can be determined from Eq. (8.9) or directly from the structural diagrams Fig. 8.3a–d; in this case

$$W(p) = \frac{\bar{y}}{\bar{Q}} = \frac{1}{\left(M + m b_1^2\right) \cdot p^2 + k}. \tag{8.10}$$

From (8.10) you can find the frequency of natural oscillations of the system:

$$\omega_{\text{nat}}^2 = \frac{k}{M + m b_1^2}. \tag{8.11}$$

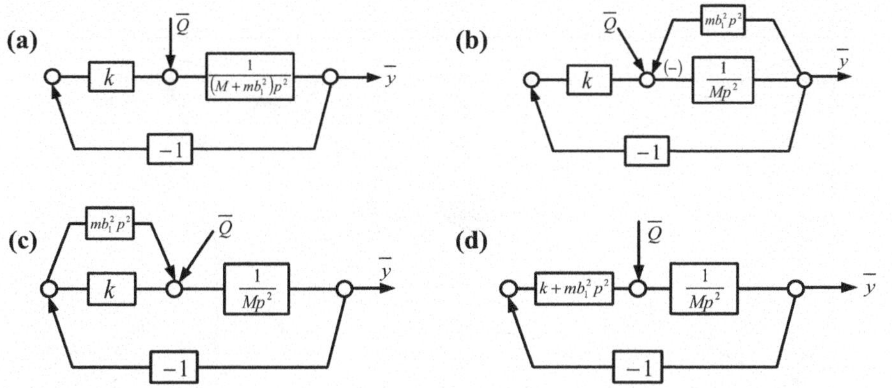

Fig. 8.3 The structural diagrams of vibration protection systems, corresponding to Eq. (8.8): **a** The system has a reduced mass $(M + mb_1^2)$; **b** the MTD is an additional feedback on acceleration with respect to the object of protection; **c** the MTD as a parallel unit; **d** the elastic element and MTD can be connected according to the rules of parallel connection of springs

Introduction of MTD can be considered as a way to control the frequency of natural oscillations. For these purposes, the parameters of the levers, the additional mass m and the angles of installation of the units may vary. If the installation angle α will take small values, then the reduced mass

$$M_{\text{red}} = M + mb_1^2 \tag{8.12}$$

may take large values due to the increase in the parameter b_1.

The disturbance caused by the motion of the bearing surface. With a kinematic perturbation, the original system (Fig. 8.1c) will behave differently, since the motion of the base ($z \neq 0$, $Q = 0$) forms additional forces of inertia of the transportation motion. In particular, the speed of point B will be the sum of the speeds of motion created by the oscillations of the base $(\dot{y}_{B_{\text{pri}}})$, as well as the motion of the object of protection $((\dot{y}_{B_{\text{reg}}}))$. Similarly, the speeds of point B_1 and B_1' should be considered. For kinematic calculations we construct a schematic diagram (Fig. 8.4).

Determination of velocity points. Find the speed of point B, while

$$\bar{\dot{y}}_B = \bar{\dot{y}}_{B_{\text{reg}}} + \bar{\dot{y}}_{B_{\text{pri}}}. \tag{8.13}$$

$\dot{y}_{Breg}$ can be determined from the expression (8.1)

$$\left|\dot{y}_{B_{reg}}\right| = y \cdot a,$$

while a is used in accordance with the expression (8.2).

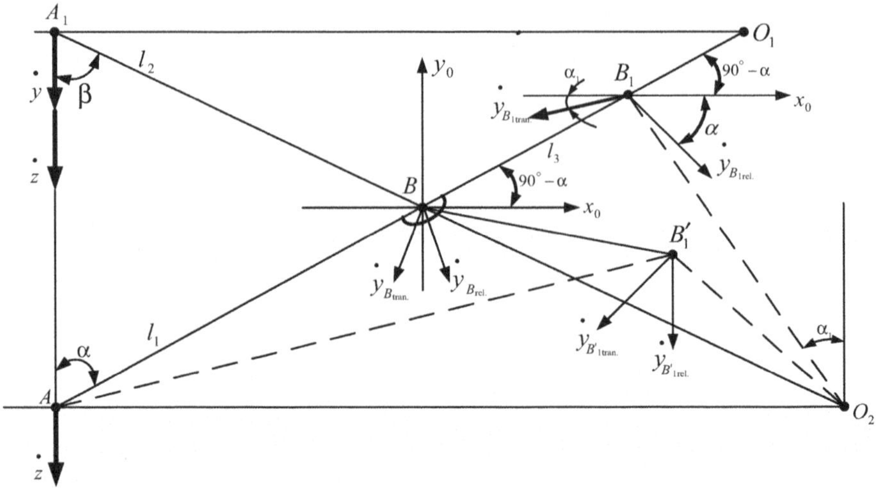

Fig. 8.4 The kinematic scheme for calculating the ratio of parameters with a kinematic perturbation

The velocity vector $\dot{y}_{B_{reg}}$ is perpendicular to AO_1 (O_1 is the instantaneous center of velocity at $\dot{z} = 0$, $\dot{y} \neq 0$). In turn, the velocity vector $\dot{y}_{Bpri}$ is perpendicular to A_1O_2 (O_2 is the instantaneous velocity center at $\dot{z} \neq 0$, $\dot{y} = 0$). The value of $\dot{y}_{B_{\text{pri}}}$ is determined by the expression

$$\dot{y}_{B_{\text{pri}}} = \frac{\dot{z} \cdot BO_2}{AO_2}. \tag{8.14}$$

We show that $AO_2 = (l_1 \cdot \cos\,\alpha + l_2 \cdot \cos\,\beta)$; $BO_2 = A_1O_2 - l_2$; $BO_2 = l_1 \frac{\cos\alpha}{\cos\beta}$; $A_1O_2 = \frac{l_1 \cos\boldsymbol{\alpha} + l_2 \cos\boldsymbol{\beta}}{\cos\boldsymbol{\beta}}$ and get

$$\dot{y}_{Bpri} = \frac{\dot{z} \cdot l_1 \cdot \cos\alpha}{\sin\beta \cdot (l_1 \cos\alpha + l_2 \cos\beta)} = \frac{\dot{z} \cdot \cos\alpha}{\sin\beta \cdot (\cos\alpha + i \cos\beta)}. \tag{8.15}$$

In this case

$$\dot{y}_{B_{\text{pri}}} = \dot{z} \cdot a_1, \tag{8.16}$$

where

$$a_1 = \frac{\cos\boldsymbol{\alpha}}{\sin\boldsymbol{\beta} \cdot (\cos\boldsymbol{\alpha} + i \cos\boldsymbol{\beta})}. \tag{8.17}$$

Since $a = \frac{i \cdot \cos \beta}{\sin \alpha \cdot (\cos \alpha + i \cos \beta)}$, then, introducing the relations, we find

$$\frac{i \cdot \cos \beta}{\sin \alpha \cdot a} = \frac{\cos \alpha}{\sin \beta \cdot a_1}; \frac{a_1 \cdot i \cdot \cos \beta}{\sin \alpha} = \frac{\cos \alpha \cdot a}{\sin \beta}; \quad a_1 = a \frac{\cos \alpha \cdot \sin \alpha}{\sin \beta \cdot i \cdot \cos \beta}.$$

The velocity vector of point B is projected on the coordinate axes y_0 and x_0 in accordance with Fig. 8.4 (in the projections along the y_0 and x_0 axes), then

$$\dot{y}_{B_{x0}} = \dot{y}_{B_{reg\,x0}} - \dot{y}_{B_{pri\,x0}} = \dot{y}_{B_{reg}} \cdot \cos \alpha - \dot{y}_{B_{pri}} \cdot \cos \beta; \tag{8.18}$$

$$\dot{y}_{B_{y0}} = -\dot{y}_{Breg} \cdot \sin \alpha - \dot{y}_{B_{pri}} \cdot \sin \beta, \tag{8.19}$$

wherein

$$\dot{y}_{B_{pri}} = a_1 \cdot \dot{z}; \tag{8.20}$$

$$\dot{y}_{B_{reg}} = a \cdot \dot{y}, \tag{8.21}$$

where $a = \frac{i \cdot \cos \beta}{\sin \alpha \cdot (\cos \alpha + i \cos \beta)}$; $a_1 = \frac{\cos \alpha}{\sin \beta \cdot (\cos \alpha + i \cos \beta)}$.

From (8.18) it follows that

$$\dot{y}_{B_{x0}} = a \cdot \dot{y} \cdot \cos \alpha - a_1 \cdot \dot{z} \cdot \cos \beta; \tag{8.22}$$

$$\dot{y}_{B_{y0}} = -a \cdot \dot{y} \cdot \sin \alpha - a_1 \cdot \dot{z} \cdot \sin \beta, \tag{8.23}$$

then you can get

$$\dot{y}_B = \sqrt{a^2 \cdot (\dot{y})^2 cos^2\alpha - 2aa_1\dot{y}\dot{z} \cdot cos\alpha \cdot cos\beta + a_1^2 \cdot (\dot{z})^2 cos^2\beta + a^2 \cdot (\dot{y})^2 sin^2\alpha}$$
$$\sqrt{+ 2aa_1\dot{y}\dot{z} \cdot sin\alpha \cdot sin\beta + a_1^2 \cdot (\dot{z})^2 sin^2\beta}$$

$$V_B^2 = a^2 \cdot (\dot{y})^2 + a_1^2 \cdot (\dot{z})^2 + 2aa_1\dot{y}\dot{z} \cdot (\sin \alpha \cdot \sin \beta - \cos \alpha \cdot \cos \beta). \tag{8.24}$$

If you accept that

$$b_2 = \sin \alpha \cdot \sin \beta - \cos \alpha \cdot \cos \beta,$$

then in this case

$$(\dot{y}_B)^2 = a^2 \cdot (\dot{y})^2 + a_1^2 \cdot (\dot{z})^2 + 2aa_1\dot{y}\dot{z} \cdot b_2. \tag{8.25}$$

8.1.3 Features of the Dynamic Properties of the System with Kinematic Perturbations

We construct a mathematical model of the system with kinematic perturbation. Find the expression for the kinetic and potential energy of the system

$$\begin{aligned} \mathrm{T} &= \frac{1}{2}M(\dot{y}-\dot{z})^2 + \frac{1}{2}m\cdot(\dot{y}_B)^2 \\ &= \frac{1}{2}M(\dot{y}-\dot{z})^2 + \frac{1}{2}m\cdot\left[a\cdot(\dot{y})^2 + a_1\cdot(\dot{z})^2 + 2aa_1b_2\dot{y}\dot{z}\right], \end{aligned} \tag{8.26}$$

$$\Pi = \frac{1}{2}k\cdot(y-z)^2. \tag{8.27}$$

Using the Lagrange equation of the second kind and the Laplace transform, we obtain the expression for the transfer function for the input action $\bar{z}$ and output—$\bar{y}$

$$W(p) = \frac{\bar{y}}{\bar{z}} = \frac{aa_1b_2mp^2 + k}{(M + a^2m)p^2 + k}, \tag{8.28}$$

where $b_2 = \cos\alpha\cdot\cos\beta - \sin\alpha\cdot\sin\beta$.

The frequency of natural oscillations of the system does not change as compared with the case of a force perturbation; however, two dynamic effects can be detected in the system. The first is the mode of dynamic oscillation damping at the frequency

$$\omega^2_{\mathrm{dyn}} = \frac{k}{aa_1b_2m}. \tag{8.29}$$

As $p\to\infty$ ($p = j\omega$ is a complex variable $j = \sqrt{-1}$) the transfer function (8.28) transforms to

$$\left|W(p)\right|_{p\to\infty} = \frac{\bar{y}}{\bar{z}} = \frac{aa_1b_2m}{M + a^2m}. \tag{8.30}$$

If $\alpha = \beta$, then $b_2 = \cos^2\alpha - \sin^2\alpha$. For $\alpha = 45°$, $b_2 = 0$. If $\alpha < 45°$, then: $b_2 > 0$. For $\alpha > 45°$, b_2.

Thus, a mechanism with mass at point B can be interpreted as an additional unit created by a second-order differentiating unit. The unit capabilities are determined by the parameters m, a, a_1, b_2. If the point of attachment of additional mass is at point B_1, then the speed of point B_1 is determined by analogy with (8.4)

$$\bar{\dot{y}}_{B_1} = \bar{\dot{y}}_{B_{1\mathrm{reg}}} + \bar{\dot{y}}_{B_{1\mathrm{pri.}}}. \tag{8.31}$$

If we project (8.31) on the y_0, x_0 axis, we obtain the following relations:

$$\left|\bar{\dot{y}}_{B_{1\text{reg.}}}\right| = \omega_2 \cdot AB_1; \quad \omega_2 = \frac{\dot{y}_{B_{1\text{reg}}}}{l_1}. \tag{8.32}$$

$$\left|\bar{\dot{y}}_{B_{1\text{reg.}}}\right| = \frac{\dot{y}_{B_{\text{reg}}}}{l_1} \cdot (l_1 + l_3). \left|\bar{\dot{y}}_{B_{1regx_0.}}\right| = \frac{\dot{y}_{B_{1reg}}}{l_1} \cdot (l_1 + l_3) \cdot \cos\alpha. \tag{8.33}$$

Here $\bar{\dot{y}}_{B_{1_{\text{pri.}}}}$ is the velocity vector $\perp$ O_2B_1. $\dot{y}_{B_{1_{\text{пep}.x_0}}} = \dot{y}_{B_{1_{\text{пep.}}}} \cdot \cos\alpha_1$. $\left|\dot{y}_{B_{1nep.}}\right| = \frac{\dot{z}\cdot O_2B_1}{AO_2}$; $AO_2 = \left(\frac{l_1 \cos\alpha + l_2 \cos\beta}{\text{tg}\beta}\right)$. With that, the magnitude of O_2B_1 is determined by the scheme in Fig. 8.4. From (8.28) it follows that the sign of b_2 determines the form of the amplitude-frequency characteristic of the system. If $b_2 > 0$, then the numerator (8.28) will be positive for any value of p. In the case when $p \to \infty$, the "locking" is possible, and, with that,

$$\left|W(p)\right|_{p\to\infty} = \frac{maa_1 b_2 p^2 + k}{(M + ma^2)p^2 + k} \to \frac{maa_1}{M + ma^2}. \tag{8.34}$$

The situation is similar when $b_2 < 0$ if $p \to \infty$. However, in this case, in the system, the mode of dynamic oscillation damping at the frequency determined by expression (8.29) is possible.

The peculiarity of the dependence between the parameters α and β (Fig. 8.5) is such that the graph divides the space of the parameters α and β into two regions, indicated by the symbols I and II. The pairs of values from region I correspond to $b_2 > 0$, and the pairs of values α and β from region II correspond to $b_2 < 0$. On the graph itself, for which the condition $b_2 = 0$ is satisfied, the pairs of values α and β determine the configuration of the units of the system in such a way that the amplitude-frequency characteristic of the system will have the same form as with a force external influence.

Thus,

$$\dot{y}_{B_{1x_0}} = \dot{y}_{B_{1\text{reg.}}} \cdot \cos\alpha - \dot{y}_{B_{1_{\text{pri}}}} \cdot \cos\alpha_1; \quad \dot{y}_{B_{1y_0}} = -\dot{y}_{B_{1\text{reg.}}} \cdot \sin\alpha - \dot{y}_{B_{1_{\text{pri.}}}} \cdot \sin\alpha_1. \tag{8.35}$$

$$(\dot{y}_{B_1})^2 = \left[\dot{y}_{B_{1\text{reg.}}} \cdot \cos\alpha - \dot{y}_{B_{1\text{pri.}}} \cdot \cos\alpha_1\right]^2 + \left[\dot{y}_{B_{1\text{reg.}}} \cdot \sin\alpha + \dot{y}_{B_{1_{\text{pri.}}}} \cdot \sin\alpha_1\right]^2. \tag{8.36}$$

We introduce a series of relations

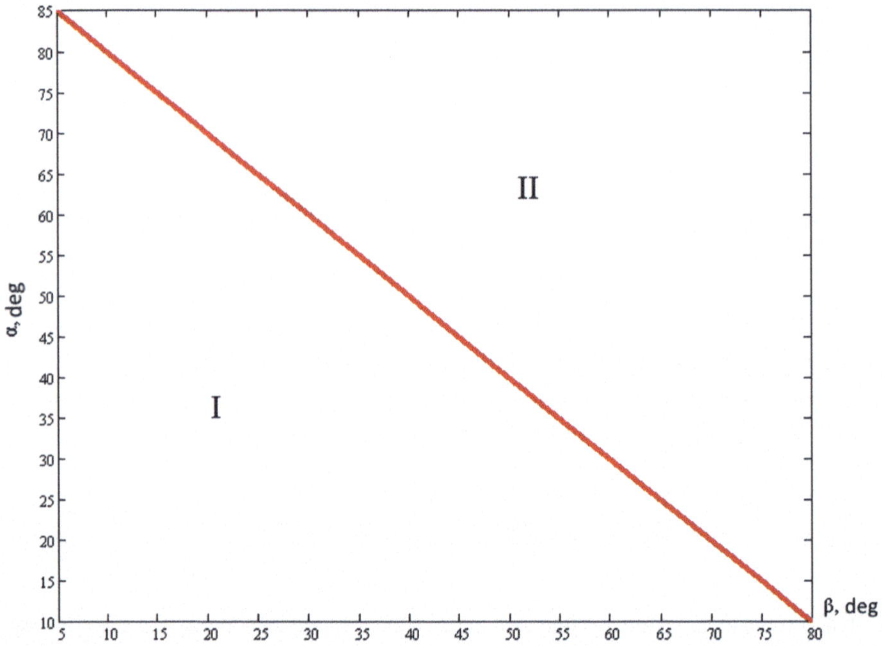

Fig. 8.5 The relationship between the angles α and β when $b_2 = 0$

$$
\begin{aligned}
&\dot{y}_{B_{1\text{reg.}}} = \frac{\dot{y}_{B_{\text{reg.}}}}{l_1}\cdot(l_1+l_3);\quad \dot{y}_{B_{\text{reg.}}} = a\cdot\dot{y};\quad \dot{y}_{B_{\text{reg.}}} = a_1\cdot\dot{z};\\
&\dot{y}_{B_{1\text{reg.}}} = \frac{\dot{y}_{B_{\text{reg.}}}\cdot B_1O_2}{BO_2} = \frac{\dot{z}a_1\cdot B_1O_2}{BO_2};\quad b_3 = \frac{B_1O_2}{BO_2};\\
&\frac{\dot{z}}{AO_2} = \frac{\dot{y}_{B_{\text{pri.}}}}{BO_2};\quad \dot{y}_{B_{\text{pri.}}} = \dot{z}\cdot\frac{BO_2}{AO_2};\quad \dot{y}_{B_{1\text{pri.}}} = \dot{z}\cdot b_3\cdot a_1;\quad BO_2 = A_1O_2 - l_2\\
&BO_2 = l_1\cdot\frac{\cos\alpha}{\cos\beta}.
\end{aligned}
\tag{8.37}
$$

We obtain that

$$
\dot{y}_{B_{\text{pri.}}} = z\cdot\frac{l_1\cdot\cos\alpha}{\cos\beta\cdot(l_1\cdot\cos\alpha+l_2\cdot\cos\beta)\cdot\text{tg}\beta} = z\cdot\frac{l_1\cdot\cos\alpha}{\sin\beta\cdot(l_1\cdot\cos\alpha+l_2\cdot\cos\beta)}.
\tag{8.38}
$$

Further construction of the equation of motion and the transfer function of the system, taking into account the particularities of the location of additional masses, is performed in the same way as for the location of the mass at point B. It is possible to consider the joint action of additional masses, which is carried out according to the above method. The introduction of additional constraints, which are mechanical chains in the configuration of the mechanism with zero degree of freedom or the Assur two-arm group, expands the possibilities of vibration protection systems. The latter is achieved by choosing the configuration of the mechanical system, depending on the angles of installation of the levers, as well as their lengths. The choice of parameters can significantly affect the properties of the system through changes in the amplitude-frequency characteristics. Thus, the type of external influence is essential to determine the capabilities of the system. The kinematic effect is characterized by the fact that it generates additional interaction forces determined by the transportation motion of the elements created by the vibrations of the bearing surface. Under certain conditions, the influence of such effects can be compensated by the configuration and features of the dynamic interactions between the elements of the system. When a force is applied, additional constraints in absolute acceleration also arise in the system. The developed approach and the proposed method for constructing mathematical models make it possible to evaluate the possibilities of searching and developing new constructive and engineering solutions associated with the removal of additional masses to different zones of the VPS, which can be used in the problems of intelligent design of vibration protection systems.

8.2 Motion Transformation Devices in Lever Structures

Similar in terms of dynamic content, the problems in which methodological concepts are developed in the search and development of methods and means of changing the state of oscillating systems are the problems of analyzing and dynamically synthesizing the systems of vibration protection of technical objects. The proximity of the problems of dynamics substantially predetermines the expediency of developing generalized approaches, including those that involve attempts to introduce into consideration new typical elements and constraints.

8.2.1 Features of the Construction of Mathematical Models

The system of vibration protection is considered (according to patent [5]), which consists of a main unit consisting of an object of protection M with a spring k_0 and a special mechanical oscillatory circuit formed by articulation linkages. Mass-and-inertia elements m are fixed in joints A and B. In addition, points A and B are interconnected by an elastic element with stiffness k (Fig. 8.6).

The articulation linkages are symmetrical and have, respectively, units in the form of weightless rigid rods l_1 and l_2. The angles α and β are determined by the configuration of the system in relation to the position of static equilibrium. All motions of system elements are considered as small; the resistance forces in the relative motions of the elements are considered small. In addition to the elastic

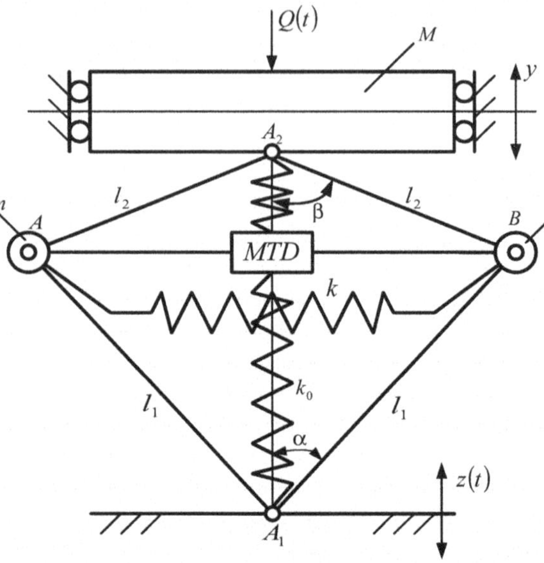

Fig. 8.6 The computational scheme of the vibration protection system with adjustable elastic-mass parameters

element with stiffness *k*, points *A* and *B* are connected by a motion transformation device (MTD); *L* is the reduced mass of the motion transformation device, which forms inertial interactions between points *A* and *B*. As such a device, the above-described non-locking screw mechanism can be used. The arising inertial forces are determined by the parameters of the relative motion of points *A* and *B*. It is assumed that the mass of the motion transformation device (*L*) is small and is not taken into account when determining the kinetic energy of the system. As external disturbances, we consider the force harmonic action *Q*, which is applied to the object of protection *M*, as well as kinematic one—in the form of the well-known law of the harmonic motion of the base or the bearing surface $z(t)$. When considering the features of the motion of the elements of the system (Fig. 8.6), it is taken into account that it simultaneously implements one of the external influences: force or kinematic. As for the joint action of two perturbations, the problem can be solved in a linear formulation based on the superposition method. However, in this case, it is necessary to establish certain ratios determining the conditions for the interaction of two factors.

This section illustrates the capabilities of the developed method for constructing mathematical models for vibration protection systems, which include mechanisms that create additional contours for the formation of dynamic interactions, new physical effects, dynamic state modes.

Force perturbation of the system ($Q \neq 0$, $z(t) = 0$). A schematic diagram of the kinematic relations in the relative motions of the elements of the system is shown in Fig. 8.7.

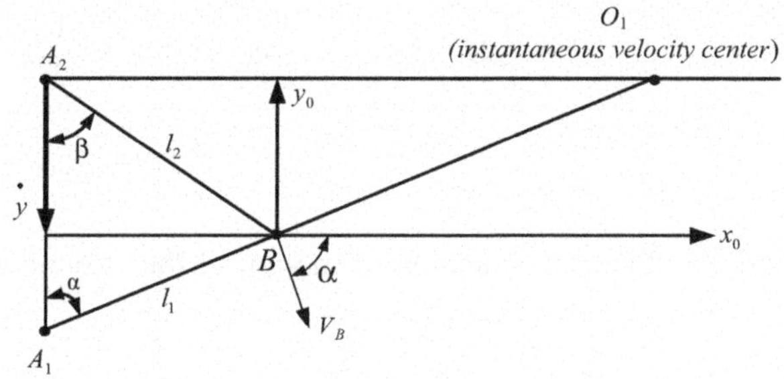

Fig. 8.7 The schematic diagram of the kinematic relations at force perturbation

In this case, the $\dot{y}_B$ velocity vector will be $\perp O_1B$. We introduce the auxiliary coordinate system x_0By_0 (Fig. 8.7), which has parallel axes with the coordinate system of a fixed basis. From the kinematic relations Fig. 8.7 it follows that

$$\overline{V}_B = \dot{y} \cdot a; \tag{8.39}$$

$$a = \frac{i \cdot \cos\beta}{\sin\alpha \cdot (\cos\alpha + i \cdot \cos\beta)}. \tag{8.40}$$

Here $i = l_2 / l_1$ is the transfer ratio of lever linkages.

To determine the horizontal deviations of points A and B in relative displacements in the direction of the x_0 axis, one can introduce the relations:

$$x_{0B} = y \cdot a \cdot \cos\alpha. \tag{8.41}$$

The total deformation of the springs k in this case will be determined

$$x_{\text{reg}_{AB}} = y \cdot 2a \cdot \cos\alpha. \tag{8.42}$$

The speeds of the relative motions of points A and B are related by the relations:

$$\dot{x}_{\text{reg}\,AB} = \dot{y} \cdot 2a \cdot \cos\alpha. \tag{8.43}$$

The kinetic and potential energy with a force perturbation can be represented

$$\mathrm{T} = \frac{1}{2}M \cdot (\dot{y})^2 + 2 \cdot \frac{1}{2}m \cdot V_B^2 + \frac{1}{2}L \cdot \left(\dot{x}_{\text{reg}_{AB}}\right)^2; \tag{8.44}$$

$$\Pi = \frac{1}{2}k_0 y^2 + \frac{1}{2}kx^2_{\text{reg}_{AB}}. \tag{8.45}$$

Taking into account (8.42) and (8.43), the resulting expressions (8.44) and (8.45) can be converted to

$$\mathrm{T} = \frac{1}{2} M \cdot (\dot{y})^2 + 2 \cdot \frac{1}{2} m \cdot (\dot{y} \cdot a)^2 + \frac{1}{2} L \cdot (2\dot{y}a \cdot \cos\alpha)^2; \tag{8.46}$$

$$\Pi = \frac{1}{2} k_0 y^2 + \frac{1}{2} k (2\dot{y}a \cdot \cos\alpha)^2 \tag{8.47}$$

and we obtain the equation of motion of the system in the form

$$\ddot{y} \cdot \left(M + 2ma^2 + 2La^2 \cos^2\alpha\right) + y \cdot \left(k_0 + 2ka^2 \cdot \cos^2\alpha\right) = Q. \tag{8.48}$$

After the Laplace transform, the transfer function of the source system is determined

$$W(p) = \frac{\bar{y}}{\bar{Q}} = \frac{1}{[M + 2a^2 \cdot (m + L \cdot \cos^2\alpha)] + k_0 + 2ka^2 \cos^2\alpha}. \tag{8.49}$$

To evaluate the possibilities of changing the dynamic properties of the system, we provide options for constructing structural diagrams (Fig. 8.8).

The element with the transfer function $2ma^2p^2$ reflects the mass-and-inertia properties of the added masses in the rotators *A* and *B*. Elastic and dissipative, as well as mass-and-inertia elements that are not objects of vibration protection, are considered as elements of the generalized spring. Elements like that in structural interpretations represent negative feedbacks in relation to the object of protection. Variants of those constraints (Fig. 8.8) can be expanded, which ultimately would create certain possibilities in the problems of dynamic synthesis of vibration protection systems.

The transfer function of the system in case of force perturbation, represented by the expression (8.50), allows you to determine the natural frequency:

$$\omega_{\text{nat}}^2 = \frac{k_0 + 2ka^2 \cos^2\alpha}{M + 2a^2(m + L\cos^2\alpha)}. \tag{8.50}$$

From (8.50) it follows that the natural frequency of the system depends essentially on the configuration of the mechanisms (angles α and β), the ratio of the lengths of the units l_1 and l_2, as well as the mass-and-inertia properties of the additional masses and the motion transformation device. With a force perturbation, the original computational scheme essentially reduces to a mechanical oscillatory system of the usual form. The parameters of that system have the ability to be adjusted by changing the shape of the mechanisms and the conditions of dynamic interaction of their elements.

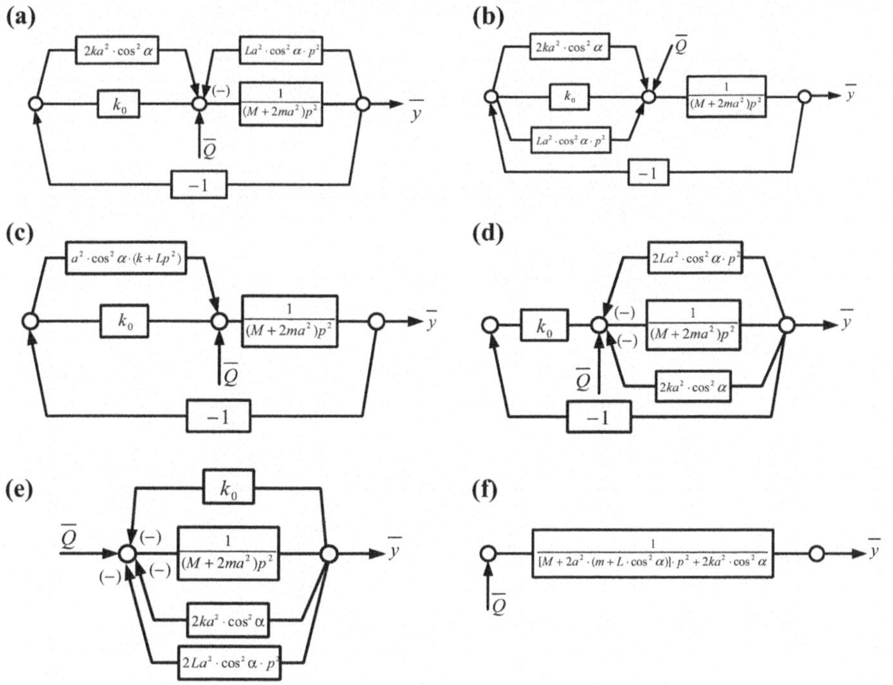

Fig. 8.8 Variants of the relative position of elements on the structural diagram of the mechanical oscillatory system in Fig. 8.6: **a** The motion transformation device as an inverse negative relationship with respect to the absolute deviation of the object ($M + ma^2$); **b** the transformation device for as a parallel unit k_0 in the base loop; **c** the motion transformation device and a spring 2 $ka^2 \cos^2\alpha$ as a parallel circuit for k_0 in the base loop; **d** elements of the system in parallel negative units with respect to the object of protection ($M + 2ma^2$); **e** The generalized structural diagram

8.2.2 *Kinematic External System Perturbation*

In this case, the schematic diagram of the kinematic relations (Fig. 8.9) will differ from the diagram in Fig. 8.7 by using another instantaneous O_2 velocity center and the notion that points A and B will participate in two types of motion: relative, which is associated with the motion of the object of protection M, and transportation—created by the motion of the base ($z(t) \neq 0$).

Using the diagram in Fig. 8.9, we find that

$$V_{B_{\text{pri}}} = \dot{z} \cdot a_1, \tag{8.51}$$

where the coefficient a_1 is determined by the expression

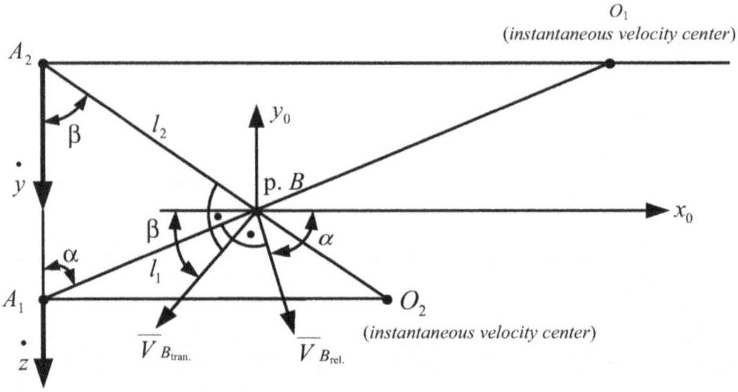

Fig. 8.9 The schematic diagram of the kinematic relations for the kinematic perturbation $z(t)$

$$a_1 = \frac{\cos \alpha}{\sin \beta \cdot (\cos \alpha + i \cdot \cos \beta)}. \tag{8.52}$$

The total velocity of point B at $\overline{V}_{Bpri} \perp A_2O_2$, $\overline{V}_{Breg} \perp A_1O_1$ can be found through the projections on the coordinate axes y_0Bx_0

$$V_{B_{x0}} = V_{B_{\text{reg}x0}} + V_{B_{\text{pri}x0}} = \dot{y} \cdot a \cdot \cos \alpha - \dot{z} \cdot a_1 \cdot \cos \beta; \tag{8.53}$$

$$V_{B_{y0}} = V_{B_{\text{reg}y0}} + V_{B_{\text{pri}y0}} = -\dot{y} \cdot a \cdot \sin \alpha - \dot{z} \cdot a_1 \cdot \sin \beta. \tag{8.54}$$

Converting (8.53) and (8.54) to the form

$$V_B^2 = (\dot{y}a \cos \alpha - \dot{z} \cdot a_1 \cdot \cos \beta)^2 + (\dot{y} \cdot a \cdot \sin \alpha + \dot{z} \cdot a_1 \cdot \sin \beta)^2, \tag{8.55}$$

after a series of calculations we get

$$V_B^2 = (\dot{y}a)^2 + (\dot{z}a_1)^2 - 2\dot{y}\dot{z}aa_1 \cdot (\sin \alpha \cdot \sin \beta - \cos \alpha \cdot \cos \beta). \tag{8.56}$$

Accept that

$$b = \sin \alpha \cdot \sin \beta - \cos \alpha \cdot \cos \beta, \tag{8.56'}$$

then:

$$V_B^2 = (\dot{y}a)^2 + (\dot{z}a_1)^2 - 2\dot{y}\dot{z}aa_1 \cdot b. \tag{8.56''}$$

To determine the speeds of the relative motion of the MTD, we use horizontal components. To determine the speeds of the relative motion of the MTD, we use the horizontal components of the velocities of the motion of points A and B. As for the horizontal displacements of the points A and B, which form the deformation of the elastic element k, then this shift will be

$$x_{\text{sum.}} = 2 \cdot (ya \cdot \cos\alpha - za_1 \cdot \cos\beta). \tag{8.57}$$

We write the expressions for the kinetic and potential energy of the system according to Fig. 8.6 with kinematic perturbation

$$\mathrm{T} = \frac{1}{2}M \cdot (\dot{y})^2 + 2 \cdot \frac{1}{2} \cdot m \cdot \left[(\dot{y}a)^2 + (\dot{z}a_1)^2 - 2\dot{y}\dot{z}aa_1 \cdot b\right] + 2 \cdot \frac{1}{2} \cdot L \cdot (\dot{y} \cdot a \cdot \cos\alpha - \dot{z} \cdot a_1 \cdot \cos\beta)^2. \tag{8.58}$$

$$\Pi = \frac{1}{2}k_0 \cdot (y - z)^2 + 2 \cdot \frac{1}{2}k \cdot (ya \cdot \cos\alpha - za_1 \cos\beta)^2, \tag{8.59}$$

we obtain the equation of motion of the system with kinematic perturbation

$$\ddot{y} \cdot \left(M + 2ma^2 + 2L \cdot a^2 \cdot \cos^2\alpha\right) + y \cdot \left(k_0 + ka^2 \cdot \cos\alpha\right) = \ddot{z} \cdot (2L \cdot aa_1 \cdot \cos\alpha \cdot \cos\beta - 2maa_1b) + z \cdot (k_0 + 2kaa_1 \cdot \cos\alpha \cdot \cos\beta). \tag{8.60}$$

After the Laplace transform (8.60) takes the form

$$\bar{y} \cdot \left[\left(M + 2ma^2 + 2L \cdot a^2 \cdot \cos^2\alpha\right) \cdot p^2 + k_0 + ka^2 \cdot \cos\alpha\right] = \bar{z} \cdot \left[(2L \cdot aa_1 \cdot \cos\alpha \cdot \cos\beta - 2maa_1b) \cdot p^2 + k_0 + 2kaa_1 \cdot \cos\alpha \cdot \cos\beta\right]. \tag{8.61}$$

The transfer function of the system with kinematic perturbation from the side of the bearing surface (Fig. 8.6) will be determined

$$W(p) = \frac{\bar{y}}{\bar{z}} = \frac{2maa_1 \cdot \left(\frac{L}{m}\cos\alpha \cdot \cos\beta - b\right) \cdot p^2 + k_0 + 2kaa_1 \cdot \cos\alpha \cdot \cos\beta}{\left[(M + 2ma^2 + 2L \cdot a^2 \cdot \cos^2\alpha) \cdot p^2 + k_0 + ka^2 \cdot \cos\alpha\right]}. \tag{8.62}$$

From (8.62) we can give findings about the dynamic properties of the system. In particular, the mode of dynamic oscillation damping is possible in the system, as well as "locking" at high frequencies at certain ratios of parameters.

8.2.3 Evaluation of the Dynamic Properties of the System with Kinematic Perturbation

From (8.62) it follows that the system has a natural oscillation frequency:

$$\omega_{\text{nat}}^2 = \frac{k_0 + 2ka^2 \cdot \cos^2\alpha}{M + 2a^2 \cdot (m + L \cdot \cos^2\alpha)}, \tag{8.63}$$

which coincides with the expression (8.50). In contrast to the force perturbation, in this case, the frequency of dynamic oscillation absorbing can be determined in the system.

$$\omega_{\text{dyn}}^2 = \frac{k_0 + 2kaa_1 \cdot \cos\alpha \cdot \cos\beta}{2maa_1 \cdot \left(\frac{L}{m} \cdot \cos\alpha \cdot \cos\beta - b\right)}. \tag{8.64}$$

With $p \to \infty$, the system, having a kinematic perturbation, "locks" and the ratio of the amplitudes of oscillations at the input and output of the system takes the value

$$\underset{p\to\infty}{W(p)} = \frac{\bar{y}}{\bar{z}} = \frac{2maa_1 \cdot \left(\frac{L}{m}\cos\alpha \cdot \cos\beta - b\right)}{M + 2ma^2 + 2L \cdot a^2 \cdot \cos^2\alpha} \tag{8.65}$$

Consider the ratio $\frac{L}{m}\cos\alpha \cdot \cos\beta - b$, which, taking into account (8.56′), can be denoted as

$$R = \frac{L}{m}\cos\alpha \cdot \cos\beta + \cos\alpha \cdot \cos\beta - \sin\alpha \cdot \sin\beta, \tag{8.66}$$

or

$$R = \left(\frac{L}{m} + 1\right)\cos\alpha \cdot \cos\beta - \sin\alpha \cdot \sin\beta. \tag{8.66′}$$

Figure 8.10 presents the dependencies between the function R, the installation angles α and β, and the values of the parameter L/m.

The function R can take a critical value of zero. If $R = 0$, then there is a condition

$$\left(\frac{L}{m} + 1\right) = \frac{\sin\alpha \cdot \sin\beta}{\cos\alpha \cdot \cos\beta} = \text{tg}\alpha \cdot \text{tg}\beta. \tag{8.67}$$

If the condition is satisfied, then it determines a type of the transfer function (8.62) such that eliminates the dynamic oscillation damping mode. Since the system under consideration correlates with the problems of searching for and developing methods and means of describing vibration disturbances, it is of interest to compare the amplitude-frequency characteristics for different values of the R parameter (Fig. 8.11).

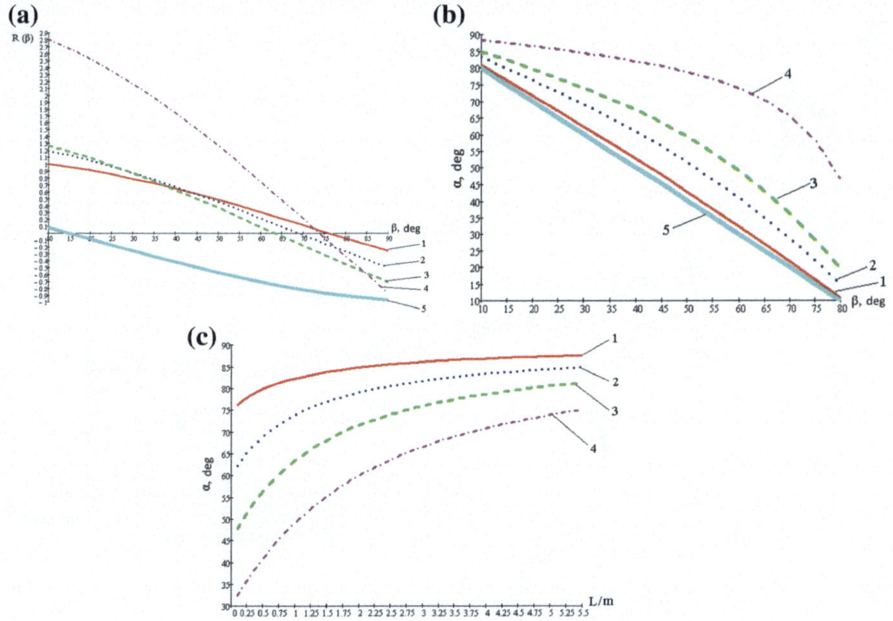

Fig. 8.10 The dependencies between the function R, the installation angles α and β, and the parameter L/m. **a** A family of curves R with fixed values of the angle α and L/m. 1—$\alpha = 15°$, $L/m = 0.1$; 2—$\alpha = 30°$, $L/m = 0.5$; 3—$\alpha = 45°$, $L/m = 1$; 4—$\alpha = 60°$, $L/m = 5$; 5—$\alpha = 75°$, $L/m = 0$; **b** dependencies between the angles α and β in the case of $R = 0$ with fixed values of L/m: 1—$L/m = 0.1$; 2—$L/m = 0.5$; curve 3—$L/m = 1$; 4—$L/m = 5$; 5—$L/m = 0$; **c** dependencies between the angle α and the L/m value at a fixed value of angle β: 1—$\beta = 15°$; 2—$\beta = 30°$; 3—$\beta = 45°$; 4—$\beta = 60°$

The amplitude-frequency characteristics significantly depend on the ratio of the parameters of the additional constraints introduced, as dynamic modes of the vibration protection systems can be chosen, and their parameters can be regulated by the appropriate selection of values characterizing the system configuration and the presence of certain elements in the structure. In the physical sense, this situation reflects the manifestations of dynamic interactions created by the transportation inertial forces that create conditions for the compensation of external influences.

The dynamic properties, introduced by additional constraints, depend on the type of external disturbances. With kinematic perturbation, it is possible to create partial compensation modes for external disturbances due to inertia forces of moving space. Motion transformation devices, as well as additional masses, can act as adjustment elements.

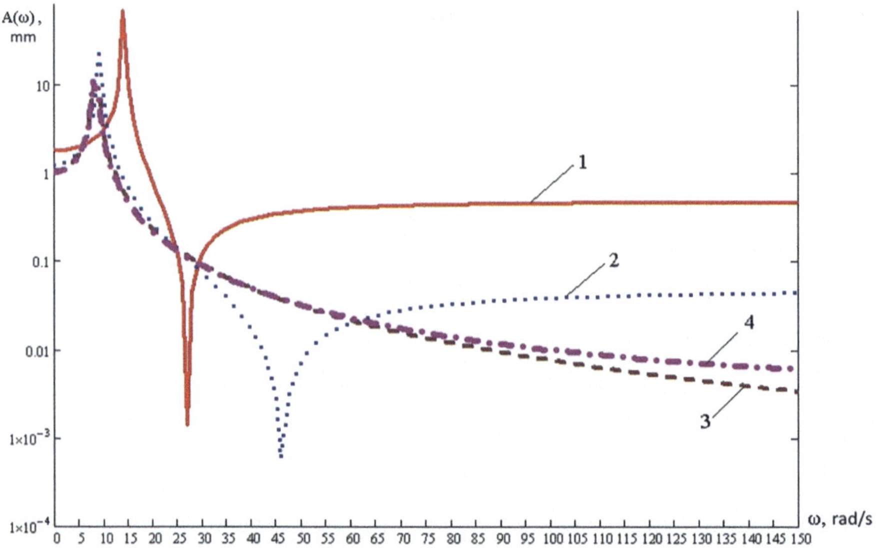

Fig. 8.11 Comparison of types of amplitude-frequency characteristics for different values of the parameter R. 1, 2—corresponds to $R > 0$; 3—$R = 0$; 4—$R < 0$

8.3 Some Constructive and Engineering Forms of Using Lever Linkages

The introduction of lever mechanisms into the structures of oscillatory systems is due to the need to take into account a number of features that arise when choosing the location of the support points of the levers, their configuration and forms of use of connecting elements. A system is considered (by patent [5]), which has an object of protection with mass m_0, which performs (Fig. 8.12) vertical motions (y_0) relative to the position of static equilibrium.

At points A and B, two lever mechanisms of inverse T-shape are attached to the object of protection. Each of the lever mechanisms has mass-and-inertia elements fixed respectively at points A_2, A_1 (m_1, m_{10}) and points B_2, B_1 (m_2, m_{20}). The elements of the system are fixed at points A_3, B_3: a spring with stiffness k_3 and a motion transformation device (MTD) in the form of a screw non-locking mechanism. In addition, points A_1 and B_1 are connected by an elastic element with stiffness k_{00}. In turn, the object also rests on a spring of stiffness k_0.

The mass-and-inertia elements m_1 and m_2 are based on surfaces I and II, the law of motion of which is known (z_1, z_2). In addition, the element k_0 is supported by the surface with the law of motion $z_0(t)$. It is assumed that the system makes small oscillations without taking into account the resistance forces. To describe the motion, the coordinate systems y_0, φ_1, φ_2 and y_0, y_1, y_2 are used. All necessary geometrical parameters are shown in Fig. 4.12.

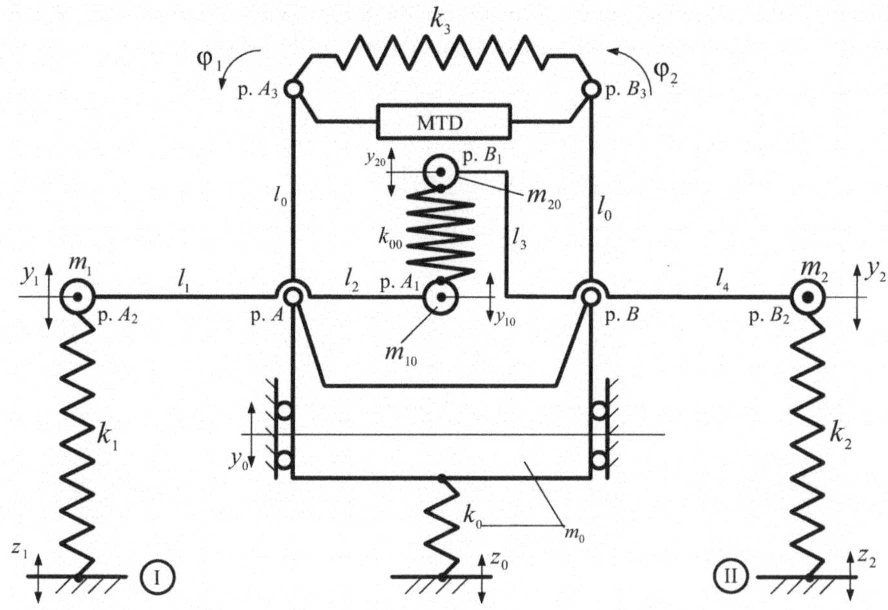

Fig. 8.12 The computational scheme of the robotic device suspension unit

Kinematic perturbations at the preliminary study stage are considered harmonic. Issues of the relationship of motion parameters z_0, z_1, z_2 are considered separately.

Among the problems to be solved is the development of a mathematical model of the system and the evaluation of the possibilities for changing the dynamic state through structural transformations and the selection of the adjustment parameters of the original design.

8.3.1 Construction of a Mathematical Model of the System

We write the expressions for the kinetic and potential energy under the assumption that the system performs small oscillations relative to the position of static equilibrium

$$
\begin{aligned}
T = {} & \frac{1}{2}\cdot(m_0+m_1+m_2+m_{10}+m_{20})\cdot(\dot{y})^2+\frac{1}{2}\cdot L\cdot(\dot{\varphi}_2-\dot{\varphi}_1)^2\cdot l^2+\frac{1}{2}\cdot(\dot{\varphi}_1)^2 \\
& \cdot\left(m_1 l_1^2+m_{10} l_2^2\right)+\frac{1}{2}\cdot(\dot{\varphi}_2)\cdot\left(m_2 l_4^2+m_{20} l_3^2\right),
\end{aligned} \tag{8.68}
$$

where φ_1 and φ_2 are the angles of rotation of the lever devices relative to the object m_0; L is the reduced mass-and-inertia characteristic of the MTD. In turn,

$$\Pi = \frac{1}{2}k_0 \cdot (y_0 - z_0)^2 + \frac{1}{2}k_1 \cdot (y_1 - z_1)^2 + \frac{1}{2}k_2 \cdot (y_2 - z_2)^2 + \frac{1}{2}k_3 \cdot (\varphi_2 - \varphi_1)^2 l^2 + \frac{1}{2}k_{00} \cdot (\varphi_2 l_3 - \varphi_1 l_2)^2. \quad (8.69)$$

We introduce a series of relations

$$\left.\begin{array}{ll} y_{10} = \varphi_1 \cdot l_2, & y_{20} = \varphi_2 \cdot l_3 \\ y_1 = y_0 + \varphi_1 \cdot l_1, & y_2 = y_0 + \varphi_2 \cdot l_4 \end{array}\right\} \quad (8.70)$$

With (8.70) taken into account, the potential energy is written as

$$\Pi = \frac{1}{2}k_0 \cdot (y_0 - z_0)^2 + \frac{1}{2}k_1 \cdot (y_1 - \varphi_1 l_2 - z_1)^2 + \frac{1}{2}k_2 \cdot (y_2 - \varphi_2 l_4 - z_2)^2 + \frac{1}{2}k_3 \cdot (\varphi_2 - \varphi_1)^2 l^2 + \frac{1}{2}k_{00} \cdot (\varphi_2 l_3 - \varphi_1 l_2)^2. \quad (8.71)$$

Find the equations of motion in the coordinates y_0, φ_1 and φ_2

$$\ddot{y}_0 \cdot (m_0 + m_1 + m_2 + m_{10} + m_{20}) + y_0 \cdot (k_0 + k_1 + k_2) - \varphi_1 k_1 l_2 - \varphi_2 k_2 l_4 = k_0 z_0 + k_1 z_1 + k_2 z_2; \quad (8.72)$$

$$\ddot{\varphi}_1 \cdot \left(m_1 l_1^2 + m_{10} l_2^2 + L \cdot l^2\right) + \ddot{\varphi}_1 \cdot \left(k_1 l_2^2 + k_2 l^2 + k_{00} l_2^2\right) + \ddot{\varphi}_2 \left(-L l^2\right) + \ddot{\varphi}_2 (-k_3 l_2 - k_{00} l_2 l_3) - k_1 l_2^2 y_0 = -k_1 z_1 l_2; \quad (8.73)$$

$$\ddot{\varphi}_2 \cdot \left(m_2 l_4^2 + m_{20} l_3^2 + L \cdot l^2\right) + \varphi_2 \cdot \left(k_2 l_4^2 + k_3 l^2 + k_{00} l_3^2\right) + \ddot{\varphi}_1 \left(-L l^2\right) + \varphi_1 \left(-k_3 l^2 - k_{00} l_2 l_3\right) - k_2 l_4 y_0 = -k_2 z_2 l_4. \quad (8.74)$$

Figure 8.13 shows the structural diagram of the initial system (see Fig. 8.12) in the coordinates y_0, φ_1, φ_2.

The peculiarity of the system is that between the partial systems (φ_1) and (φ_2) the connection has an inertial-elastic form. With the frequency of external impact

$$\omega_{\text{par}}^2 = \frac{k_{00} l_2 l_3 + k_3 l_2^2}{L l^2 p^2} \quad (8.75)$$

interpartial constraints φ_1 and φ_2 may be zeroed (as well as between y_1 and y_2). In other cases, i.e. when considering y_0 and φ_1, as well as φ_2 and y_0, the constraints are elastic (there is no zeroing).

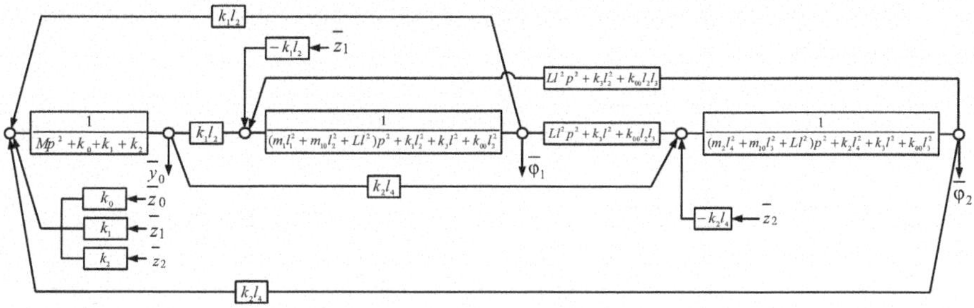

Fig. 8.13 The structural diagram of a system with T-shaped levers (see Fig. 8.12) in coordinates y_0, φ_1, φ_2

8.3.2 *Evaluation of the Dynamic Properties of the System*

To estimate the dynamic properties of the system, Kramer's formulas [8] can be used:

$$\bar{y}_0 = \frac{1}{A_0}\left[\overline{Q}_1 \cdot (a_{22}a_{33} - a_{23}a_{32}) + \overline{Q}_2 \cdot (a_{13}a_{32} - a_{12}a_{33}) + \overline{Q}_3 \cdot (a_{12}a_{23} - a_{13}a_{22})\right], \tag{8.76}$$

$$\overline{\varphi}_1 = \frac{1}{A_0}\left[\overline{Q}_1 \cdot (a_{23}a_{31} - a_{21}a_{33}) + \overline{Q}_2 \cdot (a_{11}a_{33} - a_{13}a_{31}) + \overline{Q}_3 \cdot (a_{13}a_{21} - a_{11}a_{23})\right], \tag{8.77}$$

$$\overline{\varphi}_2 = \frac{1}{A_0}\left[\overline{Q}_1 \cdot (a_{21}a_{32} - a_{22}a_{31}) + \overline{Q}_2 \cdot (a_{12}a_{31} - a_{11}a_{32}) + \overline{Q}_3 \cdot (a_{11}a_{22} - a_{12}a_{21})\right], \tag{8.78}$$

where A_0 is the characteristic equation of the system shown in Fig. 8.13,

$$A_0 = a_{11}a_{22}a_{23} - a_{11}a_{23}^2 - a_{22}a_{13}^2 - a_{33}a_{12}^2 + 2a_{12}a_{23}a_{31}. \tag{8.79}$$

Using formulas (8.67)–(8.69), one can find the transfer functions that determine the interaction between the input external disturbances from the set $\bar{z}_0$, $\bar{z}_1$, $\bar{z}_2$ and the output signals $\bar{y}_0$, $\overline{\varphi}_1$, $\overline{\varphi}_2$. If several signals act at the input of the partial system, then the principle of superposition can be used. Of particular interest is the situation in which certain relationships are established between the external signals z_0, z_1, z_2. For example, suppose that the system relies simultaneously on one surface, then it can be assumed that $z = z_0 = z_1 = z_2$ with the corresponding features of the definition of transfer functions on the basis of expressions (8.67)–(8.69). If $k_0 = 0$, then the object m_0 is not directly connected to the bearing surface, the contact is made through the elastic elements k_1 and k_2 (see Fig. 8.12). When $z = z_1 = z_2$, you can obtain the transfer function for the input signal (or generalized external action $\bar{z}$) and the output signal $\bar{y}_0$:

$$W_1(p) = \frac{\bar{y}_0}{\bar{z}} = \frac{(a_{13}a_{32} - a_{12}a_{33}) + (a_{13}a_{21} - a_{11}a_{23})}{A_0}. \tag{8.80}$$

Since a_{23} is in accordance with the structural diagram in Fig. 8.13 looks like

$$a_{23} = Ll^2p^2 + k_3l_2^2 + k_{00}l_2l_3, \tag{8.81}$$

then the numerator (8.80) with

$$a_{11} = (m_0 + m_1 + m_2 + m_{10} + m_{20}) \cdot p^2 + k_0 + k_1 + k_2 \tag{8.82}$$

is a biquadratic equation with respect to the square of the frequency of external influence. If we exclude the contact of m_0 with the bearing surface, assuming $k_0 = 0$, then with respect to the object of protection, we can expect the emergence of two modes of dynamic oscillation absorbing under the action of vibrations from the base ($z = z_1 = z_2$). Fundamentally, the situation will not change if the condition $z = z_0 = z_1 = z_2$ ($k_0 \neq 0$) is fulfilled. However, the parameters (i.e., the frequency of dynamic absorbing) will change. The amplitude-frequency characteristics of the system according to the variable $\bar{y}_0$ will be conventional. With the characteristic equation A_0, one can expect the presence of three resonant frequencies with a corresponding increase in the amplitudes of oscillations along the coordinate y_0. In this case, the two dynamic absorbing frequencies will be located between the three resonant frequencies. The suspension scheme (see Fig. 8.12) has the potential to change the dynamic state of the object of protection m_0 by selecting the adjustment parameters accordingly. These parameters can be the lengths of the units of levers, the stiffnesses of the elastic elements k_3 and k_{00}, as well as the mass-and-inertia parameters of the MTD (we mean that L is the reduced mass of the MTD and l is the length of the lever arm). It was noted above that, if the relation (8.67) is observed, partial constraints' decoupling is possible.

In suspensions with lever linkages, if there are several input signals, local dependencies between external influences are possible. For example, the relations $z = z_0 = z_1 = z_2$ reflect the possibility of vertical oscillations of all the bearing surfaces at the same time. However, situations are quite possible when motions along individual surfaces (z_0, z_1, z_2) are considered as fragments of some general motion. A motion like that can be the swinging of a common surface relative to a certain center, etc. In the spectral theory of the suspension of transport systems, more complex functional dependencies between the actions of several input signals are considered. Simplifications can be obtained by analyzing the possibilities of linear relations in the parameters of simultaneously acting perturbations. It should also be noted that the arising statement of the problem of estimating the dynamic state of a system under the action of several force factors can be considered quite "natural" under the action of an external force that is displaced relative to the center of gravity, for example, with flat vibrations of a solid body on elastic supports. The use of adjustment parameters k_3 and k_{00} in this scheme (see Fig. 8.12) is complicated by some factors. In particular, with $k_{00} \to \infty$ the connection between the elements m_{10} and m_{20} is blocked, which simultaneously leads to blocking of the relative motion of the levers (with respect to the object of protection m_0). In this regard, suspensions with gearing capabilities of lever devices are very promising.

Features of the dynamic properties of the system in the coordinates y_0, y_1, y_2. To build a mathematical model of the system, a number of relations are introduced:

$$\varphi_1 = \frac{y - y_0}{l_1} = a \cdot (y - y_0); \quad \varphi_2 = b \cdot (y - y_0); \quad a = \frac{1}{l_1}; \quad b = \frac{1}{l_4}. \tag{8.83}$$

We write the expression for the kinetic and potential energy:

$$\begin{aligned} \mathrm{T} &= \frac{1}{2} M \cdot \left(\dot{y}_0\right)^2 + \frac{1}{2} L l^2 \cdot \left(b\dot{y}_2 - b\dot{y}_0 - a\dot{y}_1 + a\dot{y}_0\right)^2 \\ &+ \frac{1}{2} \cdot \left(m_1 l_1^2 + m_{10} l_2^2\right) \cdot a^2 \cdot \left(\dot{y}_1 - \dot{y}_0\right)^2 + \frac{1}{2} \cdot \left(m_2 l_4 + m_{20} l_3^2\right) \cdot b^2 \cdot \left(\dot{y}_2 - \dot{y}_0\right)^2; \end{aligned} \tag{8.84}$$

$$\begin{aligned} \Pi &= \frac{1}{2} k_0 \cdot (y_0 - z_0)^2 + \frac{1}{2} k_1 \cdot (y_1 - z_1)^2 + \frac{1}{2} k_2 \cdot (y_2 - z_2)^2 \\ &+ \frac{1}{2} k_3 l^2 \cdot \left[b y_2 - a y_1 + y_0 (a - b)^2\right] \\ &+ \frac{1}{2} k_{00} \cdot \left[y_2 \cdot b l_3 - y_1 \cdot a l_2 + y_0 \cdot (a l_2 - b l_3)\right]^2. \end{aligned} \tag{8.85}$$

We introduce a number of symbols:

$$a - b = a_0; \quad b l_3 = a_1; \quad a l_2 = a_2; \quad a_2 - a_1 = a_3, \tag{8.86}$$

then (8.85) is converted to

$$\begin{aligned} \Pi &= \frac{1}{2} k_0 \cdot (y_0 - z_0)^2 + \frac{1}{2} k_1 \cdot (y_1 - z_1)^2 + \frac{1}{2} k_2 \cdot (y_2 - z_2)^2 \\ &+ \frac{1}{2} k_3 l^2 \cdot \left[b y_2 - a y_1 + y_0 a_0^2\right] + \frac{1}{2} k_{00} \cdot \left[y_2 \cdot a_1 - y_1 \cdot a_2 + y_0 \cdot a_3\right]^2. \end{aligned} \tag{8.87}$$

We construct a system of equations of motion in the coordinates y_0, y_1. y_2, taking

$$M = m_0 + m_1 + m_2 + m_{10} + m_{20}, \tag{8.88}$$

that will be

$$\begin{aligned}&\ddot{y}_0 \cdot \left[M + Ll^2 a_0^2 + \left(m_1 l_1^2 + m_{10} l_2^2\right) \cdot a^2 + \left(m_2 l_4^2 + m_{20} l_3^2\right) \cdot b^2\right] + y_0 \cdot \left[k_0 + k_3 l^2 a_0^2 + k_{00} a_3^2\right] \\ &\quad + \ddot{y}_1 \cdot \left[-a_0 a L l^2 - \left(m_1 l_1^2 + m_{10} l_2^2\right) a^2\right] + y_1 \cdot \left[-k_3 l^2 a - k_{00} a_3 a_2\right] \\ &\quad + \ddot{y}_2 \cdot \left[Ll^2 a_0 b - \left(m_2 l_4^2 + m_{20} l_3^2\right) b^2\right] + y_2 \cdot \left[k_3 l^2 b + k_{00} a_3 a\right] = k_0 z_0;\end{aligned} \tag{8.89}$$

$$\begin{aligned}&\ddot{y}_0 \cdot \left[-Ll^2 a a_0 - \left(m_1 l_1^2 + m_{10} l_2^2\right) \cdot a^2\right] + y_0 \cdot \left[-k_3 l^2 a a_0 - k_{00} a_3 a_2\right] \\ &\quad + \ddot{y}_1 \cdot \left[L l_1 a_1^2 - \left(m_1 l_1^2 + m_{10} l_2^2\right) a^2\right] \\ &\quad + y_1 \cdot \left[k_1 + k_3 l^2 a^2 + k_{00} a_2^2\right] + \ddot{y}_2 \cdot \left[-Ll^2 ab\right] + y_2 \cdot \left[-k_3 l^2 ab - k_{00} a_1 a_2\right] \\ &\quad = k_1 z_1;\end{aligned} \tag{8.90}$$

$$\begin{aligned}&\ddot{y}_0 \cdot \left[Ll^2 b a_0 - \left(m_2 l_4^2 + m_{20} l_3^2\right) \cdot b^2\right] + y_0 \cdot \left[k_3 l^2 b a_0 + k_{00} a_3 a_1\right] \\ &\quad + \ddot{y}_1 \cdot \left[-Ll^2 ba\right] + y_1 \cdot \left[-k_3 l^2 ab - k_{00} a_1 a_2\right] \\ &\quad + \ddot{y}_2 \cdot \left[Ll^2 b^2 + \left(m_2 l_4^2 + m_{20} l_3^2\right) \cdot b^2\right] + y_2 \cdot \left[k_2 + k_3 l^2 b^2 + k_{00} a_1^2\right] = k_2 z_2.\end{aligned} \tag{8.91}$$

Analysis of system properties using (8.89) – (8.91) can be built on the transfer function of the form

$$\underset{(z_0=z_1=z_2=z)}{W_2(p)} = \frac{\bar{y}_0}{\bar{z}} = \frac{k_0 \cdot \left(a_{22} a_{33} - a_{23}^2\right) + k_1 \left(a_{13} a_{32} - a_{12} a_{23}\right) + k_2 \left(a_{12} a_{23} - a_{13} a_{33}\right)}{A_0} \tag{8.92}$$

From (8.92), the conditions for the occurrence of dynamic absorbing modes along the coordinate $\bar{y}_0$ at various ratios of parameters can be obtained. Since the observed $\bar{y}_0$ coordinate in the y_0, φ_1, φ_2 and $\bar{y}_0$, $\bar{y}_1$ coordinate system is the same, the dynamic absorbing frequencies will be the same.

However, in analyzing (8.92), the values of the coefficients a_{ij} should be taken from the system of equation (8.89)–(8.91). As for the characteristic equation, its general construction scheme remains unchanged, however, when passing from the coordinate system y_0, φ_1, φ_2 to the system y_0, y_1, y, constant factors will be formed between the characteristic equations that determine the matching of the dimensions of the state variables (linear and angular).

8.3.3 Consideration of the Toothed Coupling of the Lever Sectors

Consider a system with two degrees of freedom, as shown in Fig. 8.14.

In the scheme (Fig. 8.14), a gearing is implemented, which ensures the connection of the angular velocities of rotation φ_1 and φ_2 $\varphi_1 r_1 = \varphi_2 r_2$, $\varphi_2 = \varphi_1 \frac{r_1}{r_2}$; at the same time, rotations are executed in different directions. The expression for potential energy is

$$\Pi = \frac{1}{2}k_0 \cdot (y_0 - z_0)^2 + \frac{1}{2}k_1 \cdot (y_1 - z_1)^2 + \frac{1}{2}k_2 \cdot (y_2 - z_2)^2 + \frac{1}{2}k_3 \cdot (\varphi_2 + \varphi_1)^2 l^2. \tag{8.93}$$

The last term of the expression (8.93) can be transformed:

$$\varphi_1 + \varphi_2 = \varphi_1 + \varphi_1 \frac{r_1}{r_2} = \varphi_1 \left(\frac{r_1 + r_2}{r_2}\right) = \varphi_1(1 + i). \tag{8.93'}$$

Then (8.93) can be written in the form

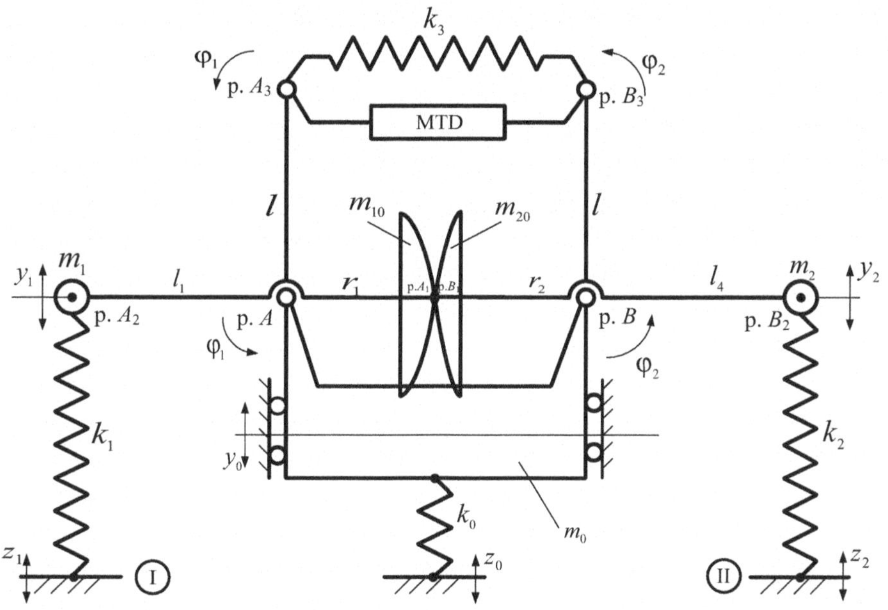

Fig. 8.14 The computational scheme of the system with a toothed gearing in the contact of two lever sectors

$$\Pi = \frac{1}{2}k_0 \cdot (y_0 - z_0)^2 + \frac{1}{2}k_1 \cdot (y_1 - z_1)^2 + \frac{1}{2}k_2 \cdot (y_2 - z_2)^2 + \frac{1}{2}k_3 \cdot \varphi_1^2 \cdot (1+i)^2 l^2. \tag{8.94}$$

Lever sectors have mass-and-inertia properties

$$J_1 = m_1 l_1^2 + m_{10} r_1^2, \tag{8.95}$$

$$J_2 = m_2 l_4^2 + m_{20} r_2^2. \tag{8.96}$$

The expression for the kinetic energy of the system is

$$\mathrm{T} = \frac{1}{2}M \cdot (y_0)^2 + \frac{1}{2}\left(J_1 + J_2 i^2\right) \cdot (\dot{\phi}_1)^2 + \frac{1}{2}Ll^2(1+i)^2 \cdot (\dot{\phi}_1)^2. \tag{8.97}$$

As

$$y_1 = y_0 - \varphi_1 l_1; \quad y_2 = y_0 - \varphi_2 l_4,$$

Equation (8.97) can be written as

$$\mathrm{T} = \frac{1}{2}(m_1 + m_2 + m_{10} + m_{20} + m_0) \cdot (\dot{y}_0)^2 + \frac{1}{2}J_1(\dot{\phi}_1)^2 + \frac{1}{2}J_2(\dot{\phi}_1)^2 i^2 + \frac{1}{2}Ll^2\dot{\phi}_1(1+i)^2. \tag{8.98}$$

We introduce a series of relationships:

$$y_1 = y_0 + \varphi_1 l_1; \quad y_2 = y_0 + \varphi_1 l_4 i; \; y_1 - y_2 = \varphi_1(l_1 - l_4 i); \; \varphi_1 = \frac{y_1 - y_2}{(l_1 - l_4 i)}, \tag{8.99}$$

which allows us to put (8.94) to a form

$$\Pi = \frac{1}{2}k_0 \cdot (y_0 - z_0)^2 + \frac{1}{2}k_1 \cdot (y_0 + \varphi_1 l_1 - z_1)^2 + \frac{1}{2}k_2 \cdot (y_0 + \varphi_1 l_4 - z_2)^2 + \frac{1}{2}k_3 \cdot \varphi_1^2 \cdot (1+i)^2 l^2. \tag{8.100}$$

We write the equations of motion of the system in the coordinates y_0 and φ_1:

$$\ddot{y}_0 M + y_0[k_0 + k_1 + k_2] + \varphi_1[k_1 l_1 + k_2 l_4 i] = k_0 z_0 + k_1 z_1 + k_2 z_2. \tag{8.101}$$

$$\dot{\varphi}_1 \cdot \left[J_1 + J_2 i^2 + Ll^2(1+i)^2\right] + \dot{\varphi}_1 \cdot \left[k_1 l_1^2 + k_4 l_4^2 i^2 + k_3 l^2(1+i)^2\right] + y_0 \cdot \left[k_1 l_1^2 + k_2 l_4 i\right] = k_1 l_1 z_1 + k_2 l_4 i z_2. \tag{8.102}$$

Table 8.1 The coefficients of the system of equations in the coordinates y_0, φ_1

a_{11}	a_{12}
$Mp^2 + k_1 + k_2 k_0$	$k_1 l_1 + k_2 l_4 i$
a_{21}	a_{22}
$k_1 l_1 + k_2 l_4 i$	$J_1 + J_2 i + Ll^2(1+i)^2 p^2 + k_1 l_1^2 + k_2 l_4^2 i^2 + k_3 l^2 (1+i)^2$
Q_1''	Q_2''
$k_0 z_0 + k_1 z_1 + k_2 z_2$	$k_1 l_1 z_1 + k_2 l_4 i z_2$

Note Q_1'', Q_2'' are generalized forces corresponding to the coordinates y_0 and φ

The coefficients of equations (8.101) and (8.102) are given in Table 8.1.

Figure 8.15 shows a structural diagram of the system in coordinates y_0 and φ. In this system, interpartial constraints are elastic.

Let $z_1 = 0$, $z_2 = 0$. Then the transfer function of the system (Fig. 8.15) takes the form

$$W_1(p) = \frac{\bar{y}_0}{\bar{z}_0} = \frac{k_0 \cdot \left\{\left[J_1 + J_2 i^2 + Ll^2 \cdot (1+i)^2\right]p^2 + k_1 l_1^2 + k_2 l_4^2 (1+i)^2 + k_3 l^2 (1+i)^2\right\}}{[a_{11}] \cdot [a_{22}] - (k_1 l_1 + k_2 l_4 i)^2}. \tag{8.103}$$

From (8.103) you can determine the dynamic absorbing frequency ($z_0 = z$; $z_1 = 0$; $z_2 = 0$):

$$\omega^2 = \frac{k_1 l_1^2 + k_2 l_4^2 (1+i)^2 + k_3 l^2 (1+i)^2}{J_1 + J_2 i^2 + Ll^2 \cdot (1+i)^2}. \tag{8.104}$$

Similarly, other dynamic absorbing modes can be found, which occur with different combinations of external influences. The possibilities of setting up the

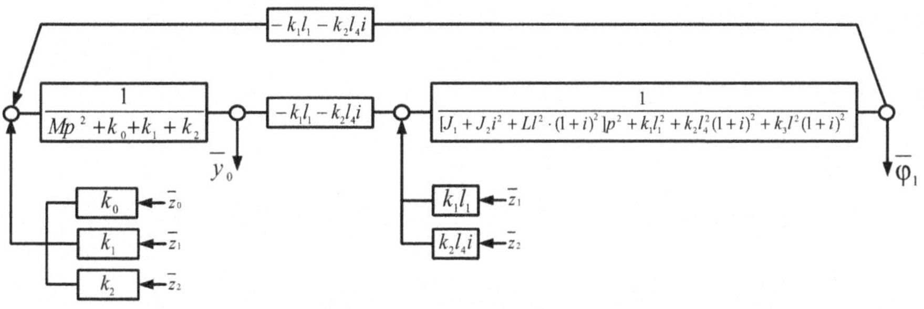

Fig. 8.15 The structural diagram of the system (Fig. 8.12) with toothed sectors

system are determined by such parameters as: stiffness k_3, length of the arms of the levers (l), transfer ratios i, and mass-and-inertia properties of the toothed sectors. The system has two resonant frequencies (or self-oscillation frequencies), which can be found from the characteristic Eq. (8.103). Thus, a mechanical system with lever linkages, considered in dynamic interactions with a vibrating base, has an extended set of adaptive properties. When providing elastic constraint k_{00} between the adjustment masses (m_1, m_2, m_{10}, m_{20}), the system can have two modes of dynamic absorbing. The stiffness of the spring k_3 and the inertial properties of the MTD (via Ll^2) can be used as adjustment parameters.

With the introduction of a toothed gearing between the lever sectors, the system loses one degree of freedom; however, adaptation capabilities in the damping of external kinematic effects can be supported by the introduction of transfer ratios between the lever sectors and the choice of other adjustment parameters. The method developed by the author for constructing and analyzing the dynamic properties of the original vibration protection system is based on taking into account the lever linkages implemented by the lever sectors. Lever sectors can be considered as a new element of mechanical oscillatory systems, possessing enhanced dynamic effects, in particular, due to the simplicity of handling and using active means by applying controlled moments of force to change the dynamic state.

8.4 Transport Suspensions. Mathematical Models. Selection of Coordinate Systems

In the dynamics of transport systems, passive suspensions with an extended set of links are widely used [9]. However, the lever linkages and their capabilities have not yet received adequate attention, although developments in this direction are quite intensive. The relaxation suspensions used in practice have a number of characteristic features—a consecutive connection “damper–elastic element”. Details, such as the location of the element in the serial connection circuit, are also taken into account.

The motion of the protected object m in the case of kinematic perturbation is determined by the coordinate y in the conditionally fixed coordinate system. As shown in [10], the equivalent (or reduced) stiffness of two consecutive standard elements (bp is a first-order differentiating unit, and k is a reinforcing unit), is determined by the formula

$$k_{\text{equ.}} = \frac{bp \cdot k}{bp + k}. \tag{8.105}$$

Note that the sequence of the consecutive arrangement of elements does not matter. The reduced spring stiffness can be found for the circuit Fig. 8.16 as follows:

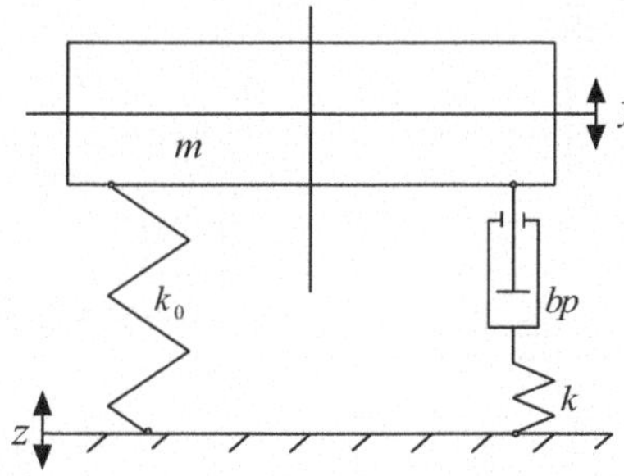

Fig. 8.16 Design suspension scheme with a relaxation chain. k_0—the stiffness of the main spring; M—the mass of the object of protection, bp—the damping element (b—the coefficient of viscous friction); k—the spring constant in the relaxation chain, $p = j\omega$—the complex variable

$$k_{\text{red}} = k_0 + k_{\text{equ}} = k_0 + \frac{pbk}{pb + k} = \frac{kk_0 + bp(k + k_0)}{bp + k}. \tag{8.106}$$

In the case under consideration, instead of purely elastic support, the object is based on a structure representing a certain combination of three elementary units (two springs k_0 and k, as well as a differentiating unit bp). At zero frequency ($p = 0$), the object is based on an elastic element with stiffness k_0; as $p \to \infty$, the object "works" with an elastic element with stiffness $k_0 + k$. Thus, the relaxation circuit makes it possible in the system in Fig. 8.16 to turn the usual spring k_0 into a passive unit with a transfer function of an aperiodic unit of general form (8.106). The properties of that unit are well studied [11–15]. Relaxation suspensions of this type make it possible to ensure smooth running, for which it is necessary to minimize the amplitude of oscillations. One of the ways to improve the dynamic properties of relaxation suspensions is considered to be damping control, which involves turning off the damper at those times when the damping force impedes the motion of the object to the equilibrium position. Such approaches make it possible to control the processes of development of oscillations and can level out resonance phenomena. At the same time, the physics of the process is more complex, since the manifestations of motion are of an integration nature, which, in turn, is determined by taking into account other factors related to the specific design of the suspension. Examples of structural solutions for relaxation suspensions are shown in Fig. 8.17. In further calculations it is assumed that the center of masses of the levers, the interacting units coincides with the point of attachment with the source of oscillations, and the "masses" of the damper bp and the spring k_2 are rather small.

Figure 8.18 shows the schemes of suspensions which include lever linkages.

The presence of constraints of that kind, implemented by lever mechanisms, allows changing the dynamic properties of suspensions, creating effects of dynamic absorbing and locking at high frequencies.

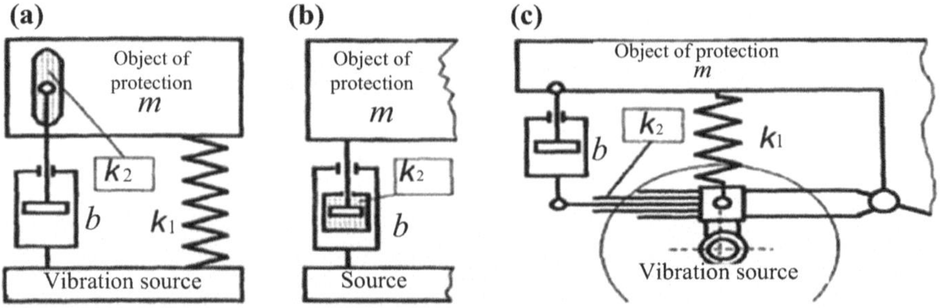

Fig. 8.17 Examples of an arrangement. **a** Serial connection with spring k_2; **b** taking account of the elasticity of the damper; **c** the elastic-damping leaf spring

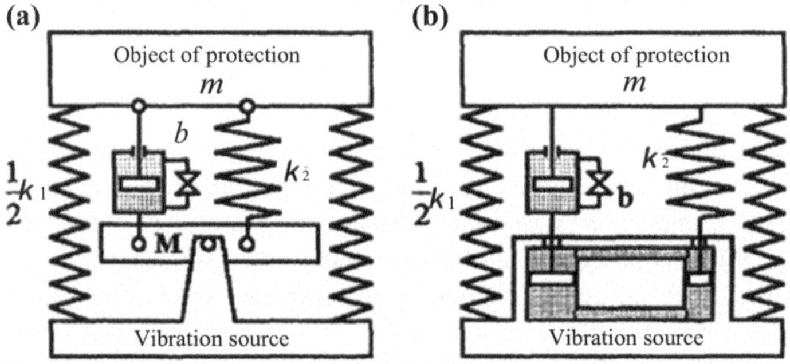

Fig. 8.18 Schemes of lever relaxation suspensions. **a** Mechanical; **b** hydromechanical

8.4.1 Formulation of the Problem. Construction of a Mathematical Model

The computational scheme of the transport suspension in general form is shown in Fig. 8.19 and is a system with three degrees of freedom and a lever mechanism.

It is assumed that the motion of the lever does not lead to noticeable deviations of the lines of action of the forces of elastic and damping elements from the vertical. We write down a number of auxiliary relations: $y_1 = y - L_1$, $y_2 = y + L_2\varphi$, $\frac{l_2}{l_1} = i$. Note that in the scheme (see Fig. 8.17) a lever of the second kind is used; therefore, the relative velocities of points A_1 and A_2 will be directed in different directions. If we take the direction of turning the lever φ_1 counterclockwise for positive, then the relative position of points A_1 and A_2 is defined respectively as $z - x_1$ and $x_2 - z$ (Fig. 8.19). For preliminary calculations, we can assume that $l_1 \approx l'_1$ and $l_2 \approx l''_2$. We also denote the distances to the attachment points A and B on the object of protection.

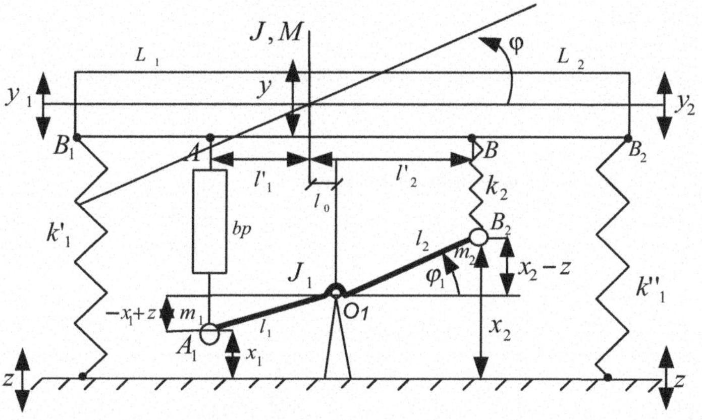

Fig. 8.19 The computational scheme of the suspension with a lever mechanism. *O*—the object gravity center; O_1—the lever attachment point; *M*, *J*—mass-and-inertia parameters of the protected object; m_1, m_2, J_1—mass-and-inertia parameters of the lever suspension; $L_1 = OB_1$, $L_1 = OB_1$, $l_0 = OO_1$, $l'_1 = OA$, $l'_2 = OB$, $l_1 = O_1A_1$, $l_2 = O_1A_2$; *z*—kinematic perturbation; k'_1, k''_1—elastic elements supporting the object on the base; *bp*—damper; k_2—the stiffness of the elastic suspension element; *y*, φ—generalized coordinates of the center of mass and rotation around the center of mass; y_1, y_2—coordinates of the motion of the object of protection relative to the fixed frame of reference; φ_1—the angle of the sweep of the lever in motion relative to the base; x_1, x_2 – coordinates of relative motion (to the base)

We consider the suspension motion in the coordinate system *y*, φ and φ_1 write the expressions for the kinetic and potential energy:

$$\mathrm{T} = \frac{1}{2}M\dot{y}^2 + \frac{1}{2}J\dot{\varphi}^2 + \frac{1}{2}J_1\dot{\varphi}_1^2; \tag{8.107}$$

$$\Pi = \frac{1}{2}k'_1(y_1 - z)^2 + \frac{1}{2}k''_1(y_2 - z)^2 + \frac{1}{2}k_2(y_B - x'_2)^2. \tag{8.108}$$

The energy dissipation function [16] can be taken in the form

$$\Phi = \frac{1}{2}b(\dot{y}_A - \dot{x}'_1)^2. \tag{8.109}$$

We introduce a number of auxiliary relations:

$$\varphi = \frac{1}{2}\frac{y_2 - y_1}{L_1 + L_2} = c(y_2 - y_1); \quad c = \frac{1}{L_1 + L_2}; \quad \varphi = \frac{z - x_1}{l_1}; \quad \varphi_1 = \frac{x_2 - z}{l_2};$$
$$x_2 = \varphi_1 + l_2 = i(z - x_1); \tag{8.110}$$

let us assume that

$$x_2 = \varphi_1 l_2;\quad x_1' = y_1 - z - x_1 = y_1 - z - l_1\varphi,\quad x_1' = y_1 - z + \varphi_1 l_1,\\ x_2' = y_2 - z + l_2\varphi_1. \tag{8.111}$$

Then (8.108) can be converted to the form

$$\Pi = \frac{1}{2}k_1'(y_1 - z)^2 + \frac{1}{2}k_1''(y_2 - z)^2 + \frac{1}{2}k_2(y + l_2\varphi - y_2 + z - l_2\varphi_1)^2. \tag{8.112}$$

Since $y_2 = y_0 + l_2\varphi$ (8.112) is written as

$$\Pi = \frac{1}{2}k_1'(y_1 - z)^2 + \frac{1}{2}k_1''(y_2 - z)^2 + \frac{1}{2}k_2[\varphi(l_1 + l_2) + (z - l_1\varphi_1)]^2. \tag{8.113}$$

We introduce into the expression for the potential energy of the coordinate system of the relative motion of elastic elements

$$\Pi = \frac{1}{2}k_1'(y_1 - z)^2 + \frac{1}{2}k_1''(y_2 - z)^2 + \frac{1}{2}k_2(y_B - x_2')^2, \tag{8.114}$$

where $y_B = y_0 + l_2\varphi$; $x_2' = y_2 - z + l_2\varphi_1$.

In turn, the dissipation function will take the form

$$\Phi = \frac{1}{2}b(\dot{y}_A - \dot{x}_1')^2,$$

where

$$\dot{y}_A = \dot{y}_0 - l_1\dot{\varphi};\quad x_1' = \dot{y}_1 - \dot{z} - l_1\dot{\varphi}_1. \tag{8.115}$$

Making a series of intermediate calculations, we write the equations of motion of the system:

$$\begin{aligned}&\ddot{y}(M) + y(k_1' + k_1'') + \varphi(-k_1'L_1 + k_1''L_2) = k_1'z + k_2''z;\\ &y(-k_1'L_1 + k_1''L_2) + \ddot{\varphi}(J) + \varphi[k_1'L_1^2 + k_1''L_2^2 + b\dot{\varphi}(L_1 - l_1)^2 + k_2(l_2 - L_2)^2]\\ &\quad + \dot{\varphi}_1(bl_1(L_1 - l_1)) + \varphi_1(-k_2(l_2 - L_2)) = -k_1'L_1z + k_1''L_2z - k_2(l_2 - L_2)z;\end{aligned} \tag{8.116}$$

$$\varphi(-k_2l_2(l_2 - L_2)) + \ddot{\varphi}_1(J_1) + \varphi_1(k_2l_2^2) = k_2l_2z.$$

The values of the coefficients of Eqs. (8.116) in the coordinates y, φ_1 are presented in Table 8.2.

Note that in the coordinates y, φ, φ_1 the system is under the action of a kinematic perturbation and does not contain information about the action of the inertia forces

Table 8.2 The coefficients of Eqs. (8.116) in the coordinate system y, φ, φ_1

a_{11}	a_{12}	a_{13}
$Mp^2 + (k_1'' + k_1')$	$-k_1'L_1 + k_1''L_2$	0
a_{21}	a_{22}	a_{23}
$(-k_1'L_1 + k_1''L_2)$	$Jp^2 + b(L_1 - l_1)^2 p + k_1''L_1^2 + k_1''L_2^2 + k_2(l_2 - L_2)^2$	$bl_1(L_1 - l_1)p - k_2(l_2 - L_2)$
a_{31}	a_{32}	a_{33}
0	$pbl_1(L_1 - l_1) - k_2 l_2(l_2 - L_2)$	$J_1p^2 + k_2 l_2^2 + bl_1^2 p$
Q_1	Q_2	Q_3
$(k_1' + k_1'')z$	$-bp(L_1 - l_1) - k_1'L_1 z + k_1''L_2 z - k_2(l_2 - L_2)z$	$k_2 l_2 z - bl_1 pz$

Note Q_1, Q_2, Q_3 are the generalized forces corresponding to the coordinates y, φ, φ_1 are determined from (8.116)

of moving space in a direct form. Of greater interest is the coordinate system in which the motion of an object is considered with respect to the base. In this case, one can introduce the coordinate $y_0 = y_{01} + z$, then $y_{01} - z = y$, and then introduce y into all the other relations.

8.4.2 *Accounting for Inertia Forces of Moving Space*

Imagine the computational scheme of the trailer suspension (see Fig. 8.19) in the coordinate system, which allows us to proceed to obtain a more convenient mathematical model. For these purposes, we introduce the additional degree of freedom y_{01}. Figure 8.20 shows the computational scheme in which the coordinate system y_1, y_2, y_1', y_2' as well as y_0, φ, y_{01}, φ_1 are used. That kind of system has four degrees of freedom. As an external influence, the base oscillations are considered. The remaining designations in Fig. 8.20 coincide with Fig. 8.19 (but k_{01} and k_1 are introduced instead of bp).

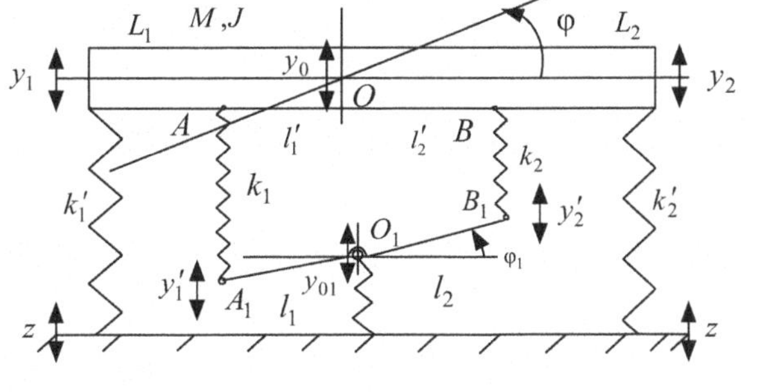

Fig. 8.20 The computational scheme of the trailer suspension

$$\mathrm{T} = \frac{1}{2}M\dot{y}_0^2 + \frac{1}{2}J\dot{\varphi}^2 + \frac{1}{2}m_1(\dot{y}_1')^2 + \frac{1}{2}m_2(\dot{y}_2')^2 + \frac{1}{2}J_1\dot{\varphi}_1^2; \tag{8.117}$$

$$\begin{aligned}\Pi = &\frac{1}{2}k_1'(y_1 - z)^2 + \frac{1}{2}k_2'(y_2 - z)^2 + \frac{1}{2}k_1(y_A - y_1')^2 + \frac{1}{2}k_2(y_B - y_2')^2 \\ &+ \frac{1}{2}k_{01}(y_{01} - z)^2.\end{aligned} \tag{8.118}$$

Let $y_{01} = (y_{01}') + z$; $y_0 = ay_1 + by_2$; $\varphi = c(y_2 - y_1)$; $y_{01} = a_1y_1' + b_1y_2'$; $\varphi_1 = c_1(y_2' - y_1')$;

$$\begin{aligned}&y_A = y_0 - l_1'\varphi; \quad y_B = y_0 + l_2'\varphi; \quad a = \frac{L_2}{L_1 + L_2}; \quad b = \frac{L_1}{L_1 + L_2}; \\ &c = \frac{1}{L_1 + L_2}; \\ &a_1 = \frac{l_2}{l_1 + l_2}; \quad b_1 = \frac{l_1}{l_1 + l_2}; \quad c_1 = \frac{1}{l_1 + l_2}.\end{aligned} \tag{8.119}$$

We introduce $y_1' = \frac{y_{01} - b_1y_2'}{a_1}$; $y_2' = \frac{y_{01} - a_1y_1'}{b_1}$, and also $y_{01} = a_1y_1' + b_1y_2'$, $y_1' = \frac{y_{01} - b_1y_2'}{a_1} = a_0y_{01} - iy_2'$, where $i = \frac{l_2}{l_1}$; $\frac{1}{b_1} = a_0$. Express φ_1 through variable y_0' $\varphi_1 = c_1(y_2' - \frac{y_{01}}{a_1} + \frac{b_1y_2'}{a_1})$,
where

$$y_1 = y_0 - l_1\varphi; \quad y_2 = y_0 + l_2\varphi. \tag{8.120}$$

Determine $y_1' = a_0y_{01} - iy_2'$, then

$$\varphi = c_1(y_2' - a_0y_{01} + iy_2') = c_1[(i+1)y_2' - a_0y_{01}], \tag{8.121}$$

and find the kinetic energy

$$\begin{aligned}\mathrm{T} = &\frac{1}{2}M\dot{y}_0^2 + \frac{1}{2}Jc^2(\dot{y}_2 - \dot{y}_1)^2 + \frac{1}{2}m_1[a_0\dot{y}_{01} - i(\dot{y}_2')^2]^2 \\ &+ \frac{1}{2}m_2(\dot{y}_2')^2 + \frac{1}{2}J_1c_1^2[\dot{y}_2' - a_0\dot{y}_{01}],\end{aligned} \tag{8.122}$$

where $i = \frac{l_2}{l_1}$, $a_0 = \frac{1}{b_1}$.

We introduce a series of relations

$$a_2 = a + l_1'c; \quad b_2 = b - l_1'c; \quad a_3 = a - l_2'c; \quad b_3 = b + l_2'c; \tag{8.123}$$

in the final form of the equation of motion of the system will take the form

$$\begin{aligned}&\ddot{y}_1(Ma^2+Jc^2)+y_1(k_1'+k_1a_2^2+k_2a_3^2)+\ddot{y}_2(Mab-Jc^2)\\&\quad+y_2(k_1a_2b_2+k_1a_2i+k_2a_3b_3)+y(k_1a_2a_0)+y_2'(-k_2a_3)=k_1'z;\end{aligned} \tag{8.124}$$

$$\begin{aligned}&\ddot{y}_1(Mab-Jc^2)+y_1(k_1a_2b_2+k_2a_3b_3)+\ddot{y}_2(Mb^2+Jc^2)\\&\quad+y_2(k_2'+k_2b_2^2+k_2b_3^2)+y_{01}(-k_2a_0b_2)+y_2'(k_1b_2i-k_2b_3)=k_2'z;\end{aligned} \tag{8.125}$$

$$\begin{aligned}&\ddot{y}_1(ma_0^2+Jc^2a_0^2)+y_1(-k_1a_2a_0)+y_2(-k_1a_0b_2)+\ddot{y}_{01}(m_1a_0^2+Jc_1^2a_0)\\&\quad+y_{01}(k_1a_0^2+k_{01})+\ddot{y}_2'[-m_1a_0i-Jc^2(1+i)]+y_2'(-k_1a_0i)=k_{01}z;\end{aligned} \tag{8.126}$$

$$\begin{aligned}&y_1(k_1a_2i-k_2a_3)+y_2(k_1b_2i-k_2b_3)+\ddot{y}_{01}[-m_1i-Jc_1a_0(1+i)]\\&\quad-y_{01}(-k_1a_0i)+\ddot{y}_2'[m_1i^2+Jc_1^2(1+i)^2]+y_2'(k_1i^2+k_2)=0.\end{aligned} \tag{8.127}$$

Table 8.3 presents the coefficients of Eqs. (8.124)–(8.127).

Using the coordinates of the Table 8.3, it is possible to construct a system of transfer functions with respect to any of the y_1, y_2, y_{01}, y_2' coordinates of the perturbation of the base; we apply the Kramer rule for this purpose. In this case, you can obtain the necessary expressions to construct amplitude-frequency characteristics, however, in this case, some dynamic properties can be also determined from the matrix of coefficients. In particular, it can be concluded that the system has four frequencies of its natural oscillations, and therefore four possible resonance peaks on the amplitude-frequency characteristics. Between the y_1, y_2, y_{01}, y_2' coordinates, intercoordinate constraints at certain frequencies may be "zeroed", which suggests the possibility of two modes of dynamic absorbing. Given the fact that the generalized forces act in three y_1, y_2, y_{01} coordinates, we can expect a more complex picture of the interdependence of motions between partial systems. However, these comments are only preliminary, since the subsequent transformation of generalized coordinates implies the introduction of joints that reduce the total number of degrees of freedom.

To account for possible joints at points A and O_1, it is necessary to introduce corresponding coordinates reflecting relative motion, which in subsequent actions could be considered as a vanishingly small. Assuming that $y_2''=y_A-y_2'$, and $y_{01}=y_{01}'+z$, we rewrite the expression for the kinetic energy in the form

$$\begin{aligned}\mathrm{T}=&\frac{1}{2}M(a\dot{y}_1+b\dot{y}_2)^2+\frac{1}{2}Jc(\dot{y}_2-\dot{y}_1)^2+\frac{1}{2}m_1[a_0\dot{y}_{01}-i(\dot{y}_2')]^2\\&+\frac{1}{2}m_2(\dot{y}_2')^2+\frac{1}{2}J_1c_1[\dot{y}_2'(1+i)-a_0\dot{y}_{01}]^2.\end{aligned} \tag{8.128}$$

The potential energy of the system in this case will take the form

Table 8.3 Coefficients of equations in the y_1, y_2, y_{01}, y_2' coordinate system

a_{11}	a_{12}	a_{13}	a_{14}
$(Ma^2 + Jc^2)p^2 + k_1' + k_1 a_2^2 + k_2 a_3^2$	$(Mab - Jc^2)p^2 + k_1 a_2 b_2 + k_1 a_2 i + k_2 a_3 b_3$	$-k_1 a_0 a_2$	$-k_2 a_3$
a_{21}	a_{22}	a_{23}	a_{24}
$(Mab - Jc^2)p^2 + k_1 a_2 b_2 + k_2 a_3 b_3$	$(Mb^2 + Jc^2)p^2 + k_2' + k_2 b_2^2 + k_2 b_3^2$	$-k_2 a_0 b_2$	$k_1 b_2 i - k_2 b_3$
a_{31}	a_{32}	a_{33}	a_{34}
$-k_1 a_2 a_0$	$-k_1 a_0 b_2$	$(m_1 a_0^2 + Jc_1^2 a_0)p^2 + k_1 a_0^2 + k_{01}$	$-[m_1 a_0 i + Jc^2(1+i)]p^2 - k_1 a_0 i$
a_{41}	a_{42}	a_{43}	a_{44}
$-k_2 a_3$	$+k_1 b_2 i - k_2 b_3$	$-[m_1 i + Jc_1 a_0(1+i)] - k_1 a_0 i$	$[m_1 i^2 + Jc_1^2(1+i)^2]p^2 + k_1 i^2 + k_2$
Q_1	Q_2	Q_3	Q_4
$Q_1 = k_1' z$	$Q_2 = k_2' z$	$Q_3 = k_{01} z$	$Q_4 = 0$

Note Q_1–Q_4 are generalized forces on the coordinates y_1, y_2, y_{01}, y_2'

$$\Pi = \frac{1}{2}k_1'(y_1 - z)^2 + \frac{1}{2}k_2'(y_2 - z)^2 + \frac{1}{2}k_1(y_A - y_1')^2 + \frac{1}{2}k_2(y_B - y_2')^2 + \frac{1}{2}k_{01}(y_{01} - z)^2. \tag{8.129}$$

Let us turn to the coordinate system in which the expressions for the kinetic and potential energy can be written as

$$\mathrm{T} = \frac{1}{2}m(\dot{y}_1 a + \dot{y}_2 b)^2 + \frac{1}{2}Jc^2(\dot{y}_2 - \dot{y}_1)^2 + \frac{1}{2}m_1(\dot{y}_1')^2 + \frac{1}{2}m_2(\dot{y}_2')^2 - \frac{1}{2}J_1c_1^2(\dot{y}_2' - \dot{y}_1')^2; \tag{8.130}$$

$$\Pi = \frac{1}{2}k_1'(y_1 - z)^2 + \frac{1}{2}k_2'(y_2 - z)^2 + \frac{1}{2}k_1(y_1'')^2 + \frac{1}{2}k_2(y_2'')^2 + \frac{1}{2}k_{01}(y_{01}'')^2; \tag{8.131}$$

where $y_{01} = y - z$; $y_1'' = y_A - y'$; $y_2'' = y_B - y_2'$; $y_A = y_0 - l_1'\varphi$; $y_1' = y_{01} - l_1\varphi_1$; $y_A = a_2 y_1 + b_2 y_2'$; $y_1' = y_{01} - l_1\varphi_1$; $y_B = y_0 + l_2'\varphi$; $y_2' = y_{01} + l_2\varphi_1$; $y_B = a_2' y_1 + b_2' y_2$; $y_2' = y_{01} + l_2\varphi$; $y_{01} = y_{01}' + z$; $y_1 = y_0 - l_1\varphi$; $y_2 = y_0 + l_2\varphi$.

We introduce a coordinate y_2'', (let $y_B - y_2' = y_2''$), and denote y_{01}' by relations $y_{01} = y_{01}' + z$, $y_B = a_3 y_1 + b_3 y_2$. We continue the further construction of the mathematical model of the suspension in the coordinate system. This coordinate system allows you to build a model of a system having a joint at the point O, i.e. in the center of gravity, which further simplifies the analysis of dynamic interactions. Let us write the initial data for the expressions of kinetic and potential energy. We turn to the y_1, y_2, y_0, y_2' coordinate system with the intention to introduce $y_2' = y_B - y_2''$. Then, taking $y_2'' = y_B - y_2' \to 0$, we can proceed to a simplified computational scheme of the suspension. We define the terms of the expression for the kinetic energy:

$$\mathrm{T} = \mathrm{T}_1 + \mathrm{T}_2 + \mathrm{T}_3 + \mathrm{T}_4 + \mathrm{T}_5, \tag{8.132}$$

where

$$\begin{aligned}
\mathrm{T}_1 &= \frac{1}{2}M(a\dot{y}_1 + b\dot{y}_2)^2 = \frac{1}{2}M_0\dot{y}_0^2; \quad \mathrm{T}_2 = \frac{1}{2}Jc^2(\dot{y}_2 - \dot{y}_1)^2;\\
\mathrm{T}_3 &= \frac{1}{2}m_1(a_0\dot{y}_{01} - i\dot{y}_1 a_3 - i\dot{y}_2 b_3 - i\dot{y}_2'')^2;\\
\mathrm{T}_4 &= \frac{1}{2}m_2\dot{y}_2 = \frac{1}{2}m_2(a_3\dot{y}_1 + b_3\dot{y}_2 - \dot{y}_2''); \quad \mathrm{T}_5 = \frac{1}{2}Jc_1^2[(i+1)^2 a_3^2\dot{y}_1^2\\
&+ 2(1+i)^2 a_3 b_3 \dot{y}_1\dot{y}_2 + (1+i)^2 b_3^2\dot{y}_2^2 + (i+1)^2(\dot{y}_2'')^2\\
&+ 2(i+1)a_0\dot{y}_2''\dot{y}_{01} + a_0^2\dot{y}_{01}^2 - 2(i+1)^2 a_3\dot{y}_1\dot{y}_2'' - 2(i+1)^2 b_3\dot{y}_2''\\
&- 2a_0 a_3(1+i)\dot{y}_1\dot{y}_{01} - 2a_0(1+i)b_3\dot{y}_2\dot{y}_{01}].
\end{aligned}$$

Let us accept $a_3(1+i) = a_6$, $b_3(1+i) = b_6$. In a similar way, we extend the expressions for the potential energy:

$$\Pi = \Pi_1 + \Pi_2 + \Pi_3 + \Pi_4 + \Pi_5, \tag{8.133}$$

where

$$\Pi_1 = \frac{1}{2}k_1'(y_1 - z)^2; \quad \Pi_2 = \frac{1}{2}k_2'(y_2 - z)^2; \quad \Pi_3 = \frac{1}{2}k_1[y_1^2 a_7^2 + 2a_7 b_7 y_1 y_2 + y_2^2 b_7^2 \\ + y_{01}^2 a_0^2 + 2a_0 i y_{01} y_2'' + i^2(\dot{y}_2'')^2 - a_7 y_1 a_0 y_{01} - y_2 b_7 a_0 y_{01} - y_1 a_7 i y_2'' - y_2 b_7 i \dot{y}_2''].$$

$\Pi_4 = \frac{1}{2}k_2(\dot{y}_2'')^2; \Pi_5 = \frac{1}{2}k_{01}(y_{01} - z)^2$, wherein $a_7 = a_2 + ia_3$. $b_7 = b + ib_3$, The equations of motion of the system in this case will take the form

$$\begin{aligned} &\ddot{y}_1(Ma^2 + Jc^2 + m_1 i^2 a_3^2 + m_2 a_3^2 + J_1 c_1^2 a_6^2) + y_1(k_1' + k_1 a_7^2) \\ &\quad + \ddot{y}_2(Mab - Jc^2 + m_1 i^2 a_3 b_3 + m_2 a_3 b_3 - m_2 a_3 + J_1 c_1^2 a_6 b_6) \\ &\quad + y_2(k_1 a_7 b_7) + \ddot{y}_{01}(-m_1 a_0 a_3 - J_1 c_1^2 a_0 a_6) + y_{01}(-k_1 a_7 a_0) \\ &\quad + \ddot{y}_2(-m_1 i^2 a_3 - m_2 a_3 - J_1 c_1^2 a_6(1 + i)) + \ddot{y}_2(-k_1 a_7 i) = k_1' z; \end{aligned} \tag{8.134}$$

$$\begin{aligned} &\ddot{y}_1(Mab - Jc^2 + m_1 i^2 a_3 b_3 + m_2 a_3 b_3 + J_1 c_1^2 a_6 b_6) + y_1(k_1 a_7 b_7) \\ &\quad + \ddot{y}_2(Mb^2 + Jc^2 + m_1 i^2 b_3^2 + m_2 b_3^2 + J_1 c_1^2 b_6^2) \\ &\quad + y_2(k_2' + k_1 b_7^2) + \ddot{y}_{01}[-m_1 a_0 i b_3 - J_1 c_1^2 a_0 b_6) + y_{01}(-k_1 b_7 a_0) \\ &\quad + \ddot{y}_2''(-m_1 i^2 b_3 - J_1 c_1^2 b_6(1 + i)] + y_2(-k_7 a_7 i) = k_2' z; \end{aligned} \tag{8.135}$$

$$\begin{aligned} &\ddot{y}_1(-m_1 a_3 a_0 - J_1 c_1^2 a_0 a_6) + y_1(-k_1 a_7 a_0) + \ddot{y}_2(-m_1 a_0 b_3 i - J_1 c_1^2 a_0 b_6) \\ &\quad + \ddot{y}_2(-k_1 b_7 a_0) + \ddot{y}_{01}(m_1 a_0^2) + y_{01}(k_1 a_0^2 + k_{01}) \\ &\quad + \ddot{y}_2(m_1 i a_0 \; + J_1 c_1^2(1 + i) a_0) + \ddot{y}_2(k_1 a_0 i) = k_{01} z; \end{aligned} \tag{8.136}$$

$$\begin{aligned} &\ddot{y}_1[-m_1 i^2 a_3 - m_2 a_3 - J_1 c_1^2(1 + i) a_6] + y_1(-k_1 a_7 i) \\ &\quad + \ddot{y}_2[-m_1 i^2 b_3 - m_2 b_3 - J_1 c_1^2 b_6(1 + i)] + y_2(-k_1 b_7 i) \\ &\quad + \ddot{y}_{01}[m_1 i a_0 + J_1 c_1^2(1 + i) a_0] + y_{01}(k_1 a_0 i) \\ &\quad + \ddot{y}_2''[m_1 i^2 + m_2 + J_1 c_1^2(1 + i)] + y_2''(k_2 + k_1 i^2) = 0. \end{aligned} \tag{8.137}$$

Table 8.4 shows the corresponding values of the coefficients of the equations of the system.

Further research is to find the transfer functions $W_1(p) = \frac{\bar{y}_1(p)}{\bar{z}}$ and $W_2(p) = \frac{\bar{y}_2(p)}{\bar{z}}$, this can be done using Kramer's formulas. Note that a change in the system of generalized coordinates leads to a change in cross-couplings between partial systems. In this case, between all partial systems it is possible to "zero out" the constraints, which will create conditions for reducing dynamic interactions in the suspension.

Table 8.4 The coefficients of Eqs. (8.134) – (8.137) in the y_1, y_2, y_0, y_2'' coordinate system

a_{11}	a_{12}	a_{13}	a_{14}
$(Ma^2 + Jc^2 + m_1 i^2 a_3^2 + m_2 a_3^2 + J_1 c_1^2 a_6^2)p^2 + k_1' + k_1 a_7^2$	$(Mab - Jc^2 + m_1 i^2 a_3 b_3 + m_2 a_3 b_3 + J_1 c_1^2 a_6 b_6)p^2 + k_1 a_7 b_7$	$(-m_1 a_0 a_3 - J_1 c_1^2 a_0 a_6)p^2 - k_1 a_7 a_0$	$(-m_1 i^2 a_3 - m_2 a_3 - J_1 c_1^2 a_6 (a+1)]p^2 - k_1 a_7 i$
a_{21}	a_{22}	a_{23}	a_{24}
$(Mab - Jc^2 + m_1 i^2 a_3 b_3 + m_2 a_3 b_3 + J_1 c_1^2 a_6 b_6)p^2 + k_1 a_7 b_7$	$(Mb^2 + Jc^2 + m_1 i^2 b_3^2 + J_1 c_1^2 b_6^2)p^2 + k_2' + k_1 b_7^2$	$(-m_1 a_0 i b_3 - J_1 c_1^2 a_0 b_6)p^2 - k_1 b_7 a_0$	$(-m_1 i^2 b_3 - m_2 b_3 - J_1 c_1^2 b_6 \times (1+i))p^2 + (-k_7 a_7 i)$
a_{31}	a_{32}	a_{33}	a_{34}
$(-m_1 a_3 a_0 - J_1 c_1^2 a_0 a_6)p^2 - k_1 a_7 a_0$	$(-m_1 a_0 b_3 i - J_1 c_1^2 a_0 b_6)p^2 - k_1 b_7 a_0$	$(m_1 a_0)^2 p^2 + k_1 a_0^2 + k_{01}$	$(m_1 i a_0 + J_1 c_1^2 \times (i+1) a_0)p^2 + k_1 a_0 i$
a_{41}	a_{42}	a_{43}	a_{44}
$[-m_1 i^2 a_3 - m_2 a_3 - - J_1 c_1^2 (1+i) a_6]p^2 - k_1 a_7 i$	$[-m_1 i^2 b_3 - m_2 b_3 - J_1 c_1^2 b_6 (1+i)]p^2 - k_1 b_7 i$	$[m_1 i a_0 + J_1 c_1^2 (1+i) a_0]p^2 + k_1 a_0 i$	$[m_1 i^2 + m_2 + J_1 c_1^2 \times (1+i)]p^2 + k_2 + k_1 i^2)$
Q_1	Q_2	Q_3	Q_4
$Q_{y_1} = k_1' z$	$Q_{y_2} = k_2' z$	$Q_{y_{01}} = k_{01} z$	$Q_{y_2''} = 0$

Note Q_1–Q_4 are generalized forces

Since a system with four degrees of freedom is used and there are no joints with the base in the system, then there are no peculiarities in the frequency characteristics, compared with the traditional forms of the systems. However, the mathematical model gives an idea of the possible spectrum of the dynamic properties of the suspension in a wide frequency range (dynamic absorbing modes, natural oscillation frequencies). Further research can be continued in the coordinate system, the choice of which implies the possibility of the formation of joints of solid bodies.

8.4.3 The Influence of the Choice of Coordinate Systems

Let us proceed to the $(y_{01} = y'_{01} + z)$, $y''_2(y_B - y'_2)$. coordinate system. In this case

$$\mathrm{T} = \mathrm{T}_1 + \mathrm{T}_2 + \mathrm{T}_3 + \mathrm{T}_4 + \mathrm{T}_5, \tag{8.138}$$

where

$$T_1 = \frac{1}{2} M \dot{y}_0^2 = \frac{1}{2} M (a\dot{y}_1 + b\dot{y}_2)^2; \quad T_2 = \frac{1}{2} Jc^2 (\dot{y}_2 - \dot{y}_1)^2;$$

$$\mathrm{T}_3 = \frac{1}{2} m_1 \left[(a_0 \dot{y}'_{01} - i\dot{y}_1 a_3 - i\dot{y}_2 b_3 + i\dot{y}''_2) + a_0 \dot{z} \right]^2;$$

$$T_4 = \frac{1}{2} m_2 \dot{y}_2^2 = \frac{1}{2} m_2 (a_3 \dot{y}_1 + b_3 \dot{y}_2 - \dot{y}''_2)^2;$$

$$\mathrm{T}_5 = \frac{1}{2} J_1 c_1^2 \left[(1+i) a_3 \dot{y}_1 + (1+i) b_3 \dot{y}_3 - (i+1) \dot{y}''_2 - a_0 \dot{y}_{01} \right]^2;$$

$$\mathrm{T}_5 = \frac{1}{2} J_1 c_1 (a_6 \dot{y}_1 + b_6 \dot{y}_2 - (1+i) \dot{y}''_2 - a_0 \dot{y}'_{01} - a_0 \dot{z})^2.$$

In turn, the potential energy of the system has the form

$$\Pi = \Pi_1 + \Pi_2 + \Pi_3 + \Pi_4 + \Pi_5, \tag{8.139}$$

where

$$\Pi_1 = \frac{1}{2} k'_1 (y_1 - z)^2; \Pi_2 = \frac{1}{2} k'_2 (y_2 - z)^2; \Pi_3 = \frac{1}{2} k_1 \left[y_1 a_7 + y_2 b_7 - i y''_2 - a_0 y_{01} \right]^2; \Pi_4 = \frac{1}{2} k_2 (y''_2)^2;$$

$$\Pi_5 = k_{01} (y'_{01})^2,$$

wherein $a_7 = a_2 + ia_3, b_7 = b_2 + ib_3$.

The given diagram of the derivation of equations provides an idea of the effect on the level of the mathematical model of the parameter values of the constituent elements of the suspension. With that, the details of the formation of joints are known. The equations of motion (more precisely, the coefficients of the equations of motion) and the generalized forces are given in Table 8.5.

Table 8.5 The values of the coefficients of equations in the y_1, y_2, y'_{01}, y''_2 coordinate system

a_{11}	a_{12}	a_{13}	a_{14}
$(Ma^2 + Jc^2 + m_1 i^2 a_3^2 + m_2 a_3^2 + J_1 c_1^2 a_6^2)p^2 + k'_1 + k_{01} a_7^2$	$(Mab - Jc^2 + m_2 a_3 b_3 + m_1 i^2 a_3 b_3 + J_1 c_1^2 a_6 b_6)p^2 + k_{01} a_7 b_7$	$(-m_1 a_0 a_3 i - J_1 c_1^2 a_0 a_6)p^2 - k_{01} a_7 a_0$	$(-J_1 c_1^2 a_6 (1+i) - m_1 i^2 a_3 - m_2 a_3)p^2 - k_1 a_7 i$
a_{21}	a_{22}	a_{23}	a_{24}
$Mab - Jc^2 + m_1 i^2 a_3 b_3 + m_2 a_3 b_3 + J_1 c_1^2 a_6 b_6 + k_{01} a_7 b_7$	$(Mb^2 + Jc^2 + m_1 i^2 b_3^2 + m_2 b_3^2 + J_1 c_1^2 b_6^2) + k'_2 + k_1 b_7^2$	$(J_1 c_1^2 b_6 a_0 - m_1 i a_0 b_3)p^2 - k_1 b_7 a_0$	$(-m_1 i^2 b_3 - m_2 b_3 - J_1 c_1^2 (1+i) b_6)p^2 - k_1 b_7 i)$
a_{31}	a_{32}	a_{33}	a_{34}
$(-m_1 a_0 i a_3 - J_1 c_1^2 a_6 a_0)p^2 - k_1 a_7 a_0$	$(-m_1 i a_0 b_3 + J_1 c_1^2 a_0 b_6)p^2 - k_{01} b_7 a_0$	$(m_1 a_0^2 + J_1 c_1^2 a_0^2)p^2 + k_1 a_0^2 + k_{01}$	$[J_1 c_1^2 (i+1) a_0 + m_1 i a_0]p^2 + k_1 a_0 i$
a_{41}	a_{42}	a_{43}	a_{44}
$[-m_1 i^2 a_3 - m_2 a_3 - J_1 c_1^2 a_6 (1+i)]p^2 - k_1 a_7 i$	$[-m_1 i^2 b_3 - J_1 c_1^2 b_6 (1+i) - m_2 b_3]p^2 - k_1 b_7 i$	$[J_1 c_1^2 (1+i) a_0 + k_1 i a_0 + m_1 a_0 i]p^2 + k_{01} a_0 i$	$[m_1 i^2 + J_1 c_1^2 (1+i)^2]p^2 + k_2$
Q_1	Q_2	Q_3	Q_4
$Q_{1_1} = m_1 a_0 a_3 z + J_1 c_1^2 a_0 a_6 z + k'_1 z + k_1 a_0 a_7$	$Q_2 = m_1 a_0 i b_3 z + J_1 c_1^2 a_0 b_6 z + z k'_2 + k_1 a_0 b_7 z$	$Q_3 = m_1 a_0 z + J_1 c_1^2 a_0^2 z$	$Q_4 = a_0 i z + J_1 c_1^2 a_0 (1+i) z$

Note Q_1–Q_4 are generalized forces

From Table 8.5 it follows that the influence of the inertial forces of moving space when choosing the corresponding coordinates is reflected through the structure of generalized forces. It can be seen that the generalized forces Q_1–Q_4 contain the members formed by the motion of the base, while inertial forces are meant. The system can be simplified by introducing joints by the coordinates y'_{01}, y''_2. To do this, you need the initial system of equations (Table 8.5) to appropriately convert.

Since the kinetic energy does not change, we write the general expressions for the potential energy in the y_1, y_2, y'_{01}, y''_2 coordinate system. Accept that

$$\Pi = \Pi_1 + \Pi_2 + \Pi_3 + \Pi_4 + \Pi_5,$$

where $\Pi_1 = \frac{1}{2}k'_1(y_1 - z)^2$; $\Pi_2 = \frac{1}{2}k'_2(y_2 - z)^2$; $\Pi_3 = \frac{1}{2}k_1[y_1a_7 + y_2b_7 - iy''_2 - a_0y'_{01} - a_0z]^2$; $\Pi_4 = k_2(y''_2)^2$; $\Pi_5 = k_{01}(y'_{01})^2$.

The equations we need to consider the dynamics of the lever suspension, i.e. the VPS, in which joints are introduced through y'_{01} and y''_2, have the following form:

$$\begin{aligned}&\ddot{y}_1(Ma^2 + Jc^2 + m_1i^2a_3^2 + m_2a_3^2 + J_1c_1^2a_6^2) + y_1(k'_1 + k_1a_7^2) + \ddot{y}_2(Mab - Jc^2 + m_1i^2a_3b_3\\ &\quad + m_2a_3b_3 + + J_1c_1a_6b_6) + y_2(k_1a_7b_7) = \ddot{z}(m_1a_0a_3 + J_1c_1^2a_0a_6) + (k'_1 + k_1a_0a_7);\end{aligned} \tag{8.140}$$

$$\begin{aligned}&\ddot{y}_1(Mab - Jc^2 + m_1i^2a_3b_3 + m_2a_3b_3 + J_1c_1^2a_6b_6) + y_1(k_1a_7b_7)\\ &\quad + \ddot{y}_2(Mb^2 + Jc^2 + m_1i^2b_3^2 + m_2b_3^2 + + J_1c_1^2b_6^2)\\ &\quad + y_2(k'_2 + k_1b_7^2) = \ddot{z}(m_1a_0ib_3 + J_1c_1^2a_0b_6) + k_2 + k_1a_0b_7.\end{aligned} \tag{8.141}$$

The computational scheme of a suspension with two joints is a system with two degrees of freedom; it can be written that

$$\begin{aligned}a_{11} &= (Ma^2 + Jc^2 + m_1i^2a_3^2 + m_2a_3^2 + J_1c_1^2a_6^2)p^2 + k'_1 + k_1a_7^2;\\ a_{12} = a_{21} &= (Mab - Jc^2 + m_1i^2a_3b_3 + m_2a_3b_3 + J_1c_1^2a_6b_6)p^2 + k'_1 + k_1a_0a_7;\\ a_{22} &= (Mb^2 + Jc^2 + m_1i^2b_3^2 + m_2b_3^2 + J_1c_1^2b_6^2)p^2 + k'_2 + k_1b_7^2.\end{aligned} \tag{8.142}$$

The values of the generalized forces are determined by the right-hand sides (8.141) and (8.142). Now you can find the corresponding transfer functions: $W_1(p) = \frac{\bar{y}_1(p)}{\bar{z}} = W_2(p) = \frac{\bar{y}_2(p)}{\bar{z}}$. In general, the relationship between the coordinates and the generalized forces is determined by the formulas:

$$\bar{y}_1 = \frac{\overline{Q}_1a_{22} - \overline{Q}_2a_{12}}{a_{11}a_{22} - a_{12}^2}, \tag{8.143}$$

$$\bar{y}_2 = \frac{-\bar{Q}_1 a_{12} - \bar{Q}_2 a_{11}}{a_{11}a_{22} - a_{12}^2}. \quad (8.144)$$

Thus, the mathematical model of the suspension can be obtained from the matrix, the elements of which are presented in Table 8.5, by eliminating two columns a_{i3}, a_{i4} and two rows of the matrix $3i,\ 4i\ (i = \overline{i,3})$. The system of equations (8.141) and (8.142) is determined by the remaining 2×2 matrix.

Mathematical models of transportation vehicle suspensions include a sufficiently large number of elements, including lever linkages forming joints. The proposed approach to the construction of mathematical models is based on the choice of systems of generalized coordinates, within the framework of which the focus is on the coordinates of relative motions. With an appropriate choice of the coordinates of the relative motion, their "zeroing" is possible, which leads to a decrease in the number of degrees of freedom of motion of the system, but it allows obtaining a mathematical model of a system with joints. In this case, a corresponding transformation of the generalized forces takes place, taking into account the equality of virtual works on possible displacements. Subsequent actions to evaluate the dynamic properties are associated with the use of the transfer functions of the system in the problems of dynamic synthesis. It seems rational to search for parametric spaces in which, at certain frequencies, a dynamical damping would be simultaneously possible in several or all coordinates of the system.

8.5 Features of the Dynamic Interactions of Elements in Transportation Vehicle Suspension Schemes

Methodological aspects of the construction of mathematical models of mechanical oscillatory systems, incorporating various mechanisms, whose mass-and-inertia and elastic properties can be adjusted in advance or in motion, are developed.

Consider a system comprising an object of protection (*M*), which relies on a certain supporting block consisting of double articulation linkages, in the form of two-arm Assur groups with rotational pairs at A_1, A_2; B_1, B_2; C_1, C_2; D_1, D_2. The computational scheme of the system is shown in Fig. 8.21a; the object of protection executes vertical oscillations, which are initiated by the action of force and kinematic perturbations, respectively, $Q(t)$ and $z(t)$.

External influences are represented by harmonic functions, which implies the further construction of structural mathematical models in the form of dynamically equivalent automatic control systems and the determination of transfer functions [17]. It is assumed that the oscillatory motions are small, and the resistance forces are small. It is assumed that the oscillatory motions are small, and the resistance forces are small. In this approach, the linear model can be used to preliminarily evaluate the dynamic capabilities of constructive and engineering forms of

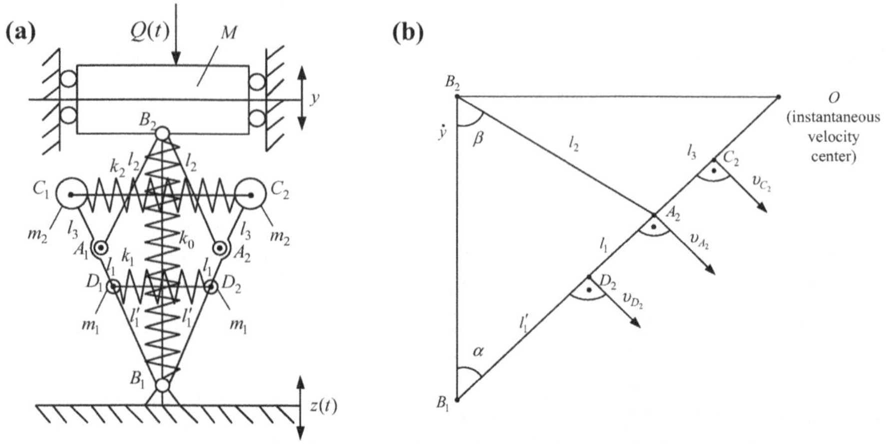

Fig. 8.21 The computational scheme (**a**) and kinematic diagram (**b**) of the supporting device

interaction of elements, as well as calculations related to the evaluation of dynamic responses in the hinges.

At points C_1, C_2 and D_1, D_2, additional masses $m_1(D_1, D_2)$ and $m_2(C_1, C_2)$ are fixed, the position of which can synchronously change. In Fig. 8.21a the following notation is adopted: $B_1A_1 = B_1A_2 = l_1$; $B_2A_1 = B_2A_2 = l_2$; $B_1C_1 = B_1C_2 = l_3$. In turn, the points D_1 and D_2 are connected by a spring with stiffness k_1, the points C_1 and C_2 are connected by the spring k_2. The protection object (M) rests on a spring with stiffness k_0. In addition, the symbols $A_1C_1 = A_2C_2 = l_3$ are introduced, and the angles α and β determine the inclinations of the rods l_1 and l_2 relative to the vertical. The instantaneous velocity center of the B_2A_2 unit at the speed of the object $\dot{y}$ and at $Q(t) \neq 0$ and $z(t) = 0$ is shown by the point O in Fig. 8.21b.

The speed of point A_2 at $z = 0$ (i.e. with a fixed bearing surface) is determined:

$$\upsilon_{A2} = \frac{\dot{y}}{B_2O} A_2O. \tag{8.145}$$

$$B_1B_2 = l_1\cos(\alpha) + l_2\cos(\beta); \quad B_1B_2 = B_1O\cos(\alpha); \quad B_1O = \frac{B_1B_2}{\cos(\alpha)};$$

$$A_2O = B_1O - l_1 = \frac{l_1\cos(\alpha) + l_2\cos(\beta)}{\cos(\alpha)} - l_1 = \frac{l_2\cos(\beta)}{\cos(\alpha)}; \frac{B_2O}{B_1B_2} \tag{8.146}$$

$$= \mathrm{tg}(\alpha), B_2O = B_1B_2\mathrm{tg}(\alpha).$$

Using (8.145) and (8.146), we find:

$$\upsilon_{A2} = \dot{y}\frac{l_2\cos(\beta)}{\cos(\alpha)(l_1\cos(\alpha) + l_2\cos(\beta))\mathrm{tg}(\alpha)}$$

or

$$\upsilon_{A2} = \dot{y}\frac{l_2\cos(\beta)}{(l_1\cos(\alpha) + l_2\cos(\beta))\sin(\alpha)}. \quad (8.147)$$

We introduce the concept of the transfer ratio $i = l_1/l_2$, then

$$\upsilon_{A2} = \dot{y}\frac{i\cos(\beta)}{(\cos(\alpha) + i\cos(\beta))\sin(\alpha)}. \quad (8.148)$$

The absolute velocity vector under force perturbation of p. $A_2 \perp B_1O$, since $z(t) = 0$. With a kinematic external disturbance, when $z(t) \neq 0$ and $Q(t) = 0$ there will be another distribution of the velocities of the elements of the system, which is physically explained by the formation of another scheme for the transmission of dynamic interactions. In this regard, it is problematic to take into account the joint action of several external factors of influence, although the superposition principle can also be used when considering linear systems.

8.5.1 Features of Power External Disturbance of the System

If the concept of the center of instantaneous velocities is used to determine the speeds of motion of the points of the elements, then the kinematic scheme in Fig. 8.21b can be taken as a basis for determining the velocities of the points D_2 and C_2.

The speed of the point D_2 will be perpendicular to the unit B_1D_2, then

$$\upsilon_{D2} = \frac{\upsilon_{A2}l_1'}{l_1} = \frac{\dot{y}ii_1\cos(\beta)}{\sin(\alpha)(\cos(\alpha) + i\cos(\beta))}, \quad (8.149)$$

where i_1 is the transfer ratio determined by $i_1 = B_1D_2/B_1A_2 = l_1'/l_1$.

The transfer ratio i_1 (by definition) will be less than unity, since $B_1D_2 < B_1A_2$.

In turn, the absolute velocity of the point C_2 can be found (at $z = 0$). The speed of the point C_2 will be $\perp B_1C_2$:

$$\upsilon_{C2} = \frac{\dot{y}i\cos(\beta)(l_1 + l_3)}{\sin(\alpha)(\cos(\alpha) + i\cos(\beta))l_1} = \frac{\dot{y}ii_2\cos(\beta)}{\sin(\alpha)(\cos(\alpha) + i\cos(\beta))}. \quad (8.150)$$

Here i_2 is the transfer ratio, which is defined by:

$$i_2 = \frac{l_1 + l_3}{l_1}. \quad (8.151)$$

Knowing the absolute velocities at points C_1, C_2, D_1 and D_2, taking into account the symmetry of the mechanism, we find the kinetic energy, assuming that the mass-and-inertia properties of the rods are of little importance:

$$\mathrm{T} = \frac{1}{2}M\dot{y}^2 + 2\frac{1}{2}m_1(\upsilon_{D\,1})^2 + 2\frac{1}{2}m_2(\upsilon_{C\,1})^2, \tag{8.152}$$

or in expanded form

$$\begin{aligned}\mathrm{T} &= \frac{1}{2}M\dot{y}^2 + m_1\left\{\frac{\dot{y}ii_1\cos(\beta)}{(\cos(\alpha)+i\cos(\beta))\sin(\alpha)}\right\}^2 + m_2\left\{\frac{\dot{y}ii_2\cos(\beta)}{(\cos(\alpha)+i\cos(\beta))\sin(\alpha)}\right\}^2\\ &= \frac{1}{2}\dot{y}^2\left\{M + \frac{2i\cos(\beta)(m_1i_1+m_2i_2)}{(\cos(\alpha)+i\cos(\beta))\sin(\alpha)}\right\}.\end{aligned} \tag{8.153}$$

We introduce the reduced mass

$$m_{\text{пр}} = \frac{2i\cos(\beta)(m_1i_1+m_2i_2)}{(\cos(\alpha)+i\cos(\beta))\sin(\alpha)}. \tag{8.154}$$

Find an expression for potential energy. If we assume that a deviation $\dot{y}$ will be implemented at p. B_2, then at p. A_2, the horizontal displacement will be

$$\delta_{A2} = \frac{yi\cos(\alpha)\cos(\beta)}{(\cos(\alpha)+i\cos(\beta))\sin(\alpha)} = \frac{yi\cos(\beta)}{tg(\alpha)(\cos(\alpha)+i\cos(\beta))}; \tag{8.155}$$

in turn for pp. D_2 and C_2:

$$\delta_{D2} = \delta_{A2}\frac{l_1'}{l_1} = \frac{yii_1\cos(\beta)}{tg(\alpha)(\cos(\alpha)+i\cos(\beta))}, \tag{8.156}$$

$$\delta_{C2} = \frac{yii_2\cos(\beta)}{tg(\alpha)(\cos(\alpha)+i\cos(\beta))}. \tag{8.157}$$

If oscillations are considered relative to the position of static equilibrium, then the potential energy will be determined

$$\Pi = \frac{1}{2}k_0\dot{y}^2 + \frac{1}{2}k_1\left\{\frac{yii_1\cos(\beta)}{tg(\alpha)(\cos(\alpha)+i\cos(\beta))}\right\}^2 + \frac{1}{2}k_2\left\{\frac{yii_2\cos(\beta)}{tg(\alpha)(\cos(\alpha)+i\cos(\beta))}\right\}^2 \tag{8.158}$$

or

$$\Pi = \frac{1}{2} y^2 \left\{ k_0 + \frac{(k_1 i_1^2 + k_2 i_2^2) i^2 (\cos(\beta))^2}{tg(\alpha)(\cos(\alpha) + i\cos(\beta))} \right\}. \tag{8.159}$$

Thus, the elastic system will be characterized by the reduced stiffness

$$k_{\text{red}} = k_0 + \frac{(k_1 i_1^2 + k_2 i_2^2) i^2 (\cos(\beta))^2}{tg(\alpha)(\cos(\alpha) + i\cos(\beta))}. \tag{8.160}$$

The original computational scheme (Fig. 8.21a) can be reduced to a generalized form, as shown in Fig. 8.22a–c. For the construction of structural diagrams used the method described in [17].

The differential equation of motion on the basis of the expressions (8.152), (8.154), (8.159) and (8.160) obtained above will be written in operator form

$$\bar{y}[(M + m_{\text{red}})p^2 + k_0 + k_{\text{red}}] = \overline{Q}, \tag{8.161}$$

where $p = j\omega$ is a complex variable, and the sign – (tilde) means the Laplace transform image.

The transfer function of the system under the force of external influence is

$$W(p) = \frac{\bar{y}}{\overline{Q}} = \frac{1}{[(M + m_{\text{пр}})p^2 + k_0 + k_{\text{пр}}]}, \tag{8.162}$$

where $\bar{y}$, $\overline{Q}$ correspond to the Laplace transformation of functions.

Figure 8.22a–c gives the computational scheme of the initial system and structural models-analogues of the differential Eq. (8.161).

The dynamic properties of the system in this case are determined by well-known relations. For example, the natural oscillation frequency is determined by the expression

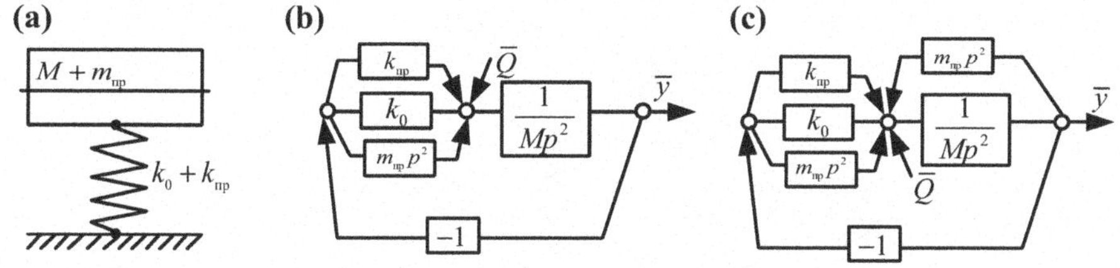

Fig. 8.22 Equivalent computational schemes and structural diagrams of the system: **a** a system with one degree of freedom of the traditional type; **b** a system with the allocation of the basic structure with the addition of the above parameters; **c** a structural diagram with a mechanism in the form of an additional negative relationship in absolute acceleration

$$\omega_{\mathrm{nat}}^2 = \frac{k_0 + k_{\mathrm{red}}}{M + m_{\mathrm{red}}}. \tag{8.163}$$

In this case, the adjustment capabilities of the system are of interest due to the appropriate choice of parameters m_1, m_2, i_1, i_2 and angles α and β, as well as additional springs k_1 and k_2.

8.5.2 Forced Oscillations of the System with External Kinematic Disturbance

The type of impact on the object of protection depends on whether there are mechanisms in the composition of the system that transform the motions of the constituent elements. In this case, the parameters of the mathematical model change, which requires taking into account a number of features. Figure 8.23 shows a schematic diagram of the kinematic relations.

To carry out kinematic calculations, an additional coordinate system x_0, y_0 is introduced in Fig. 8.23. As for the consideration of the motion of the object (*M*), its position is determined by the coordinate *y* in a fixed basis. In the case of kinematic perturbation, the elements of the mechanism participate in two motions, which are initiated by the vertical motion of pp. B_1 and B_2, forming a transportation motion (the designations $\upsilon'_{A2}, \upsilon'_{C2}, \upsilon'_{D2}$), as well as the relative motion caused by

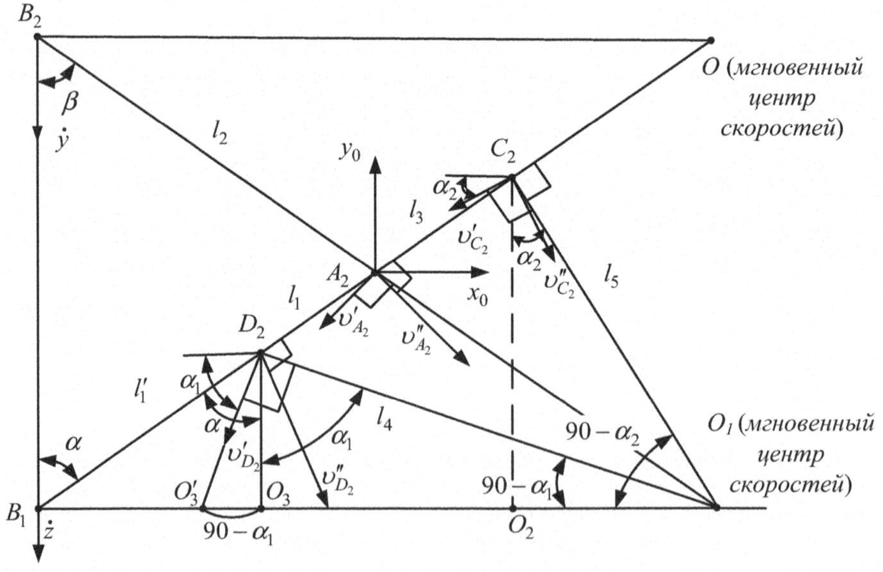

Fig. 8.23 The kinematic diagram of relations of motion parameters with kinematic perturbation

y—$\upsilon''_{A2}, \upsilon''_{C2}, \upsilon''_{D2}$. The corresponding components are distinguished when considering the displacements at pp. D_1, D_2; C_1, C_2. As for displacements, when determining the potential energy of the elastic elements k_1 and k_2, only horizontal components on the x_0 axis are taken into account. The components of the velocities in relative and figurative motions are determined in projections on the x_0 and y_0 axes with the subsequent determination of the velocities in absolute motion.

Using the approach presented above, we obtain that

$$\upsilon'_{A2} = \frac{\dot{z}\cos(\alpha)}{\sin(\beta)(\cos(\alpha) + i\cos(\beta))}. \tag{8.164}$$

Accordingly, for pp. D_2 and C_2 speeds can be determined if we find the angular velocity of rotation around p. O, then

$$\omega_{O1} = \frac{\upsilon'_{A2}}{A_2O_1}. \tag{8.165}$$

If $B_2O_1 = \frac{l_1\cos(\alpha) + l_2\cos(\beta)}{\cos(\beta)}$, then $A_2O_1 = B_2O_1 - l_2 = l_1\frac{\cos(\alpha)}{\cos(\beta)}$.

The angular velocity is thus determined:

$$\omega_{O1} = \frac{\dot{z}\cos(\alpha)}{\sin(\beta)(\cos(\alpha) + i\cos(\beta))l_1}. \tag{8.166}$$

Denote $O_1D_2 = l_4$, $O_1C_2 = l_5$. The values of l_4 and l_5 can be found from geometric constructions (Fig. 8.23), then

$$\upsilon'_{D2} = \omega_{O1}l_4 = \frac{l_4\dot{z}\cos(\alpha)}{l_1\sin(\beta)(\cos(\alpha) + i\cos(\beta))}, \tag{8.167}$$

respectively

$$\upsilon'_{C2} = \omega_{O1}l_5 = \frac{l_5\dot{z}\cos(\alpha)}{l_1\sin(\beta)(\cos(\alpha) + i\cos(\beta))}. \tag{8.168}$$

In determining the speeds of pp. C_2 and D_2 in relative motion use expressions (8.148)–(8.150), then

$$\upsilon''_{A2} = \frac{\dot{y}i\cos(\beta)}{\sin(\alpha)(l_1\cos(\alpha) + i\cos(\beta))}, \tag{8.169}$$

$$\upsilon''_{C2} = \frac{\dot{y}ii_2\cos(\beta)}{\sin(\alpha)(l_1\cos(\alpha) + i\cos(\beta))}, \tag{8.170}$$

$$\upsilon''_{D2} = \frac{\dot{y}ii_1\cos(\beta)}{\sin(\alpha)(l_1\cos(\alpha) + i\cos(\beta))}. \tag{8.171}$$

To determine the parameters in absolute motion, we use the auxiliary system of coordinates x_0, y_0:

$$\upsilon_{C2\,x0} = \upsilon''_{C2\,x0} + \upsilon'_{C2\,x0} = \dot{y}aii_2\cos(\alpha) - \dot{z}\frac{l_5}{l_1}a_1\cos(\alpha_2), \tag{8.172}$$

$$\upsilon_{C2\,y0} = -\dot{y}aii_2\sin(\alpha) - \dot{z}\frac{l_5}{l_1}a_1\sin(\alpha_2), \tag{8.173}$$

$$\upsilon_{D2\,x0} = \dot{y}aii_1\cos(\alpha) - \dot{z}\frac{l_4}{l_1}a_1\cos(\alpha_1), \tag{8.174}$$

$$\upsilon_{D2\,y0} = -\dot{y}aii_1\sin(\alpha) - \dot{z}\frac{l_4}{l_1}a_1\sin(\alpha_1), \tag{8.175}$$

where

$$a = \frac{\cos(\beta)}{\sin(\alpha)(\cos(\alpha) + i\cos(\beta))}, \tag{8.176}$$

$$a_1 = \frac{\cos(\alpha)}{\sin(\beta)(\cos(\alpha) + i\cos(\beta))}, \tag{8.177}$$

The kinetic energy of the system is determined by the expression

$$\begin{aligned} \mathrm{T} = {} & \frac{1}{2}M\dot{y}^2 + 2\frac{1}{2}m_1\left\{[\dot{y}aii_1\cos(\alpha) - \dot{z}i_3a_1\cos(\alpha_1)]^2\right. \\ & \left. + [-\dot{y}aii_1\sin(\alpha) - \dot{z}i_3a_1\sin(\alpha_1)]^2\right\} + 2\frac{1}{2}m_2\left\{[\dot{y}aii_2\cos(\alpha) - \dot{z}i_4a_1\cos(\alpha_2)]^2\right. \\ & \left. + [-\dot{y}aii_2\sin(\alpha) - \dot{z}i_4a_1\sin(\alpha_2)]^2\right\}, \end{aligned} \tag{8.178}$$

where $i_3 = l_4/l_1$, $i_4 = l_5/l_1$, the angles α_1 and α_2 are determined by the kinematic scheme Fig. 8.23.

Find an expression for potential energy

$$\begin{aligned} \Pi = {} & \frac{1}{2}k_0(y - z)^2 + 2\frac{1}{2}k_1[\dot{y}aii_1\cos(\alpha) - \dot{z}i_3a_1\cos(\alpha_1)]^2 \\ & + 2\frac{1}{2}k_2[-\dot{y}aii_2\cos(\alpha) - \dot{z}i_4a_1\cos(\alpha_2)]^2. \end{aligned} \tag{8.179}$$

Let us write the expression of the kinetic energy (8.178) in detail:

$$\begin{aligned}
\mathrm{T} = &\frac{1}{2}M\dot{y}^2 + m_1\left[a^2\dot{y}^2(ii_1)^2(\cos^2(\alpha) + \sin^2(\alpha)) + a_1^2\dot{z}^2 i_3^2(\cos^2(\alpha_1) + \sin^2(\alpha_1))\right] \\
&+ 2[aa_1\dot{y}\dot{z}ii_1\sin(\alpha)\sin(\alpha_1) - aa_1\dot{y}\dot{z}ii_1\cos(\alpha)\cos(\alpha_1)] \\
&+ m_2\left[a^2\dot{y}^2(ii_2)^2(\cos^2(\alpha) + \sin^2(\alpha)) + a_1^2\dot{z}^2 i_4^2(\cos^2(\alpha_2) + \sin^2(\alpha_2))\right] \\
&+ 2[aa_1\dot{y}\dot{z}ii_2(\sin(\alpha)\sin(\alpha_2) - \cos(\alpha)\cos(\alpha_2)],
\end{aligned} \tag{8.180}$$

Using the Lagrange equation of the second kind, we represent the expression in the Laplace transform images

$$\begin{aligned}
&\bar{y}\left[M + 2m_1a^2(ii_1)^2 + 2m_2a^2(ii_2)^2\right]p^2 + k_0 \\
&\quad + 2k_1[a^2(ii_1)^2\cos^2(\alpha)] + 2k_2[a^2(ii_2)^2\cos^2(\alpha)] \\
&= \bar{z}\cdot 2\cdot\left[\begin{array}{l} aa_1im_1i_1i_3(-\sin(\alpha)\sin(\alpha_1) + \cos(\alpha)\cos(\alpha_1)) \\ \quad - im_2i_2i_4(-\sin(\alpha)\sin(\alpha_2) + \cos(\alpha)\cos(\alpha_2)) \end{array}\right]p^2 \\
&\quad + k_0 + 2k_1aa_1ii_1i_3\cos(\alpha)\cos(\alpha_1) - 2k_2aa_1ii_1i_4\cos(\alpha)\cos(\alpha_2)
\end{aligned} \tag{8.181}$$

The transfer function of the system is

$$W(p) = \frac{\bar{y}}{\bar{z}} = \frac{2aa_1i[m_1i_1i_3r + m_2i_2i_4r_1]p^2 + (k_1i_1i_3\cos(\alpha)\cos(\alpha_1) -}{[M + 2a^2i^2(m_1i_1^2 + m_2i_2^2)]p^2 +}\cdots$$
$$\cdots\frac{-k_2i_2i_4\cos(\alpha)\cos(\alpha_2)) + k_0}{+2a^2\cos^2(\alpha)i^2(k_1i_1^2 + k_2i_2^2) + k_0}. \tag{8.182}$$

where

$$r = \cos(\alpha)\cos(\alpha_1) - \sin(\alpha)\sin(\alpha_1), \tag{8.183}$$

$$r_1 = \cos(\alpha)\cos(\alpha_2) - \sin(\alpha)\sin(\alpha_2). \tag{8.184}$$

Dynamic absorbing modes are possible in the system at frequency

$$\omega_{\text{dyn}}^2 = \frac{2aa_1i\cos(\alpha)[k_2i_2i_4\cos(\alpha_2) - k_1i_1i_3\cos(\alpha_1)] - k_0}{2aa_1i(m_1i_1i_3r + m_2i_2i_4r_1)}. \tag{8.185}$$

Natural oscillation frequency

$$\omega_{\text{nat}}^2 = \frac{k_0 + 2a^2\cos^2(\alpha)i(k_1i_1^2 + k_2i_2^2)}{M + m_1a^2\dot{y}^2(ii_1)^2 + m_2a^2(ii_2)^2}. \tag{8.186}$$

Of interest is the analysis of the numerator of the transfer function.

$$\begin{aligned} R = m_1 i_1 i_3 r + m_2 i_2 i_4 r_1 &= m_1 i_1 i_3 (\cos(\alpha)\cos(\alpha_1) - \sin(\alpha)\sin(\alpha_1)) \\ &\quad + m_2 i_2 i_4 (\cos(\alpha)\cos(\alpha_2) - \sin(\alpha)\sin(\alpha_2)) \\ &= \cos(\alpha)[m_1 i_1 i_3 \cos(\alpha_1) + m_2 i_2 i_4 \cos(\alpha_2)] \\ &\quad - \sin(\alpha)[m_1 i_1 i_3 \sin(\alpha_1) + m_2 i_2 i_4 \sin(\alpha_2)]. \end{aligned} \tag{8.187}$$

If $R = 0$, then

$$\mathrm{tg}(\alpha)\frac{m_1 i_1 i_3 \cos(\alpha_1) + m_2 i_2 i_4 \cos(\alpha_2)}{m_1 i_1 i_3 \sin(\alpha_1) + m_2 i_2 i_4 \sin(\alpha_2)} = 1 \tag{8.188}$$

determines the ratio between the parameters of the system in which the dynamic absorbing mode does not occur, and the system with a kinematic perturbation will behave in the same way as a conventional system when exposed to the same type.

8.5.3 Evaluation of the Dynamic Properties of the System

Let us find the dependences between the parameters of the system α, β, etc., which determine the values of the angles α_1 and α_2, as well as the distances l_1 and l_5 to the instantaneous center of velocities O_2.

Using the circuit diagram in Fig. 8.23 we find by the theorem of sines that

$$\frac{\sin(\alpha + \alpha_2)}{B_1 O_1} = \frac{\sin\left(\frac{\pi}{2} - \alpha\right)}{l_5} = \frac{\cos(\alpha)}{l_5}. \tag{8.189}$$

Whereas $l_5 \cos\alpha_2 = C_2 O_2$, to $l_5 = C_2 O_2 / \cos(\alpha_2)$. After substitution of l_5 in (8.189) we obtain

$$\frac{\sin(\alpha + \alpha_2)}{B_1 O_1} = \frac{\cos(\alpha)\cos(\alpha_2)}{C_2 O_2}. \tag{8.190}$$

In turn

$$B_1 O_1 = (l_1 \cos(\alpha) + l_2 \cos(\beta)) tg(\beta), \tag{8.191}$$

that after substitution gives in (8.191)

$$\frac{\sin(\alpha + \alpha_2)}{\cos(\alpha_2)} = \frac{\cos(\alpha)(l_1 \cos(\alpha) + l_2 \cos(\beta)) tg(\beta)}{D_2 O_2}. \tag{8.192}$$

Since

$$C_2O_2 = (l_1 + l_3)\cos(\alpha), \tag{8.193}$$

then after substitution in (8.192), we obtain

$$\mathrm{tg}(\alpha_2) = \frac{(l_1\cos(\alpha) + l_2\cos(\beta))\mathrm{tg}(\beta)}{(l_1 + l_3)}. \tag{8.194}$$

Assuming that $\sin(\alpha+\alpha_2) = \sin(\alpha)\cos(\alpha_2) + \cos(\alpha)\sin(\alpha_2)$, we find that

$$\mathrm{tg}(\alpha_2) = \frac{(l_1\cos(\alpha) + l_2\cos(\beta))\mathrm{tg}(\beta) - (l_1 + l_3)\sin(\alpha)}{(l_1 + l_3)\cos(\alpha)}. \tag{8.195}$$

As for the definition of l_5 values,

$$l_5 = \frac{(l_1 + l_3)\cos(\alpha)}{\cos(\alpha_2)}. \tag{8.196}$$

Using a similar approach with respect to the point where the additional mass m_1 is fixed (points D_2, D'_2), we obtain

$$\mathrm{tg}(\alpha_1) = \frac{(l_1\cos(\alpha) + l_2\cos(\beta))tg(\beta) - l'_1\sin(\alpha)}{l'_1\cos(\alpha)}. \tag{8.197}$$

In this case, l_4 can be determined by the formula

$$l_4 = \frac{l'_1\cos(\alpha)}{\cos(\alpha_1)}. \tag{8.198}$$

The expression for the transfer function (8.182) can be converted to

$$W(p) = \frac{\bar{y}}{\bar{z}} = \frac{R_1p^2 + R_2}{R_3p^2 + R_4}, \tag{8.199}$$

Here

$$R_1 = 2aa_1i(m_1i_1i_3r + m_2i_2i_4r_1), \tag{8.200}$$

$$R_2 = 2aa_1i(k_1i_1i_3\cos(\alpha)\cos(\alpha_1) - k_2i_2i_4\cos(\alpha)\cos(\alpha_2)) + k_0, \tag{8.201}$$

$$R_3 = M + 2a^2i^2(m_1i_1^2 + m_2i_2^2), \tag{8.202}$$

$$R_4 = 2a^2\cos^2(\alpha)(k_1 i_1^2 + k_2 i_2^2) + k_0, \tag{8.203}$$

where $i_3 = l_4/l_1$, $i_4 = l_5/l_1$, the values of r and r_1 are represented by expressions (8.183) and (8.184).

The combination of the values of R_1–R_4 determines the features of the transfer function.

8.5.4 Comparative Analysis of the Dynamic Properties of the System

The features of the transfer function (8.199) are such that the values of the coefficients R_1–R_4 determine the dynamic properties of the original system. For example, when $R_1 = 0$, the transfer function for $R_2 > 0$ will reflect such properties of the system that are typical of a conventional mechanical oscillatory system with one degree of freedom with kinematic perturbation. In this case, $R_1 = 0$ determines the boundary conditions for the selection of adjustment parameters that can be obtained from (8.200) using expressions (8.195)–(8.198). Thus, when $R_1 = 0$, one can obtain a series of relations.

1. We substitute (8.183) and (8.184) into (8.200), we obtain a relation characterizing the relationship of the system parameters at $R_1 = 0$.

$$\begin{aligned} R_1 &= m_1 i_1 i_3 r + m_2 i_2 i_4 r_1 = m_1 i_1 i_3 \cos(\alpha)\cos(\alpha_1) - m_1 i_1 i_3 \sin(\alpha)\sin(\alpha_1) \\ &\quad + m_2 i_2 i_4 \cos(\alpha)\cos(\alpha_2) - m_2 i_2 i_4 \sin(\alpha)\sin(\alpha_2), \end{aligned} \tag{8.204}$$

Convert (8.204) to the expanded form:

$$\begin{aligned} R_1 &= \sin(\alpha)(m_1 i_1 i_3 \sin(\alpha_1) + m_2 i_2 i_4 \sin(\alpha_2)) \\ &= \cos(\alpha)(m_1 i_1 i_3 \cos(\alpha_1) + m_2 i_2 i_4 \cos(\alpha_2)). \end{aligned} \tag{8.205}$$

The expression (8.205) can be transformed to the form

$$tg(\alpha) = \frac{m_1 i_1 i_3 \cos(\alpha_1) + m_2 i_2 i_4 \cos(\alpha_2)}{m_1 i_1 i_3 \sin(\alpha_1) + m_2 i_2 i_4 \sin(\alpha_2)}. \tag{8.206}$$

Note that $i_3 = l_4/l_1$, $i_4 = l_5/l_1$. In turn

$$l_4 = \frac{l_1' \cos(\alpha)}{\cos(\alpha_1)}, \quad l_5 = \frac{(l_1 + l_3)\cos(\alpha)}{\cos(\alpha_2)}.$$

Denote $l_1\cos(\alpha)$ + $l_2\cos(\beta)$= L, $i_1i_3 = i_1\frac{l_1'\cos(\alpha)}{\cos(\alpha_1)}$, $i_2i_4 = i_2\frac{(l_1+l_3)\cos(\alpha)}{\cos(\alpha_2)}$, then

$$tg(\alpha) = \frac{m_1i_1\frac{l_1'\cos(\alpha)}{\cos(\alpha_1)}\cos(\alpha_1)+m_2i_2\frac{(l_1+l_3)\cos(\alpha)}{\cos(\alpha_2)}\cos(\alpha_2)}{m_1i_1\frac{l_1'\cos(\alpha)}{\cos(\alpha_1)}\sin(\alpha_1)+m_2i_2\frac{(l_1+l_3)\cos(\alpha)}{\cos(\alpha_2)}\sin(\alpha_2)} = \frac{m_1i_1l_1'+m_2i_2(l_1+l_3)}{m_1i_1tg(\alpha_1)+m_2i_2(l_1+l_3)tg(\alpha_2)}. \tag{8.207}$$

Denote $tg(\alpha_1) = \frac{tg(\beta)L-l_1'\sin(\alpha)}{l_1'\cos(\alpha)}$, and $tg(\alpha_2) = \frac{tg(\beta)L-(l_1+l_3)\sin(\alpha)}{(l_1+l_3)\cos(\alpha)}$. Assume

$$\begin{aligned} tg(\alpha_1)m_1i_1l_1'+tg(\alpha_2)m_2i_2(l_1+l_3) &= \frac{tg(\beta)L}{\cos(\alpha)}\left[\frac{m_1i_1l_1'}{l_1'}+\frac{m_2i_2(l_1+l_3)}{l_1+l_3}\right] \\ &- \frac{\sin(\alpha)}{\cos(\alpha)}\left[\frac{m_1i_1l_1'}{l_1'}+\frac{m_2i_2(l_1+l_3)(l_1+l_3)}{l_1+l_3}\right] \\ &= \frac{tg(\beta)L}{\cos(\alpha)}(m_1i_1+m_2i_2)-tg(\alpha)(m_1i_1l_1'+m_2i_2(l_1+l_3)). \end{aligned} \tag{8.208}$$

Thus, expression (8.206) can be in the form of a trigonometric equation:

$$R_1 = tg(\alpha) - \frac{m_1i_1l_1'+m_2i_2(l_1+l_3)}{\frac{tg(\beta)L}{\cos(\alpha)}(m_1i_1+m_2i_2)-tg(\alpha)(m_1i_1l_1'+m_2i_2(l_1+l_3))}. \tag{8.209}$$

Let $m_1i_1l_1'+m_2i_2(l_1+l_3) = N$, then

$$R_1 = tg(\alpha) - \frac{1}{\frac{tg(\beta)L}{Ncos(\alpha)}(m_1i_1+m_2i_2)-tg(\alpha)N} = tg(\alpha) - \frac{N\cos(\alpha)}{tg(\beta)L(m_1i_1+m_2i_2)-N\sin(\alpha)}. \tag{8.210}$$

2. When R_1 = 0, you can also investigate the possibility of "zeroing out" of the R_2 coefficient, since when the equality of that kind is achieved, the transfer function numerator may become zero, which corresponds to the ideas about the system's capabilities to absorb external influences. The protected object remains stationary.

By substituting (8.176) and (8.177) into (8.201), and also by i = \sin(α) / \sin(β), we obtain the relation for R_2 = 0.

$$R_2 = 2\left(\frac{\cos(\beta)}{\sin(\alpha)(\cos(\alpha) + i\cos(\beta))}\right) \cdot \left(\frac{\cos(\alpha)}{\sin(\beta)(\cos(\alpha) + i\cos(\beta))}\right)$$
$$\times \frac{\sin(\alpha)}{\sin(\beta)}(k_1 i_1 i_3 \cos(\alpha)\cos(\alpha_1) - k_2 i_2 i_4 \cos(\alpha) cos(\alpha_2)) + k_0. \quad (8.211)$$

Using the relations from the previous section $i_1 i_3 = i_1 \frac{l_1' \cos(\alpha)}{\cos(\alpha_1)}$, $i_2 i_4 = i_2 \frac{(l_1 + l_3)\cos\alpha}{\cos\alpha_2}$, we transform the expression (8.199)

$$R_2 = \frac{\cos(\alpha)}{tg(\beta)(\cos(\alpha)\sin(\beta) + \sin(\alpha)\cos(\beta))}\left(k_1 i_1 \frac{l_1' \cos(\alpha)}{\cos(\alpha_1)}\cos(\alpha)\cos(\alpha_1)\right.$$
$$\left. - k_2 i_2 \frac{(l_1 + l_3)\cos(\alpha)}{\cos(\alpha_2)}\cos(\alpha)\cos(\alpha_2)\right) + k_0 \quad (8.212)$$
$$= \frac{\cos^3(\alpha)}{tg(\beta)(\cos(\alpha)\sin(\beta) + \sin(\alpha)\cos(\alpha))}(k_1 i_1 l_1' - k_2 i_2 (l_1 + l_3)) + k_0.$$

3. If R_1 and R_2 in the numerator of the transfer function (8.199) are simultaneously zero at a certain ratio of the parameters of the mechanical oscillatory system, then the amplitude-frequency characteristic (AFC) is not implemented. In this case, the motion of the base leads to interactions of the mass-and-inertia elements of the symmetric mechanism. At the same time, the object does not move, as a dynamic absorbing mode occurs. It is difficult to analytically define these conditions because of the complexity of trigonometric expressions, but the problem can be solved by numerical simulation. Figure 8.24 presents the frequency response of the system, for which values of R_1 and R_2 are close to zero, in particular, $R_1 = 0.006$, $R_2 = 0.002$. The amplitude-frequency characteristic in Fig. 8.24 reflects the dynamic properties of the system, including resonant effects, which will also be observed with small values of R_1 and R_2. Expression

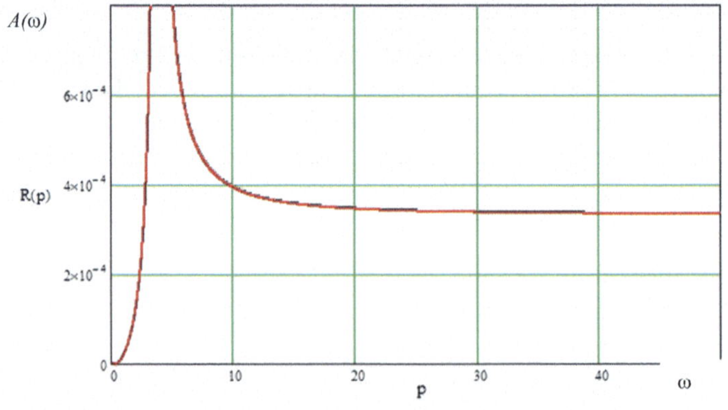

Fig. 8.24 Amplitude-frequency characteristic of the system with equality $R_1 = R_2 = 0$

(8.199) is a detailed transfer function. If R_1 and R_2 are small but not zero at the same time, then it is necessary to take into account the occurrence of the resonance mode and “locking” of the system at high frequencies, then the zone of steady reduction in the amplitude of natural oscillations will be in the superresonance region, in which the oscillation transfer coefficient will be a constant small number, that can be interpreted as a specific mode of dynamic absorbing.

$$
\begin{aligned}
W(p) = \frac{\bar{y}}{\bar{z}} = \frac{\left[tg(\alpha) - \frac{N\cos(\alpha)}{tg(\beta)L(m_1 i_1 + m_2 i_2) - N\sin(\alpha)}\right]p^2 + }{\left[M + 2\left[\frac{\cos(\beta)}{\sin(\alpha)\left(\cos(\alpha) + \frac{\sin(\alpha)}{\sin(\beta)}\cos(\beta)\right)}\right]^2 \times \right.} \cdots \\
\cdots \frac{+\left[\frac{\cos^3(\alpha)}{tg(\beta)(\cos(\alpha)\sin(\beta) + \sin(\alpha)\cos(\beta))}\right] \times}{\left.\times \left(\frac{\sin(\alpha)}{\sin(\beta)}\right)^2 (m_1 i_1^2 + m_2 i_2^2)\right] p^2 + 2\left[\frac{\cos(\beta)}{\sin(\alpha)\left(\cos(\alpha) + \frac{\sin(\alpha)}{\sin(\beta)}\cos(\beta)\right)}\right]^2 \times} \cdots \\
\cdots \frac{\times (k_1 i_1 l'_1 - k_2 i_2 (l_1 + l_3)) + k_0]}{\times (\cos(\alpha))^2 \left(\frac{\sin(\alpha)}{\sin(\beta)}\right)(k_1 i_1^2 + k_2 i_2^2) + k_0}.
\end{aligned}
\tag{8.213}
$$

Parameter values are: M = 20 kg; m_1 = 4.21 kg; m_2 = 0.5 kg; k_1 = 10 N/m; k_2 = 140 N/m; k_3 = 98.36 N/m; l_1 = 0.5 m; l_2 = 0.5 m; l_3 = 0.5 m; l'_1 = 0.3 m; $\alpha = \beta = = 45°$.

Analysis of the AFC shows that it is possible to reduce the level of vibrations to very small values. When $R_1 = 0$ and $R_2 = 0$, the specific dynamic absorbing mode extends over the entire frequency range, which will have a common point at which the amplitude ratio point can take large values. Thus, the object of protection can be stationary in a wide frequency range of external disturbances. Adjustment of the dynamic state of the system can be carried out by changing the stiffness of the pneumatic element k_2 using the system of automatic tracking of the state of the object.

An introduction to mechanical oscillatory systems of additional constraints that have constructive and engineering forms in the form of mechanisms creates the possibility of changing the dynamic state by choosing the parameters of the adjustment contour. That circuit can be formed by special mass-and-inertia elements, which are placed on the units of the articulation linkages. Placing additional elements in a certain way on the lower arm of the mechanism, it is possible to ensure the possibility of implementing the frequency characteristics of a system of various types, including the possibilities of implementing dynamic oscillation damping modes. It is shown that the introduction of mechanisms with additional masses corresponds to the formation in the system of inverse negative constraints by absolute deviation. The features of the influence on the frequency characteristics of external influences are revealed. Approaches to the construction of mathematical models are proposed. The necessary relations for setting up systems for the implementation of the modes of the required type are obtained.

8.6 Motion Transformation Devices in the Suspension with Two Degrees of Freedom

Protection of transportation vehicles and technological equipment from the action of vibration disturbances is based on the use of a wide range of special elastic-dissipative elements in the form of shock absorbers, dampers, vibration isolators for various purposes. At the same time, a number of questions related to the study of the possibilities of introducing special motion transformation devices into the structures of mechanical oscillatory systems have not yet been duly considered in detail, which predetermines interest in searching for non-traditional means of increasing the effectiveness of vibration protection systems.

The possibilities of changing the dynamic properties of mechanical oscillatory systems are considered when special mechanisms are introduced into their structure that use lever linkages and effects that arise in the meantime.

8.6.1 Description of System Properties

One of the options for using the mass-and-inertia mechanism as part of a mechanical oscillatory system with the object of protection against vibrations in the form of a solid body is shown in Fig. 8.25.

Two lever mechanisms (l_3, l_5 and l_4, l_6 respectively) at pp. A_1 and B_1 are connected by a screw non-locking mechanism, which has a reduced moment of inertia L. In the process of changing the position of pp. A_1 and B_1, lying on the same axis, the longitudinal relative displacements are converted into rotating and reciprocating motions of the elements of the screw connection. The arising inertial forces form additional efforts at pp. A_1 and B_1. The frictional forces in the screw pairs, as well as in all the joints at the preliminary stage of studying the dynamic properties, are small. The screw mechanism is considered as a motion transformation device and can be made on another structural basis, for example, using toothed gearings.

The features of the scheme under consideration are the introduction of a new type of coupling between an oscillating system and a solid body. Such relationships determine the interplay between partial systems that are unusual for ordinary systems.

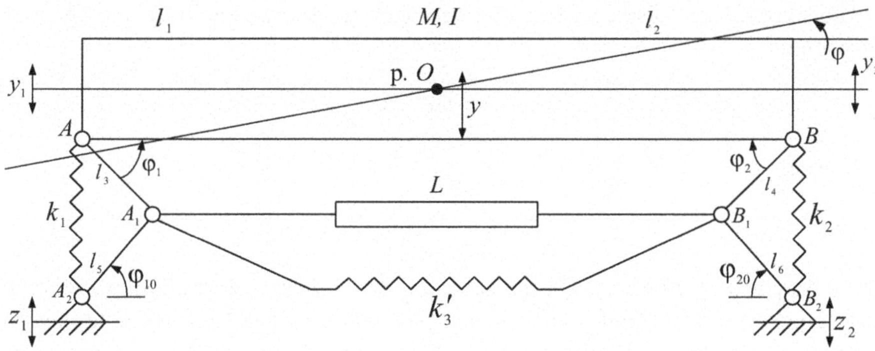

Fig. 8.25 The computational scheme of the vibration protection system with two degrees of freedom with lever interacting mechanisms

The following notation is introduced in Fig. 8.25: l_1, l_2 are the distances to the center of gravity of p. *O*; y_1, y_2, y and φ are, respectively, the systems of generalized coordinates with respect to a fixed basis; z_1, z_2 are kinematic perturbations; *M*, *I* are the mass and the moment of inertia of the object of protection (solid body); k_1, k_2, k_3' are the coefficients of the elastic elements; l_3–l_6 are the lengths of the units of the lever mechanisms.

It is assumed that the system makes small oscillations relative to the position of static equilibrium; coordinates y_1 and y_2 are associated with a fixed coordinate system; in the position of static equilibrium, the angles of inclination of the units of the lever mechanism, according to Fig. 8.25, are determined by the angles φ_1, φ_{10} and φ_2, φ_{20}. In this case, the solid body performs only vertical motions; points A_1 and B_1 are connected by a motion transformation device, in which the slope of the A_1B_1 line has no significant effect. It is supposed that $l_3 = l_4 = l_5 = l_6 = l$, $\varphi_{10} = \varphi_1$, $\varphi_{20} = \varphi_2$. We also believe that the rods of the articulation linkage are weightless.

Figure 8.26 presents a schematic diagram of the relative position of the rods of the articulation linkage.

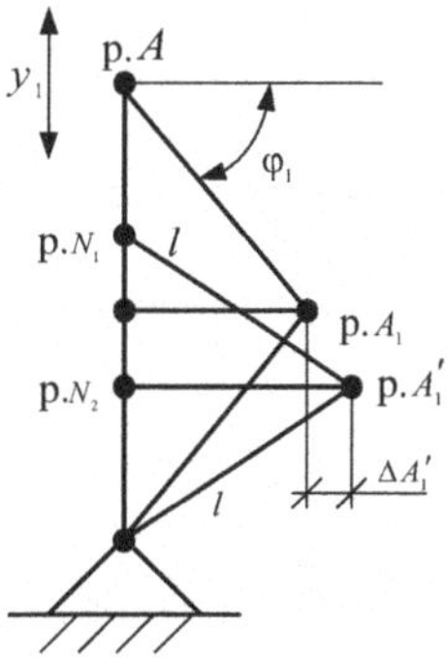

Fig. 8.26 The layout of the rods with small changes in the angle φ_1

The longitudinal displacement of p. A_1 can be determined from the obvious correlation:

$$\Delta A_1 = A_1' N_2 - A_1 N_1.$$

In turn:

$$y_1 = l \cdot \sin \varphi_1 - l \cdot \sin(\varphi_1 - \Delta\varphi_1)$$

where $\Delta\varphi_1$ is a small change in the angle φ_1 determining the configuration of the relative position of the rods [18]. Omitting the geometric details, we note that, in accordance with Fig. 8.26, the point A_1 offset when changing the coordinate y_1 is:

$$\Delta A_1 = b_1 \times y_1.$$

where b_1 is a geometric parameter.

In addition, when moving along the y_2 coordinate, you can obtain that:

$$\Delta B_2 = b_2 \times y_2,$$

where b_2 is also determined from geometric relationships.

If we assume that $\varphi_1 = \varphi_2$, то $b_1 = b_2$; the transformation device can work depending on the method of arrangement of the articulation linkages (symmetrically or unidirectionally), then the change in length A_1B_1 will be determined:

$$\Delta(A_1B_1) = b_1(y_1 \mp y_2).$$

As for b_1 and b_2, then with $z_1 = 0$, $z_2 = 0$ and $l = l_i$ $(i = \overline{3,6})$, $b_1 = b_2 = \mathrm{tg}\varphi_1$.

8.6.2 Construction of Mathematical Models

Find expressions for the kinetic and potential energies of the system

$$\mathrm{T} = \frac{1}{2} M(\dot{y})^2 + \frac{1}{2} J(\dot{\varphi})^2 + \frac{1}{2} L b_1^2 (\dot{y}_1 \pm \dot{y}_2)^2,$$

where y, φ is the coordinate of the center of gravity and the angle of rotation of the solid body;

L is the reduced mass of the motion transformation device;

$$\Pi = \frac{1}{2} k_1 \cdot y_1^2 + \frac{1}{2} k_2 \cdot y_2^2 + \frac{1}{2} k_3 \cdot (y_1 - y_2)^2.$$

We assume that $y = ay_1 + by_2$, $\varphi = c(y_2 - y_1)$,

where

$$a = \frac{l_2}{l_1 + l_2}, \quad b = \frac{l_1}{l_1 + l_2}, \quad c = \frac{1}{l_1 + l_2}, \quad L_1 = Lb_1^2, \quad k_3 = k_3' b_1^2. \tag{8.214}$$

When considering the conditions for the symmetric arrangement of weightless rods, we can assume:

$$\Delta A_1 = (y_1 - y_2) \cdot b_1, \tag{8.214'}$$

then the kinetic energy of the motion transformation device will take the form:

$$\mathrm{T}_3 = \frac{1}{2} L_1 (\dot{y}_1 - \dot{y}_2), \tag{8.214''}$$

which corresponds to the case (8.214′), when the layout of the rods is symmetric. Thus, the kinetic and potential energies can be written:

$$\mathrm{T} = \frac{1}{2} M (a\dot{y}_1 + b\dot{y}_2)^2 + \frac{1}{2} Jc^2 (\dot{y}_2 - \dot{y}_1)^2 + \frac{1}{2} Lb_1^2 (\dot{y} + \dot{z}_1 - \dot{y}_2)^2; \tag{8.215}$$

$$\Pi = \frac{1}{2} k_1 \cdot (y_1 - z_1)^2 + \frac{1}{2} k_2 \cdot y_2 + \frac{1}{2} k_3 \cdot (y_1 + z_1 - y_2)^2. \tag{8.215'}$$

In (8.215) and (8.215′), the kinematic perturbation $z_1(t) \neq 0$ (for $z_2(t) = 0$) is taken into account. The system of differential equations of motion in the coordinates y_1 and y_2 takes the form:

$$\begin{aligned} &\bar{y}_1 \left[\left(Ma^2 + Jc^2 + L_1 \right) p^2 + k_1 + k_3 \right] + \bar{y}_2 \left[\left(Mab - Jc^2 - L_1 \right) p^2 - k_3 \right] \\ &\quad = k_1 \overline{z_1} - k_3 \overline{z_1} - L_1 \overline{z_1} p^2; \end{aligned} \tag{8.216}$$

$$\bar{y}_2 \left[\left(Mb^2 + Jc^2 + L_1 \right) p^2 + k_2 + k_3 \right] + \bar{y}_1 \left[\left(Mab - Jc^2 - L_1 \right) p^2 - k_3 \right] = k_3 \overline{z_1} - L_1 \overline{z_1}. \tag{8.217}$$

The structural diagram of the system in accordance with (8.216) and (8.217) has the form, as shown in Fig. 8.27.

Note that the introduction of motion transformation devices with the transfer function $L_1 p^2$ (here $p = j\omega$ is a variable [9], and $\bar{y}_1$, $\bar{y}_2$, $\overline{z_1}$ are Laplace functions' transform images) reflected in the form of transfer functions of partial systems, crosslinks, and transformations of external influence $\overline{z_1}$ In this case, due to mechanical constraints, the perturbation $\overline{z_1}$ is the input signal y_1 and y_2.

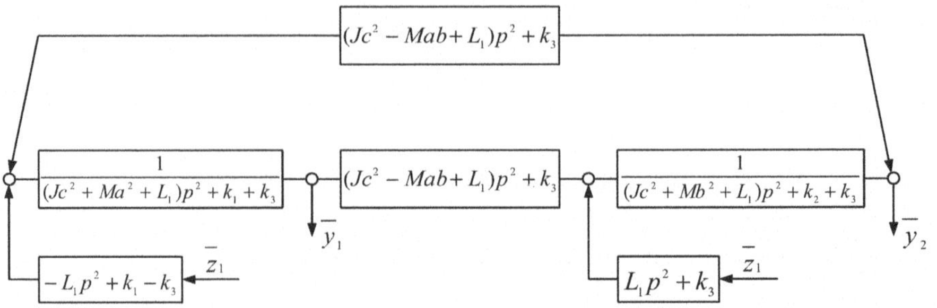

Fig. 8.27 The structural diagram of the system with new constraints, reflecting the effect of the inserted device (L_1)

8.6.3 *The Analysis of Dynamic Properties*

The structural diagram of the system shown in Fig. 8.27, differs from the known ones in that, given the initial restrictions, interpartial relations are not inertial, but elastic-inertial in nature. This suggests the possibility of the emergence of the "decoupling" effect of partial systems at the frequency of external influence:

$$\omega^2 = \frac{k_3}{Jc^2 + L_1 - Mab}. \tag{8.218}$$

Note also that the reduced mass of inertia L_1 of a motion transformation device can be considered as an adjustment parameter. In particular, the parameter L_1 is included in the expression for determining the partial frequencies:

$$\omega_1^2 = \frac{k_1 + k_3}{Jc^2 + L_1 + Ma^2}; \tag{8.219}$$

$$\omega_2^2 = \frac{k_2 + k_3}{Jc^2 + L_1 + Mb^2}. \tag{8.220}$$

The larger the value of L_1, the smaller the values of the partial frequencies. In addition, L_1 is included in the frequency characteristic equation, changing its coefficients, and this is due to the values of natural frequencies, which will also take the corresponding underestimated values.

Since the external disturbance acts simultaneously on the inputs on the coordinates $\bar{y}_1$ and $\bar{y}_2$, the transfer functions are determined based on the principle of superposition:

$$W_1(p) = \frac{\bar{y}_1}{\bar{z}_1} = \frac{1}{A_0} \cdot \{(k_1 - k_3 - L_1 p^2) \cdot [(Mb^2 + Jc^2 + L_1)p^2 + k_2 + k_3] \\ + (L_1 p^2 + k_3) \cdot [(Jc^2 - Mab + L_1)p^2 + k_3]\}, \tag{8.221}$$

$$W_2(p) = \frac{\bar{y}_2}{\bar{z}_1} = \frac{1}{A_0} \cdot \{(k_1 - k_3 - L_1 p^2) \cdot [(Jc^2 - Mab - L_1)p^2 + k_3] \\ + (L_1 p^2 + k_3) \cdot [(Jc^2 + Ma^2 + L_1)p^2 + k_1 + k_3]\}, \tag{8.222}$$

where

$$A_0 = [(Ma^2 + Jc^2 + L_1)p^2 + k_1 + k_3] \cdot [(Mb^2 + Jc^2 + L_1)p^2 + k_2 + k_3] \\ - [(Jc^2 - Mab + L_1)p^2 + k_3]^2. \tag{8.223}$$

The same order of numerator and denominator can be considered characteristic of the transfer functions defined by the expressions (8.221) and (8.222), which is due to the presence of a number of features.

The introduction of the L_1, k_3 constraint leads to the fact that the external action $z_1(t)$ is transformed into a simultaneous perturbation in two partial systems.

In accordance with the structure of the characteristic Eq. (8.223), in the system we can expect the possibility of two resonances at the frequencies of the natural oscillations of the system ω_{1nat} and ω_{2nat}, determined from Eq. (8.223). Since L_1 can be considered as a freely selectable parameter, it is possible to find that value of L_1 at which the partial frequencies can be equal, which is determined by:

$$L_1 = \frac{1}{(k_1 - k_2)} \cdot \{M \cdot [(k_1 b^2 - k_2 a^2) + k_3(b^2 - a^2)] + Jc^2 \cdot (k_1 - k_3)\}. \tag{8.224}$$

When moving along the coordinates y_1 and y_2, it follows from (8.221) and (8.222) that, in the general case, two modes of dynamic oscillation damping are possible for each of the coordinates y_1 and y_2. If we take $p = 0$, which corresponds to a static effect, then the position is determined by the ratios of oscillation amplitudes:

$$W_1(p)_{p=0} = \frac{\bar{y}_1}{\bar{z}_1} = \frac{(k_1 - k_3) \cdot (k_2 + k_3) + k_3^2}{(k_1 + k_3) \cdot (k_2 + k_3) - k_3^2} = \frac{k_1 k_2 - k_2 k_3 + k_1 k_3}{k_1 k_2 + k_2 k_3 + k_1 k_3}, \tag{8.225}$$

$$W_2(p)_{p=0} = \frac{\bar{y}_2}{\bar{z}_1} = \frac{(k_1 - k_3) \cdot k_3 + k_3 \cdot (k_1 + k_3)}{(k_1 + k_3) \cdot (k_2 + k_3) - k_3^2} = \frac{2k_1 k_3}{k_1 k_2 + k_2 k_3 + k_1 k_3}. \tag{8.226}$$

In turn, as $p \to \infty$:

$$W_1''(p)_{p\to\infty} = \frac{(-L_1)\cdot(Mb^2+Jc^2+L_1)+L_1(Jc^2-Mab+L_1)}{(Ma^2+Jc^2+L_1)\cdot(Mb^2+Jc^2+L_1)-(Jc^2-Mab+L_1)^2};\tag{8.227}$$

$$W_2''(p)_{p\to\infty} = \frac{(-L_1)\cdot(Jc^2-Mab+L_1)+L_1(Ma^2+Jc^2+L_1)}{(Ma^2+Jc^2+L_1)\cdot(Mb^2+Jc^2+L_1)-(Jc^2-Mab+L_1)^2}.\tag{8.228}$$

From the analysis (8.227) it follows that:

$$|W_1''(p)| = \frac{L_1 b}{(Jc^2+L_1)},\tag{8.227'}$$

and the expression (8.228) has a module:

$$|W_2''(p)| = \frac{L_1 a}{(Jc^2+L_1)}.\tag{8.228'}$$

Note that by definition, $a < 1$ and $b < 1$, therefore, the values of (8.227′) and (8.228′) will also be less than unity.

As for the influence of the parameter k_3, then for $k_3 = 0$ the following relation holds:

$W_1(p)_{p=0} = \frac{\bar{y}_1}{\bar{z}_1} = 1$; on the y_2 coordinate, we obtain, respectively:

$$W_2(p)_{p=0} = \frac{\bar{y}_2}{\bar{z}_1} = 0.$$

If $L_1 = 0$, then:

$$\underset{L_1=0}{W_1(p)} = \frac{\bar{y}_1}{\bar{z}_1} = \frac{(k_1-k_3)\cdot[(Mb^2+Jc^2)p^2+k_2+k_3]+}{[(Ma^2+Jc^2)p^2+k_1]\cdot[(Mb^2+Jc^2)p^2+k_2]-}\cdots$$
$$\cdots\frac{+k_3\cdot[(Jc^2-Mab)p^2+k_3]}{-k_3\cdot[(Jc^2-Mab)p^2+k_3]^2},\tag{8.229}$$

$$\underset{L_1=0}{W_2(p)} = \frac{\bar{y}_2}{\bar{z}_1} = \frac{(k_1-k_3)\cdot[(Jc^2-Mab)p^2+k_3]+}{[(Ma^2+Jc^2)p^2+k_1+k_3]\cdot[(Mb^2+Jc^2)p^2+k_2+k_3]-}\cdots$$
$$\cdots\frac{+k_3\cdot[(Ma^2+Jc^2)p^2+k_1+k_3]}{-k_3\cdot[(Jc^2-Mab)p^2+k_3]^2}.\tag{8.230}$$

In this case, the connection between the partial systems is through the elastic element k_3. If we assume that $k_3 = 0$, $L_1 = 0$, then:

$$\underset{\substack{L_1=0\\k_3=0}}{W_1(p)} = \frac{\bar{y}_1}{\bar{z}_1} = \frac{k_1 \cdot [(Mb^2 + Jc^2)p^2 + k_2]}{[(Ma^2 + Jc^2)p^2 + k_1] \cdot [(Mb^2 + Jc^2)p^2 + k_2] - (Jc^2 - Mab)p^4}, \tag{8.231}$$

$$\underset{\substack{L_1=0\\k_3=0}}{W_2(p)} = \frac{\bar{y}_2}{\bar{z}_1} = \frac{k_1 \cdot (Jc^2 - Mab)p^2}{[(Ma^2 + Jc^2)p^2 + k_1] \cdot [(Mb^2 + Jc^2)p^2 + k_2] - (Jc^2 - Mab)p^4}, \tag{8.232}$$

which corresponds to the results given in [9].

8.6.4 Asymmetrical Case of the Arrangement of Mechanisms

We will assume that the kinetic energy introduced into the mechanical system by a motion transformation device in an asymmetrical arrangement is:

$$\mathrm{T}_3 = \frac{1}{2} L_1 (\dot{y}_1 + \dot{y}_2), \tag{8.233}$$

then (8.214) transforms:

$$T = \frac{1}{2} M (a\dot{y}_1 + b\dot{y}_2)^2 + \frac{1}{2} Jc^2 (\dot{y}_2 - \dot{y}_1)^2 + \frac{1}{2} Lb_1^2 (\dot{y}_1 + \dot{z}_1 + \dot{y}_2)^2. \tag{8.234}$$

In turn:

$$\Pi = \frac{1}{2} k_1 \cdot (y_1 - z_1)^2 + \frac{1}{2} k_2 \cdot y_2 + \frac{1}{2} k_3 \cdot (y_1 + z_1 + y_2)^2. \tag{8.235}$$

Using the usual methods of generating differential equations, we obtain:

$$\begin{aligned} &\bar{y}_1 \left[(Ma^2 + Jc^2 + L_1) p^2 + k_1 + k_3 \right] \\ &\quad + \bar{y}_2 \left[(Mab - Jc^2 + L_1) p^2 + k_3 \right] = \bar{z}_1 \cdot (k_1 - k_3 - L_1 p^2); \end{aligned} \tag{8.236}$$

$$\begin{aligned} &\bar{y}_2 \left[(Mb^2 + Jc^2 + L_1) p^2 + k_2 + k_3' \right] \\ &\quad - \bar{y}_1 \left[(Mab - Jc^2 + L_1) p^2 + k_3 \right] = -k_3 \bar{z}_1 - L_1 \bar{z}_1 p^2. \end{aligned} \tag{8.237}$$

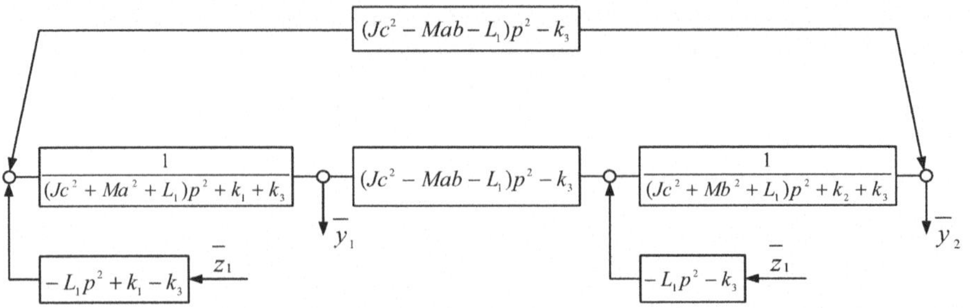

Fig. 8.28 The structural diagram of the system with an asymmetric motion transformation mechanism

The system of equations (8.236) and (8.237) describes the motion of the system in the case when in the computational scheme (Fig. 8.25) the right bundle of rods is rotated through 180°, which makes the system asymmetric.

The structural diagram of the mechanical oscillatory system in this case takes the form, as shown in Fig. 8.28.

It should be noted that the case of asymmetric arrangement coincides with the structural diagram in Fig. 8.27:

1. The partial systems are the same: interpartial communication operators in a symmetric scheme have the form:

$$W_{12} = \left(Jc^2 - Mab + L_1\right)p^2 + k_3, \tag{8.238}$$

and with asymmetrical one, respectively:

$$W'_{12} = \left(Jc^2 - Mab - L_1\right)p^2 - k_3 \tag{8.239}$$

2. The form of the action of external forces has changed: if for y_1 with a symmetric scheme, the operator for transforming external effects z_1 had the form:

$$W_{z_1} = L_1p^2 + k_1 - k_3, \tag{8.240}$$

then with an asymmetric scheme, respectively:

$$W_{z_2} = -L_1p^2 + k_1 - k_3. \tag{8.241}$$

3. In turn, by coordinate $\bar{y}_2$, the external influence had a transformation operator z_1:

$$W'_{z_1} = L_1p^2 + k_3, \tag{8.242}$$

It has become:

$$W''_{z_1} = -L_1 p^2 - k_3. \tag{8.243}$$

In general, additional constraints predetermine changes in dynamic properties in a sufficiently large range.

If we determine the characteristic equation A'_0 reflecting the properties of the system in Fig. 8.28, we get:

$$\begin{aligned} A'_0 = {} & \left[\left(Ma^2 + Jc^2 + L_1\right)p^2 + k_1 + k_3\right] \cdot \left[\left(Mb^2 + Jc^2 + L_1\right)p^2 + k_2 + k_3\right] \\ & - \left[\left(Jc^2 - Mab + L_1\right)p^2 - k_3\right]^2. \end{aligned} \tag{8.244}$$

When comparing (8.223) and (8.244), it can be noted that the dynamic properties of a system with an asymmetric arrangement of mechanisms will be different, and $A_0 = A'_0$.

The transfer functions of the system along the coordinates y_l and y_2 with the input disturbance $\overline{z_1}$ can be, in accordance with the structural diagram in Fig. 8.28, write:

$$\begin{aligned} W'_1(p) = \frac{\bar{y}_1}{\overline{z_1}} = \frac{1}{A'_0} \cdot {} & \left\{\left(k_1 - k_3 - L_1 p^2\right) \cdot \left[\left(Mb^2 + Jc^2 + L_1\right)p^2 + k_2 + k_3\right] \right. \\ & \left. + \left(-L_1 p^2 - k_3\right) \cdot \left[\left(Jc^2 - Mab - L_1\right)p^2 - k_3\right]\right\}, \end{aligned} \tag{8.245}$$

$$\begin{aligned} W'_2(p) = \frac{\bar{y}_2}{\overline{z_1}} = \frac{1}{A'_0} \cdot {} & \left\{\left(k_1 - k_3 - L_1 p^2\right) \cdot \left[\left(Jc^2 - Mab - L_1\right)p^2\right] \right. \\ & \left. + \left(-L_1 p^2 - k_3\right) \cdot \left[\left(Jc^2 + Ma^2 + L_1\right)p^2 + k_1 + k_3\right]\right\}. \end{aligned} \tag{8.246}$$

Although the partial systems in the structural diagrams Figs. 8.27 and 8.28 have the same partial systems, yet the modes of dynamic oscillation damping will be different.

The structural diagram of the system shown in Fig. 8.27 differs from the known ones by the fact that, given the initial constraints, interpartial relations are not inertial, but elastic-inertial in nature. This suggests the possibility of the emergence of the "decoupling" effect of partial systems at the frequency of external influence:

$$\omega^2 = \frac{k_3}{Jc^2 + L_1 - Mab}. \tag{8.247}$$

Note also that the reduced moment of inertia L_1 of a motion transformation device can be considered as an adjustment parameter. In particular, the parameter L_1 is included in the expression for determining the partial frequencies:

$$\omega_1^2 = \frac{k_1 + k_3}{Jc^2 + L_1 + Ma^2}, \tag{8.248}$$

$$\omega_2^2 = \frac{k_2 + k_3}{Jc^2 + L_1 + Mb^2}. \tag{8.249}$$

The larger the value of L_1, the smaller the values of the partial frequencies. In addition, L_1 is included in the frequency characteristic equation, changing its coefficients, and this is due to the values of natural oscillation frequencies, which will also take the corresponding underestimated values.

8.6.5 Possible Forms of Development of Ideas About the Introduction of Additional Constraints

Figure 8.29a–d provides some computational schemes of mechanical oscillatory systems in which the beam-type constraints are implemented, ensuring the use of transformation effects of relative motions.

Various options of transforming schemes in accordance with Fig. 8.29a are considered in [19]. The schematic solution (Fig. 8.29b) of the suspension of the object of protection from the vibration of the base allows the interpartial constraints to be converted into inertial-elastic, which is reflected in the results of studies cited [3]. The variant of protection of instrumentation is shown in Fig. 8.29c, in which the elastic properties of the suspension are provided by the elastic-lever system. In the diagram of the vibration protection system in Fig. 8.29d are represented by lever supports (pp. *A* and *B*), which can be interconnected through additional elastic constraints or through toothed gearings.

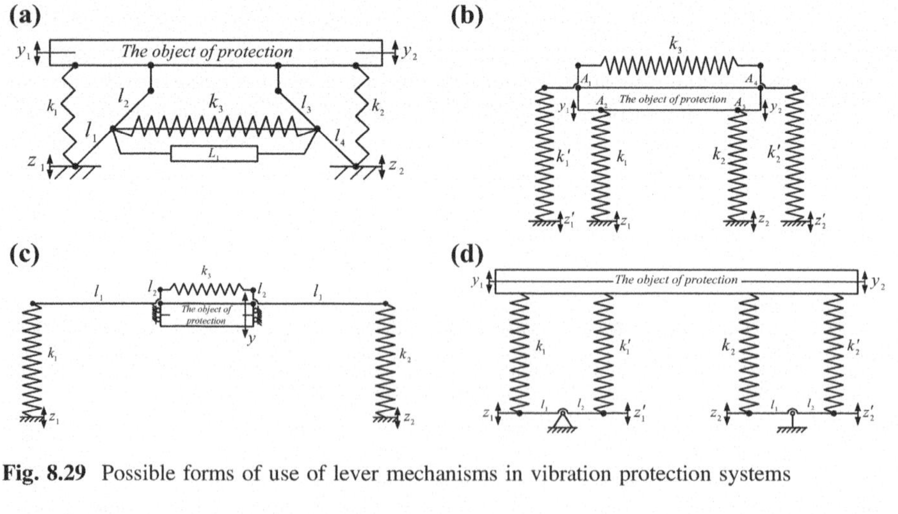

Fig. 8.29 Possible forms of use of lever mechanisms in vibration protection systems

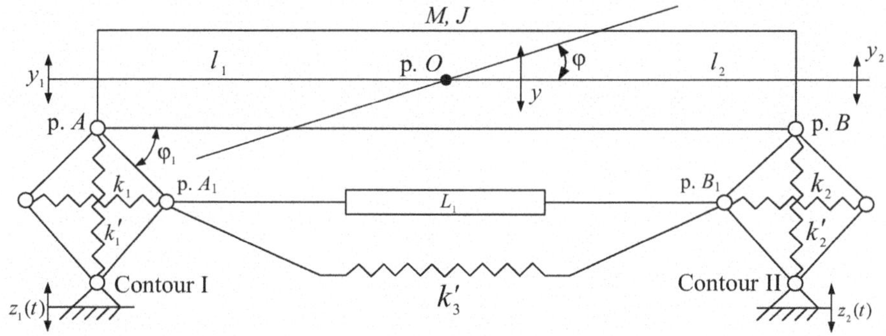

Fig. 8.30 The computational scheme of the system with the introduction of additional (corrective) elastic elements (k_1' and k_2')

The introduction of corrective constraints. Figure 8.30 shows vibration protection system correction options by building an elastic system that has the ability to provide functionality even if one of the elements fails.

In solving this problem (Fig. 8.30), as compared with the previous one (Fig. 8.25), a more complex mechanical system is used. It consists of an object of protection (in the form of a solid body *M*, *J*), based on a mechanism formed by two contours (each of 4 identical weightless rods of length *l*), which are interconnected by a motion transformation device with a reduced moment of inertia *L*: the elastic element with stiffness k_3' joins in parallel to MTD at pp. A_1 and B_1.

The reduced mass-and-inertia and elastic parameters of the system depend, in particular, not only on the length of weightless rods, but also on the angle of inclination φ_1 (Fig. 8.30). Through φ_1, the parameter b_1 is determined, which depends nonlinearly on φ_1. However, considering the small oscillations of the system relative to the position of static equilibrium, we can assume that the parameters of the system are determined by the values of $L_1 = Lb_1^2$, $k_3 = k_3' b_1^2$.

Taking account of kinematic perturbations $z_1(t)$ and $z_2(t)$, as shown above, has its own characteristics, manifested in the fact that through a mechanical system each of the external perturbations $z_1(t)$ and $z_2(t)$, taken separately, creates effects at the inputs of each partial system. The solution to the problem of determining the total response, in this case, to the harmonic external influence, can be found on the basis of the superposition principle.

If two perturbations act simultaneously, an acceptable solution can be found if some connection is defined between $z_1(t)$ and $z_2(t)$. In a particular case, we can assume that $z_1(t) = z_2(t)$. Otherwise, the output reaction should be the result of the vector summation of two dynamic processes.

When considering the dynamic properties of the scheme (Fig. 8.30), vertical elastic elements with stiffnesses k_1 and k_2 play a certain role in the evaluation of conditions. In the first section of the article, where k_1 and k_2 were directly related to the bearing surfaces and points *A* and *B* on a solid body, stability issues were provided in accordance with the Routh-Hurwitz criteria. However, when $k_1 = 0$,

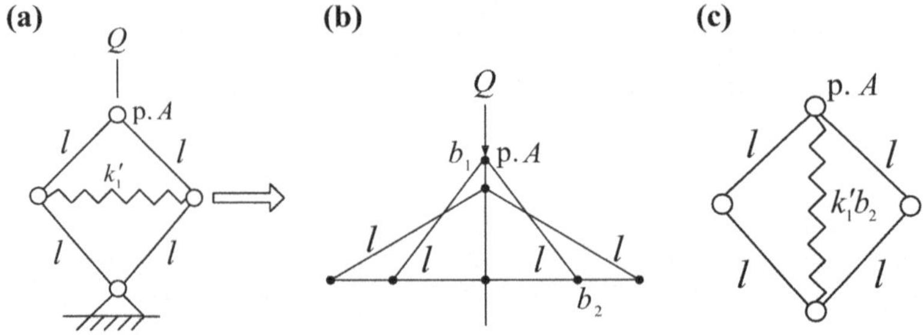

Fig. 8.31 Computational (**a**), principal (**b**) and equivalent (**c**) schemes for determining the elastic properties of a closed contour

$k_2 = 0$, the stability of the system is violated due to the emergence of a cyclic coordinate in the system.

In the case under consideration (Fig. 8.30), when the elements k_1 and k_2 are inside a closed contour, it can be assumed that the elastic element is reduced to an equivalent spring. Figure 8.31 presents a schematic diagram for determining the equivalent stiffness of the spring.

The kinetic energy of the system (Fig. 8.30) is determined by the expression:

$$\mathrm{T} = \frac{1}{2}M(\dot{y})^2 + \frac{1}{2}J(\dot{\varphi})^2 + \frac{1}{2}L(\dot{y}_1 - \dot{y}_2 + \dot{z}_1)^2 \tag{8.250}$$

where $L_1 = L \cdot b_1^2$ (b_1 is the geometric parameter of the relations between the elements of the system) with $z_1(t) \neq 0$ and $z_2(t) = 0$.

Potential energy of an elastic element in contours *I* and *II*:

$$\begin{gathered}\Pi_1 = \frac{1}{2}k_1' \cdot b_2^2 \cdot (y_1 + z_1)^2, \\ \Pi_2 = \frac{1}{2}k_2 \cdot b_3^2 \cdot y_2^2, \;\; \Pi_3 = \frac{1}{2}k_3' \cdot (b_2 y_1 + b_2 z_1 - b_3 y_2)^2.\end{gathered} \tag{8.251}$$

Assuming that $b_2 = b_3$, after a series of transformations, we obtain the system of differential equations of motion in the coordinates y_1, y_2:

$$\begin{aligned}&\bar{y}_1\left[\left(Ma^2 + Jc^2 + L_1\right)p^2 + k_1'b_2^2 + k_3'b_2^2\right] + \bar{y}_2\left[\left(Mab - Jc^2 - L_1\right)p^2 - k_3'b_2^2\right] \\ &\quad = -k_1'z_1b_2^2 - k_3'b_2^2z_1 - L_1z_1p^2;\end{aligned} \tag{8.252}$$

$$\begin{aligned}&\bar{y}_2\left[\left(Ma^2 + Jc^2 + L_1\right)p^2 + k_2'b_2^2 + k_3'b_2^2\right] + \bar{y}_1\left[\left(Mab - Jc^2 - L_1\right)p^2 - k_3'b_2^2\right] \\ &\quad = k_3'b_2^2z_1 - L_1z_1p^2.\end{aligned} \tag{8.253}$$

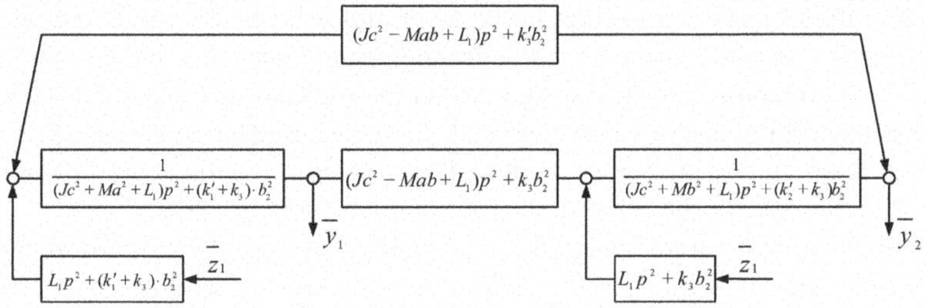

Fig. 8.32 The structural diagram of the system in Fig. 8.30 without elastic elements k_1 and k_2

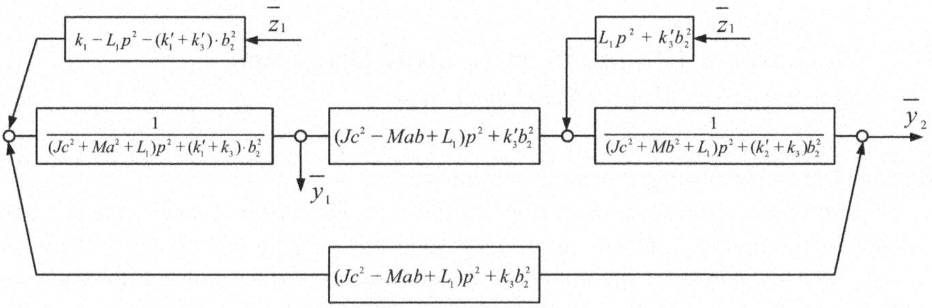

Fig. 8.33 The structural diagram of a system with corrective elastic elements k_1 and k_2

The structural diagram in accordance with (8.252) and (8.253) has the form, as shown in Fig. 8.32.

If the vertical springs k_1 and k_2 were not introduced in the supports (pp. *A* and *B*), the system would have been unstable. Therefore, in the structural diagram in Fig. 8.33 elastic elements k_1 and k_2 are introduced.

From Fig. 8.33 it follows that in partial systems, in addition to the springs shown, having stiffnesses $(k_1' + k_3) \cdot b_2^2$ and $(k_2' + k_3) \cdot b_2^2$, respectively, the springs with stiffnesses k_1 and k_2 work in parallel.

Note that in the considered constructive version of the inclusion of mechanisms in the structure of the mechanical oscillatory system, the elastic elements k_1', k_2', k_3' are connected not in series (as it might seem in the computational scheme), but in parallel. Further research is carried out on the option of assessing the dynamic properties of a system with a symmetric scheme of mechanisms (Fig. 8.25).

The proposed method of changing dynamic properties is based on the introduction into the oscillatory system of mechanical rod structures that are connected by a mechanism for motion transformation. This technique allows you to change the nature of interpartial relations and thereby adjust the system to specific tasks of vibration protection. The construction of mathematical models of systems uses the possibilities of transforming the original mathematical model in the form of a system of differential equations into a structural model that has the form of a dynamically equivalent automatic control system. Thus, in mechanical oscillatory systems, comprising the object of protection in the form of a solid body with two degrees of freedom, it is possible to introduce additional constraints that induce significant changes in the dynamic properties. The evaluation of possible dynamic properties is based on the technology of frequency analysis using transfer functions. The peculiarities of the dynamic properties are that the numerator and the denominator of the transfer functions have the same order for kinematic perturbation.

As the adjustment parameters of the system, the values of the reduced moment of inertia L, the spring stiffness of the additional constraint k_3, and parameters of the geometric nature b, depending on the system configuration, which can change independently, can be used.

8.7 Mechanical Chains in Structural Diagrams of Vibration Protection Systems

Method for determining dynamic responses

Vibration protection systems (VPS) of technical objects are quite diverse and can contain various elements that are often combined into mechanical chains. If there are inertial elements in the circuits and periodic external influences, dynamic interactions form constraint reactions that determine the state of the protected object, contacts with bearing surfaces and the level of force interactions between the elements.

Dynamic constraint reactions are quite simply defined in planar mechanisms, which are closed mechanical chains with one or several fixed units. Kinetostatic calculations and methodical basis of calculations are well developed and are reflected in [20]. At the same time, the definition of constraint reactions in mechanical oscillatory systems is less developed. Although the issues of the dynamics of oscillatory motions in the mechanisms and vibration protection systems obtain considerable attention. The specificity of vibration protection systems is in the predominance of periodic motions of the object of protection and the units of the VPS, which in practice often results in opening of contacts, sampling of gaps and the occurrence of vibro-impact processes, which are undesirable effects in terms of ensuring the reliability and safety of operation of machines, equipment and apparatus [10, 21–23]. A methodological basis is proposed for determining dynamic responses in mechanical oscillatory systems using structural mathematical models-analogues obtained by means of the Laplace transform of the original linear or linearized differential equations of motion.

8.7.1 Description of the Original Positions

Consider as an initial computational scheme of a vibration protection system in the form of a mechanical oscillatory system with three degrees of freedom (Fig. 8.34).

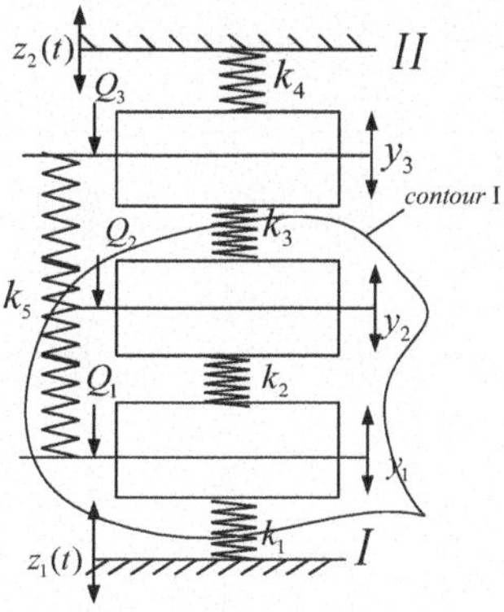

Fig. 8.34 The computational scheme of the vibration protection system with three degrees of freedom

The system executes small motions under the action of periodic force and kinematic perturbations that can be applied to all mass-and-inertia elements $-Q_i\left(i=\overline{1,3}\right)$. In addition, it is possible to take into account the action of kinematic perturbations $z_k(k=\overline{1,2})$ associated with the bearing surfaces *I* and *II* (Fig. 8.34). The fixed coordinate system y_1, y_2, y_3 is selected as the main one; friction forces are considered negligibly small, and the motions of elements with masses m_1, m_2, m_3 occur vertically; k_1, k_2, k_3, k_4, k_5 denote the stiffness coefficients of linear elastic elements. Contour I in Fig. 8.34 determines the location of the vibration protection device (VPD); the object of protection has mass m_3. The bearing surface *II* is introduced to build a more complete scheme: if necessary, k_4 and k_5, as well as Q_i and z_k, can take zero values. We write the expression for the kinetic and potential energies:

$$T=\frac{1}{2}m_1\dot{y}_1^2+\frac{1}{2}m_2\dot{y}_2^2+\frac{1}{2}m_3\dot{y}_3^2, \tag{8.254}$$

$$\Pi=\frac{1}{2}k_1y_1^2+\frac{1}{2}k_2(y_2-y_1)^2+\frac{1}{2}k_3(y_3-y_2)^2+\frac{1}{2}k_4y_3^2+\frac{1}{2}k_5(y_3-y_1)^2. \tag{8.255}$$

To build a mathematical model, the standard methodology [24] based on the Lagrange formalism is used:

$$m_1\ddot{y}_1+y_1(k_1+k_2+k_3)-k_2y_2-k_5y_3=k_1z_1; \tag{8.256}$$

$$m_2\ddot{y}_2+y_2(k_2+k_3)-k_2y_1-k_5y_3=0; \tag{8.257}$$

$$m_3\ddot{y}_3+y_3(k_3+k_4+k_5)-k_3y_2-k_5y_1=Q_3. \tag{8.258}$$

If we accept that $z_1=0, z_2=0$, and $Q_1\neq 0$, $Q_2\neq 0$, $Q_3\neq 0$, the relations between the variables y_1, y_2, y_3 and Q_1, Q_2, Q_3 in the operator form can be represented as the following expressions:

$$\bar{y}_1 = \frac{\overline{Q}_1(a_{22}a_{33} - a_{23}a_{32}) + \overline{Q}_2(a_{13}a_{32} - a_{33}a_{12}) + \overline{Q}_3(a_{12}a_{23} - a_{13}a_{22})}{A_0}, \quad (8.259)$$

$$\bar{y}_2 = \frac{\overline{Q}_1(a_{23}a_{31} - a_{31}a_{33}) + \overline{Q}_2(a_{11}a_{33} - a_{13}a_{31}) + \overline{Q}_3(a_{13}a_{21} - a_{11}a_{23})}{A_0}, \quad (8.260)$$

$$\bar{y}_3 = \frac{\overline{Q}_1(a_{21}a_{32} - a_{22}a_{31}) + \overline{Q}_2(a_{12}a_{31} - a_{11}a_{32}) + \overline{Q}_3(a_{11}a_{22} - a_{12}a_{21})}{A_0}, \quad (8.261)$$

where

$$A_0 = a_{11}a_{22}a_{33} - a_{11}a_{23}a_{32} - a_{33}a_{12}a_{21} - a_{22}a_{13}a_{31} + a_{13}a_{21}a_{32} + a_{12}a_{23}a_{31}, \quad (8.262)$$

is the characteristic equation of the system (the system is symmetric: $a_{12} = a_{21}$, $a_{13} = a_{23}$, $a_{23} = a_{32}$); as in relation to (8.256)–(8.258), the Laplace transform is used for further calculations.

Using (8.259)–(8.261), one can construct a sufficiently large number of transfer functions. For further consideration, it is assumed that an external periodic (harmonic) force $Q_3 \neq 0$ acts on the object of protection, the remaining force factors are equal to 0. The structural diagram of the system corresponding to Eqs. (8.256)–(8.258) is shown in Fig. 8.35.

Under the action of external disturbances, dynamic responses occur between the object of protection m_3 and support *II*, as well as between the inertial intermediate element m_1 of the VPD and the bearing surface *I*. Dynamic reactions will also occur between the individual elements of the system.

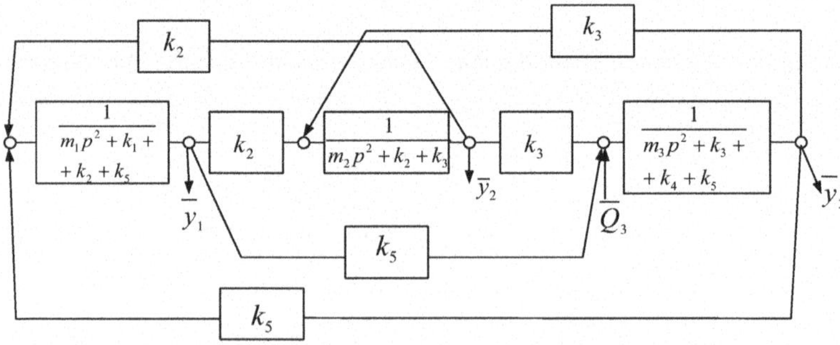

Fig. 8.35 The structural diagram for the vibration protection system in Fig. 8.34 when $\overline{Q}_3 \neq 0$

8.7.2 Evaluation of the Dynamic Properties of the System

Using expressions (8.259)–(8.261), we write transfer functions for the structural system in Fig. 8.35:

$$W_1(p) = \frac{\bar{y}_1}{\overline{Q}_3} = \frac{a_{12}a_{23} - a_{13}a_{22}}{A_0}; \tag{8.263}$$

$$W_2(p) = \frac{\bar{y}_2}{\overline{Q}_3} = \frac{a_{13}a_{21} - a_{11}a_{23}}{A_0}; \tag{8.264}$$

$$W_3(p) = \frac{\bar{y}_3}{\overline{Q}_3} = \frac{a_{11}a_{22} - a_{12}a_{21}}{A_0}, \tag{8.265}$$

where A_0 is defined by the expression (8.260). Since the structural diagram in Fig. 8.35 has a relationship between the coordinates y_1 and y_3 ($k_5 \neq 0$), then the case of loading $Q_2 \neq 0$ is of particular interest. We write the transfer functions of the system for this case:

$$W_1'(p) = \frac{\bar{y}_1}{\overline{Q}_2} = \frac{a_{13}a_{32} - a_{12}a_{33}}{A_0}; \tag{8.266}$$

$$W_2'(p) = \frac{\bar{y}_2}{\overline{Q}_2} = \frac{a_{11}a_{33} - a_{13}a_{31}}{A_0}; \tag{8.267}$$

$$W_3'(p) = \frac{\bar{y}_3}{\overline{Q}_2} = \frac{a_{12}a_{31} - a_{11}a_{32}}{A_0}. \tag{8.268}$$

Expressions (8.263)–(8.264) for transfer functions can be used to transform the structural diagram in Fig. 8.35.

1. Consider the case $\overline{Q}_3 \neq 0$, then (8.265) can be transformed as

$$W_3(p) = \frac{\bar{y}_3}{\overline{Q}_3} = \frac{1}{a_{33} - \frac{a_{11}a_{23}^2}{A_1} - \frac{a_{22}a_{13}^2}{A_1} + \frac{2a_{12}a_{23}a_{31}}{A_1}}, \tag{8.269}$$

where

$$A_1 = a_{11}a_{22} - a_{12}^2. \tag{8.270}$$

In detailed form (8.269) is shaped into:

$$W_3^*(p) = \frac{\bar{y}_3}{\bar{Q}_3} = \frac{1}{m_3p^2 + k_3 + k_4 + k_5 - \frac{k_5^2\left(m_2p^2 + k_2 + k_3\right)}{A_2} - \ldots} \ldots \tag{8.271}$$

$$\ldots \frac{}{-\frac{k_3^2\left(m_1p^2 + k_1 + k_2 + k_5\right)}{A_2} - \frac{2k_5k_2k_3}{A_2}}.$$

Here,

$$A_2 = \left(m_1p^2 + k_1 + k_2 + k_5\right) \cdot \left(m_2p^2 + k_2 + k_3\right) - k_2^2, \tag{8.271$'$}$$

and $p = j\omega$ is the complex variable ($j = \sqrt{-1}$).

In turn, by the expression (8.271) a structural diagram can be constructed, as shown in Fig. 8.36. In this diagram, the structural element corresponding to the object of protection m_3 is allocated. All other elements are grouped into a feedback circuit, which, by its physical nature, is a generalized spring with stiffness, defined by the expression:

$$\bar{k}'_{\text{red}} = (k_3 + k_4 + k_5)A_2 - \frac{k_5^2(m_2p^2 + k_2 + k_3) + k_3^2(m_1p^2 + k_1 + k_2 + k_5) - 2k_2k_3k_5}{A_2}. \tag{8.272}$$

Note that the feedback in the structural diagram (Fig. 8.36) will have a positive sign. The dimension of the input signal in the feedback circuit corresponds to the displacement, and the output signal has the dimension of force. That is, the feedback, in the physical sense, reflects the interaction with the bearing surface *II*, which corresponds to the dynamic reaction

$$\overline{R}_B = k_4\bar{y}_3 = k_4W_3(p)\overline{Q}_3. \tag{8.272$'$}$$

The second part of the dynamic reaction is determined by the interaction of the elastic elements k_4 and k_5, as well as by the additional term, which in expression (19) has a negative sign:

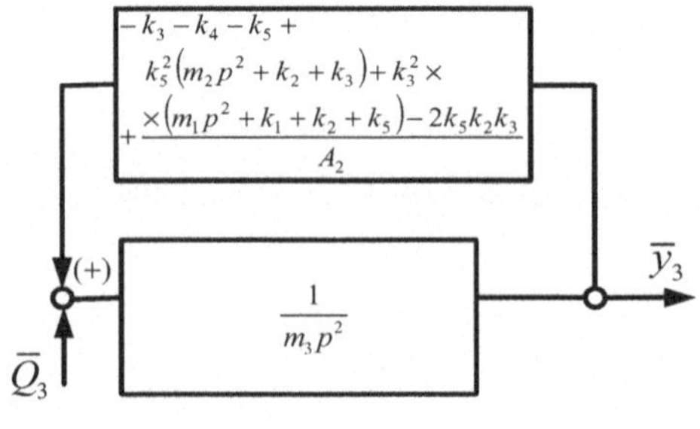

Fig. 8.36 The reduced structural diagram for determining the dynamic reaction at the object of protection

$$\overline{R}_{m_3} = \left[\frac{(k_4+k_3+k_5)A_2 - \left[k_5^2\begin{pmatrix} m_2p^2 \\ +k_2+k_3 \end{pmatrix} + k_3^2\begin{pmatrix} m_1p^2+k_1 \\ +k_2+k_5 \end{pmatrix} - 2k_2k_3k_5\right]}{A_2}\right]\bar{y}_3. \tag{8.272''}$$

The transfer function at the force input Q_3 and output – $\overline{R}_{m_3}$ is as follows:

$$W_{m_3}(p) = \frac{\overline{R}_{m_3}}{\overline{Q}_3} = \frac{(k_3+k_4+k_5)A_2 - \left[k_5^2\begin{pmatrix} m_2p^2 \\ +k_2+k_3 \end{pmatrix} + k_3^2\begin{pmatrix} m_1p^2+k_1 \\ +k_2+k_5 \end{pmatrix} - 2k_2k_3k_5\right]}{A_0}W_3(p),$$

where

$$W_3(p) = \frac{1}{A_0}\left[(k_3+k_4+k_5)A_2 - k_5^2\begin{pmatrix} m_2p^2 \\ +k_2+k_3 \end{pmatrix} - k_3^2\begin{pmatrix} m_1p^2 \\ +k_1 \\ +k_2+k_5 \end{pmatrix} - 2k_2k_3k_5\right]. \tag{8.272'''}$$

In turn:

$$W_B(p) = \frac{\overline{R}_{II}}{\overline{Q}_3} = \frac{k_4A_2}{A_0}. \tag{8.272''''}$$

In general, it follows from (8.272''') and (8.272'''') that the amplitude-frequency characteristics have three resonances, as well as two frequencies at which the dynamic reactions $\overline{R}_{m_3}$ will be zero. This corresponds to the mode when the dynamic part of the overall response at the object of protection m_3 will be zero, but the load remains static, determined by the gravity of the elements of the vibration protection system and preliminary preloads of the springs. As for the dynamic response on the bearing surface *II*, in accordance with (8.272''''), the reaction R_{II} will also have two "zeroing" frequencies. Note that these modes coincide with the dynamic absorbing modes along the coordinate y_3 when a force is applied to an object of protection, which follows from the expression (8.265) or (8.271).

In this situation, special cases are possible, for example, k_4 and k_5 can be taken equal to zero. In this case, the expression (8.272) is converted to the form:

$$\bar{k}'_{\text{red}_1} = \frac{k_3A_3 - k_3^2(m_1p^2+k_1+k_2)}{A_3}, \tag{8.273}$$

where

$$A_3 = \left(m_1p^2 + k_1 + k_2\right)\left(m_2p^2 + k_3\right) + k_2\left(m_1p^2 + k_1\right). \tag{8.274}$$

Knowing (8.273), one can find a dynamic response at the protected object m_3:

$$\overline{R}'_{m_3} = \overline{k}'_{\text{red}_1}\overline{y}'_3, \tag{8.275}$$

where $\overline{y}'_3 = W_3(p)\overline{Q}_3(k_4 = 0, k_5 = 0)$.

Thus, for the object of protection, dynamic responses can be found both for contacts with the bearing surface and for the complete reaction, which is the sum of the static and dynamic components. If $k_4 = 0, k_5 = 0$, then the reaction value can also be found using (8.273).

2. In order to find the dynamic response on the intermediate mass m_1 of the VPD, it is necessary to use the expression (8.269):

$$W_1(p) = \frac{\overline{y}_1}{\overline{Q}_3} = \frac{\left[k_2k_3 + k_5\left(m_2p^2 + k_2 + k_3\right)\right]/A_4}{\left(m_1p^2 + k_1 + k_2 + k_5\right) - \frac{k_5^2\left(m_2p^2 + k_2 + k_3\right) + \left(m_3p^2 + k_3 + k_4 + k_5\right)k_2^2 - 2k_2k_3k_5}{A_4}}. \tag{8.276}$$

where

$$A_4 = \left(m_2p^2 + k_2 + k_3\right) \cdot \left(m_3p^2 + k_3 + k_4 + k_5\right) - k_3^2. \tag{8.277}$$

The structural diagram for the case of reducing the general system to an intermediate element m_1 has the form, as shown in Fig. 8.37.

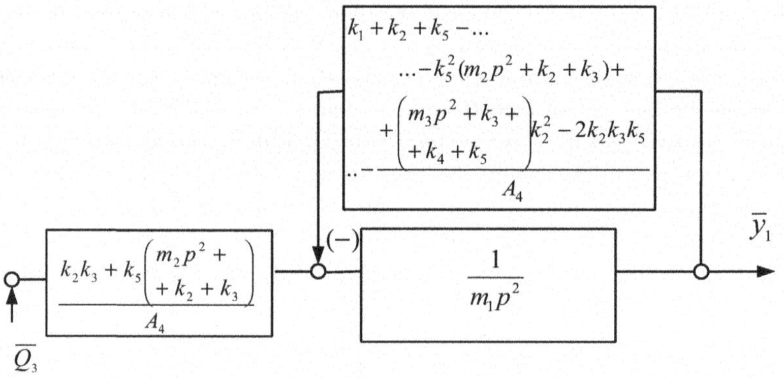

Fig. 8.37 The structural diagram for determining the dynamic reaction on the intermediate element VPD—m_1

Note that when determining the transfer function (8.276), it is possible to present some external force $\overline{Q_1'}$ at the system input, which has the form:

$$\overline{Q_1'} = \overline{Q_3'}\frac{k_2k_3 + k_5(m_2p^2 + k_2 + k_3)}{A_4}. \tag{8.278}$$

In this case, $\overline{Q_1'}$ is the equivalent external disturbance force transferred from the object of protection m_3 to the intermediate mass of the VPD—m_1. In turn, if we assume that the perturbation in the system has the form (8.278), then we can determine the transfer function from $\overline{Q_1'}$ along the coordinate $\bar{y}_3$. The reduced stiffness of the generalized spring, and such is the physical essence of the negative feedback circuit in the structural diagram in Fig. 8.37, is determined (at $k_5 = 0$):

$$\bar{k}''_{\text{red}} = \frac{(k_1 + k_2)A_5 - k_2^2(m_3p^2 + k_3 + k_4)}{A_5}, \tag{8.279}$$

where

$$A_5 = \left(m_2p^2 + k_2 + k_3\right)\left(m_3p^2 + k_3 + k_4\right) - k_3^2. \tag{8.279$'$}$$

As regards the definition of a reaction $\bar{R}_I$, it can be found.

$$\bar{R}_I = k_1\bar{y}_1 = k_1W_1^*(p)\bar{Q}_1, \tag{8.280}$$

here $W_1^*(p)$ is determined from (8.269) with $k_5 = 0$; i.e. $W_1^*(p) = W_1(p)_{k_5=0}$.

The dynamic response $\bar{R}_I$ is determined by the level of interaction with the bearing surface *I* (Fig. 8.34). Dynamic response on the intermediate element can be found

$$\bar{R}_{m_1}(p) = \bar{k}''_{\text{red}}\bar{y}_1$$

$$= \frac{(k_1 + k_2)\left[\begin{pmatrix} m_2p^2 + \\ + k_2 + k_3 \end{pmatrix}\begin{pmatrix} m_3p^2 + \\ + k_3 + k_4 \end{pmatrix} - k_3^2\right] - k_2^2\begin{pmatrix} m_3p^2 + \\ + k_3 + k_4 \end{pmatrix}}{A_5} W_1^*(p)\bar{Q}_1'. \tag{8.281}$$

To define $W_{m_1}(p)$, $\overline{Q}_1$ is introduced taking into account the defined (8.278), and taking $k_5 = 0$.

Similarly, according to the dynamic responses on the bearing surface *I* and the intermediate element m_1, transfer functions can be obtained from external force $\overline{Q}_3$ by the parameters of the reactions $\bar{R}_I$ and $\bar{R}_{m_1}$, as well as transfer functions on the corresponding dynamic responses with a force input action $\overline{Q}_3$.

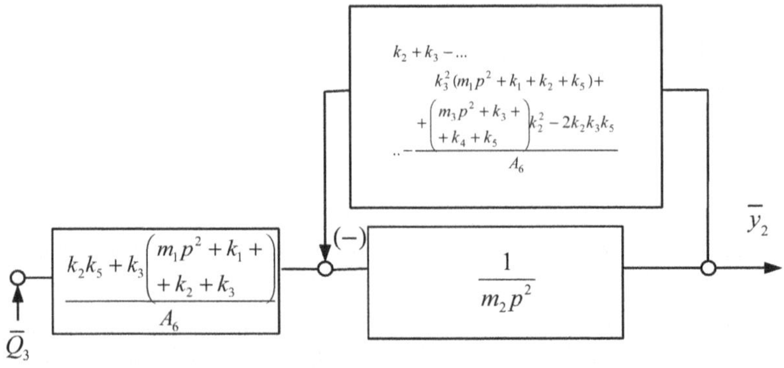

Fig. 8.38 The reduced structural diagram to determine the dynamic response from $\overline{Q}_3$ along the coordinate $\bar{y}_2$

3. We proceed to the definition of the dynamic response from $\overline{Q}_3$ along the coordinate $\bar{y}_2$ associated with the intermediate mass m_2. In this case, the expression (8.270) takes the form:

$$W_2(p) = \frac{\bar{y}_2}{\overline{Q}_3} = \frac{[k_5 + k_2 + k_3(m_1p^2 + k_1 + k_2 + k_5)]/A_6}{(m_2p^2 + k_2 + k_3) - \frac{\begin{pmatrix} m_1p^2 + k_1 \\ + k_2 + k_5 \end{pmatrix} k_3^2}{A_6} - \frac{\begin{pmatrix} m_3p^2 + k_3 \\ + k_4 + k_5 \end{pmatrix} k_2^2 + 2k_2k_3k_5}{A_6}}, \tag{8.282}$$

where $A_6 = (m_1p^2 + k_1 + k_2 + k_5) \cdot (m_3p^2 + k_2 + k_3) - k_5^2$.

The structural diagram of the system for determining the dynamic response from on the intermediate element m_2 is shown in Fig. 8.38.

The reduced stiffness of the generalized spring is determined by the scheme Fig. 8.38; at $k_5 = 0$

$$\bar{k}''_{\text{red}} = \frac{(k_2 + k_3)A_7 - k_3^2(m_1p^2 + k_1 + k_2) - (m_3p^2 + k_3 + k_4)k_2^2}{A_7}, \tag{8.283}$$

where

$$A_7 = (m_1p^2 + k_1 + k_2) \cdot (m_3p^2 + k_3 + k_4). \tag{8.283'}$$

We will find the values of $\bar{k}'_{\text{red}}, \bar{k}''_{\text{red}}, \bar{k}'''_{\text{red}}$ directly from the computational scheme Fig. 8.34 with $k_5 = 0$, using the method described in [8], and we obtain:

$$\bar{k}'^{*}_{\text{red}} = k_4 + \frac{k_3[(m_1p^2 + k_1)k_2 + m_2p^2(m_1p^2 + k_1 + k_2)]}{(k_1 + m_1p^2)k_2 + m_2p^2(m_1p^2 + k_1 + k_2) + k_3^2(m_1p^2 + k_1 + k_2)}, \tag{8.284}$$

$$\bar{k}''^{*}_{\text{red}} = k_1 + \frac{k_2[(m_3p^2 + k_4)k_3 + m_2p^2(m_3p^2 + k_3 + k_4)]}{(k_4 + m_3p^2)k_3 + m_2p^2(m_3p^2 + k_3 + k_4) + k_2(m_3p^2 + k_3 + k_4)}, \tag{8.285}$$

$$k^{\bar{\prime\prime\prime}*}_{\text{red}} = \frac{k_2[(m_1p^2 + k_1)(m_3p^2 + k_3 + k_4)) + k_3(m_1p^2 + k_1 + k_2)(m_3p^2 + k_4)]}{(k_1 + k_2 + m_1p^2)(m_3p^2 + k_3 + k_4)}. \tag{8.286}$$

8.7.3 *Method of Direct Transformations of the Computational Scheme*

Making the appropriate transformations in expressions (8.272) with $k_5 = 0$ (8.283), (8.286), it can be shown that the expressions obtained on the basis of the proposed method and using direct convolution methods are the same. Computational schemes for obtaining the reduced stiffnesses are shown in Fig. 8.39a, b.

Note that the computational scheme Fig. 8.39b includes a system with two elastic elements with stiffnesses coefficients k_4 and {{}}, respectively. Note that stiffness {{}} is formed in the following sequence of actions:

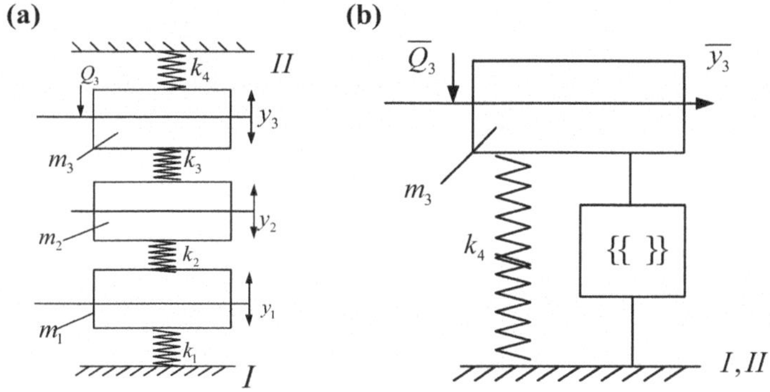

Fig. 8.39 A schematic diagram of the formation of the reduced stiffness according to the computational scheme (**a**) in the case of force perturbation $\bar{Q}_3$ at the object of protection; **b** reduced scheme

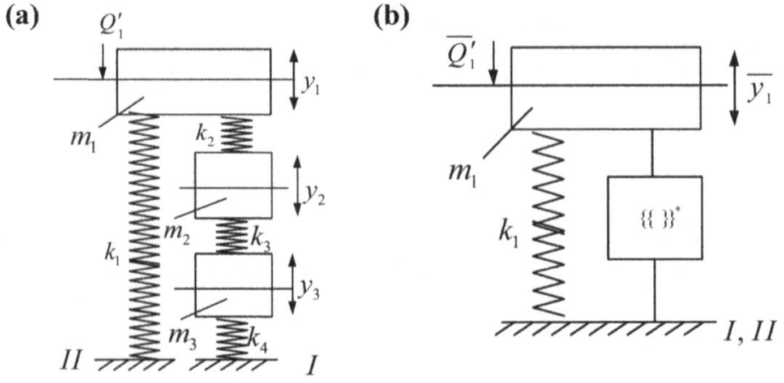

Fig. 8.40 Schematic diagram of the formation of the reduced stiffness according to the computational scheme (**a**) with an equivalent force perturbation $\overline{Q}_1'$ at the element m_1; **b** reduced scheme

$$\left(m_1 p^2\right) \to \left[\frac{(m_1 p^2 + k_1)k_2}{m_1 p^2 + k_1 + k_2}\right] \to \{[\,] + m_2 p^2\} \to \left\{\left\{\frac{\{\}k_3}{\{\} + k_3}\right\}\right\}, \tag{8.287}$$

which leads to the expression (8.285).

Figure 8.40a, b shows how the reduced stiffness is formed, that is, the stiffness coefficient of the generalized spring (8.285) is determined. This is related to the definition of an equivalent external force $\overline{Q}_1'$ by the formula (8.278), which is shown in Fig. 8.37.

In determining the reduced stiffness of the generalized spring $\{\{\}\}^*$ (Fig. 8.40b), the following sequence of operation of the convolution of the structure is used

$$\left(m_3 p^2 + k_4\right) \to \left[\frac{(m_3 p^2 + k_4)k_3}{m_3 p^2 + k_3 + k_4}\right] \to \{[\,] + m_2 p^2\} \to \left\{\left\{\frac{\{\}k_2}{\{\}^* + k_2}\right\}\right\}^*, \tag{8.288}$$

which leads to the expression (8.286), which is shown schematically Fig. 8.40b. The total stiffness is determined by summing k_1 and $\{\{\}\}^*$.

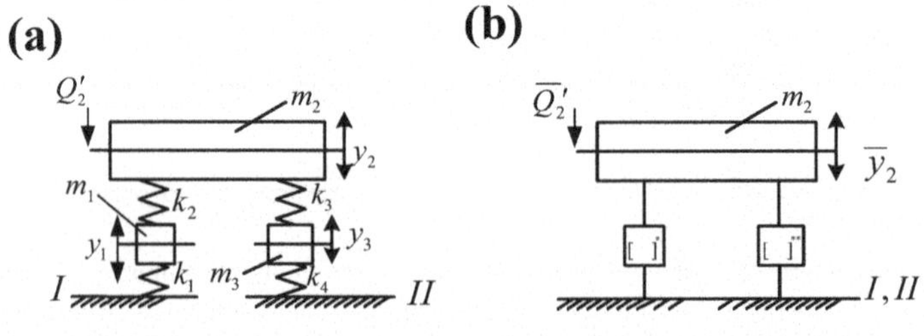

Fig. 8.41 The schematic diagram of the formation of the reduced stiffness of the generalized spring according to the computational scheme (**a**); **b** reduced scheme

With regard to the definition of the reduced stiffness associated with the intermediate element m_2, the equivalent force (Fig. 8.38) is determined by the expression

$$\overline{Q}_2' = \overline{Q}_3 \frac{k_2 k_5 + k_3(m_1 p^2 + k_1 + k_2 + k_5)}{(m_1 p^2 + k_1 + k_2 + k_5)(m_3 p^2 + k_2 + k_3) - k_5^2}. \tag{8.289}$$

With further calculations accept $k_5 = 0$.

The basic computational scheme and the possibility of reduction to a simplified form are shown in Fig. 8.41a, b.

In this case, the equivalent external force $\overline{Q}_2'$ is determined from the expression (8.289). In turn, the overall stiffness coefficient is formed by the sum of two stiffnesses $[\,]^*$ and $[\,]^{**}$ (Fig. 8.41b) based on the implementation of the following sequences of actions: $(m_1 p^2 + k_1) \rightarrow \left[\frac{(m_1 p^2 + k_1)k_2}{m_1 p^2 + k_1 + k_2}\right]^*$, and

$$(m_3 p^2 + k_4) \rightarrow \left[\frac{(m_3 p^2 + k_4)k_3}{m_3 p^2 + k_3 + k_4}\right]^{**}, \tag{8.290}$$

which ultimately leads to the expression (8.288).

Thus, the definition of dynamic responses is associated with the evaluation of the reduced stiffnesses of generalized springs, which create dynamic stiffness, depending on the frequency of external disturbances. Two approaches are shown, based on the proposed method of forming a negative feedback chain on a dedicated mass-and-inertia element, as well as on consecutive convolution procedures using the original computational schemes (Figs. 8.39a, 8.40a and 8.41a). Note that convolution is simple enough only for planar systems; therefore, in this case, it was assumed that $k_5 = 0$. In cases where non-planarity has to be taken into account, it is necessary to use a number of preliminary transformations, which complicates the method for determining the reduced stiffnesses, but leads to similar results.

Thus, the basis of the method of determining dynamic responses on the object of protection, as well as on the inertial elements of the vibration protection device (VPD) and on the bearing surfaces *I* and *II* is the transformation of the structural diagram of the system to a form corresponding to the second-order integration unit with the formation of additional negative feedback, which, in its essence, is a dynamic response. In a system with three degrees of freedom, such transformations are made, if necessary, not only for the object of protection, but also for intermediate masses. The method is also applicable to systems with a larger number of degrees of freedom; however, this results in a noticeable increase in the complexity of calculations. The structure of additional negative feedback can be obtained more simply by using the characteristic equation of the system included in the corresponding transfer function. The equation is transformed with the allocation of the desired partial system with subsequent transformations of the structure.

When constructing the reduced structural diagrams, a force, initially applied to the object of protection, can be transferred to the intermediate inertial element. A force like that is equivalent in the sense that the transfer functions obtained using, when transferring a force to a new point, give the same result as in the usual order. The proposed method allows us to obtain the transfer function of the dynamic response of an element or a support by multiplying the corresponding reduced stiffness coefficient of the generalized spring (or dynamic stiffness) by the corresponding displacement of the element in the coordinate system.

References

1. Belokobylsky, S.V., Eliseev, S.V., Sitov, I.S.: Dynamics of Mechanical Systems. Lever Linkages and Inertial-Elastic Constraints, 319 p. Polytekhnika Publ., St. Petersburg (2013)
2. Khomenko, A.P., Eliseev, S.V., Artyunin, A.I., et al.: Mechanisms in Elastic Oscillatory Systems: Features of Taking into Account Dynamic Properties, Problems of Vibration Protection of Machines, Instruments and Equipment, 187 p. IrGUPS Publ., Irkutsk (2013). Dep. in VINITI 08.15.13, No. 243-B 2013
3. Eliseev, S.V., Artyunin, A.I., Khomenko, A.P., Kaimov, E.V., Eliseev, A.V.: A Device for Damping Oscillations. Pat. 136112 Russian Federation. Applicant and patent holder: IrGUPS; Applied 07.25.2013; publ. 12.27.2013, Bull. No. 36
4. Eliseev, S.V., Artyunin, A.I., Kaimov, E.V., Eliseev, A.V.: A Device for Regulating the Elastic-Dissipative Properties of a Vibration Protection System. Pat. 142137 Russian Federation. Applicant and patent holder: IrGUPS. Appl.: 01/09/2014; publ. 06/20/2014, Bull. No. 17
5. Khomenko, A.P., Eliseev, S.V., Artyunin, A.I., Eliseev, A.V., Kaimov, E.V. A Device for Damping Oscillations. Pat. 150331 Russian Federation. Applicant and patent holder: IrGUPS. Appl.: 09/25/2014; publ. 02/10/2015, Bull. No. 4
6. Khomenko, A.P., Eliseev, S.V., Milovanov, A.I., Bityukova, S.M., Ermoshenko, Y.V., Gozbenko, V.E.: The Method of Controlling the Characteristics of Linear Oscillations and the Device for Its Implementation, 6 p. Pat. 2246647 Russian Federation. Applicant and patent holder: IrGUPS. Appl.: 15.11.2002; publ. 20.02.2005, Bull. No. 5
7. Khomenko, A.P., Eliseev, S.V., Bolshakov, R.S.: Dynamic properties of the design of an elastic contacting block in vibration protection problems. Bulletin of Irkutsk State Technical University **5**(100), 33–43
8. Druzhinsky, I.A.: Mechanical chains, p. 240. Mashinostroenie Publ., Moscow (1977)
9. Eliseev, S.V., Reznik, YuN, Khomenko, A.P., et al.: Dynamic synthesis in generalized problems of vibration protection and vibration isolation of technical objects, p. 523. ISU Publ., Irkutsk (2008)
10. Rothenberg, R.V.: Fundamentals of the "driver-car-road-environment" system reliability, p. 216. Mashinostroenie Publ., Moscow (1986)
11. Verigo M.F., Kogan A.Ya. The interaction of the track and rolling stock. In Verigo M.F. (ed.). Moscow: Transport Publ., 1986. 559 p
12. Dembaremdiker, A.D.: Shock absorbers for vehicles, p. 199. Mashinostroenie Publ., Moscow (1985)
13. Klimov, A.V.: Dynamics of a lever relaxation suspension with intermittent damping. Ph.D. (Engineering) thesis, 186 p. Orlov. State Tech. University Publ., Oryol (2001)
14. Logunov, A.S.: Dynamics of pneumatic elements and motion transformation devices in the systems of vibration protection of objects. Author's abstract of Ph.D. (Engineering) thesis, 20 p. IrGUPS Publ., Irkutsk (2009)

15. Khachaturov, A.A., Afanasyev, V.L., Vasilyev, V.S., et al.: Dynamics of the road—bus—car —driver system. In: Khachaturov, A.A. (ed.). Mashinostroenie Publ., Moscow, 536 p (1976)
16. Loitsyansky, L.G., Lurie, A.I.: The Course of Theoretical Mechanics (2 vols.), vol. 2: Dynamics, 640 p. Nauka Publ., Moscow (1983)
17. Eliseev, S.V., Reznik, YuN, Khomenko, A.P.: Mechatronic approaches in the dynamics of mechanical oscillatory systems, p. 384. Nauka Publ., Novosibirsk (2011)
18. Chuprakov, YuI: Hydraulic systems of protection of the human operator from the general vibration, p. 224. Mashinostroenie Publ., Moscow (1987)
19. Belokobylsky, S.V., Eliseev, S.V., Kashuba, V.B.: Applied problems of the structural theory of vibration protection systems, p. 363. Polytekhnika Publ., St. Petersburg (2013)
20. Artobolevsky, I.I.: Theory of mechanisms and machines, p. 638. Nauka Publ., Moscow (1988)
21. Kolesnikov, V.I., Vorobiev, V.B., Shapovalov, V.V., et al.: Improving the interaction of the track and rolling stock, p. 365. Marshrut Publ., Moscow (2006)
22. Kruglov, YuA, Tumanov, YuA: Shock protection of machines, equipment and equipment, p. 222. Mashinostroenie Publ., Leningrad (1986)
23. Khokhlov, A.A.: Dynamics of complex mechanical systems, p. 172. MIIT Publ., Moscow (2002)
24. Lurie, A.I.: Analytical mechanics, p. 516. Nauka Publ., Moscow (1986)

The manufacturer's authorised representative in the EU is Springer Nature Customer Service Centre GmbH, Europaplatz 3, 69115 Heidelberg, Germany. If you have any concerns regarding our products, please contact ProductSafety@springernature.com

Printed and bound by CPI Group (UK) Ltd, Croydon, CR0 4YY
30/09/2026
02225657-0016